Atlas of Mammal Distribution through Africa from the LGM (~18 ka) to Modern Times

The zooarchaeological record

Hélène Jousse

ARCHAEOPRESS PUBLISHING LTD
13-14 Market Square
Bicester
Oxfordshire OX26 6AD
www.archaeopress.com

ISBN 978 1 78491 540 7
ISBN 978 1 78491 541 4 (e-Pdf)

This book is available direct from Archaeopress or from our website www.archaeopress.com

Table of Contents

Acknowledgements

The author is grateful to the Lise Meitner Program from the Austrian FWF Der Wissenschaftsfonds (Fonds zur Förderung der wissenschaftlichen Forschung) that supported two years of post-doctoral research during 2007-2008 when this atlas was composed.

This work was hosted by the Mammals Collection at the Naturhistorisches Museum Wien. Dr. Barbara Herzig supervised the study, helped for the administrative work and facilitated all research in the collection and is thoughtfully acknowledged. The team at the Naturhistorisches Museum Wien is thanked for their help and collaboration and especially Dr. Spitzenberger, M. Bibl and Ms. Windl are thanked for their availability.

The post-doctorate was also reviewed by a senior researcher and referee, Prof. Christian Sturmbauer (University of Graz, Austria). Prof. Fiona Marshall (University of California), Joris Peters (Univeristy of Munich), Wim Van Neer (Institute for Natural Sciences, Brussels) and Jean-Denis Vigne (MNHN, Paris) are acknowledged for their scientific and friendly support. The author also addresses her thanks to Kris Kovarovic, who kindly commented and improved the English text, and to Jean-Marc Richard, who archived the formatting work.

Chapter 1 Introduction

This work provides the first overview of mammal species distributions in Africa since the Last Glacial Maximum (LGM, 18 ky) to modern time. It is derived from data published mainly in the zooarchaeological literature until 2009. During a post-doctoral project hosted in the zoological department of mammal collection at the Naturhistoriches Museum in Vienna (Austria), the occurrences of taxa in archaeological sites on the African continent were recorded in a database, integrating geographical and chronological information. This record offers the opportunity to produce a chronological atlas of mammalian distributions by presenting their occurrences on successive maps over the last 18 ky.

Similar works were conducted previously at a smaller scale, including the macromammals of the southern subcontinent (Plug and Badenhorst, 2001) and studies restricted to individual countries (eg. Gautier and Van Neer, 2005 for Ghana). The aim of the present project is to take all of the African mammalian fauna into account without taxonomic or geographic restrictions. Every effort has been made to record the fauna as accurately as possible, however there may be some omissions or inconsistencies due to a lack of data, missing publications, poorly referenced ages or faunal lists or incorrect ages attributed to excavation levels.

This work is useful for zooarchaeologists dealing with one particular species by providing a bibliographical work that documents its past locations. It must be noted that fauna are mainly documented through their presence at archaeological sites and are therefore tied to the presence of humans and their activities. This may only partially reproduce their true past distribution. However, the sites offer a good coverage throughout space and time and generally reflect the extent of mammalian distributions, although the limits of their distributions may be further refined.
The atlas will aid in the investigation of palaeoecological issues, such as the capacity of mammals to adapt to climatic change and respond to human disturbance in the recent past of Africa. Since the LGM, drastic climatic changes occurred on the African continent, with alternating periods of aridity (~18-10 ky and 5-0 ky) and humidity (~10-5 ky). Concomitantly, human economy and behaviour changed as populations migrated and adapted their dietary patterns, mainly with food production (agriculture and pastoralism), diet specialisation, social organisation, semi-nomadic or sedentary way of life. All together, climatic and anthropogenic impacts have an influence on the natural environment, and directly or indirectly on mammals (reduction of distribution area, of population size, of individual size, etc). Other questions than can be investigated through this atlas are: What was the mammal's former distribution? Did refuge areas exist and, if so, what were their geographical barriers and the potential migration routes of species between them? Did some species face extinction during the hyper-arid LGM? How rapidly did distribution areas decrease for each taxa? Finally, these data presented here can indicate which species suffered more from the recent climatic degradation, and thus which are potentially more endangered by climatic change today.

The database also provides information that is fundamental to a better understanding of what influenced the present-day distribution, dynamism and structure of mammalian communities in Africa. By incorporating a larger temporal scale to modern ecological studies, it may help control their conservation since desiccation and human disturbance in Africa is still a worrying question for their future.

Chapter 2 Data and Methods

Database structure and working tools

The database counts more than 850 sites and dated site levels, corresponding to about 9 000 occurrences of mammals. The structure of the database was built using the software Microsoft ACCESS. It includes the following zooarchaeological and chronological data:

- Geography: country, coordinates, detailed location and publication reference.
- Cultural context: general attribution to Palaeolithic/Epipalaeolithic/Late Stone Age/ Iron Age, or to the pastoral phases identified for each region, and then precise information about local culture/ facies/traditions defined usually by ceramic and/or lithic industries.
- Chronological context, which is often problematic when radiometric ages are rare or absent, or when they date not directly human occupation but environmental phases (molluscs,…).
- Occurrences of taxa integrated through published faunal lists, with or without quantification (Number of Remains, Minimum Number of Individuals, are present in the database, but not provided here).
- Mammal biometrics, through published dimensions taken on fossil bones, facilitating the study of morphological and body size changes during the LGM.
- Historical maps are obtained by exporting the data in GIS software. The DIVA-GIS software, available freely online, plots together the spatial and temporal data. The maps are then composited with Adobe Illustrator and Photoshop softwares.

Taxonomy

Taxa mentioned in the zooarchaeological literature follow different classifications depending on the authors and the evolution of systematics over time. To homogenise the dataset, only one classification system has been used : Wilson and Reeder (1992), in accordance with that used in the NHM Wien mammals collections and in many institutions. An updated publication is available in Wilson and Reeder (2005), but the former is more consistent with the one found in the majority of the publications. It considers, for example, that antelopes are classifed into subfamilies (e.g. Cephalophinae, Tragelaphinae), while others place this level of taxonomic differentiation below, in the level of the tribe (e.g. Cephalophini, Tragelaphini). This classification is available online at www.bucknell.edu/MSW3/.

All the mammals are not well known in the zooarchaeological record. There are few natural sites; most of the recorded sites result from the accumulation of human deposition of food remains. Therefore fauna that mirror the human diet are documented: mainly herbivores, but a few rodents, while carnivores are rare. Domesticated species are included as well as wild. Compared with the larger bodied indigenous African fauna, our present knowledge of past micromammals - Afrosoricids, Macroscelids, Insectivora and bats - are very weak. Even among the rodents the data are not very consistent.

Identifications and quantifications were made by many authors whose technical skills and involvement vary widely. Some taxonomic attributions may seem questionable to specialists. Identifications to species level may have been based on poorly diagnostic skeletal elements and indirect attributions can be based on the current distribution of the same species, on environmental conditions, etc.

However, when compiling data for this atlas, the published identifications are provided although this introduces a “subjective filter”. Each user of this database is therefore responsible for the sorting and selection of the data for his or her own use. When an identification comes with cf., af. or ? to show uncertainty, this is added in a note. Abundance data are not plotted on maps since they are more likely to further introduce a selective filter than to illustrate real ecological abundance.

Geography

The DIVA-GIS software integrates a number of spatial data, including a topographic map covering the continent, which is found in the files of “SRTM30 dataset built on CGIAR-SRTM data aggregated to 30 seconds”. A coarse topographic cutting (0/500/1000/5000/4000 m) was chosen to locate the main reliefs (Fig. 1). The main river systems of Africa are also indicated. The spatial distribution of sites is heterogeneous throughout the continent, which relates to research efforts interests, field research difficulties, accessibility, or political conflicts. The dataset shows how some regions are favoured by researchers, such as South Africa where a rich record of Iron Age sites is known (Plug and Badenhorst, 2001) and the Nile Valley in Egypt and Sudan, while other countries are almost devoid of data.

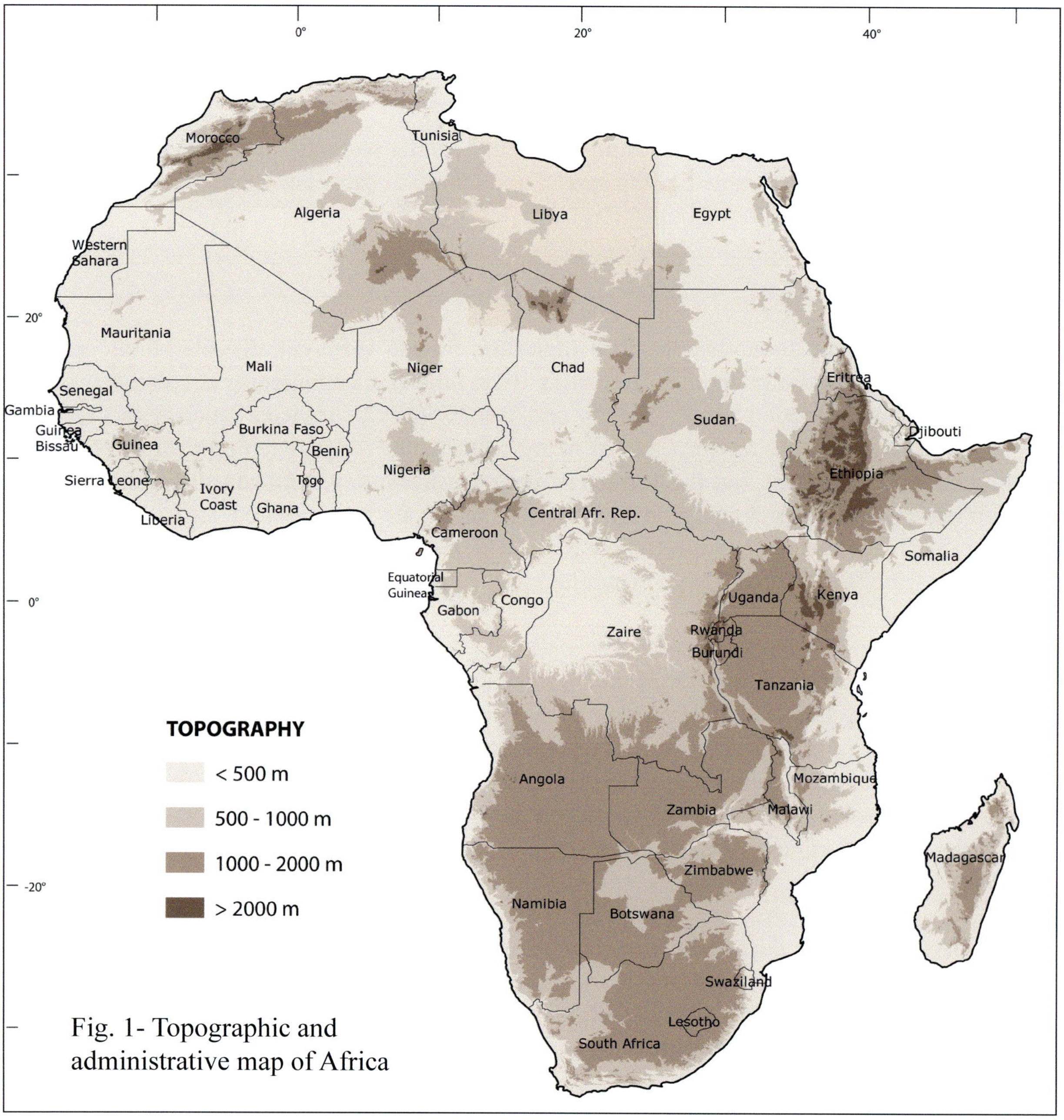

Fig. 1- Topographic and administrative map of Africa

Chronology

The problem of resolving different site ages in the literature has been addressed by providing all the radiometric ages available, even if there was only one for a site. Here again, it will be up to each user to select or reject this data for his or her own use. The time periods with their associated cultural facies include Late Paleolithic (eg. Aterian), Epipaleolithic (eg. Capsian), Mesolithic (eg, Khartoum), Holocene communities (eg: Later Stone Age, Neolitic, Elmenteitan, Pastoral Phases), Iron Age, and historical times up until the 20th century.

All radiocarbon ages were calibrated using the online software Oxcal 4.1 (Bronk Ramsey, 2001), with the IntCal04 calibrating curve for Northern Hemisphere (Reimer et al., 2004) and SHCal04 calibrating curve for the Southern Hemisphere until ~11 ka (McCormac et al., 2004).

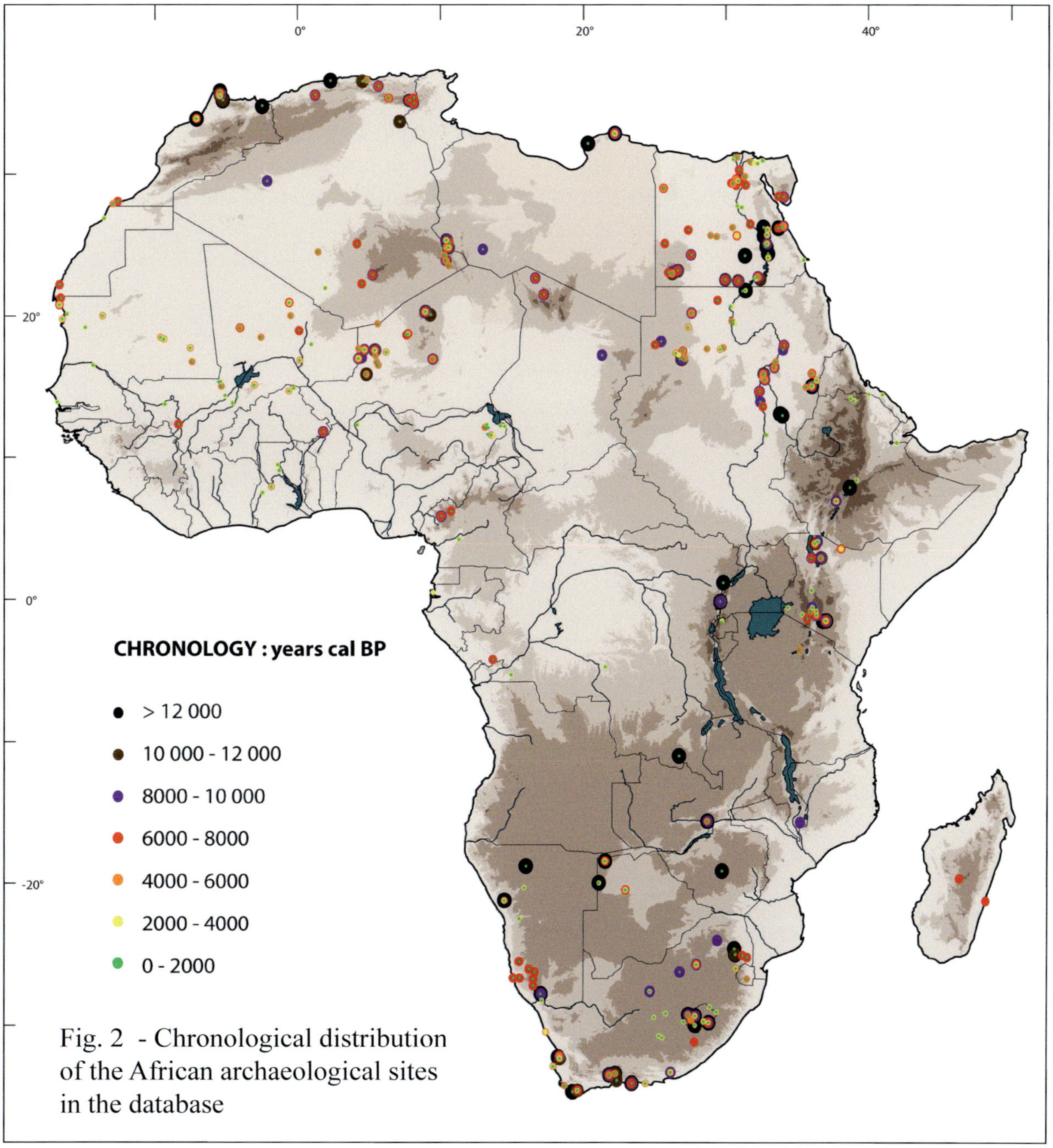

Fig. 2 - Chronological distribution of the African archaeological sites in the database

Chronology is another heterogeneity in the database, as seen on figure 2. There are few sites prior to 10 ka. This is mainly due to the climato-historical fact that human occupation was scarse during the hyper-arid period of LGM and nomadic people left few traces of their presence during this time. From 10 ky onward, the sites become more frequent and the spatial distribution illustrates the dispersion of people through time across the continent, with sub-Saharan Africa mainly colonised after 3 ky. Therefore, the time period between ~20 and 10 ka is grouped as one single interval, and then different colors are used to illustrate successive 2 ka intervals after that. The size of the points does not signify an abundance of sites or species, but it is a way to indicate "stacked" occurrences, when a species is present in the same place at different times. Each site is attributed to a time interval using the calibrated age. To simplify, the midpoint of the calibration interval (95% confidence) is calculated, and this midpoint value is used to allocate the adequate chronological color.

Atlas manual

Figures 1 and 2 present the topography, countries, and the chronological dataset used in this work. Chapter 3 provides the site data including the geography and archaeological context. Each site has a reference number, following the alphabetic order within each country. The radiocarbon ages are listed in Chapter 4. Individual maps and species occurrences are given in Chapter 5. All of the bibliographical references are listed with a number by alphabetical order at the end of the book in Chapter 6, and these numbers are used in the data tables.

Chapter 3 Site Data :
GEOGRAPHY, ARCHAEOLOGICAL CONTEXT

Abbreviations found in the table are the following:

Abbrev.	Meaning
*	Approximative coordinates
E	Epipaleolithic
EIA	Early Iron Age
Hi	Historic
Ho	Holocene
IA	Iron Age
LIA	Late Iron Age
LP	Late Paleolithic
LPl	Late Pleistocene
LSA	Late Stone Age
MIA	Middle Iron Age
MSA	Middle Stone Age
N	Neolithic
P	Paleolithic
Pl	Pleistocene
AT	Atérian
Ib	Iberomaurusian
pcW	Post-Classic Wilton
Iwc	Interior Wilton Classic
PN	Pastoral Neolithic
SPN	Savannah-Oriented Pastoral Neolithic

N site	Site	SITE N	Latitude			Longitude			*	Culture, facies, industry, chronology, notes		Citation
Algeria												
1	A-88-15	89	24	30	N	1	30	E		LSA	Ho, ~3rd mill. BC	290
2	A-88-28	90	22	0	N	2	0	E		LSA	Ho, ~3rd-1st mill. BC	290
3	A-88-35	88	33	45	N	7	10	E		LSA	Ho, ~Early N and E	290
4	Ain Mistehayia	384	35	11,65	N	7	45,9	E		LSA		150, 186, 187
5	Amekni	286	22	53	N	5	14	E		LSA	N	24, 35, 214
6	Columnata	370	35	29	N	1	16	E		LSA	Ib, N, Capsian Tradition	44
7	Damous el Ahmar	497	35	21	N	8	5	E		LSA	N, Capsian Tradition	320
8	Djurdjura	701	36	28	N	4	10	E		LSA		135
9	Dra-Mta-el-Ma-el-Abiod	897	34	55	N	8	5	E		LSA		36
10	Grotte Capeletti	498	35	20	N	6	26	E		LSA	N, Capsian Tradition	320
11	Gueldaman Akbou	369	36	40	N	4	55	E		N	N, Capsian Tradition	44
12	In Guezzam South	368	19	29	N	5	42	E	*	N		6
13	Kef Zoura D	386	35	8,56	N	7	47,8	E		LSA		186, 187
14	Mankhor	910	23	30	N	10	40	E		LSA	N	45, 281
15	Medjez II	377	36	8	N	5	40	E		LSA		25
16	Meniet	287	25,1	1	N	4,18	1	E				35, 214, 327
17	Oued Adjemet	470	22	20	N	4	30	E		LSA	N?	214
18	Oued Telidjene A	385	35	12	N	7	47	E	*	LSA		186, 187
19	Takouatz Guerrissène	702	36	32	N	4	30	E		LSA		135
20	Tamar Hat	757	36	40	N	2	25	E		LSA	Ib	185, 210
21	Ti-n-Hanakaten	371	23	52	N	10	22	E		LSA		44
22	Zmeilet el-Barka	350	29	25	N	2	3	W		LSA		3
Angola												
23	Benfica	119	8	57	S	13	9	E		LSA	Lupembo-Tshitolien	291
24	Kamabanga	117	8	51	S	13	14	E	*			291
25	Kitala	118	8	51	S	13	14	E	*			291
Botswana												
26	Bosutwe-Mapungubwe	357	21	57,15	S	26	36,65	E		EIA-LIA		241
27	Bosutwe-Taukome	355	21	57,15	S	26	36,65	E		EIA		241
28	Bosutwe-Toutswe	356	21	57,15	S	26	36,65	E		EIA		241
29	Bosutwe-Zimbabwe	358	21	57,15	S	26	36,65	E		LIA		241
30	Broadhurst	807	24	32	S	25	57	E		EIA		78

N site	Site	SITE N	Latitude			Longitude			*	Culture, facies, industry, chronology, notes		Citation
31	Commando Kop	526	22	8	S	29	10	E		EIA	Leopard´Kopje A	336
32	Divuyu	686	18	55	S	21	44	E	*	EIA		79
33	Drotsky's Cave-H	767	19	55	S	21	5	E	*	Ho		260
34	Drotsky's Cave-LP	768	19	55	S	21	5	E	*	LPl		260
35	Drotsky's Cave-P	769	19	55	S	21	5	E	*	Pl		260
36	Matanga	858	20	52,16	S	27	37,33	E		IA	Zimbabwe	319
37	Matlapaneng	688	19	53	S	23	23	E		EIA		79
38	Nqoma	687	18	26	S	21	20	E	*	EIA		79
39	Qogana	689	18	55	S	22		E	*	EIA		79
40	Taukome	645	21	53	S	26	56	E		EIA	Zhizo, Toutswe Tradition	254
41	Toteng1-SU1	770	20	25	S	22	57	E		LSA		258
42	Toteng1-SU2	771	20	25	S	22	57	E		LSA		258
43	Toteng1-SU3	772	20	25	S	22	57	E		LSA		258
44	Toteng1-SU4	773	20	25	S	22	57	E		LSA		258
45	Toteng3-SU3	776	20	25	S	22	57	E		LSA		258
46	Toteng3-SU4	777	20	25	S	22	57	E		LSA		258
47	White Paintings RS-1	490	18	25	S	21	30	E		Hi?		259
48	White Paintings RS-2ab	491	18	25	S	21	30	E		LSA-EIA		259
49	White Paintings RS-2b-3ab	492	18	25	S	21	30	E		LSA		259
Burkina Faso												
50	Cocorba-IA	304	14	42	N	0	29	W		IA		183
51	Cocorba-LSA	303	14	42	N	0	29	W		late LSA		183
52	Kissi 22	311	14	39	N	0	9	W		MIA		183
53	Kissi 40	312	14	38	N	0	8	W		LIA		183
54	Oursi hu-beero	310	14	42	N	0	28	W		IA		183
55	Oursi Village-EIA	307	14	43	N	0	29	W		EIA		183
56	Oursi Village-LIA	309	14	42	N	0	27	W		LIA		183
57	Oursi Village-MIA	308	14	42	N	0	27	W		MIA		183
58	Oursi-IA	306	14	43	N	0	29	W		EIA		183
59	Oursi-LSA	305	14	43	N	0	29	W		LSA		183
60	Péntènga 89/1-spit 13	21	11	50	N	1	50	E		LSA		92
61	Péntènga 89/1-spits 15-21	20	11	50	N	1	50	E		LSA		92
62	Péntènga 89/1-spits 4-10	22	11	50	N	1	50	E		IA		92

N site	Site	SITE N	Latitude			Longitude			*	Culture, facies, industry, chronology, notes		Citation
63	Saouga 94/120	313	14	22	N	0	11	W		Hi		183
64	Saouga 95/7	314	14	22	N	0	10	W		late LSA/IA		183
65	Tin Akof	300	14	55	N	0	17	W		late LSA		183
Burundi												
66	Ruzizi Plain	116	3	15	S	29	20	E	*			291
Cameroon												
67	Abeke	32	6	11	N	10	42	E		LSA		74
68	Aissa Dugjé	678	11	5	N	24	12	E	*			197
69	Aissa Dugjé-b	692	11	5	N	24	12	E	*			28
70	Blé A	346	11	58,75	N	14	59,63	E		IA		183
71	Blé B	347	11	58,75	N	14	59,63	E		IA		183
72	Blé C	348	11	58,75	N	14	59,63	E		IA		183
73	Blé E	349	11	58,75	N	14	59,63	E		IA		183
74	Djaba Hosséré	911	8	19	N	13	57,5	E		IA		180
75	Nkang	92	4	16	N	11	19	E		IA	Obobogo Tradition	208, 296
76	Shum Laka II c	30	5	51	N	10	4	E		LSA		74
77	Shum Laka II d	31	5	51	N	10	4	E		LSA		74
78	Shum Laka IIa+b	29	5	51	N	10	4	E		LSA		74
Central African Republic												
79	Mondongué	120	4	15	N	18	30	E	*			296
80	Lingbangbo	121	4	15	N	18	30	E	*			296
Chad												
81	Delebo	887	17	12	N	21	15	E				108
82	Enneri Bardagué	883	22	36	N	16	37	E		LSA		108
83	Gabrong-b	880	21	30	N	17	8	E		LSA		108
84	Gabrong-d	879	21	30	N	17	8	E		LSA		108
85	Koyom	33	9	48	N	15	52	E		IA	Recent	256, 257
Congo												
86	Kawezi	122	3	15	S	29	20	E	*	IA		296
87	Ntadi Yom ba-level B	35	4	15	S	13	46	E		LSA	Tshitolien	308, 309
88	Ntadi Yomba-level A	34	4	15	S	13	46	E		LSA	Tshitolien	308, 309
89	Tongo	123	1	10	S	29	0	E		EIA		296

N site	Site	SITE N	Latitude			Longitude			*	Culture, facies, industry, chronology, notes		Citation
Djibouti												
90	Asa Koma	205	11	4	N	42	5	E	*	LSA		131, 132
Egypt												
91	Abu Ballas - 83/39	38	24	14	N	27	25	E		LSA		314
92	Abu Ballas - 85/50-1	40	24	14	N	27	25	E		LSA		314
93	Abu Ballas - 85/51-3	41	24	14	N	27	25	E		LSA		314
94	Abu Ballas - 85/56	39	24	14	N	27	25	E		LSA		314
95	Abu Sha´ar-4C	74	27	22	N	33	41	E		IA	Late Roman	313
96	Abu Sha´ar-5C	75	27	22	N	33	41	E		IA	Late Roman	313
97	Abusir	664	29	54	N	31	12	E		LSA	1st Dynastie	23
98	Adaima-burials	26	25	10	N	32	40	E	*	LSA	End Naqada I to Naqada III	301
99	Adaima-habitat	65	25	10	N	32	40	E	*		Predynastic, Naqada IIc to III A1	212
100	Adaima-habitat	25	25	10	N	32	40	E	*	LSA	End Naqada I to Naqada III	301
101	Ain Tirghi cemetery	395	29	0	N	25	30	E		LSA	About 2-3 centry AD	58
102	Al-Ma´abda	182	27	20	N	31	1	E			Roman	235
103	Al-Ma´abda	288	27	20	N	31	1	E			Roman	235
104	Baboon Catacomb	816	29	51	N	31	14	E				126, 127
105	Balat-Mastaba V	677	25	34,9	N	29	14,78	E		LSA	End of VI th Dynastie	54
106	Berenike 95/97- Early Roman	67	23	55	N	35	27	E		IA	Early Roman	310
107	Berenike 95/97- Late Roman	68	23	55	N	35	27	E		IA	Late Roman	310
108	Berenike 95/97-Ptolemaic	69	23	55	N	35	27	E		IA	Ptolemaic	305
109	Bir Kiseiba-EN	135	22	34	N	29	53	E		LSA	Early Neolithic	103
110	Bir Kiseiba-LN	137	22	34	N	29	53	E		LSA	Late Neolithic	103, 353
111	Bir Kiseiba-MN	136	22	34	N	29	53	E		LSA	Middle Neolithic	103, 353
112	Buto/1-3	293	31	11	N	30	45	E		LSA	Naggada Iib	344
113	Buto-6	294	31	11	N	30	45	E		LSA	Old Kingdom	344
114	Buto-9-12	295	31	11	N	30	45	E		LSA	Ptolemaic to Graeco-Roman	344
115	Catfish Cave-L1	489	22	43	N	32	7	E		LSA	Late Ho	355
116	Catfish Cave-L2	488	22	43	N	32	7	E		LSA	Ho	355
117	Catfish Cave-L3	487	22	43	N	32	7	E		LSA	Aceramic Ho	355
118	Catfish Cave-L4	486	22	43	N	32	7	E		LSA	Early Holocene	355
119	Catfish Cave-L5	485	22	43	N	32	7	E		LSA?	Early Holocene	355
120	Dahschûr	699	31	2	N	30	28	E		LSA	12e Dynasty	19

N site	Site	SITE N	Latitude			Longitude			*	Culture, facies, industry, chronology, notes		Citation
121	Dakhleh Oasis, SMU	394	29	0	N	25	30	E		LSA	Sheikh Muftah Unit	59
122	Dakhleh Oasis, Ba	396	29	0	N	25	30	E		LSA	Bashendi	59
123	Dakhleh Oasis, Roman	398	29	0	N	25	30	E		IA	Roman	59
124	Dakhleh Oasis,OK	397	29	0	N	25	30	E		LSA	Old Kingdom	59
125	Djara 90/1	694	27	20	N	30		E	*			118
126	Eastpan 95-1/1	695	25		N	28	30	E	*			118
127	Eastpan 95-2	696	25		N	28	30	E	*			118
128	Ein el Gazzareen	871	25	39,15	N	28	52,03	E		LSA	Old Kingdom	60
129	El Abadiya 2 North sector	14	25	51	N	32	42	E		LSA	Naqada I	323
130	El Abadiya 2 South sector	13	25	51	N	32	42	E		LSA	Naqada I	323
131	El Abadyia-A-B	64	25	51	N	32	42	E	*	MSA	Late Paleolithic	322
132	El Abadyia-C1-C2	63	25	51	N	32	42	E	*	MSA	Shuwikhatian	322
133	El Omari	499	29	15	N	31	19	E		LSA	Late Neolithic, Predynastic	320
134	El Salamuni	693	26	27	N	31	40	E	*			324
135	Elephantine-Necropolis	494	24	5	N	32	53	E		LSA	Late and Middle Kindgom	19, 341, 342
136	Elephantine-NW Chnum temple	666	24	5	N	32	53	E		ISA	25-26th dynasty	22
137	Elephantine-Satet temple	495	24	5	N	32	53	E		LSA	4th-11th Dynasties	19, 341, 342
138	Elkab-Epipaleo	130	25	7	N	32	48	E		MSA	E, Elkabien	100
139	Elkab-Necropolis	129	25	7	N	32	48	E			New Kingdom	111
140	El-Tell el Iswid-Early Dynastic	390	30	50	N	31	41	E			Early dynastic - Naqada III	287
141	El-Tell el Iswid-Predynastic	389	30	50	N	31	41	E			Late predynastic period	287
142	El-Tell el-Iswid A	670	30	50	N	31	42	E	*	LSA	Late predynastic	287
143	El-Tell el-Iswid B	671	30	50	N	31	42	E	*	LSA	Naqada III-Early Dynastic	287
144	El-Zarkeh	73	26	0	N	33	47	E		IA	Early Roman	294
145	Fayoum G1-H1	145	29	20	N	30	30	E		LP		99
146	Fayoum G3-H2	144	29	20	N	30	30	E		LSA	N	99
147	Fayoum G5-G4	143	29	20	N	30	30	E		LSA	Predynastic and old Kingdom	99
148	Fayoum Site 1	424	29	35	N	30	40	E		LSA	Fayoum A - Neolithic	31
149	Fayoum Site 2	425	29	35	N	30	47	E		LP	Fayoum B	31
150	Fayoum Site 3	426	29	34	N	30	53	E		LSA	Fayoum A - Neolithic	31
151	Fayoum Site 4	427	29	34	N	30	38	E		LSA	Fayoum A - Neolithic	31
152	Fayoum Site 5	428	29	23	N	30	23,5	E		LSA	Fayoum A - Neolithic	31
153	Gilf Kebir	42	23	4	N	25	52	E	*	LSA		314

N site	Site	SITE N	Latitude			Longitude			*	Culture, facies, industry, chronology, notes		Citation
154	Hierakonpolis HK6 - Naqada IC-IIA	16	25	2	N	32	49	E	*	LSA	Predynastic, Early Dynastic, Naqada IC-IIA	311
155	Hierakonpolis HK6 - Naqada IIA-B	17	25	2	N	32	49	E	*	LSA	Predynastic, Early Dynastic, Naqada IIA-B	311
156	Hierakonpolis HK6 - Naqada III	18	25	2	N	32	49	E	*	LSA	Predynastic, Early Dynastic, Naqada IIIc	311
157	Hierakonpolis HK6 - Naqada I-II	15	25	2	N	32	49	E	*	LSA	Predynastic, Early Dynastic, Naqada I-II	311
158	Idfu, Isna	383	24	59	N	32	52	E		MSA	LP	129
159	Karnak North	496	25	44	N	32	42	E		LSA	New Kindgom	341, 342
160	Kharga depression-graeco-roman	204	25	39	N	30	38	E			Graeco roman	101
161	Kharga depression-Neolithic	203	25	39	N	30	38	E		LSA	N	101
162	Kom Ombo	183	24	28	N	32	57	E				61, 229
163	Lobo - 81/55	37	26	4	N	27	17	E	*	LSA		314
164	Maadi	55	29	49	N	31	19	E	*	LSA	Late Predynastic	136, 347
165	Mahgar Dendera 2	19	26	3	N	32	48	E		LSA	End of Badarian	139, 299
166	Makhadma 2	61	26	12	N	32	38	E		MSA	LP	312, 321
167	Makhadma 4	62	26	12	N	32	38	E		MSA	LP	312, 321
168	Merimde Benisâlame	291	30	19	N	30	51	E		LSA	N	342, 347
169	Minshat Abu Omar	665	30	50	N	32	6	E	*	LSA		18, 342
170	Mons Claudianus	72	26	51	N	33	32	E	*	IA	Early Roman	294
171	Napta Playa 77-1/4/5/6	201	22	30	N	30	45	E		LSA	N, Early Khartoum	101
172	Napta Playa 77-8	202	22	30	N	30	45	E		LSA	Middle to Late Neolothic	101
173	Napta Playa E77-3/7	200	22	30	N	30	45	E		LSA	LP	101
174	Qantir	700	30	48	N	31	50	E		LSA	19/20 Dynasties	19, 342
175	Qasr el-Sagha-1	296	29	34	N	30	38	E		LSA	Early N	342
176	Qasr el-Sagha-2	297	29	34	N	30	38	E		LSA	Middle N	342, 343
177	Qasr el-Sagha-3	298	29	34	N	30	38	E		IA	12th Dynasty, about 1990-1800 BC	342
178	Qasr Ibrim	222	22	37	N	31	59	E		LSA	Napatan Period	
179	Quseir al-Qadim	71	26	6	N	34	17	E		IA	Early Roman	294, 345
180	Selima Sandsheet - 85/78	43	21	5	N	29	20	E		LSA		314
181	Selima Sandsheet - 85/79	44	21	5	N	29	20	E		LSA		314
182	Shenshef	124	23	43	N	35	22	E	*		Early to Late Roman	306
183	Shuwikhat 1	66	26	11	N	32	36	E		MSA	Shuwikhatian	297
185	Sodmein Cave older backfill	56	26	14	N	33	38	E		MSA	QSR-44	325
186	Sodmein Cave younger backfill	57	26	14	N	33	38	E		LSA	QSR-44	325
187	Tell el-Dab´a	289	30	44	N	32	1	E		LSA	Early 18th Dynasty	349

N site	Site	SITE N	Latitude			Longitude			*	Culture, facies, industry, chronology, notes		Citation
188	Tell el-Dab´a - b	667	30	44	N	32	1	E		LSA	Middle Kingdom to 2nd intermed	17, 342
189	Tell Heboua	256	30	57	N	32	27	E			1786-1552 cal BC : New Kindgom	50
190	Tell Mashkuta	668	30	45	N	32	20	E	*			342
191	Toukh	367	25	56	N	32	45	E		N		93
192	Tree Shelter-1	862	26	16,332	N	33	57,379	E		IA		202
193	Tree Shelter-3	24	26	16,332	N	33	57,379	E		LSA		326
194	Tree Shelter-4	863	26	16,332	N	33	57,379	E		LSA		202
195	Tree Shelter-5	23	26	16,332	N	33	57,379	E		LSA		326
196	Tuna el-Gebel	292	27	46	N	30	44	E			Saitic-Ptolemaic	20, 346, 348
197	Umm el-Qaab, Friedhof U	290	26	11	N	30	25	E			Naqada Ia/IIIa	81
198	Wadi el Akhdar	199	23	1	N	26	6	E				227
199	Wadi el Bakht	198	23	12	N	26	28	E				227
200	Wadi Kalalat	125	23	55	N	35	27	E	*		Early Roman	306
201	Wadi Kubbaniya - E-78-3	60	24	15	N	31	19	E		MSA	LP	110, 114
202	Wadi Kubbaniya - E-78-4	58	24	15	N	31	19	E		MSA	LP	110, 114
203	Wadi Kubbaniya - E-81-1	59	24	15	N	31	19	E		MSA	LP	110, 114
204	Western Thebes- Tomb 10	663	16	30	N	14	13	E			Late Ptolemaic-Roman	21
205	Willmann´s Camp - 81/61	36	25	5	N	25	37	E	*	LSA		314
206	Workmen Village	387	27	40	N	31	0	E		LSA		188
Ethiopia												
207	Axum-D	211	14	8	N	38	43	E			Pre-Later Axumite	34
208	Axum-K	212	14	8	N	38	43	E			Post-Axumite	34
209	Gademotta	142	7	55	N	38	42	E	*		MSA	98
210	Gobedra	889	14	0	N	39	0	E	*	LSA		207
211	Laga Oda-a	892	14	30	N	41	0	E	*	LSA		207
212	Laga Oda-b	893	14	30	N	41	0	E	*	IA		207
213	Lake Besala	888	8,37	0	N	39,16	0	E	*	LSA		207
214	Lalibela	891	12	30	N	38,7	0	E	*	IA		207
215	Moche Borago-1a	912	6	53	N	37	44,5	E		LSA	Aceramic Early Phase 1	181
216	Moche Borago-1b	913	6	53	N	37	44,5	E		LSA	Aceramic Intermediate Phase I	181
217	Moche Borago-1c	914	6	53	N	37	44,5	E		LSA	Aceramic Late Phase 1	181
218	Moche Borago-2	915	6	53	N	37	44,5	E		LSA	Ceramic Holocene	181
219	Moche Borago-3	916	6	53	N	37	44,5	E		LSA	Ceramic Holocene	181

N site	Site	SITE N	Latitude			Longitude			*	Culture, facies, industry, chronology, notes		Citation
220	Quiha	890	14	30	N	40	0	E	*	LSA		207
Gabon												
221	Okala	253	0	29,43	N	9	24,75	E		N	Okala Grup	63
222	Oveng	93	0	28,72	N	9	31	E		EIA		304
Ghana												
223	Anyinam-Area A-B-C	176	6	25	N	1	34	W				116
224	Anyinam-Area D	175	6	25	N	1	34	W				116
225	Asantamanso-ASII/III	172	6	25	N	1	34	W				116
226	Asantamanso-ASIV	173	6	25	N	1	34	W				116
227	Asantamanso-ASV	174	6	25	N	1	34	W				116
228	Begho	163	7	51	N	2	28	W		IA		116
229	Bibease Plantation	167	5	46	N	0	14	W				116
230	Bono Manso	165	7	40	N	1	50	W	*			116
231	Daboya	166	9	31	N	1	22	W		IA		116
232	Dawu	162	5	59	N	0	5	W		IA		116
233	Esiease - A-C-D	177	6	25	N	1	34	W				116
234	Esiease - H	178	6	25	N	1	34	W				116
235	Esiease -recent	179	6	25	N	1	34	W				116
236	Frederiksgave plantation	168	5	45	N	0	14	W				116
237	Kintampo Rockshelter K6-IA	156	8	1	N	1	45	W		IA		116
238	Kintampo Rockshelter K6-K	157	8	1	N	1	45	W		LSA	Kintampo	42, 116
239	Kintampo Rockshelter K6-K/P	158	8	1	N	1	45	W		LSA	Kintampo/Punpun	116
240	Kintampo Rockshelter K6-P	159	8	1	N	1	45	W		LSA	Punpun	116
241	Mandara	169	9	1	N	2	35	W		IA		116
242	Mankuma	170	9	10	N	2	29	W		IA		116
243	Mumute	161	7	35	N	2	22	W	*			116
244	Ntereso	160	9	7	N	1	13	W		LSA	Kintampo	42, 116
245	Nyanawase/Akwamu Amanfoso	164	5	55	N	0	39	W				116
246	Nyanga	171	9	14	N	2	21	W		IA		116
Kenya												
247	Akira-N	718	0	14	S	36	19	E	*	LSA	Neolithic, Akira/Elmenteitan	122, 124
248	Akira-PIA	716	0	14	S	36	19	E	*	LSA	Pastoral Iron Age	122, 124
249	Akira-SPN	717	0	14	S	36	19	E	*	LSA	SPN, Akira wares	122, 124

N site	Site	SITE N	Latitude			Longitude			*	Culture, facies, industry, chronology, notes		Citation
250	Causeway Site	712	0	48	S	36	23	E		LSA	SPN, Eburrean 5	122, 124
251	Crescent Island main	444	0	49	S	36	23	E	*	LSA	SPN (SPN), Narusura wares	122, 124
252	Dongodien	723	3	56	N	36	12	E		LSA	Nderit ceramics	122
253	Ele Bor A-A	703	3	33	N	38	7	E	*	LSA		123
254	Ele Bor A-B	704	3	33	N	38	7	E	*	LSA		123
255	Ele Bor A-C	705	3	33	N	38	7	E	*	LSA		123
256	Ele Bor A-D	706	3	33	N	38	7	E	*	LSA		123
257	Engaruani	206	1	20	S	35	42	E	*	LSA	EHo?	205
258	Enkapune Ya Muto - Eburran 4a	484	0	50	S	36	9	E		LSA	Eburran 4	199, 201
259	Enkapune Ya Muto - Eburran 4b	483	0	50	S	36	9	E		LSA	Eburran 4	199, 201
260	Enkapune Ya Muto - Eburran 5a	482	0	50	S	36	9	E		LSA	Eburran 5	199, 201
261	Enkapune Ya Muto - Eburran 5b	481	0	50	S	36	9	E		LSA	Eburran 5	199, 201
262	Enkapune Ya Muto - Eburran 5c	480	0	50	S	36	9	E		LSA	Eburran 5	199, 201
263	Enkapune Ya Muto BSS-HTL	479	0	50	S	36	9	E		LSA		199, 201
264	FxJj12-a	898	4	5	N	36	20	E		LSA		280
265	FxJj12-b	899	4	5	N	36	20	E		LSA		280
266	FxJj12-N	905	4	5	N	36	20	E		LSA		280
267	GaJi1	902	3	56	N	36	16	E		LSA		280
268	GaJi11	903	3	57	N	36	23	E		LSA		280
269	GaJi12	904	3	57	N	36	25	E		LSA		280
270	GaJi2	901	3	54,5	N	36	14	E		LSA		280
271	GaJi3	900	3	57	N	36	13	E		LSA		280
272	Gogo Falls-I	477	0	34	S	34	21	E		LSA	Elmentetian	159
273	Gogo Falls-II-III	210	0	34	S	34	21	E		LSA	Elmentetian	159, 204, 207
274	Gogo Falls-sq26-29	478	0	34	S	34	21	E		LSA	Elmentetian	159
275	GvJm 19-HA	461	1	29	S	37	3	E		LSA	Holocene Aceramic	200
276	GvJm 19-HC	462	1	29	S	37	3	E		LSA	Holocene Ceramic	200
277	GvJm 19-TLPG	460	1	29	S	37	2	E		LSA	LPI Geometric	200
278	GvJm 22-LHC	464	1	29	S	37	3	E		LSA	Holocene Ceramic	200
279	GvJm 22-LPG	463	1	29	S	37	3	E		LSA	LPI Geometric	200
280	GvJm 46-LPG	465	1	29	S	37	3	E		LSA	LPI Geometric	200
281	GvJm 62-H	469	1	29	S	37	3	E		LSA	Ho	200
282	GvJm 62-LPG	468	1	29	S	37	3	E		LSA	LPI Geometric	200

N site	Site	SITE N	Latitude			Longitude			*	Culture, facies, industry, chronology, notes		Citation
283	GvJm 62-LPNG	467	1	29	S	37	3	E		MSA	LPI Non Geometric	200
284	Kulchurdo	707	2	20	N	37	59	E	*	Hi		120
285	Lemek North-East	209	1	6	S	35	23	E		LSA	Narosura wares	122
286	Lothagam	909	2	55	N	36	4	E		LSA		280
287	Lowasera-a	906	2	53,5	N	36	43	E		LSA		280
288	Lowasera-b	907	2	53,5	N	36	43	E		LSA		280
289	Lowasera-c	908	2	53,5	N	36	43	E		LSA		280
290	Maasai Gorge Shelter	714	0	43	S	36	15	E	*	LSA	Elmenteitan	122, 124
291	Maringishu	710	0	1	N	36	6	E	*	LSA	SPN, Narosura wares	122, 124
292	Marula Rock Shelter	724	0	40	S	36	10	E	*	LSA	N, Eburrean 3	122
293	Naivasha Railway Shelter	713	6	47	S	36	27	E	*	LSA	PN, Eburrean 5, Narosura and Salasun	122, 124
294	Narosura	722	1	1	S	35	20	E	*	LSA	SPN, Narosura wares	122, 124
295	Ndabibi	715	0	49	S	36	15	E	*	LSA	SPN, Narosura wares	122, 124
296	Ngamuriak	207	1	6	S	35	23	E		LSA	Elmentetian	205
297	Ngenyn	709	0	37	N	36	1	E	*	LSA	SPN, Akira, Narosura and Turkwell	122, 124
298	Prolonged Drift	711	0	27	S	36	6	E	*	LSA	SPN, Narosura wares	121, 122, 124
299	Salasun-LSA	721	1	9	S	36	21	E	*	LSA	Akira/Elmenteitan wares	122, 124
300	Salasun-PIA	719	1	9	S	36	21	E	*	LSA	Pastoral IA	122, 124
301	Salasun-PN	720	1	9	S	36	21	E	*	LSA	PN	122, 124
302	Sambo Ngige	208	1	6	S	35	23	E		LSA	Elmentetian	205
Lesotho												
303	Bolahla	748	30	4	S	28	24	E		LSA	pCW	243
304	Ha Makotoko	787	29	20	S	27	48	E		LSA	Oakhurst Industry	243
305	Lequetsoana	750	29	26	S	27	35	E		LSA	pCW and ceramic pCW (2700-100 bp)	249
306	Likoaeng-I	739	29	44	S	28	45	E		LSA- EIA	Ceramic Wilton Assemblage	249
307	Likoaeng-III-V	740	29	44	S	28	45	E		LSA- EIA	pCW	249
308	Likoaeng-VII-IX	741	29	44	S	28	45	E		LSA- EIA	pCW	249
309	Likoaeng-XI	742	29	44	S	28	45	E		LSA- EIA	pCW	249
310	Likoaeng-XIII	743	29	44	S	28	45	E		LSA	pCW	249
311	Likoaeng-XVII	744	29	44	S	28	45	E		LSA	pCW	249
312	Liphofung-CW	786	29	44	S	28	27	E		LSA	Classic Wilton	243
313	Liphofung-pCW	745	29	44	S	28	27	E		LSA	pCW	249
314	Muela	746	29	44	S	28	27	E		LSA	pCW	243

N site	Site	SITE N	Latitude			Longitude			*	Culture, facies, industry, chronology, notes		Citation
315	Ntloana Tsoana	788	29	10	S	27	49	E		LSA	Oakhurst Industry	243
316	Sehonghong-a	749	29	46	S	28	47	E		LSA	pCW and ceramic pCW	249
317	Sehonghong-b	778	29	46	S	28	47	E		LSA	Classic Wilton	248
318	Sehonghong-c	779	29	46	S	28	47	E		LSA	Oakhurst levels	248
319	Sehonghong-d	780	29	46	S	28	47	E		LSA	Robberg levels	248
320	Sehonghong-e	781	29	46	S	28	47	E		LSA	MSA/LSA levels	248
321	Tloutle-CWI	782	29	28	S	27	46	E		LSA	Classic Wilton	243
322	Tloutle-O	783	29	28	S	27	46	E		LSA	Oakhurst Industry	243
323	Tloutle-pCW	747	29	28	S	27	46	E		LSA	Late pCW	243, 249
324	Tloutle-R	784	29	28	S	27	46	E		LSA	Robberg	243
Libya												
325	Et Tera	364	32	10	N	20	18	E		LP	Ib	176
326	Hauah Fteah- Historic	354	32	50	N	22	5	E		Hi	Graeco-Roman and Pharaonic Age	176
327	Hauah Fteah-Iberomaurusian	351	32	50	N	22	5	E			IB	176
328	Hauah Fteah-Libyco-capsian	352	32	50	N	22	5	E			Libyco-capsian	176
329	Hauah Fteah-Neolithic	353	32	50	N	22	5	E		N		176
330	In Aghelachem-RT,SF	270	25	14,1	N	10	19,35	E		LSA	Garamantian	2
331	In Aghelachem-T1	269	24	14,1	N	10	19,35	E		LSA	Late Pastoral	2
332	Murzuq	882	24	40	N	13	0	E	*	LSA		223
333	Ti-n-Thora East	140	25	8	N	10	27	E		LSA		43, 106, 113
334	Ti-n-Thora North	139	25	8	N	10	27	E		LSA		43, 106, 113
335	Ti-n-Thora Two Caves	141	25	8	N	10	27	E				106, 113
336	Uan Afuda-US1-3	264	24	52,12	N	10	30,32	E		LSA	Mesolithic	68
337	Uan Afuda-US4-7	263	24	52,12	N	10	30,32	E		LSA	Mesolithic	68
338	Uan Muhuggiag (LA)	262	24	50,53	N	10	30,78	E		LSA	Late Acacus or Mesolithic	68
339	Uan Muhuggiag (LP)	261	24	50,53	N	10	30,78	E		LSA	Late Pastoral	68, 95
340	Uan Muhuggiag (MP)	260	24	50,53	N	10	30,78	E		LSA	Middle Pastoral	68, 95
341	Uan Tabu-11	266	24	51,58	N	10	31,7	E		LSA	Mesolithic	68
342	Uan Tabu-14-20	265	24	51,58	N	10	31,7	E		LSA	Mesolithic	68
343	Uan Telocat (EP)	257	24	46,12	N	10	36,07	E		LSA	Early Pastoral	68, 95
344	Uan Telocat (LP)	259	24	46,12	N	10	36,07	E		LSA	Late Pastoral	68, 94
345	Uan Telocat (MP)	258	24	46,12	N	10	36,07	E		LSA	Middle Pastoral	68, 95
346	Wadi Tannezzuft-00/195	268	25	30,23	N	10	22,28	E		LSA	Late pastoral	2

N site	Site	SITE N	Latitude			Longitude			*	Culture, facies, industry, chronology, notes		Citation
347	Wadi Tannezzuft-96/129	267	25	22,1	N	10	18,75	E		LSA	Late pastoral	2
Madagascar												
348	Fompona	422	21	15,46	S	48	20,75	E				89
349	Madagascar	423	19	40	S	46	30	E	*			89
Malawi												
350	Chikumba	588	15	55	S	35	10	E	*	EIA		331
351	Malowa-1	872	15	49	S	35	13	E		LIA		158
352	Malowa-2	873	15	49	S	35	13	E		LIA		158
353	Malowa-3	874	15	49	S	35	13	E		IA		158
354	Malowa-4	875	15	49	S	35	13	E		IA		158
355	Malowa-5	876	15	49	S	35	13	E		LSA		158
356	Matope Court-EIA	586	15	22	S	34	56	E		EIA		331
357	Matope Court-IA	584	15	22	S	34	56	E		IA		331
358	Matope Court-LIA	585	15	22	S	34	56	E		LIA		331
359	Namichimba	587	15	20	S	35	10	E	*	IA		331
Mali												
360	Akumbu-AK3	235	15	4	N	5	35	W		IA		194, 285
361	Akumbu-EIA	236	15	4	N	5	35	W		IA		194, 285
362	Akumbu-MIA	237	15	4	N	5	35	W		IA		194, 285
363	Asselar	473	19	0	N	0	10	E		LSA	N, Facies A	273
364	Dia-Shoma Horizon I	218	14	21	N	4	58	W		IA		16
365	Dia-Shoma Horizon II	219	14	21	N	4	58	W		IA		16
366	Dia-Shoma Horizon III+IV	220	14	21	N	4	58	W		IA		16
367	Dia-Shoma Horizon V	221	14	21	N	4	58	W		IA		16
368	Erg Ine Sakane	225	20	55	N	0	32	W		LSA	N	154, 307
369	Est-Hassi el Abiod	224	18	27,5	N	2	28	W		LSA	N	154, 307
370	Fanfannyégènè I	255	13	45	N	9	14	W		N		47
371	Gao Ancien	217	16	16	N	0	3	W		IA		196
372	Gao Saney	216	16	16	N	0	3	W		IA		196
373	Hamdallahi	382	14	19	N	4	6	W		IA	Hi	57, 320
374	Hassi el-Abiod	223	19	7	N	3	55	W		LSA	N	154, 307
375	In Begaouaen	474	18	0	N	1	0	E	*	LSA	N, Facies B	273
376	Jenné-Jeno - phase I/II	85	13	53	N	4	32	W		IA		190, 293

N site	Site	SITE N	Latitude			Longitude			*	Culture, facies, industry, chronology, notes	Citation
377	Jenné-Jeno - phase III	86	13	53	N	4	32	W	IA		190, 293
378	Jenné-Jeno - phase IV	87	13	53	N	4	32	W	IA		190, 293
379	Karkarichinkat KN05	864	16	52	N	0	12	E	LSA	N, Facies K	91
380	Karkarichinkat N-S	475	16	52	N	0	12	E	LSA	N, Facies K	271, 272
381	Kobadi	228	15	21,5	N	5	29	W	LSA	N	155
382	Kolima Sud-I	233	15	23	N	5	19	W	IA	1700-1300 cal BC (Phase I)	189, 195
383	Kolima Sud-II	234	15	23	N	5	19	W	N	1300-500 cal BC (Phase II)	189, 195
384	Korounkorokalé-I	215	12	24	N	8	14	W	IA	Date : transitiond II-I (after 1020+-105 bp)	192
385	Korounkorokalé-III	213	12	24	N	8	14	W	LSA	Preceramic LSA, transition III-II (before 5185 +-95 BP)	192
386	Mansa Musa	691	16	16	N	0	3	W	IA		149
387	Tessalit	226	20	0	N	0	20	W	LSA	N	154, 307
388	Tiabel Goudiodié	232	15	3	N	5	15	W	IA		189, 195
389	Tongo Maaré Diabel	231	15	1	N	2	57	W	IA	Classic Period	189
390	Windé Koroji I,III	277	15	8	N	2	56	W	LSA	Windé Koroji Faciès	191
391	Windé Koroji SO	276	15	8	N	2	56	W	LSA	Gourma facies, Kobadi trad.	191
392	Zampia-Tumulus 7	278	15	7	N	2	57	W	LSA	Zampia Faciès	191
Mauritania											
393	Berouâga	365	21	14,45	N	16	27,02	W	LSA	N	157
394	Chami-A2	373	20	7	N	16	6	W			27
395	Dhar Tichitt- highest beach	447	18	28	N	9	30	W	LSA	N, supposed to be earlier than 4500-400 bp	215
396	Dhar Tichitt- latest phase	451	18	28	N	9	30	W	LSA	N	215
397	Dhar Tichitt- lowest beach	449	18	28	N	9	30	W	LSA	N	215
398	Dhar Tichitt- middle beach	448	18	28	N	9	30	W	LSA	N	215
399	Dhar Tichitt- post beach	450	18	28	N	9	30	W	LSA	N	215
400	Dhar Tichitt Site 38	446	18	25	N	9	14	W	LSA	N	143, 145
401	Dhar Tichitt Site 46	445	18	24	N	9	15	W	LSA	N	143, 144
402	DN4-BKH-S5	227	16	45	N	7	15	W	LSA	N	225
403	Draa Malichigdane	503	20	0	N	13	35	W	LSA		40
404	Ile de Tidra	500	19	47	N	16	24	W	LSA		328
405	Khatt Lemaiteg	361	19	12	N	14	45	W	LSA	N	14
406	Khneivissa	229	17	45	N	7	28	W	LSA	N	4
407	Tegdaoust-a	378	17	25	N	10	25	W		Hi	26
408	Tegdaoust-b	379	17	25	N	10	25	W		Hi	26

N site	Site	SITE N	Latitude			Longitude			*	Culture, facies, industry, chronology, notes	Citation
409	Tegdaoust-c	380	17	25	N	10	25	W		Hi	26
410	Tegdaoust-d	381	17	25	N	10	25	W		Hi	26
411	Tichitt - Akjinjeir phase	452	18	28	N	9	30	W	LSA	N	216
412	Tichitt - Arriane phase	453	18	28	N	9	30	W	LSA	N	216
413	Tichitt - Chebka phase	454	18	25	N	9	20	W	LSA	N	216
414	Tichitt - Goungou phase	457	18	20	N	9	15	W	LSA	N	216
415	Tichitt - Khimiya phase	458	18	21	N	9	13	W	LSA	N	216
416	Tichitt - Naguez phase	455	18	25	N	9	20	W	LSA	N	216
417	Tichitt - Nkhal phase	456	18	25	N	9	20	W	LSA	N	216
418	Tintan	372	20	46	N	16	36	W	LSA		27
Morocco											
419	Bou Saria A	436	35	31	N	5	25	W	IA	P	220
420	Bou Saria B	437	35	31	N	5	25	W		Proto-campaniforme (recent Neolithic)	220
421	Bou Saria C-D	438	35	31	N	5	25	W	LSA	Cardial	220
422	El Harhoura 1	802	33	54,14	N	6	56,6	W	LSA	AT	5
423	El Harhoura 2	868	33	54,14	N	6	56,6	W	LSA	early N	217
424	El Mnasra	867	33	55,67	N	6	57,22	W	LSA	Early N, Cardial	217
425	Ghar-Khal I	439	35	52,607	N	5	25,517	W	Hi		220
426	Ghar-Khal IIIb-V	441	35	52,607	N	5	25,517	W	LSA	Ancient N	220
427	Ghar-Khal II-IIIa	440	35	52,607	N	5	25,517	W	LSA	Recent N	220
428	Ghar-Khal-Ib	865	35	52,607	N	5	25,517	W	LSA	Ib	29
429	Kaf-taht-el-Ghar-A	429	35	31	N	5	21	W	P		220
430	Kaf-taht-el-Ghar-B	430	35	31	N	5	21	W	P	Campaniforme	220
431	Kaf-taht-el-Ghar-C	431	35	31	N	5	21	W	LSA	Recent N	220
432	Kaf-taht-el-Ghar-D	432	35	31	N	5	21	W	LSA	Cardial 2	220
433	Kaf-taht-el-Ghar-E	433	35	31	N	5	21	W	LSA	Cardial 1	220
434	Kaf-taht-el-Ghar-F	434	35	31	N	5	21	W	LSA	Late N	220
435	Kaf-taht-el-Ghar-G	435	35	31	N	5	21	W	E		220
436	Kehf el Hammar	708	35	20,62	N	5	10,93	W	LP	Ib	13
437	Kehf-el-Baroud	442	33	55	N	7	0	W	IA	Campaniforme?	220
438	Kehf-el-Baroud-aven	443	33	55	N	7	0	W	IA	bronze	220
439	Rouazi necropolis	366	33	51	N	7	2	W	LSA	N	72
440	Taforalt	866	34	48,63	N	2	24,5	W	LSA	Ib	29

N site	Site	SITE N	Latitude			Longitude			*	Culture, facies, industry, chronology, notes		Citation
Mozambique												
441	Manekweni-1	656	22	10	S	34	30	E		LIA	Great Zimbabwe	12
442	Manekweni-2-ABC	659	22	10	S	34	30	E		LIA	Great Zimbabwe	12
443	Manekweni-3	660	22	10	S	34	30	E		LIA	Great Zimbabwe	12
444	Manekweni-4	661	22	10	S	34	30	E		LIA	Great Zimbabwe	12
445	Manekweni-5	662	22	10	S	34	30	E		LIA	Great Zimbabwe	12
Namibia												
446	Aar I and II	535	26	43	S	16	31	E		LSA		282
447	Apollo 11-a	540	27	45	S	17	6	E		LSA	LSA with pottery	282
448	Apollo 11-b	541	27	45	S	17	6	E		LSA	LSA Wilton	282
449	Apollo 11-c	542	27	45	S	17	6	E		Early LSA	LPI	282
450	Brandberg-BBC	97	21	9	S	14	32	E	*	IA	Brandberg Culture	303
451	Brandberg-late LSA	96	21	9	S	14	32	E	*	Late LSA		303
452	Brandberg-LSA	95	21	9	S	14	32	E	*	LSA		303
453	Brandberg-MSA	94	21	9	S	14	32	E	*	MSA		303
454	Fackelträger	544	22	30	S	15	32	E	*	LSA		282
455	Geduld-LIA	843	20	17	S	15	50	E				278
456	Geduld-LSA	844	20	17	S	15	50	E				278
457	Haalenberg	543	26	40	S	15	32	E		LSA		282
458	Namtib	536	26	2	S	16	15	E		LSA		282
459	Nos	538	25	30	S	15	30	E		LSA		282
460	Oshigambo Peninsula	727	18	42,633	S	16	3,098	E		LPI		142
461	Pokkenbank	534	27	13	S	16	31	E		LSA		282
462	Steenbras Bay	539	26	40	S	15	9	E		LSA		282
463	Tiras	537	26	13	S	16	34	E		LSA		282
Niger												
464	Adrar Bous	388	20	18	N	8	57	E		LSA		41, 264
465	Adrar Bous 10	363	20	19,83	N	9	2	E		LSA	N	265
466	Adrar Bous-K	869	20	18	N	8	57	E		LSA	Kiffian	125, 276
467	Adrar Bous-T	870	20	18	N	8	57	E		LSA	Tenerian	125, 276
468	Arlit-a	885	18	44	N	7	43	E		LSA		108
469	Arlit-b	884	18	44	N	7	43	E		LSA		108
470	Awzerkat-1	419	17	14	N	5	32	E		LSA		82

N site	Site	SITE N	Latitude			Longitude		*	Culture, facies, industry, chronology, notes		Citation
471	Chin Tafidet	285	17	26	N	6	15	E	LSA	N	90, 224
472	Gawane	421	16	30	N	5	46	E	LSA		82
473	Gobero-K	725	16	55,88	N	9	30,13	E	LSA	Kiffian	270
474	Gobero-T	726	16	55,88	N	9	30,13	E	LSA	Tenerian	270
475	In Aridal 1	411	17	42,5	N	4	21	E	LSA		82
476	Madaouéla	655	18	40	N	7	35	E	LSA	N	55
477	Mentes 1	401	16	59	N	4	18	E	LSA		82
478	Mentes 13	408	17	1	N	4	19	E	LSA		82
479	Mentes 20	409	16	58	N	4	19	E	LSA		82
480	Mentes 21	410	16	57	N	4	21	E	LSA		82
481	Mentes 4	403	16	59	N	4	18	E	LSA		82
482	Mentes 5	406	16	59	N	4	18	E	LSA		82
483	Mentes 6	404	17	1	N	4	19	E	LSA		82
484	Mentes 7	407	17	1	N	4	19	E	LSA		82
485	Mentes 7b	405	17	1	N	4	19	E			82
486	Mentes nord	402	16	58	N	4	18	E	LSA		82
487	Takene Bawat 1	412	17	33	N	4	38	E	LSA		82
488	Takene Bawat 2	413	17	33	N	4	43	E	LSA		82
489	Takene Bawat 5	414	17	34	N	4	42	E	LSA		82
490	Takene Bawat 6	415	17	35	N	4	45	E	LSA		82
491	Tarada-C	400	15	58	N	4	52	E	LSA		82
492	Tarada-T	399	15	58	N	4	52	E	LSA		82
493	Tasara-1	420	16	48	N	5	40	E	LSA		82
494	Tin Kotkon	416	17	32	N	5	20	E	LSA		82
495	Tin Ouaffadene	362	20	10,67	N	9	11,5	E	LSA	N	265
496	TMY1	417	17	34	N	5	26	E	LSA		82
497	TMY3	418	17	36	N	5	24	E	LSA		82
Nigeria											
498	Bukarkurari-I	327	11	45	N	13	25	E	LSA	Gajiganna I	183
499	Bukarkurari-II	328	11	45	N	13	25	E	LSA	Gajiganna II	183
500	Daima	886	12	12	N	14	27	E	LSA		108
501	Dorota	333	11	57	N	13	9	E	EIA	Bama Deltaic Complex	183
502	Elkido North	332	12	3	N	13	14	E	EIA	Bama Deltaic Complex, 2nd-5/6th C. AD	183

N site	Site	SITE N	Latitude			Longitude			*	Culture, facies, industry, chronology, notes		Citation
503	Gajiganna A	316	12	15,4	N	13	13	E		LSA	Gajiganna Iib	30, 183
504	Gajiganna B1-lower	317	12	15,4	N	13	12,8	E		LSA	Gajiganna IIa/b	30, 183
505	Gajiganna B2	318	12	15,4	N	13	12,8	E		LSA	Gajiganna Iic	30, 183
506	Gajiganna C	315	12	16	N	13	12,9	E		LSA	Gajiganna Iia	183
507	Galaga	334	12	20	N	13	15	E		EIA	Bama Deltaic Complex, mid 9th to recent	183
508	Gilgila-IIa	322	12	14	N	13	11	E		LSA	Gajiganna Iia	183
509	Gilgila-IIb	323	12	14	N	13	11	E		LSA	Gajiganna Iib	183
510	Gilgila-IIc	324	12	14	N	13	11	E		LSA	Gajiganna Iic	183
511	Kariari	330	11	32	N	13	37	E		LSA	Gajiganna II	183
512	Kelumeri	329	11	32	N	13	39	E		LSA	Gajiganna later phase	183
513	Kursakata	91	12	20	N	4	10	E		LSA		130
514	Kursakata-IA	336	12	19	N	14	12	E		IA		183
515	Kursakata-LSA	335	12	19	N	14	12	E		LSA		183
516	Labe Kanuri	331	12	3	N	13	24	E		EIA	Bama Deltaic Complex, 1st to 5th century AD	183
517	Labe NA 97/24	320	12	4	N	13	23	E		LSA	Gajiganna I	183
518	Mege I	337	12	15	N	14	15	E		IA	Daima I	183
519	Mege II	338	12	15	N	14	15	E		IA	Daima II	183
520	Mege III	339	12	15	N	14	15	E		IA	Daima III	183
521	Mege IV	340	12	15	N	14	15	E		IA	Recent	183
522	NA 93/10	325	12	10	N	13	14	E		LSA	Gajiganna Iic	183
523	NA 93/36	321	12	12	N	13	15	E		LSA	Gajiganna Iia	183
524	NA 93/42	319	12	10	N	13	15	E		LSA	Gajiganna Group	183
525	Ngala I	341	12	20	N	14	11	E		IA		183
526	Ngala II	342	12	20	N	14	11	E		IA		183
527	Ngala IIIa	343	12	20	N	14	11	E		IA		183
528	Ngala IIIb	344	12	20	N	14	11	E		IA		183
529	Ngala IV	345	12	20	N	14	11	E		IA		183
530	Zilum	326	12	23	N	13	21	E		LSA	Gajiganna III	183
Rhodesia												
531	Chamabvefva	813	30	4,75	S	20	13,25	E		IA		147
532	Harleigh Farm-1	840	18	25	S	32	12	E		IA	Zimbabwe complex	261
533	Harleigh Farm-2	841	18	25	S	32	12	E		IA	Inyanga Lowland culture	261
534	Mabveni	812	20	22	S	30	28	E		EIA	Gokomere Tradition	146

N site	Site	SITE N	Latitude			Longitude			*	Culture, facies, industry, chronology, notes		Citation
Rwanda												
535	Akameru-Cyinkomane	114	1	29	S	29	36	E				291
536	Muganza	113	1	31	S	29	42	E			Sediments dated, but fauna could be more recent	291
537	Ryamurari	115	2	0	S	30	0	E	*			291
Senegal												
538	Cubalel-Ia	238	16	16	N	14	0	W		IA	Dated to 0-250AD	193
539	Cubalel-Ib	239	16	16	N	14	0	W		IA	Dated to 250-400 AD	193
540	Cubalel-II	240	16	16	N	14	0	W		IA	Dated to 400-600 AD	193
541	Cubalel-III	241	16	16	N	14	0	W		IA	Dated to 600-950 AD	193
542	Ndiamon-Badat	279	13	53,85	N	16	43,49	W		IA		11, 295
543	Ogo region	675	15	33	N	13	17	W		Hi		56
544	Sintiou-Bara	271	15	42	N	13	23	W		P		283
545	Siouré-Ia	248	16	27	N	14	9	W		IA	Dated to 0-250AD	193
546	Siouré-Ib	249	16	27	N	14	9	W		IA	Dated to 250-400 AD	193
547	Siouré-II	250	16	27	N	14	9	W		IA	Dated to 400-600 AD	193
548	Siouré-IV, V	252	16	27	N	14	9	W		IA	Dated to 950-1400 AD	193
549	Tulel-Fobo	83	13	38	N	15	39	W		P		302
550	Walaldé-W1/1-19	153	16	30	N	14	13	W		IA		77
551	Walaldé-W1/20-25	154	16	30	N	14	13	W		IA		77
552	Walaldé-W2	155	16	30	N	14	13	W		IA		77
Sinai												
553	Abu Gedar	697	28	26	N	33	42	E				64
554	El´Awag	698	28	22	N	33	55	E				64
South Africa												
555	/Ai Tomas	613	28	17	S	17	4	E		LSA		140
556	2627 CD7	833	26	47,73	S	22	22,12	E		IA		184
557	Badfontein Valley-2530 AD 10	803	25	20	S	30	20	E		IA, Hi	Marateng phase	66
558	Blombos Cave	595	34	23	S	21	14	E				140, 141
559	Blydefontein I-II	621	30	50	S	25	30	E		LSA	phase 5 =Interior Wilton Ceram	169
560	Blydefontein III-IV	622	30	50	S	25	30	E		LSA	phase 4 - IWC developed	169
561	Blydefontein V	624	30	50	S	25	30	E		LSA	phase 3-4, IWC developed - IWC	169
562	Blydefontein VI-VII	623	30	50	S	25	30	E		LSA	phase 3 - IWC	169
563	Boomplass Cave-a	525	33	22	S	22	11	E		LSA	Robberg Tradition	76, 167

N site	Site	SITE N	Latitude			Longitude		*	Culture, facies, industry, chronology, notes		Citation
564	Boomplass Cave-b	524	33	22	S	22	11	E	LSA	Albany Tradition	76, 167
565	Boomplass Cave-c	523	33	22	S	22	11	E	LSA	Wilton Tradition	76, 167
566	Boomplass Cave-d	299	33	22	S	22	11	E	LSA	Herder Unit	167
567	Border Cave	528	27	14	S	32	0	E	IA		166
568	Buffelskloof RS-a	601	33	30,66	S	21	52	E	MSA	Albany	168, 218
569	Buffelskloof RS-b	602	33	30,66	S	21	52	E	LSA	Wilton	168, 218
570	Buffelskloof RS-c	603	33	30,66	S	21	52	E	LSA	Wilton	168, 218
571	Buffelskloof RS-d	604	33	30,66	S	21	52	E	LSA	Wilton	168, 218
572	Buffelskloof RS-e	605	33	30,66	S	21	52	E	LSA	Wilton	168, 218
573	Burchell´s Shelter	634	28	48	S	23	43	E	LSA		169
574	Bushman Rock Shelter-LSA	569	25	0,5	S	30	38	E	LSA		238
575	Bushman Rock Shelter-MSA	568	24	35	S	30	38	E	MSA		238
576	Byeneskranskop Cave 1a	519	34	35	S	19	28	E	LSA	Prewilton	170
577	Byeneskranskop Cave 1b	520	34	35	S	19	28	E	LSA	Prewilton	170
578	Byeneskranskop Cave 1c	521	34	35	S	19	28	E	LSA	Prewilton	170
579	Byeneskranskop Cave 1d	522	34	35	S	19	28	E	LSA	Wilton	170
580	Collingham Shelter	790	29	27	S	29	47	E	LSA	pCW	243
581	Colwinton Shelter-CW	791	31	7,75	S	27	44,37	E	LSA	Classic Wilton	243
582	Colwinton Shelter-pCW	792	31	7,75	S	27	44,37	E	LSA	pCW	243
583	Connies Limpet Bar	517	32	18	S	18	20	E	LSA		173
584	De Hoop	764	29	14	S	27	44	E	LSA	pre-ceramic, Post-classic Wilton	350
585	Deelpan	576	29	11	S	25	45	E	Natural site		269
586	Die Kelders	615	34	32	S	19	22	E	LSA		10, 140
587	Dikbosch 1	642	28	39	S	28	54	E	LSA		76, 169
588	Dikbosch 2	633	28	39	S	28	54	E	LSA		169
589	Doornfontein	628	28	12,33	S	23	2	E	LSA		15, 169
590	Drie Susters	849	33	6,06	S	18	0,05	E	IA		277
591	Drie Susters Main	850	33	6,05	S	18	0,06	E	IA		277
592	Driebos	854	33	6,03	S	19	3	E	IA		277
593	Eiland Salt Works-a	597	23	40	S	30	40	E	EIA	Matola Phase	244
594	Eiland Salt Works-b	598	23	40	S	30	40	E	EIA	Eiland Phase	244
595	Eland´s Bay Cave-Al	505	32	18	S	18	20	E	LSA	Albany	173
596	Eland´s Bay Cave-Rl	504	32	18	S	18	20	E	LSA	Robberg	173

N site	Site	SITE N	Latitude			Longitude		*	Culture, facies, industry, chronology, notes		Citation
597	Eland´s Bay Cave-Wia	506	32	18	S	18	20	E	LSA	Wilton	173
598	Eland´s Bay Cave-Wib	507	32	18	S	18	20	E	LSA	Wilton	173
599	Eland´s Bay Cave-Wic	508	32	18	S	18	20	E	LSA	Wilton	173
600	Eland´s Bay Cave-Wid	509	32	18	S	18	20	E	LSA	Wilton	173
601	Eland´s Bay Open-a	514	32	18	S	18	20	E	LSA		173
602	Eland´s Bay Open-b	515	32	18	S	18	20	E	LSA		173
603	Emambeni	827	26	45	S	31	35	E	LSA		171
604	Equus Cave 1A	795	27	37	S	24	37	E	LSA		175
605	Ficus A	647	24	10	S	29	13	E	EIA	6th century AD	254
606	Ficus B	648	24	10	S	29	13	E	EIA	9th century AD	254
607	Ficus C	649	24	10	S	29	13	E	LIA	14th century AD	254
608	Forest Hall	860	33	54	S	22	20	E			356
609	Glentyre	810	34	0	S	22	42	E	LIA		88
610	Green Elliott Shelter I-III	625	30	55	s	25	45	e	LSA	Phase 6 = Smithfield	169
611	Green Elliott Shelter IV-VI	626	30	55	s	25	45	E	LSA	Phase 5 = IWC	169
612	Haaskraal Shelter	594	31	34	S	24	13	E	IA		97
613	Hail Stone Midden	516	32	18	S	18	20	E	LSA		173
614	Hapi Pan	567	22	25	S	31	20	E	LIA		240
615	Happy Rest	596	23	30	S	31	50	E	IA		339
616	Heuningneskrans-I/O	821	34	36	S	30	39	E	LSA	Intermediate/Oakhurst complex	171
617	Heuningneskrans-R	820	34	36	S	30	39	E	LSA	Robberg Industry	171
618	Hill X	734	24	1	S	31	9	E	Hi	LIA	250
619	HKLP Ia	829	26	1	S	30	48	E	LSA		177
620	HKLP Ib	830	26	1	S	30	48	E	LSA/IA		177
621	HKLP Va	831	26	0	S	30	46	E	LSA		177
622	HKLP Vb	832	26	0	S	30	46	E	LSA/IA		177
623	Icon	582	22	25	S	29	15	E	IA		332
624	Jubilee Shelter-O	857	25	42	S	27	55	E	LSA	Oakhurst	286
625	Jubilee Shelter-PW	855	25	42	S	27	55	E	LSA	Post-Wilton	286
626	Jubilee Shelter-W	856	25	42	S	27	55	E	LSA	Wilton	286
627	Kasteelberg A	616	32	48	S	17	56	E	LSA		140, 172, 174
628	Kasteelberg B	617	32	48	S	17	56	E	LSA		172, 174
629	Kekane	652	24	10	S	29		E	LIA	19th-20th century AD	254

N site	Site	SITE N	Latitude			Longitude			*	Culture, facies, industry, chronology, notes		Citation
630	Klasies River Mouth-1DIII	531	34	5	S	24	22	E	*	LSA		165
631	Klasies River Mouth-1I	529	34	5	S	24	22	E	*	LSA		165
632	Klasies River Mouth-1II	530	34	5	S	24	22	E	*	LSA		165
633	Klasies River Mouth-5II	532	34	5	S	24	22	E	*	LSA		165
634	Klein Kliphuis	839	32	7	S	18	51	E		LSA		7, 317
635	Klingbeil 2530 AB 1	808	25	5,9	S	30	29,8	E		EIA		86
636	Klipspring	599	25	8	S	30	44	E	*	EIA		244
637	Kreeftebaai	852	33	8,06	S	17	59,08	E		IA		277
638	Langsdraai	600	25	6	S	30	42	E	*	EIA		244
639	Large Rock Shelter	806	33	19,5	S	26	8,5	E		LSA	Wilton	75
640	Le2	564	23	45	S	31	30	E		LIA		240, 252
641	Le6	550	23	50	S	31	40	E		EIA	Tsende	239, 240
642	Le7a	547	23	50	S	31	40	E		EIA	Early Eiland Matola	239
643	Le7b	551	23	50	S	31	40	E		EIA	Balule	239
644	Leliehoek	763	29	12	S	27	26	E		LSA	Pre-ceramic	85
645	Limerock 1	631	28	33	S	24	8	E		LSA		169
646	Limerock 2	632	28	33	S	24	8	E		LSA		169
647	Lydenburg Heads site	809	25	6,33	S	30	29,66	E		EIA	Lydenburg	87
648	Ma38	545	23	43	S	31	21	E		EIA	Matola	239
649	Ma4	555	23	5	S	31	30	E		EIA	Balule	239
650	Magogo-1	578	28	27	S	30	47	E		EIA		335
651	Magogo-2	579	28	27	S	30	47	E		EIA		335
652	Magogo-3	580	28	27	S	30	47	E		EIA		335
653	Malwula	828	26	15	S	26	45	E		LSA		171
654	Mamba 1	583	28	56	S	31	2	E		EIA		337
655	Mapotini Hill	732	23	57	S	31	5	E		Hi	LIA	250
656	Mapungubwe-Hill	612	22	25	S	29	15	E		EIA	Leopards´ Kopje B	336
657	Mapungubwe-K2/TS3	527	22	25	S	29	15	E		IA	Leopard´Kopje A	336
658	Mapungubwe-ST	611	22	25	S	29	15	E		EIA	Leopards´ Kopje B	333, 336
659	Marupale Hill	738	24		S	31	6	E		Hi	LIA	250
660	Mauermanshoek	765	28	51	S	27	14	E		LSA		350
661	Mgoduyanuka	572	28	43	S	29	23	E		LIA		246
662	Mhlopeni	581	28	28	S	30	47	E		EIA		335

N site	Site	SITE N	Latitude			Longitude			*	Culture, facies, industry, chronology, notes		Citation
663	Mhlwazini Cave	789	29	2	S	29	23	E		LSA	pCW	243
664	Mo8	546	23	48	S	31	38	E		EIA	Early Eiland Matola	239
665	Msuluzi Confluence	646	28	45	S	30	8	E		EIA		254
666	Muhululu Hill	736	24	1	S	31	10	E		Hi	LIA	250
667	Ndondondwane	577	28	50	S	31	2	E		EIA		340
668	Nelson Bay Cave- PW	753	34	6	S	23	24	E		LSA	Pre-Wilton	163
669	Nelson Bay Cave- R	752	34	6	S	23	24	E		LSA		163
670	Nelson Bay Cave- W	754	34	6	S	23	24	E		LSA	Wilton	163
671	Oakhurst	811	34	0	S	22	42	E		LIA		88
672	OFD1	635	29	22	S	25	0	E	*	LSA		169
673	Ol20	552	24		S	31	46	E		EIA	Balule	239
674	Ondini	859	28	15	S	31	26	E		LIA		352
675	Onink	826	26	45	S	31	35	E		LSA	Late Ho	171
676	Oudepost I	805	33,08		S	18,02		E		Hi		71
677	Pa8.1	556	22	29	S	31	21	E		EIA	Late Ziwa	239
678	Pancho's Kitchen Midden 1-2	817	32	20,22	S	18	20	E				151
679	Pancho's Kitchen Midden 3-4	818	32	20,22	S	18	20	E				151
680	Pancho's Kitchen Midden 6-7	819	32	20,22	S	18	20	E		LSA		151
681	Paternoster1-2	836	32	48	S	17	53	E		LSA		262
682	Paternoster3-4	837	32	48	S	17	53	E				262
683	Paternoster5-6	838	32	48	S	17	53	E				262
684	Ph1	566	23	45	S	31	30	E		LIA		240
685	Ph6	560	23	30	S	31	10	E		LIA		240
686	Ph9	553	23	47	S	31	21	E		EIA	Balule	239
687	Pjene Hill	737	24	1	S	31	7	E		Hi	LIA	250
688	Pont-Drift-LKA	609	22	13,867	S	29	9,05	E		EIA	Leopards´ Kopje A	247, 333, 336
689	Pont-Drift-Zhizo	608	22	13,867	S	29	9,05	E		EIA	Zhizo tradition	247, 333, 336
690	Powerhouse Cave	629	27	37	S	24	38	E		LSA		169
691	Pr34	558	25	5	S	31	15	E		LSA		240
692	QwaQwa Museum	606	28	29	S	28	44	E		LIA		32
693	Randjies	359	23	20	S	28	51	E		LIA		242
694	Ravenscraig Shelter-pCW	794	30	0	S	27	47	E		LSA	pCW	243
695	Ravenscraig Shelter-R	793	30	0	S	27	47	E		LSA	Robberg	243

N site	Site	SITE N	Latitude			Longitude			*	Culture, facies, industry, chronology, notes	Citation
696	Rhenosterkloof	651	24	53	S	27	39	E	LIA		254
697	Riversmead Shelter I-II	618	30	40	S	25	20	E	LSA	Phase 5 = IWC	169
698	Riversmead Shelter III-IV	619	30	40	S	25	20	E	LSA	Phase 4 - IWC dev	169
699	Riversmead Shelter VI-IX	620	30	40	S	25	20	E	LSA	Phase 3 - IWClassic	169
700	Rooiels	842	34	18,13	S	18	49,12	E		Late Ho	274
701	Rooikrans-CW	785	29	35	S	27	36	E	LIA	Classic Wilton	243
702	Rooikrans-pCW	650	29	35	S	27	36	E	LIA	Late pCW	243
703	Roosfontein	814	28	49	S	27	44	E	LSA	pCW	162
704	Rose Cottage Cave-A-Mn	751	29	13	S	27	28	E	LSA	pCW and ceramic pCW	243, 249
705	Rose Cottage Cave-DB2	759	29	13	S	27	28	E	LSA	Robberg	350
706	Rose Cottage Cave-OIC	760	29	13	S	27	28	E	LSA	Oakhurst Industrial Complex	243
707	Rose Cottage Cave-WI	761	29	13	S	27	28	E	LSA	Wilton	243
708	Schroda	607	22	14	S	29	25	E	EIA	Zhizo tradition	336
709	Sebatini Hill	730	23	58	S	31	7	E	Hi	LIA	250
710	Selongwe Hill	735	23	54	S	31	8	E	Hi	LIA	250
711	Serotwe Hill	731	24		S	31	5	E	Hi	LIA	250
712	SH14a	554	23	10	S	31	19	E	EIA	Balule	239
713	SH16	559	23	10	S	31	25	E	LIA		240
714	Sh27	565	23	10	S	31	25	E	LIA		240
715	Shankare Hill-a	728	23	58	S	31	10	E	Hi	IA	250
716	Shankare Hill-b	729	23	58	S	31	10	E	Hi	LIA	250
717	SK17	549	24	58	S	31	43	E	EIA	Lyndenburg	239
718	Sk4	557	25	10	S	31	30	E	LSA		240
719	Smitswinkelbaai Cave	459	34	16	S	18	28	E			198
720	Sonkoanini Hill	733	24	1	S	31	7	E	Hi	LIA	250
721	Spoegrivier-a	614	30	29,67	S	17	22	E	LSA		8, 140, 329
722	Spoegrivier-b	755	30	29,67	S	17	22	E	LSA		140, 329
723	Spoegrivier-c	756	30	29,67	S	17	22	E	LSA		140, 329
724	Spring Cave	518	32	18	S	18	20	E	LSA		173
725	Steenbokfontein-1	589	32	9,7	S	18	20	E	LSA		9, 152
726	Steenbokfontein-2	590	32	10	S	18	20	E	LSA		9, 152
727	Steenbokfontein-3	591	32	10	S	18	20	E	LSA		9, 152
728	Steenbokfontein-4a	592	32	10	S	18	20	E	LSA		9, 152

N site	Site	SITE N	Latitude			Longitude			*	Culture, facies, industry, chronology, notes		Citation
729	Steenbokfontein-4b	593	32	10	S	18	20	E		LSA		9, 152
730	Tandjesberg	762	29	5	S	28	24	E		LSA	Hi	350
731	Ti3	562	24	5	S	31	35	E		LIA		240
732	Ti6	563	24	5	S	31	35	E		LIA		240
733	Tienfontein 2	639	30	10	S	27	0	E	*	LSA		169
734	Tienfontein 4A	640	30	10	S	27	0	E	*	LSA		169
735	Tortoise cave-a	510	32	18	S	18	20	E		LSA	Wilton	173
736	Tortoise cave-b	511	32	18	S	18	20	E		LSA	Wilton	173
737	Tortoise cave-c	512	32	18	S	18	20	E		LSA	Wilton	173
738	Tortoise cave-d	513	32	18	S	18	20	E		LSA	Wilton	173
739	Tsh1	548	24	54	S	31	40	E		EIA	Mutlumuvi	239
740	Twyfelpoort	766	28	37	S	27	34	E		LSA		350
741	Umgungundlovu	360	28	32,1	S	31	31,1	E		LIA		242, 251
742	Ventershoek	627	29	45	S	27	0	E	*	LSA	Phase 6 = Smithfield	169
743	Vlaeberg	851	33	6,08	S	18	0,06	E		IA		277
744	Voëlvlei	853	33	24	S	19	4,06	E		IA		277
745	Westbury Rock Shelter	801	28	45	S	27	40	E		LSA		245, 284
746	Wildebeestfontein	653	26	27	S	29	9	E		LIA		254
747	Witklip-1	845	32	55,06	S	17	59,06	E		IA		277
748	Witklip-2	846	32	55,06	S	17	59,06	E		IA		277
749	Witklip-3	847	32	55,06	S	17	59,06	E		IA		277
750	Witklip-4	848	32	55,06	S	17	59,06	E		LSA		277
751	Wosi-early EIA	573	28	54,417	S	31	1,833	E		EIA	Msuluzi Phase	318, 338
752	Wosi-middle-late EIA	574	28	54,417	S	31	1,833	E		EIA	Msuluzi Phase	318, 338
753	Wosi-terminal EIA	575	28	54,417	S	31	1,833	E		EIA	Msuluzi Phase	318, 338
Sudan												
754	Abu Darbein	195	17	42	N	34	2	E		LSA	Mesolithic	232, 234
755	Abu Hugar	134	13	1	N	33	58	E		MSA		104
756	Abu Tabari 02/1	800	17	42	N	28	38	E	*	LSA		255
757	Aneibis	197	17	57	N	34	0	E		LSA	Mesolithic	232, 234
758	Awlad el Imam	77	15	54,1	N	32	35,7	E		LSA	Early Kharthoum	39
759	Berber	274	18	1	N	33	59	E		IA	Postmeroitic	179
760	Buhen	690	21	44	N	31	11	E		Hi	Early part of middle kingdom	65

N site	Site	SITE N	Latitude			Longitude			*	Culture, facies, industry, chronology, notes		Citation
761	Debbat Bangdit	128	11	35	N	32	46	E		LSA		115
762	Debbat El Eheima	127	11	45	N	32	46	E		LSA		115
763	Djarabona 84/13	799	17	14	N	26	29	E	*	LSA		255
764	El Damer	196	17	36	N	33	58	E		LSA	Mesolithic	232, 234
765	El Kadada	138	16	45	N	33	33	E		LSA	N	107
766	El Kenger Middle/East	273	15	57	N	32	39	E		LSA		38
767	El Mahalab	673	15	30	N	32	40	E		LSA	Mesolithic	51
768	El Nofalab	393	15	54	N	32	32	E		LSA		84
769	El Qala´a	81	15	58	N	32	35,5	E		LSA	Late Early Kharthoum	39
770	El Temeyim	80	15	55,3	N	32	36	E		LSA	Early Kharthoum	39
771	El Zakiab	391	15	46	N	32	35	E		LSA		84
772	Erg of Ennedi	798	18	10	N	25	25	E				221
773	Esh Shaheinab	181	16	2	N	32	33	E		LSA		84, 226
774	Gash Delta-AA	682	15	57	N	36	5	E		LSA	Amm Adam Group	119
775	Gash Delta-K	683	15	57	N	36	5	E	*	LSA	Kassala group	119
776	Gebel Tageru	796	17	2	N	27	7	E				221
777	Geili	131	16	1	N	32	34	E		LSA	Early N (Esh Shaheinab)	102, 109
778	Gism el-Arba	669	19	26	N	30	32	E		LSA	Kerma	52
779	Guli	680	13	34	N	32	34	E		LSA	Early Khartoum	1
780	Jebel et Tomat	684	13	45	N	32	32	E	*			62, 128
781	Jebel Umm Marrahi	76	15	59	N	32	32	E		LSA	Mesolithic, Early Kharthoum	112
782	Kabbashi	82	15	53,9	N	32	35,8	E		LSA	Early Kharthoum	39
783	Kadero	146	15	44	N	32	35	E		LSA	Central Sudanese Neolithic	105
784	Kadruka	272	19	26,25	N	30	32,75	E		N	Kerma	46
785	Kerma	254	19	38	N	30	25	E		LSA	Kerma	48
786	Khabbashi Haitah	78	15	53,7	N	32	36	E		LSA	Early Kharthoum	39
787	Khartoum Hopital	180	15	35	N	32	31	E		LSA	Early Khartoum	226
788	Khasm el Girba - Kassala	99	15	9,4	N	36	8,5	E		LSA	Kassala phase	228, 292
789	Khasm el Girba - KG14	98	14	59,5	N	35	58	E		LSA	Pre-Saroba	228, 292
790	Khasm el Girba - KG15-16	27	15	0	N	35	58	E		LP		203, 228, 292
791	Khasm el Girba - KG68	28	14	59,5	N	35	58	E		LSA		203, 228, 292
792	Lake Sidigh Complex	797	16	55	N	26	46	E				221
793	Laqiya area - 82/29	47	20	14	N	27	30	E		LSA		314

N site	Site	SITE N	Latitude			Longitude			*	Culture, facies, industry, chronology, notes		Citation
794	Laqiya area - 82/31	48	20	14	N	27	30	E		LSA		314
795	Laqiya area - 82/lake	49	20	14	N	27	30	E				314
796	Lower Wadi Howar-5	472	17	45	N	29	45	E				222
797	Mahal Teglinos K1-CG	150	15	28	N	36	21	E		LSA	Classical Gash group	117
798	Mahal Teglinos K1-EG	148	15	28	N	36	21	E		LSA	Early Gash group	117
799	Mahal Teglinos K1-JM	152	15	28	N	36	21	E		LSA	Jebel Mokram	117
800	Mahal Teglinos K1-LG	151	15	28	N	36	21	E		LSA	Late Gash group	117
801	Mahal Teglinos K1-MG	149	15	28	N	36	21	E		LSA	Middle Gash Group	117
802	Mahal Teglinos K1-PG	147	15	28	N	36	21	E		LSA	Proto-Gash Group	117
803	Palaeolake S98/S20-21	192	18	0	N	25	0	E	*	LSA		233
804	Rahib-80/87	193	17	33	N	26	57	E		LSA		233, 255
805	Saggai 1	132	15	55	N	32	36	E		LSA	Early N (Khartoum)	102
806	Sai	275	20	42	N	30	19	E		IA	New Kindgom	53
807	Sarurab	681	15	56,25	N	32	32,5	E		LSA	Ealry Khartoum Tradition	160, 161
808	Selima Sandsheet - 85/73-2	45	21	5	N	29	20	E		LSA		314
809	Selima Sandsheet - 85/80	46	21	5	N	29	20	E		LSA	situated near 85/73, a contemporaneous site	314
810	Shabona	685	14	38	N	32	16	E		LSA	Early Khartoum	62, 128
811	Shaqaduq-S1-A	189	16	20	N	33	20	E		LSA		231
812	Shaqaduq-S1-B	188	16	20	N	33	20	E		LSA		231
813	Shaqaduq-S21	187	16	15	N	33	20	E		LSA		228, 231
814	Sheikh el Amin	674	15	30	N	32	40	E	*	LSA	N	51
815	Sheikh Mustafa	672	15	30	N	32	40	E	*	LSA	Mesolithic	51
816	Singa	133	13	9	N	33	56	E		MSA		104
817	Soba	676	15	48	N	32	33	E		Hi	Transitional-early medieval	49
818	Tagra	679	13	56	N	32	22	E		LSA	Early Khartoum	1
819	Um Direiwa	392	15	40	N	32	37	E		LSA		84
820	Umm Singid	79	15	58,2	N	32	37,25	E		LSA	Early Kharthoum	39
821	Upper Atbara hinterland	191	15	6	N	36	7,7	E		LSA		228
822	Upper Atbara Valley	190	14	59,1	N	35	38	E		LSA		228
823	Wadi Halfa	284	21	48	N	31	22	E		Pl		182
824	Wadi Hariq 01/1	282	19	12	N	27	17	E		LSA		153, 255
825	Wadi Hariq 01/4	283	19	12	N	27	17	E		LSA	No dates, attributed to ~ 2200-1100 BC	153, 255
826	Wadi Hariq 97/5	280	19	12	N	27	17	E		LSA		153

N site	Site	SITE N	Latitude			Longitude			*	Culture, facies, industry, chronology, notes		Citation
827	Wadi Hariq 97/7	281	19	12	N	27	17	E		LSA		153
828	Wadi Howar - 80/73	50	17	40	N	29	30	E		LSA	Khartoum pottery tradition	314
829	Wadi Howar - 84/1	54	17	40	N	29	30	E		LSA	Shaheinab N	314
830	Wadi Howar - 84/12-1	52	17	40	N	29	30	E		LSA	Shaheinab N	314
831	Wadi Howar - 84/13	53	17	40	N	29	30	E		LSA	Shaheinab N	233, 314
832	Wadi Howar - 84/19	51	17	40	N	29	30	E		LSA	Shaheinab N	314
833	Wadi Howar S96/5-1	194	17	10	N	26	20	E		LSA		233
834	W-Gebel Rahib	471	17	30	N	26	55	E				222
Tanzania												
835	Mumba	834	3	32	S	35	16	E		LSA		209
Western Sahara												
836	Izriten	376	28	1	N	12	28	W				27
837	Sebkha Amtal	895	23	5	N	13	29	W		LSA		236
838	Sebkha Laasailia	894	26	50	N	13	29	W		LSA		236
839	Sebkha Mahariat	896	22	16	N	16	33	W		LSA		236
840	Site F	374	27	58	N	12	49	W		N		27
841	Site H	375	27	45	N	13	43	W			camel dated from 1180+-120 bp (Ly-1.103) intrusive	27
Zaire												
842	Dimba	107	4	42	S	21	28	E	*	LSA	Ngovo Group	291
843	Ishango-AB	184	0	8,3	S	29	36,2	E		LPI		230
844	Ishango-C	185	0	8,3	S	29	36,2	E		LPI		230
845	Ishango-Z-Post-Em	186	0	8,3	S	29	36,2	E		EHi		230
846	Kakontwe	105	11	0	S	26	43	E	*			291
847	Kamilamba	101	7	48	S	27	1	E		IA	Ancient to Classic Kisalien	257, 288
848	Kiantapo	112	10	4	S	25	47	E		IA		289, 291
849	Kilulu	102	7	50	S	26	59	E		IA	Classic Kasilien-Kabambien A	257, 288
850	Malemba-Nkulu-CK	103	8	4	S	26	48	E		IA	Classic Kasilien	257, 288
851	Malemba-Nkulu-KA	108	8	4	S	26	48	E		IA	Kabambien A	288, 291
852	Malemba-Nkulu-KB	109	8	4	S	26	48	E		IA	Kabambien B	288, 291
853	Mashita Mbanza	110	5	24	S	19	6	E	*	IA		291
854	Matupi-I	126	1	13,88	N	29	49,38	E		IA		289
855	Matupi-II	861	1	13,88	N	29	49,38	E		LSA		289
856	Ngovo	106	5	18,5	S	14	55,42	E		LSA	Ngovo Group	291

N site	Site	SITE N	Latitude			Longitude			*	Culture, facies, industry, chronology, notes		Citation
857	Nkile	111	0	40	S	21	0	E		IA		291
858	Sanga	100	8	10	S	26	30	E		IA	Classic Kasilien	257, 288
Zambia												
859	Bulila	643	15	40	S	27	9	E	*	EIA		254
860	Leopard's Hill-NI	823	15	36	S	28	43	E		LSA	Nachikufian I	171
861	Leopard's Hill-NII	824	15	36	S	28	43	E		LSA	Nachikufian II	171
862	Leopard's Hill-NIII	825	15	36	S	28	43	E		LSA	Nachikufian III	171
863	Leopard's Hill-PN	822	15	36	S	28	43	E		LSA	Pre-Nachikufian or Proto-LSA	171
864	Namakala	570	16	42	S	24	58	E		EIA	EIA Phase 1, ~6-7 th century AD	237
865	Nanga	571	16	39	S	25	0	E		EIA	EIA Phase 1 and 2, ~8-10 th century AD	237
866	Salumano	644	17	5	S	23	53	E		EIA	8th-10th century AD	254
Zanzibar												
867	Kizimkazi Dimbani	104	6	26	S	39	28	E		IA		298
Zimbabwe												
868	Mtanye	878	21	23	S	29	15	E		IA	Leopard's Kopje facies	148
869	Muozi	835	18	12	S	32	48	E				253
870	Redcliff Cave	804	19	1	S	29	46	E		LPI, early LSA	Tshangula horizon	70
871	Ruanga Ruin	815	17	1,58	S	31	40,7	E		LIA	Later Zimbabwe and Muzengezi wares	96
872	Simamwe	877	21	28	S	29	24	E		IA	Zhizo facies	148

Chapter 4 Chronological Data

Abbreviations found in the table are the following:

Abbrev	Meaning
ND	No date
1D	One date
Env	Environmental data
Arch	Archaeological data
Fau	Faunal data
Unk	Unknown
B	Bone
Bb	Burnt bone
BoA	Bone apatite
BoC	Bone collagen
Ca	Carbonate
Ce	Ceramic
Ch	Charcoal
Co	Coprolithe
En	Enamel
EnA	Enamel apatite
EnC	Enamel collagen
IR	Iron residue
L	Leather
MoS	Molluscs shell
OM	Organic matter
OES	Ostrich egg shell
Sed	Sediment
V	Vegetal
W	Wood

N site	Date N / Other chronological attribution	Age BP	±	cal BC		ky Mid BC/AD	ky Mid BP	Citation	Material	Note (taxon, level)	ND	1D	Env	Arch	Fau	Unk
Algeria																
1	Holocene, about third millenium BC					-2,5	4,45	290			+					
2	Holocene, about third to first millenium BC					-1,5	3,45	290			+					
3	Early Holocene					-9	10,95	290			+					
4				-9800	-9540	-9,67	11,62	150	MoS	Snail				+	+	
4				-9470	-9300	-9,39	11,34	150	MoS	Snail				+		
4				-7960	-7790	-7,88	9,83	150	MoS	Snail				+		
4				-7610	-7480	-7,55	9,5	150	MoS	Snail				+		
4				-7320	-7150	-7,24	9,19	150	MoS	Snail				+		
5	Gif-464	5500	250	-4934	-3789	-4,36	6,31	35	Ch	Middle layer 0,3-0,6 m, tr.2				+		
5	Gif-1222	6800	220	-6203	-5319	-5,76	7,71	35	Ch	Tr. 1, 0,6-0,9 m				+		
5	UW-87	8050	80	-7290	-6690	-6,99	8,94	35	Ch	West rock A, 0,8m				+		
5	MC-212	8670	150	-8226	-7501	-7,86	9,81	35	Ch	Square base				+		
6	MC-155	8280	200	-7708	-6689	-7,2	9,15	44	Ch	Columnatian				+		
6	Gif-309	6340	300	-5846	-4599	-5,22	7,17	44	Ch	Capsian				+		
6	MC-154	7300	200	-6563	-5762	-6,16	8,11	67	Ch	Columnatian				+		
6	Gif-308	6850	300	-6386	-5225	-5,81	7,76	67	Ch	Capsian				+		
7	Alg-10	5720	195	-5196	-4074	-4,64	6,59	320	MoS	150-160 cm				+		
7		5400	190	-4681	-3799	-4,24	6,19	320								+
8	OxA-6805	1550	40	422	596	0,51	1,44	135	BoC	*Ursus arctos*		+			+	
9	Mc-628	7280	120	-6414	-5922	-6,17	8,12	67	Ch					+		
9	Mc-625	7090	110	-6211	-5741	-5,98	7,93	67	Ch					+		
9	Mc-627	7000	110	-6065	-5673	-5,87	7,82	67	Ch					+		
10		6530	250	-5981	-4941	-5,46	7,41	320		3,2-3 m transition E/N						+
10		4340	200	-3621	-2474	-3,05	5	320		1,05-0,5 m						+
10		5900	150	-5208	-4458	-4,83	6,78	266		3-2 m						+
10		4900	250	-4317	-3030	-3,67	5,62	266		1,05-0,5 m						+
10		5380	140	-4521	-3825	-4,17	6,12	266		2-1,5 m						+
10		5400	140	-4533	-3955	-4,24	6,19	266		2-1,5 m						+
10		5740	140	-4934	-4337	-4,64	6,59	266		3-2 m						+
11	Ly-3657	4630	130	-3655	-2943	-3,3	5,25	44	B	*Bos taurus*						
12	Neolithic					-4	5,95	6			+					
13	AECV-1812C	7340	120	-6432	-6009	-6,22	8,17	150	Ch	Unit IIIB				+		
13	SMU-1082	7750	50	-6656	-6471	-6,56	8,51	150	MoS	Unit IIIB				+		
13	SMU-1084	6620	110	-5728	-5368	-5,55	7,5	150	Ch	Unit IIIa				+		
13	SMU-1095	7590	60	-6593	-6270	-6,43	8,38	150	Ch	Unit IIIB				+		
13	SMU-712	9390	130	-9140	-8313	-8,73	10,68	150	Ch	Unit V				+		
13	I-9838	6575	170	-5873	-5210	-5,54	7,49	150	Ch	Unit IIIa				+		
13	SMU-1081	7150	200	-6418	-5673	-6,05	8	150	Ch	Unit IIIB				+		

N site	Date N	Age BP	±	cal BC		ky Mid BC/AD	ky Mid BP	Citation	Material	Note (taxon, level)	ND	1D	Env	Arch	Fau	Unk
	Other chronological attribution															
13	SMU-1096	7210	340	-6984	-5481	-6,23	8,18	150	Ch	Unit IIIB				+		
13	SMU-1120	7350	50	-6364	-6078	-6,22	8,17	150	Ch	Unit IIIB				+		
13	SMU-1097	7442	560	-7755	-5307	-6,53	8,48	150	Ch	Unit IIIB				+		
13	AECV-1813C	7430	90	-6445	-6093	-6,27	8,22	150	Ch	Unit IIIB				+		
13	SMU-1154	6770	90	-5846	-5515	-5,68	7,63	150	Ch	Unit IIIa				+		
13	I-9836	6485	125	-5658	-5215	-5,44	7,39	150	Ch	Unit IIIa				+		
13	SMU-1099	6520	170	-5752	-5064	-5,41	7,36	150	Ch	Unit IIIa				+		
13	I-9837	6505	125	-5660	-5222	-5,44	7,39	150	Ch	Unit IIIa				+		
13	SMU-1108	9100	130	-8704	-7886	-8,3	10,25	150	MoS	Unit V				+		
13	SMU-704	8580	150	-8205	-7327	-7,77	9,72	150	Ch	Unit IV				+		
13	SMU-1121	8390	170	-7938	-7034	-7,49	9,44	150	Ch	Unit IV				+		
13	I-9835	5965	115	-5208	-4590	-4,9	6,85	150	Ch	Unit I		+		+		
14		5525-4865		-4490	-3510	-4	5,95	281						+		
14	Pa-1457	4870	120	-3945	-3375	-3,66	5,61	281	B	*Bos taurus*				+	+	
15	MC-151	6500	100	-5632	-5299	-5,47	7,42	67	Ch	100 cm				+		
15	MC-326	8550	150	-8185	-7189	-7,69	9,64	67	Ch	300-325				+		
15	MC-323	7280	120	-6414	-5922	-6,17	8,12	67	Ch	225-250 cm				+		
15	MC-327	8860	150	-8294	-7598	-7,95	9,9	67	Ch	325-350 cm				+		
15	Gif-889	8480	300	-8288	-6708	-7,5	9,45	67	Ch	230-320 cm				+		
15	Gif877	8270	185	-7606	-6698	-7,15	9,1	67	Ch	100-130				+		
15	MC-213	7860	120	-7051	-6479	-6,77	8,72	67	Ch	100-140 cm				+		
15	MC-319	7570	160	-6806	-6066	-6,44	8,39	67	Ch	125-150 cm				+		
15	MC-318	7030	120	-6203	-5670	-5,94	7,89	67	Ch	100-125 cm				+		
16	Sa-59	5410	300	-4981	-3637	-4,31	6,26	327	Sed	Soil						
17	Holocene (Neolithic)					-6	7,95	214			+					
18	About 7600 BP					-6,4	8,35	187			+					
19	9620 BP					-9	10,95	135			+					+
20	24152-25429 to 18933-19558 cal BP					-22	23,95	210			+					
21	Gif-5417	7110	140	-6328	-5716	-6,02	7,97	44	Ch	S4						
21	Mc-677	6650	90	-5730	-5469	-5,6	7,55	44	Ch	S5						
21	Gif-5467	6450	130	-5637	-5078	-5,36	7,31	44	Ch	S4						
21	Gif-5416	5800	120	-4942	-4371	-4,66	6,61	44	Ch	S4						
21	Mc-676	4100	70	-2877	-2491	-2,68	4,63	44	Ch	S3						
22	Gif1656	7700	180	-7057	-6227	-6,64	8,59	3	Ch					+		
22	Gif-1702	7200	180	-6419	-5742	-6,08	8,03	3	OES					+		
Angola																
23	Pta-212	1810	50	135	398	0,27	1,68	291								
23	?	1770	55	135	434	0,28	1,67	291								
24	Gif-6182	1120	60	782	1140	0,96	0,99	291	Ch							

N site	Date N / Other chronological attribution	Age BP	±	cal BC		ky Mid BC/AD	ky Mid BP	Citation	Material	Note (taxon, level)	ND	1D	Env	Arch	Fau	Unk
25	Gif-6011	720	60	1229	1400	1,31	0,64	291	Ch							
Botswana																
26	Early-Late Iron Age, 1200-1300 AD					1,25	0,7	241			+					
27	Weighmann	1333	35	660	860	0,76	1,19	267	Ch					+		
27	Weighmann	1251	32	708	945	0,83	1,12	267	Ch					+		
28	Early Iron Age, 800-1200 AD					1	0,95	241			+					
29	Late Iron Age, 1450 AD- XVIIe century					1,5	0,45	241			+					
30	Wits-837	590	50	1305	1448	1,38	0,57	78	Ch	Midden deposit		+		+		+
31	LKA~ 950 to 1130 AD					1	0,95	336			+					
32	Weighmean	1308	29	675	869	0,77	1,18	267	Ch			+		+		
33	Beta-50162	5470	90	-4449	4003	-0,22	2,17	260	Ch			+		+		
34	Beta-47862	12450	80	-12976	-12206	-12,59	14,54	260	Ch					+		
34	Beta-50163	11240	60	-11303	-11046	-11,17	13,12	260	Ch					+		
35	~30 000-11 000 BP					-15	16,95	260			+					
36	I-12813	580	75	1286	1482	1,38	0,57	319	Ch	Pit A, 10-20 cm				+		
36	I-12814	530	75	1300	1622	1,46	0,49	319	Ch	Pit B, 20-30 cm				+		
37	Weighmann	1241	44	695	973	0,83	1,12	79	Ch			+		+		+
38	AD850-1090					1	0,95	79								+
39	9-10th century AD					0,9	1,05	79								+
40	Weighmann	1252	57	686	972	0,83	1,12	267	Ch			+		+		
41	LSA					-5	6,95	258			+					
42	Beta-18669	2020	40	-83	131	0,02	1,93	258	B	*Ovis aries* , sq. K;, 75-80 cm+L165				+		
42	Beta-1904888	2070	40	-162	75	-0,04	1,99	258	B	*Bos taurus* , sq.L., 70-75 cm				+		
42	UGA03-120	3.8+-0.5				-3,8	5,75	258		2/75				+		
42	UGA03-119	3.7+-0.4				-3,7	5,65	258		2/78				+		
42	UGA03-117	5.0+-0.8				-5	6,95	258		2/110				+		
43	Beta-186670	1480	40	556	667	0,61	1,34	258	En	*Bos taurus* , sq. L., 55-60 cm				+		
43	Beta-184609	1560	60	427	648	0,54	1,41	258	Ch	Sq. I., 55-60 cm				+		
43	AA-58186	1639	33	405	560	0,48	1,47	258	Ch in Ce	Sq. G., 45-50 cm				+		
43		2.1+-0.5				-2,1	4,05	258		3/55		+				
44	AA-58187	643	31	1300	1408	1,35	0,6	258	Ch in Ce	Sq. F., 35-40 cm				+		
44	AA-58188	571	31	1326	1445	1,39	0,56	258	Ch in Ce	Sq. H., 40-45 cm				+		
45	Beta-182060	1610	40	418	597	0,51	1,44	258	OM	In ceramic sherd, sq.3, 30-35 cm				+		
45	Beta-186671	1130	40	888	1025	0,96	0,99	258	En	Caprinae, sq.7, 30-35 cm				+		
46	Beta-194360	1710	40	259	534	0,4	1,55	258	Ce	Toteng 3, 20-30 cm				+		
48	UCR-3370	3570	60	-2022	-1688	-1,86	3,81	259	OES	Sq.22, 60-70 cm				+		
48	Beta-47864	2260	80	-401	-50	-0,23	2,18	259	Ch	Sq. 14, 40-50 cm				+		
48	Beta-33056	1080	100	779	1209	0,99	0,96	259	Ch	Sq. 9, 40-50 cm				+		
48	Beta-33681	2640	130	-1009	-396	-0,7	2,65	259	Ch	Sq. 5, 77 cm				+		

N site	Date N	Age BP	±	cal BC		ky Mid BC/AD	ky Mid BP	Citation	Material	Note (taxon, level)	ND	1D	Env	Arch	Fau	Unk
	Other chronological attribution															
48	Beta-33052	500	60	1326	1623	1,47	0,48	259	Ch	Sq.1-4, 62 cm				+		
48		200	50	1648	1953	1,8	0,15	259	OES	Unit 1, square 17, 20/30 cm		+		+		
48	OXA-6038	1225	60	692	988	0,84	1,11	259	B	*Ovis aries* , sq.23, 40-50 cm				+		
49	UGA-6728	6840	40	-5743	-5621	-5,68	7,63	259	OES	Sq.23, 110-120 cm				+		
49	U.Botswana	5700+-600				-5,7	7,65	259	Sed	Quartz sand, sq.13, 100 cm						
49	UCR-3365	35510	560			-33	34,95	259	Ch	Sq.12, 120-130 cm				+		
49	Beta-47865	4330	160	-3357	-2480	-2,92	4,87	259	Ch	Sq,11, 110-120 cm				+		
49	U. Washington	20600+-1900				-20,6	22,55	259	Sed	Quartz sand, sq. 19, 120 cm				+		
49	UCR-3366	3900	50	-2468	-2143	-2,31	4,26	259	OES	Sq.21, 110-120				+		
49	Beta-33053	3700	130	-2457	-1691	-2,07	4,02	259	Ch	Sq.6, 90-100 cm				+		
49	UCR-3367	28360	240	-26906	-25945	-26,43	28,38	259	OES	Sq. 21, 110-120 cm				+		
49	UCR-3368	3860	50	-2460	-2050	-2,26	4,21	259	OES	Sq.22, 110-120 cm				+		
49	AA-35233	3555	45	-1946	-1691	-1,82	3,77	259	Ch	Sq.22, 110-120 cm				+		
Burkina Faso																
50						0,5	1,45	183			+					
51	UtC-7282	3722	33	-2270	-2026	-2,15	4,1	183	Ch							
52	KI-4487	1240	25	687	870	0,78	1,17	183	Ch							
52	UtC-6467	1882	34	58	227	0,14	1,81	183	V	*Vitex* sp.						
53	KI-4885	1030	25	906	1033	0,97	0,98	183	Ch							
53	KIA-10613	1097	29	889	1013	0,95	1	183	Ch							
54	KL-4861	960	30	1021	1155	1,09	0,86	183	Ch							
54	KL-4860	1092	43	828	1024	0,93	1,02	183	Ch							
55	UtC-7354	1587	35	405	553	0,48	1,47	183	Ch							
56	KI-4549	960	25	1021	1155	1,09	0,86	183	Ch							
56	KI-4862	860	40	1045	1261	1,15	0,8	183	Ch							
57	KI-4551	1130	35	781	990	0,89	1,06	183	Ch							
57	KI-4550	1070	30	895	1021	0,96	0,99	183	Ch							
58	Erl-2650	1829	52	68	333	0,2	1,75	183	Ch							
58	Erl-3126	1845	41	70	313	0,19	1,76	183	Ch							
58	UtC-8513	1864	42	58	244	0,15	1,8	183	Ch							
58	KN-4785	1907	46	3	228	0,12	1,83	183	Ch							
58	Erl-3109	1895	41	25	228	0,13	1,82	183	Ch							
58	UtC-7353	1891	33	53	221	0,14	1,81	183	Ch							
59	UtC-7281	2931	32	-1260	-1021	-1,14	3,09	183	Ch							
60	Like site 61 (Péntènga 89/1-spits 15-21)					-6	7,95	92			+					
61	UtC-1992	7140	60	-6205	-5889	-6,05	8	92		Spits 21, -170 cm						
61	KI-3352	7000	100	-6061	-5709	-5,89	7,84	92		Spits 21, -170 cm						
61	KI-3351	7590	90	-6607	-6247	-6,43	8,38	92	Ch	Spits 17, -130 cm						
61	KI-3474	6990	160	-6212	-5623	-5,92	7,87	92		Spits 15, -110 cm						

N site	Date N	Age BP	±	cal BC		ky Mid BC/AD	ky Mid BP	Citation	Material	Note (taxon, level)	ND	1D	Env	Arch	Fau	Unk
	Other chronological attribution															
61	UtC-1993	7480	60	-6441	-6235	-6,34	8,29	92		Spits 21, -170 cm						
62	KI-3350	890	38	1036	1219	1,13	0,82	92	Ch	Spit 4, -20 cm						
62	UtC-2316	1640	50	259	541	0,4	1,55	92		Spit 4, -20 cm						
62	KI-3470	1650	90	141	600	0,37	1,58	92		Spit 10, -70 cm						
62	Like site 61 (Péntènga 89/1-spits 15-21)					-6	7,95	92			+					
63	KN-4783	964	35	1015	1161	1,09	0,86	183	Ch							
63	KI-4552	1150	30	780	973	0,88	1,07	183	Ch							
63	KN-4784	1139	40	779	988	0,88	1,07	183	Ch							
63	KN-4775	1136	44	778	990	0,88	1,07	183	Ch							
64	UtC-6465	1197	29	717	937	0,83	1,12	183	V	*Vitellaria paradoxa*						
64	KN-493?	657	60	1262	1412	1,34	0,61	183	Ch							
64	KN-4940	1135	36	780	988	0,88	1,07	183	Ch							
64	KI-4362	1230	35	687	885	0,79	1,16	183	Ch							
64	KN-4939	720	36	1222	1385	1,3	0,65	183	Ch							
65	KN-4777	3479	45	-1914	-1689	-1,8	3,75	183	Ch							
65	UtC-6466	3413	37	-1876	-1617	-1,75	3,7	183	Ch							
65	KN-4776	3380	100	-1921	-1451	-1,69	3,64	183	Ch							
65	UtC-4906	2840	49	-1191	-850	-1,02	2,97	183	V	*Pennisetum americanum*						
Burundi																
66	GrN-6108	100	45	1684	1953	1,82	0,13	291		Mid 17-mid 19th C AD						
66	GrN-6109	190	45	1654	1953	1,8	0,15	291		Mid 17-mid 19th C AD						
Cameroon																
67	Hv10586	5565	120	-4708	-4074	-4,39	6,34	74	Ch	-10 -15 cm						
68	TO-7516	950	50	997	1208	1,1	0,85	197	B	Horse, B20-55				+		
68	TO-7517	680	50	1257	1401	1,33	0,62	197	B	Horse, A-B 70-120				+		
68	TO-7515	1310	60	635	876	0,76	1,19	197	B	Horse, E110-140				+		
69	Beta-116638	1100	60	778	1025	0,9	1,05	28		Unit A, level 14				+		
70	RT-1666	775	40	1185	1287	1,24	0,71	183	Ch							
70	RT-1668a	1010	45	899	1156	1,03	0,92	183	Ch							
70	RT-1667	755	45	1182	1380	1,28	0,67	183	Ch							
70	RT-1668b	1095	45	784	1024	0,9	1,05	183	Ch							
71	RT-1670	560	50	1298	1437	1,37	0,58	183	Ch							
72	RT-1673	905	45	1030	1215	1,12	0,83	183	Ch							
72	RT-1672	915	50	1024	1215	1,12	0,83	183	Ch							
72	RT-1671	1080	50	783	1032	0,91	1,04	183	Ch							
73	RT-1681	992	40	985	1155	1,07	0,88	183	Ch							
73	RT-1679	850	50	1042	1269	1,16	0,79	183	Ch							
74	18-19th century					1,9	0,05	180			+					+
75	Lv-1940	2580	70	-897	-419	-0,66	2,61	208	Ch	Pit 6						

N site	Date N / Other chronological attribution	Age BP	±	cal BC		ky Mid BC/AD	ky Mid BP	Citation	Material	Note (taxon, level)	ND	1D	Env	Arch	Fau	Unk
75	Lv-1944	2490	110	-838	-383	-0,61	2,56	208	Ch	Pit 9						
75	Lv-1943	2490	80	-790	-411	-0,6	2,55	208	Ch	Pit 9						
75	Lv-1942	2400	60	-756	-390	-0,57	2,52	208	Ch	Pit 9						
75	Lv-1939	2420	70	-764	-395	-0,58	2,53	208	Ch	Pit 3						
75	Lv-1946	2170	80	-394	-41	-0,22	2,17	208								
75	Lv-1941	2340	70	-752	-206	-0,48	2,43	208	Ch	Pit 7 bis						
75	Lv-1945	2310	90	-754	-170	-0,46	2,41	208	Ch	Pit 13						
76	Hv8965	6980	260	-6405	-5474	-5,94	7,89	74	Ch	-30 -35 cm						
77	Hv8964	8705	275	-8547	-7143	-7,85	9,8	74	Ch	-50 -55 cm						
78	Hv8963	6070	340	-5670	-4266	-4,97	6,92	74	Ch	-20-25 cm						
Central African Republic																
79	BM-2425					1,8	0,15	296	Ch	140+-240 BP						
80	Bdy-255	430	180	1261	1954	1,61	0,34	296	Ch	Between 16-20th C AD						
80	Bdy-582	559	77	1277	1457	1,37	0,58	296	Ch	Between 16-20th C AD						
Chad																
81	Gif-351	7180	300	-6653	-5512	-6,08	8,03	108	Ch							+
82	Hv-2195	6930	370	-6600	-5076	-5,84	7,79	108	B	Human						+
82	Hv-5482	5220	160	-4357	-3695	-4,03	5,98	108	OES						+	
82	Hv-2775	7455	180	-6679	-5933	-6,31	8,26	108	B	Giraffa, elephant						+
83		6130	90	-5298	-4846	-5,07	7,02	108								
84	Hv-2748	8065	100	-7319	-6688	-7	8,95	108	Ch					+		
85	Left in 1945, maximal occupation 200 years					1,9	0,05	256			+					
Congo																
86	Iron Age					0,5	1,45	296			+					
87	Gif-4392	7090	140	-6211	-5665	-5,94	7,89	308	Ch	Level B						
87	Gif-4613	6890	160	-6019	-5481	-5,75	7,7	308	MoS	Achatina, level B						
88	Gif-4391	1990	90	-175	322	0,07	1,88	308		Lower part of level A						
89	Gif-9010	1690	80	235	600	0,42	1,53	296	Ch							
89	Gif-9006	1620	90	260	651	0,46	1,49	296	Ch							
Djibouti																
90	GIF-8183	3510	70	-2029	-1667	-1,85	3,8	131, 132								
90	GIF-7404	3440	90	-1974	-1523	-1,75	3,7	131, 132	Ch							
Egypt																
91	Hv-8691	8210	115	-7534	-6834	-7,18	9,13	327	Ch					+		
92	KN-3692	7160	70	-6213	-5902	-6,06	8,01	327	OES					+		
93	Hv-8241	6405	325	-5987	-4619	-5,3	7,25	327	Ch					+		
94	SMU-1693	7790	340	-7527	-6051	-6,79	8,74	327	Ce					+		
94	KN-3634	7500	80	-6500	-6216	-6,36	8,31	327	Ch					+		
94	Hv-8694	7780	130	-7038	-6433	-6,74	8,69	327	Ch					+		

N site	Date N / Other chronological attribution	Age BP	±	cal BC		ky Mid BC/AD	ky Mid BP	Citation	Material	Note (taxon, level)	ND	1D	Env	Arch	Fau	Unk
94	Hv-8690	7380	70	-6398	-6081	-6,24	8,19	327	Ch					+		
95	Late Roman, 4th century					0,4	1,55	313								
96	Late Roman, Monastry : end of 4th to 6th century					0,4	1,55	313								
97	1st Dynasty					-3	4,95	23			+					
98				-3600	-2950	-3,28	5,23	301								
99						-3,2	5,15	212			+					
100				-3600	-2950	-3,28	5,23	301			+					
101	2-3 century AD (2190+-60 BP)					0,2	1,75	58			+					
102	Roman temple, no date					0,2	1,75	235			+					
103	Roman					0,4	1,55	235			+					
104	Upper (+-400-200 BC) and Lower Gallery (+-200-30 BC)					0,38	1,57	126, 127			+					
105	End of VI th Dynasty ~ 2200 bc					-2,2	4,15	54			+					
106						0,1	1,85	310			+					
107	End of third to 6 century AD					0,6	1,35	310			+					
108	Ptolemaic (3-2 century BC)					-0,3	2,25	305			+					
109	SMU-481	9000	100	-8456	-7821	-8,14	10,09	320						+		
109	SMU-486	8150	70	-7450	-6848	-7,15	9,1	320	OES					+		
109	SMU-489	8740	95	-8202	-7591	-7,9	9,85	320	Ch					+		
109	SMU-494	8740	70	-8172	-7594	-7,88	9,83	320	OES					+		
110	SMU-247	6010	45	-5017	-4790	-4,9	6,85	327	Ch					+		
111	7700-6500 BP					-5,8	7,75	353								
112	Naggada IIb, about 3500–3200 BC					-3,35	5,3	344			+					
113	Old Kingdom, about 2630-2150 BC					-2,4	4,35	344			+					
114	Ptolemaic to Graeco-Roman, 305 BC-600AD					0,15	1,8	344			+					
115	Late Holocene (LSA with Pottery)					-7	8,95	355		Layer 1	+					
116	Y-1680	4830	80	-3782	-3376	-3,58	5,53	355	Ch	Layer 2a		+		+		
117	Early to Middle Holocene (no pottery LSA)					-7	8,95	355		Layer 3	+					
118	Y-1646	7060	120	-6211	-5720	-5,97	7,92	355	Ch	Layer 4		+		+		
119	Layer 5 undated, probably Late Pleist.-Early Holocene					-10	11,95	355		Layer 5	+					
120	12e Dynasty					-1,9	3,85	19			+					
121	Late Neolithic, Sheikh Muftah Unit					-5	6,95	59			+					
122	S-2415	5745	80	-4788	-4401	-4,59	6,54	327	OES					+		
123	Roman					0,2	1,75	59			+					
124	Old Kingdom					-2,42	4,37	59			+					
125	UtC-9460	7913	43	-7029	-6651	-6,84	8,79	118	OES					+		
125	KN-4422	7820	90	-7029	-6474	-6,75	8,7	118	OES							
126	UtC-5459	7752	48	-6655	-6473	-6,56	8,51	118	Ch	Bottom layer				+		
126	UtC-5623	7750	150	-7057	-6274	-6,67	8,62	118	Ch	Bottom layer				+		
126	UtC-5585	7529	48	-6466	-6254	-6,36	8,31	118	Ch	Top layer/surface				+		

N site	Date N / Other chronological attribution	Age BP	±	cal BC		ky Mid BC/AD	ky Mid BP	Citation	Material	Note (taxon, level)	ND	1D	Env	Arch	Fau	Unk
126	UtC-5582	7820	50	-6816	-6506	-6,66	8,61	118	Ch	Bottom layer				+		
127	UtC-5749	6102	46	-5210	-4859	-5,03	6,98	118	Ch	Steinplatz				+		
127	UtC-5451	6170	80	-5313	-4911	-5,11	7,06	118	Ch	Steinplatz				+		
128	Old Kingdom, fourth Dynasty					-2,6	4,55	60			+					
129	GrA-20144	4910	45	-3785	-3637	-3,71	5,66	323	V	*Triticum turgidium, Hordeum vulgare* , North Sector, -45 cm						
130	KN-5187	4825	40	-3696	-3522	-3,61	5,56	92		Spit 13, - 95 cm						
130	GrA-20142	5080	45	-3971	-3775	-3,87	5,82	323	V	*Triticum turgidium* , South sector, -5 cm						
131	12500 years BP					-12,5	14,45	322			+					
132	25 000 years BP					-23	24,95	322			+					
133	C-463	5255	230	-4593	-3540	-4,07	6,02	137	Ch			+		+		
134	Lv-1237	5730	100	-4788	-4364	-4,58	6,53	324		Hearth 3		+		+		
135	Late and Middle Kindgom					-2	3,95	341			+					
136	25-26th dynasty					-0,6	2,55	22			+					
137	4th-11th Dynastys					-2,3	4,25	341			+					
138	Lv-465	7930	160	-7301	-6466	-6,88	8,83	327	Ch					+		
138	Lv-393	8340	160	-7729	-6845	-7,29	9,24	327	Ch					+		
138	Lv-464	7990	150	-7339	-6510	-6,92	8,87	327	Ch					+		
138	GrN-7189c	8090	75	-7318	-6776	-7,05	9	327	Ch					+		
139	New Kingdom					-1,25	3,2	111			+					
140	Early dynastic - Naqada III					-3,2	5,15	287			+					
141	Late predynastic period					-4	5,95	287			+					
142	Phase A: Late predynastic Period					-4	5,95	287			+					
143	Phase A: Late predynastic Period, Naqada III-Early Dynastic					-3,2	5,15	287			+					
144	first to beginning of third century AD					0,1	1,85	294								
145	6000 - 5000 BP non calib					-4,3	6,25	99			+					
146	I-4127	5810	115	-4947	-4374	-4,66	6,61	137	Ch			+				
147	3500-3200 bc (E29-G5) - 3000-2600 bc (E29-G4)					-3	4,95	99			+					
148	I-4131	5860	115	-5007	-4457	-4,73	6,68	31	Ch			+				
149	Beta-4872	7720	70	-6680	-6441	-6,56	8,51	31	Ch					+		
149	Beta-4180	7600	70	-6600	-6263	-6,43	8,38	31	Ch					+		
149	Beta-4871	8220	105	-7531	-6868	-7,2	9,15	31	Ch					+		
150		5475	70	-4462	-4071	-4,27	6,22	327	Ch					+		
151	Fayoum A date from 5500 to 6300 uncal. BP					-4,8	6,75	31			+					
152	Fayoum A date from 5500 to 6300 uncal. BP					-4,8	6,75	31			+					
153	6000-5000 BP					-4,5	6,45	314								
154	Naqada IC-IIA (c. 3800-3650 B.C.)					-3,7	5,65	311			+					
155	Naqada IIA-B					-3,5	5,45	311			+					
156	Naqada III (c. 3200-3000 B.C.)					-3,1	5,05	311								

N site	Date N / Other chronological attribution	Age BP	±	cal BC		ky Mid BC/AD	ky Mid BP	Citation	Material	Note (taxon, level)	ND	1D	Env	Arch	Fau	Unk
157	Predynastic, Early Dynastic, Naqada I-II					-4	5,95	311			+					
158	Upper Paleolithic					-20	21,95	129			+					
159	New Kingdom					-1,3	3,25	341			+					
160	SMU377	2890	70	-1302	-902	-1,1	3,05	101	Ch					+		
160	Graeco-roman (post 2000 BP)					0,5	1,45	101			+					
161	SMU-412	4650	60	-3634	-3136	-3,39	5,34	134	Ch					+		
161	SMU-408	4740	60	-3641	-3372	-3,51	5,46	134	Ch					+		
161	SMU791	5450	80	-4454	-4054	-4,25	6,2	134	OES					+		
162	Between 15,000 and 10,500 B.C					-12,5	14,45	61			+					
163	From 8650 to 6400					-6	7,95	314								
164	Beta-2804	4730	60	-3639	-3372	-3,51	5,46	183	W							
164	Beta-2805	5010	50	-3951	-3696	-3,82	5,77	183	V							
165	Lv-1312	5110	75	-4051	-3706	-3,88	5,83	137	Ch					+		
165	Lv-1313	3180	60	-1610	-1316	-1,46	3,41	137	Ch					+		
165	Lv-1236	4810	50	-3697	-3383	-3,54	5,49	137	Ch					+		
166	GrN-12031	13330	170	-14448	-13315	-13,88	15,83	321	Ch	Hearth at 13N1E						
166	GrN-12029	12060	280	-12981	-11426	-12,2	14,15	321	Ch	Sheikh Houssein silts						
166	GrN-12030	12270	100	-12767	-11929	-12,35	14,3	321	Ch	Hearth at 17N2E						
167	GrN-12984	12410	120	-13003	-12101	-12,55	14,5	312	Ch							
167	GrN-12985	12410	60	-12891	-12189	-12,54	14,49	312	Ch							
167	GrN-12034	12940	130	-13811	-12977	-13,39	15,34	312	Ch	60 cm (27N7-8E)						
167	GrN-12986	12340	70	-12794	-12078	-12,44	14,39	312	Ch							
167	GrN-12033	12570	80	-13137	-12326	-12,73	14,68	321	Ch	70 cm (19N5E)						
167	GrN-12032	12790	170	-13706	-12370	-13,04	14,99	321	Ch	50-80 cm (15-18N8E)						
167	GrN-12981	12320	100	-12873	-12024	-12,45	14,4	312	Ch							
168	KN-3275	5830	60	-4835	-4541	-4,69	6,64	347	Ch	Lev I.1		+				
168	KN-3276	5790	60	-4783	-4502	-4,64	6,59	347	Ch	Lev I.2						
168	KN-3277	5890	60	-4933	-4607	-4,77	6,72	347	Ch	Lev I.3						
168	KN-3278	5590	60	-4541	-4338	-4,44	6,39	347	Ch	Lev V.1						
168	KN-3279	5760	60	-4763	-4461	-4,61	6,56	347	Ch	Lev V.2						
168		5440	75	-4449	-4054	-4,25	6,2	347	Ch	Lev V.3						
169	Late Period, 6-4 Centuries bc					-1,2	3,15	18			+					
170	First to second century AD					0,1	1,85	294								
171	SMU208	7930	40	-7032	-6682	-6,86	8,81	327	Ch							
171	SMU189	7910	100	-7067	-6532	-6,8	8,75	327	OES							
171	SMU240	7970	70	-7060	-6679	-6,87	8,82	327	Ch							
171	SMU202	8000	110	-7294	-6614	-6,95	8,9	327	OES							
171	SMU249	8040	90	-7294	-6679	-6,99	8,94	327	Ch							
171	SMU253	8070	90	-7308	-6698	-7	8,95	327	B	Human						

N site	Date N	Age BP	±	cal BC		ky Mid BC/AD	ky Mid BP	Citation	Material	Note (taxon, level)	ND	1D	Env	Arch	Fau	Unk
	Other chronological attribution															
171	SMU252	8080	90	-7322	-6700	-7,01	8,96	327	Ch							
171	SMU219	8120	80	-7421	-6815	-7,12	9,07	327	B	Human						
171	SMU255	8130	60	-7338	-6847	-7,09	9,04	327	Ch							
171	SMU257	8290	80	-7521	-7083	-7,3	9,25	327	OES							
172	SMU362	6280	60	-5461	-5057	-5,26	7,21	327								
172	SMU421	6950	150	-6202	-5562	-5,88	7,83	327	Ch							
172	SMU373	6680	70	-5711	-5489	-5,6	7,55	327	Ch							
172	SMU424	6690	80	-5722	-5486	-5,6	7,55	327	Ch							
172	SMU352	6600	90	-5702	-5374	-5,54	7,49	327	OES							
172	SMU452	6570	70	-5632	-5379	-5,51	7,46	327	Ch							
172	SMU497	6550	80	-5633	-5362	-5,5	7,45	327	Ch							
172	SMU435	6500	80	-5616	-5319	-5,47	7,42	327	Ch							
172	SMU441	6310	90	-5474	-5057	-5,27	7,22	327	Ch							
172	SMU361	6240	70	-5357	-5011	-5,18	7,13	327	Ch							
172	SMU366	6130	80	-5295	-4848	-5,07	7,02	327	Ch							
172	SMU473	5810	80	-4843	-4464	-4,65	6,6	327	Ch							
172	SMU368	6500	90	-5619	-5315	-5,47	7,42	327	Ch							
173	SMU200	9360	60	-8794	-8457	-8,63	10,58	327	Ch					+		
174	1300-1150 avant JC					-1,2	3,15	19			+					
175	Gd-1499	6380	60	-5476	-5226	-5,35	7,3	343	Ch	QS IX/81						
175	Gd-979	6290	100	-5474	-5022	-5,25	7,2	343	Ch	QS X/81						
175	Gd-2021	6480	170	-5722	-5056	-5,39	7,34	343	Ch	QS XI/81						
176	Gd-976	4820	100	-3894	-3366	-3,63	5,58	343	Ch	QS VII A/81				+		
176	Gd-973	4580	180	-3711	-2879	-3,3	5,25	327	Ch					+		
177	12th Dynasty					-1,99	3,94	342			+					
178	OXA-1061	2690	90	-1114	-551	-0,83	2,78	193	OM	Camel dung pellets						
178	OXA-1062	2800	80	-1195	-806	-1	2,95	183	V	Barley straw						
178	OXA-1060	2470	160	-971	-196	-0,58	2,53	183	B	Camel mandible						
179	First to early third century AD					0,1	1,85	294			+					
180	SMU-766	6590	170	-5872	-5215	-5,54	7,49	327						+		
180	A-2516	6660	320	-6230	-4911	-5,57	7,52	327	OES					+		
181	SMU-766	6590	170	-5872	-5215	-5,54	7,49	327						+		
181	A-2516	6660	320	-6230	-4911	-5,57	7,52	327	OES					+		
182	5-6th century.Early Roman (few) - Late Roman					0,5	1,45	306			+					
183	OX85 - TLI 253	25000	2500	-32308	-18864	-25,59	27,54	296								
185	GrN-16782	29950	900	-29615	-25911	-27,76	29,71	325	Ch	Top of older backfill						
186	Lv-2085	6360	70	-5476	-5216	-5,35	7,3	325	Ch	Top of the layer						
186	Lv-2082	6320	100	-5481	-5046	-5,26	7,21	325	Ch	Top of the layer						
186	Lv-2083	6940	100	-6006	-5661	-5,83	7,78	325	Ch	Top of the layer						

N site	Date N / Other chronological attribution	Age BP	±	cal BC		ky Mid BC/AD	ky Mid BP	Citation	Material	Note (taxon, level)	ND	1D	Env	Arch	Fau	Unk
187	Early 18th Dynasty					-1,4	3,35	349			+					
188	Mid. Kingdom to 2nd interm. period ~ 1800 to 1750-1550 cal bc					-1,8	3,75	17, 342			+					
189	1786-1552 cal BC: New Kindgom					-1,6	3,55	50			+					
190	6th century BC - 200 AD					0,2	1,75	342			+					
191	Neolithic					-4	5,95	93			+					
192	GrN-22559	200	35	1643	1955	1,8	0,15	202	Ch			+		+		
193	GrN-12560	6630	45	-5627	-5489	-5,56	7,51	326	Ch	Feature 12 (hearth)						
193	Lv-2185	5330	60	-4328	-3999	-4,16	6,11	326	Ch	Feature 17 (hearth)						
193	GrN-22562	6770	60	-5773	-5561	-5,67	7,62	326	Ch	Feature 15 (hearth)						
193	UtC-5390	5835	30	-4786	-4611	-4,7	6,65	202	Ch					+		
193	GrN-22651	4930	30	-3772	-3651	-3,71	5,66	326	Ch	Feature 23 (hearth)						
194	Beta-229537	7630	50	-6592	-6421	-6,51	8,46	202	Ch			+		+		
195	KIK-655/UtC-5388	7790	70	-6907	-6462	-6,68	8,63	326	Ch	Lower occup. levels				+		
195	KIK-516/UtC-4193	7910	90	-7056	-6598	-6,83	8,78	326	Ch	Lower occup. levels				+		
195	KIK-656/UtC-5389	8120	45	-7304	-7039	-7,17	9,12	326	Ch	Lower occup. levels				+		
196	Saitic-Ptolemaic, Late period, about 600 to 300 BC					-0,45	2,4	20			+					
197	Naqada Ia/IIIa					-4	5,95	81			+					
198	SMU273	6890	80	-5976	-5641	-5,81	7,76	227	OES	Uppermost playa sed						
198	KN-2878	7700	60	-6642	-6452	-6,55	8,5	227	Ch	Human occupation						
198	Hv-8242	3830	365	-3343	-1444	-2,39	4,34	327	Ch	Human occupation				+		
199	SMU273	6890	80	-5976	-5641	-5,81	7,76	227	OES	Uppermost playa sed			+			
200	1-2th century					0,1	1,85	306			+					
201	AA-2028	18100	140	-20077	-19022	-19,55	21,5	354	V	*Cyperus rotundus* l. 20						
201	AA-2029	18000	200	-20081	-18790	-19,44	21,39	354	V	*Cyperus rotundus* l. 22						
201	AA-2030	17870	140	-19712	-18666	-19,19	21,14	354	V	*Cyperus rotundus* l.24						
202	AA-2033	17810	150	-19588	-18591	-19,09	21,04	354	V	*Cyperus rotundus* l.f						
202	AA-2032	17950	225	-20072	-18685	-19,38	21,33	354	V	*Cyperus rotundus* l.c						
202	AA-2031	17300	160	-18966	-18101	-18,53	20,48	354	V	*Cyperus rotundus* l.b						
203	AA-2934	17490	190	-19271	-18255	-18,76	20,71	354	V	*Cyperus rotundus* l.b						
203	AA-2935	17530	170	-19277	-18336	-18,81	20,76	354	V	*Cyperus rotundus* l.c						
203	AA-2939	17210	160	-18832	-18019	-18,43	20,38	354	V	*Cyperus rotundus* l.g						
203	AA-2938	17990	150	-19987	-18850	-19,42	21,37	354	V	*Sirpus* l.f						
203	AA-2937	17430	280	-19418	-18029	-18,72	20,67	354	V	*Cyperus rotundus* l.e						
203	AA-2936	17560	220	-19435	-18266	-18,85	20,8	354	V	*Cyperus rotundus* l.d						
204	1st C BC - 4th C AD, Late Ptolemaic-Roman Empire					0,15	1,8	21			+					
205	Between 8860 and 6400 BP					-6	7,95	314								
206	1350 cal BC					-1,35	3,3	188			+					
Ethiopia																
207	Pre-Later Axumite					-0,5	2,45	34			+					

N site	Date N / Other chronological attribution	Age BP	±	cal BC		ky Mid BC/AD	ky Mid BP	Citation	Material	Note (taxon, level)	ND	1D	Env	Arch	Fau	Unk
208	Post-Axumite					0,6	1,35	34			+					
209	Upper Pleistocene : 35 000 to 5 000 BP					-15	16,95	98			+					
210	BM-1153	CAL b.c. 856+		-803	-909	-0,86	2,81	207			+					
211	Gif-3269	2740	110	-1261	-571	-0,92	2,87	236	B	Human				+		
212	14th-17th C AD					1,5	0,45	207			+					
212	~3500 BP					-1,8	3,75	207			+					
213	~3500 BP					-1,8	3,75	207			+					
214	14th C AD					1,4	0,55	207			+					
215	About 7,5 ka BP (arid phase)					-6,4	8,35	181			+					
216	About 5 ka BP : wet phase					-3,8	5,75	181			+					
217	GIF 11246	4370	70	-3331	-2885	-3,1	5,05	181				+				
218	GIF 11244	2180	45	-381	-111	-0,25	2,2	181				+				
219	GIF 11242	1480	60	433	656	0,54	1,41	181				+				
220	~2000 -4000 BP					-1,2	3,15	207			+					
Gabon																
221	Gif-8966	1870	50	25	253	0,14	1,81	63		Fosse XVI						
221	Beta-25546	2290	80	-747	-115	-0,43	2,38	63		Fosse VIII						
221	Beta-25549	2460	60	-767	-408	-0,59	2,54	63		Fosse XV						
221	Beta-25548	2120	60	-360	2	-0,18	2,13	63		Fosse XIII						
221	Gif-8612	2170	50	-379	-61	-0,22	2,17	63		Fosse XI						
221	Beta-25582	2450	70	-767	-404	-0,59	2,54	63		Fosse X						
222	Gif-6424	1650	70	238	560	0,4	1,55	304	MoS							
222	Beta-14833	1740	70	87	504	0,3	1,65	304	Ch, V	Palm nuts						
222	Beta-14832	1970	70	-166	213	0,02	1,93	304	MoS	*Anadara senilis*						
222	Gif-8151	1860	40	66	242	0,15	1,8	304	MoS	*Anadara senilis*						
222	Arc-343	1900	50	3	236	0,12	1,83	304	MoS	*Anadara senilis*						
Ghana																
223	A : 1650-today, B: 1830-today, C: 1850-today					1,9	0,05	116			+					
224	Gd-6546	170	70	1533	1955	1,74	0,21	116	Ch							
224	Gd-5998	550	50	1299	1442	1,37	0,58	116	Ch							
225		1020	130	716	1260	0,99	0,96	279						+		
225	Beta-25084	600	70	1280	1432	1,36	0,59	279						+		
225	Beta-25090	670	40	1268	1395	1,33	0,62	279						+		
225	Beta-25087	690	40	1257	1394	1,33	0,62	279						+		
225	Beta-25092	700	80	1174	1413	1,29	0,66	279						+		
225	Beta-25081	1070	80	775	1155	0,97	0,98	279						+		
226	Beta-25086	490	40	1324	1465	1,39	0,56	279						+		
226	Beta-25085	310	40	1470	1655	1,56	0,39	279						+		
226	Beta-25089	550	40	1304	1438	1,37	0,58	279						+		

N site	Date N	Age BP	±	cal BC		ky Mid BC/AD	ky Mid BP	Citation	Material	Note (taxon, level)	ND	1D	Env	Arch	Fau	Unk
	Other chronological attribution															
227	Beta-25083	190	70	1523	1955	1,74	0,21	279						+		
227	Beta-25091	230	70	1487	1954	1,72	0,23	279						+		
227	Beta-25082	210	50	1524	1955	1,74	0,21	279						+		
228	N-932	500	100	1284	1634	1,46	0,49	279						+		
228	N-2142	905	80	995	1265	1,13	0,82	279						+		
228	N-930	520	100	1277	1633	1,46	0,49	279						+		
228	N-929	240	100	1465	1954	1,71	0,24	279						+		
228	N-2288	430	75	1327	1644	1,49	0,46	279						+		
228	GX-2030	400	100	1307	1952	1,63	0,32	279						+		
228	N-931	285	100	1437	1954	1,7	0,25	279						+		
228	N-2287	355	60	1444	1647	1,55	0,4	279						+		
228	N-2141	830	80	1030	1285	1,16	0,79	279						+		
228	N-1438	410	75	1403	1649	1,53	0,42	279						+		
229	19 th century AD					1,9	0,05	116			+					
230	BLN-1731	190	40	1644	1955	1,8	0,15	279						+		
230	N-2493	715	75	1169	1405	1,29	0,66	279						+		
230	N-2491	240	75	1470	1954	1,71	0,24	279						+		
230	N-1803	570	75	1281	1447	1,36	0,59	279						+		
230	N-2628	335	70	1437	1797	1,62	0,33	279						+		
230	N-2490	430	75	1327	1644	1,49	0,46	279						+		
230	N-2629	365	70	1429	1655	1,54	0,41	279						+		
230	BLN-1729	315	30	1484	1648	1,57	0,38	279						+		
230	BLN-1728	420	30	1426	1618	1,52	0,43	279						+		
231	S-1858	2340	100	-767	-198	-0,48	2,43	279						+		
231	SFU-391	520	100	1277	1633	1,46	0,49	279						+		
231	Beta-7855	1100	170	644	1255	0,95	1	279						+		
231	SFU-388	280	80	1443	1954	1,7	0,25	279						+		
231	5CX	330	50	1454	1649	1,55	0,4	279						+		
231	S-1856	2710	245	-1498	-234	-0,87	2,82	279						+		
231	GX-6133	2010	140	-386	318	-0,03	1,98	279						+		
231	SFU-274	1870	100	-91	398	0,15	1,8	279						+		
231	S-2372	1285	450	-378	1450	0,54	1,41	279						+		
231	SFU-12	1670	190	-96	765	0,33	1,62	279						+		
231	3CX	630	70	1267	1426	1,35	0,6	279						+		
231	Beta-7858	890	160	780	1395	1,09	0,86	279						+		
232	N-4091	380	70	1425	1649	1,54	0,41	279						+		
232	N-4089	450	60	1324	1634	1,48	0,47	279						+		
232	N-4090	395	70	1419	1645	1,53	0,42	279						+		
232	N-4092	395	75	1410	1650	1,53	0,42	279						+		

N site	Date N	Age BP	±	cal BC		ky Mid BC/AD	ky Mid BP	Citation	Material	Note (taxon, level)	ND	1D	Env	Arch	Fau	Unk
	Other chronological attribution															
233	Gd-6543	190	60	1529	1955	1,74	0,21	116		A : 1850-1920, C: 1750-1850 , D: 1870-1950 AD				+		
233	Gd-5997	230	50	1513	1955	1,73	0,22	116						+		
234	Gd-6542	450	80	1316	1642	1,48	0,47	116				+		+		
235	Recent					1,9	0,05	116			+					
236	Recent (18-19th century AD?)					1,8	0,15	116			+					
237	Iron Age					1,5	0,45	116			+					
238	I-2697	3220	110	-1771	-1213	-1,49	3,44	42						+		
238	I-2698	3560	100	-2198	-1641	-1,92	3,87	42						+		
238	Birm-30	3339	35	-1733	-1526	-1,63	3,58	42						+		
239	UCR-1692	3550	127	-2278	-1536	-1,91	3,86	300	V	Palm nuts, unit 1-2 l.3a-b				+		
239	Birm-29	3750	84	-2458	-1951	-2,2	4,15	351		Layer 3				+		
239	UCR-1693	3605	100	-2279	-1691	-1,99	3,94	351	V	Palm nuts, unit 1-3 l.4				+		
239	UCR-1691	3700	90	-2456	-1831	-2,14	4,09	351	V	Palm nuts, unit 1 l.2b-d				+		
239	UCR-1690	3495	100	-2128	-1535	-1,83	3,78	351	V	Palm nuts, unit 2 l.2b-d				+		
240	Punpun phase = after 3500 BP -?					-2	3,95	116			+					
241	Beta-12178	140	80	1645	1955	1,8	0,15	117	Ch		+					
242	Beta-12181	120	60	1668	1949	1,81	0,14	116	Ch							
242	Beta-12180	400	90	1324	1795	1,56	0,39	116	Ch							
242	Beta-12182	170	60	1648	1953	1,8	0,15	116	Ch							
243	N-1984	3350	100	-1888	-1433	-1,66	3,61	351		Pit				+		
244	SR-61	3190	120	-1747	-1129	-1,44	3,39	42	OM	Layer 1b (pit)				+		
244	SR-81	3270	100	-1873	-1316	-1,59	3,54	42		Layer 2				+		
244	SR-52	3580	130	-2336	-1609	-1,97	3,92	42	W	Layer 3				+		
245	Mid 17th-20th century AD					1,8	0,15	116			+					
246	Beta-12184	180	60	1643	1954	1,8	0,15	116	Ch	1730-1790 AD						
246	Beta-12183	160	60	1655	1953	1,8	0,15	116	Ch	1730-1790 AD						
Kenya																
247	Akira/Elmenteitan wares					0,5	1,45	122			+					
248	GX-4386	1965	140	-351	422	0,04	1,91	124		Gelatin						+
248	GX-4384	1255	140	580	1145	0,86	1,09	124	Ch							+
248	GX-4385	1775	150	-1	634	0,32	1,63	124	Ch							+
249	Savannah-oriented Pastoral Neolithic, Akira wares					0,5	1,45	122			+					
250		2045	125	-361	319	-0,02	1,97	124				+				+
251	GX-4588	2405	150	-802	-60	-0,43	2,38	124	BoA							+
251	GX-4585	2660	120	-1001	-406	-0,7	2,65	124	BoA							+
251	GX-4587	2795	155	-1374	-421	-0,9	2,85	124		Gelatin						+
251	Gx-4586/A	2215	120	-510	119	-0,2	2,15	206								+
251	Gx-4586/B	2535	140	-911	-209	-0,56	2,51	124		Gelatin						+
251	GX-4589	2660	160	-1193	-382	-0,79	2,74	124		Gelatin						+

N site	Date N	Age BP	±	cal BC		ky Mid BC/AD	ky Mid BP	Citation	Material	Note (taxon, level)	ND	1D	Env	Arch	Fau	Unk
Other chronological attribution																
252	SUA-637-b	4100	125	-3006	-2291	-2,65	4,6	122	Ch	Acid humic				+		
252	SUA-637	3945	135	-2875	-2051	-2,46	4,41	122	Ch					+		
252	P-2610	3960	60	-2829	-2235	-2,53	4,48	33	Ch	Unit 5C						
252	GX-4642-IA	3405	130	-2035	-1421	-1,73	3,68	33	BoA	Mammal, Unit 5C						
252	GX-4642-IIA	4580	170	-3693	-2892	-3,29	5,24	33	BoA	Fish, Unit 5C						
252		3860	60	-2476	-2141	-2,31	4,26	122	Ch					+		
253	P-2613	1470	50	436	659	0,55	1,4	123	Ch	Lower section				+		
253	P-2614	1700	50	220	529	0,37	1,58	123	Ch	Lower section				+		
253	P-2612	630	50	1281	1409	1,35	0,6	123	Ch	Upper section				+		
254	GX-5744-A	3820	170	-2861	-1778	-2,32	4,27	123	BoA					+		
254	GX-5743-A	3495	130	-2196	-1504	-1,85	3,8	123	BoA					+		
254	GX-5745-A	5150	170	-4334	-3645	-3,99	5,94	123	BoA					+		
255	GX-5158-A	6595	170	-5873	-5217	-5,55	7,5	123	BoA					+		
255	GX-5157-A	6340	170	-5621	-4910	-5,27	7,22	123	BoA					+		
255	GX-5157-G	5240	285	-4711	-3380	-4,05	6	123	BoC					+		
256	GX-5156-A	5935	220	-5321	-4357	-4,84	6,79	123	BoA					+		
256	GX-5156-G	4895	240	-4257	-3031	-3,64	5,59	123	BoA					+		
257	No date - early Holocene? (Later stone Age)					-6	7,95	205			+					
258	GX-9947	5860	200	-5214	-4265	-4,74	6,69	199	Ch	RBL3.1				+		
258	ISGS-1750	6350	150	-5539	-4851	-5,2	7,15	199	Ch	RBL3.1				+		
259	GX-9399	5365	235	-4680	-3647	-4,16	6,11	199	Ch	RBL2.3				+		
259	GX-9944	5265	220	-4504	-3532	-4,02	5,97	199	Ch	RBL2.2						
259	GX-9946	5785	200	-5205	-4072	-4,64	6,59	199	Ch	DBS1.6				+		
259	ISGS-1732	5220	70	-4229	-3787	-4,01	5,96	199	Ch	DBS1.1				+		
259	GX-9945	5470	215	-4724	-3768	-4,25	6,2	199	Ch	DBS1.2				+		
260	GX-9942	4475	210	-3637	-2506	-3,07	5,02	199	Ch	RBL2.1A				+		
260	ISGS-1946	3030	70	-1406	-1006	-1,21	3,16	199	Ch	RBL2.1A1				+		
260	ISGS2308	3990	70	-2623	-2201	-2,41	4,36	199	Ch	RBL2.1B2				+		
260	GX-9943	4535	170	-3639	-2698	-3,17	5,12	199	Ch	RBL2.1B				+		
260	ISGS-1742	4860	70	-3711	-3373	-3,54	5,49	199	Ch	RBL2.1B5				+		
260	ISGS-2040	3110	70	-1493	-1060	-1,28	3,23	199	Ch	RBL2.1A1				+		
261	GX-9941	3125	185	-1740	-845	-1,29	3,24	199	Ch	BVA1				+		
261	ISGS-1745	3390	70	-1869	-1445	-1,66	3,61	199	Ch	RBL1				+		
262	GX-9940	3280	190	-2011	-1006	-1,51	3,46	199	Ch	BS1A				+		
262	ISGS-1746	2710	80	-1012	-540	-0,78	2,73	199	Ch	BS1B				+		
262	ISGS-1733	2860	70	-1192	-810	-1	2,95	199	Ch	BS1A				+		
263	ISGS-1756	2610	70	-827	-415	-0,62	2,57	199	Ch	BSS				+		
263	GX-9939	2570	175	-1056	-194	-0,63	2,58	199	Ch	BSS				+		
263	ISGS-1737	2560	70	-795	-414	-0,6	2,55	199	Ch	HTL				+		

N site	Date N	Age BP	±	cal BC		ky Mid BC/AD	ky Mid BP	Citation	Material	Note (taxon, level)	ND	1D	Env	Arch	Fau	Unk
	Other chronological attribution															
263	ISGS-1757	2820	70	-115	-801	-0,46	2,41	199	Ch	HTL				+		
264	GX-5481	8395	270	-8206	-6704	-7,46	9,41	280	BoA	Mammal, spits 11-13				+		
264	GX-5480	8355	235	-8165	-6687	-7,43	9,38	280	BoA	Mammal, spits 4-7				+		
265	RI-954	9540	260	-9761	-8260	-9,01	10,96	280	MoS	Spits 14-15				+		
265	GX-5479	9260	235	-9249	-7871	-8,56	10,51	280	MoS	Spits 14-15				+		
266	GX-4733	3215	155	-1889	-1056	-1,47	3,42	280	B	Human. Too young?				+		
267	About 9000-7000 BP					-6	7,95	280			+					
268	GX-5476	7855	160	-7176	-6427	-6,8	8,75	280	BoA	Fish				+		
268	Hel-1276	8520	130	-7956	-7189	-7,57	9,52	280	MoS	*Mutela, Melanoides*				+		
268	Hel-1277	8710	130	-8216	-7570	-7,89	9,84	280	MoS	*Etheria elliptica*				+		
269	LSA					-7	8,95	280			+					
270	P-2609	3970	60	-2833	-2288	-2,56	4,51	280	Ch					+		
270	SUA-634	4160	110	-3015	-2468	-2,74	4,69	280	Ch					+		
271	GX-5475	4560	185	-3712	-2763	-3,24	5,19	280	BoA	Fish, middle section		+		+		
272	HAR-6262	3020	100	-1425	-921	-1,17	3,12	159	Ch	Trench I, lev M3				+		
273	HAR-6261	1710	70	231	566	0,4	1,55	159	Ch	Trench II- lev FA4				+		
273	HAR-6254	1990	80	-165	253	0,04	1,91	159	Ch	Base of trench III - midd F12				+		
273	HAR-6260	1900	70	5	378	0,19	1,76	159	Ch	Trench II- lev A3				+		
273	HAR-6252	1770	80	130	534	0,33	1,62	159	Ch	Top of trench III-midd F2				+		
274	OxA-3497	3480	75	-1923	-1527	-1,73	3,68	159	Ch	Sq 29, 130-140 cm				+		
274	OxA-3496	2030	65	-168	210	0,02	1,93	159	Ch	Sq 28, 130-140 cm				+		
274	OxA-3493	2030	70	-169	213	0,02	1,93	159	Ch	Sq 26, 100-110 cm				+		
274	OxA-3494	3170	70	-1529	-1131	-1,33	3,28	159	Ch	Sq 26, 160-170 cm				+		
274	OxA-3495	2000	70	-151	238	0,04	1,91	159	Ch	Sq 27, 120-130 cm				+		
275	GX-6757b	8010	340	-7655	-6099	-6,88	8,83	200						+		
276	GX-6755	6040	225	-5462	-4370	-4,92	6,87	200						+		
276	GX-6754	4145	240	-3353	-1985	-2,67	4,62	200						+		
276	GX-6756	6770	225	-6058	-5221	-5,64	7,59	200						+		
277	GX-6758	13705	430	-15802	-13221	-14,51	16,46	200				+		+		
278	N-1076	1307	122	575	1019	0,8	1,15	200						+		
278	UCLA-1709D	1490	131	265	889	0,58	1,37	200						+		
278	UCLA-1709C	2311	127	-756	-1	-0,38	2,33	200						+		
279	UCLA-1709b	17670	800	-21356	-17470	-19,41	21,36	200						+		
279	GX-3698	13730	430	-15840	-13253	-14,55	16,5	200						+		
279	UCLA-1709a	17700	760	-21198	-17510	-19,35	21,3	200						+		
279	GX-3699	15320	450	-17526	-15300	-16,41	18,36	200						+		
280	GX-5350a	19330	945	-23653	-18997	-21,33	23,28	200	Bb	51-54 cm				+		
280	GX-5349a	20780	1050	...	-20816	-20,8	22,75	200	Bb	87-90 cm				+		
281	Undated Holocene					-4	5,95	200			+					

N site	Date N / Other chronological attribution	Age BP	±	cal BC		ky Mid BC/AD	ky Mid BP	Citation	Material	Note (taxon, level)	ND	1D	Env	Arch	Fau	Unk
282	GX-1774	21535	980	-21869	-17811	-19,84	21,79	200						+		
282	GX1773a	12195	330	-13221	-11466	-12,34	14,29	200						+		
282	GX1773b	11950	460	-13272	-11032	-12,15	14,1	200						+		
283	40 to 21 ka					-30	31,95	200			+					
284	14-15 th century AD					1,4	0,55	120			+					
285	GX-8532	2225	140	-740	129	-0,31	2,26	205								
286		6300	800	-7183	-3531	-5,36	7,31	280		Upper excavation				+		
286	UCLA-1247E	7160	80	-6220	-5890	-6,06	8,01	280	MoS	Base				+		
287	Hel-867	9420	200	-9256	-8286	-8,77	10,72	280	MoS			+		+		
288	GX-4346	7735	150	-7048	-6268	-6,66	8,61	280	BoA	Fish, unit 7		+		+		
289		4460	110	-3498	-2891	-3,19	5,14	280	BoA	Unit 4				+		
289	GX-4349	4110	110	-2919	-2348	-2,63	4,58	280	BoA	Fish, unit 4				+		
290		1560	135	223	810	0,52	1,43	124								+
290		1545	135	239	858	0,55	1,4	124								+
290		2325	145	-768	-1	-0,38	2,33	124								+
290		2595	135	-997	-378	-0,69	2,64	124								+
290	ZX-4471	2515	140	-896	-207	-0,55	2,5	124	BoA							+
291	GX-4466	1695	105	86	570	0,33	1,62	124	BoA			+				+
292	GX-6763	7195	260	-6505	-5554	-6,03	7,98	122				+				+
293		2000	135	-354	387	0,02	1,93	124				+				+
294		2660	115	-997	-407	-0,7	2,65	124								+
294		2760	115	-1211	-523	-0,87	2,82	124								+
294		2640	115	-974	-402	-0,69	2,64	124								+
294		2360	110	-762	-110	-0,44	2,39	124								+
295	GX-4465	2225	155	-752	132	-0,31	2,26	124		Gelatin				+		
295	GX-4463	1665	145	85	669	0,38	1,57	124	BoA					+		
295	GX-4463	1415	150	387	991	0,69	1,26	124		Gelatin				+		
296	Gx-8533	2135	140	-406	240	-0,08	2,03	205	Ch							
296	GX-8534	1940	140	-349	506	0,08	1,87	205	Ch							
297	Birm-770	2020	130	-378	240	-0,07	2,02	124	Ch							+
297	Birm-767	1970	130	-358	335	-0,01	1,96	124	BoC							+
298	GX-5753-G	2530	160	-974	-195	-0,58	2,53	124	EnC	Elephant					+	
298	GX-5753-A	2315	150	-768	5	-0,38	2,33	121	EnA	Elephant					+	
299	GX-4469	6595	235	-5978	-4983	-5,48	7,43	124	BoA							+
299	GX-4422	7255	225	-6477	-5658	-6,07	8,02	124	BoA					+		
300	GX-4420	1185	140	650	1176	0,91	1,04	124	Ch			+				+
301	GX-4468	2990	170	-1601	-795	-1,2	3,15	124	BoA							+
301	GX-4421	2680	150	-1189	-394	-0,79	2,74	124	BoA					+		
302	Elmentetian					0,1	1,85	205								

N site	Date N	Age BP	±	cal BC		ky Mid BC/AD	ky Mid BP	Citation	Material	Note (taxon, level)	ND	1D	Env	Arch	Fau	Unk
	Other chronological attribution															
Lesotho																
303	Pta-5413	90	45	1685	1955	1,82	0,13	213	Ch	S2 sq., - 300 mm				+		
303	Pta-5549	910	80	1025	1280	1,15	0,8	213	Ch	Q12 sq., -200-300 mm				+		
304	Oakhurst Complex : 12000-8300 BP					-8	9,95	243			+					
305	pCW: 2700-100 BP					-1	2,95	249			+					
306	Pta-7877	1310	80	582	894	0,74	1,21	249	Ch	Layer I		+		+		
307	Pta-7865	1830	15	140	339	0,24	1,71	249	Ch	Layer III				+		
307	Pta-7092	1850	40	90	377	0,23	1,72	249	Ch	Layer V				+		
307	Pta-7097	1850	15	134	322	0,23	1,72	249	Ch	Layer III				+		
308	Pta-7098	2060	45	-164	116	-0,02	1,97	249	Ch	Layer VII				+		
308	Pta-7870	2100	80	-355	128	-0,11	2,06	249	Ch	Layer VII				+		
309	Pta-7101	2390	60	-749	-204	-0,48	2,43	249	Ch	Layer XI		+		+		
310	Pta-7093	2650	60	-899	-523	-0,71	2,66	249	Ch	Layer XIII		+		+		
311	GrA-13535	3110	50	-1433	-1130	-1,28	3,23	249	Ch	Layer XVII		+		+		
312	CWI: 8000-3200 BP					-3,5	5,45	243			+					
313	pCW: 2700-100 BP					-1	2,95	249			+					
314	pCW: 2700-100 BP					-1	2,95	243			+					
315	Oakhurst Complex : 12000-8300 BP					-8	9,95	243			+					
316	Pta-885	1400	50	597	779	0,69	1,26	267	Ch					+		
316	Pta-6084	1240	50	694	976	0,84	1,11	267	Ch					+		
316	Pta-6063	1710	20	261	529	0,4	1,55	267	Ch					+		
317	Classic Wilton levels : ~6000 y BP					-4,8	6,75	248			+					
318	Oakhurst level : ~ 7000-9800 y BP					-7	8,95	248			+					
319	Robberg levels : ~11000-12500 y BP					-11,5	13,45	248			+					
320	MSA/LSA levels : 15700-26000 y BP					-20	21,95	248			+					
321	CWI: 8000-3200 BP					-3,5	5,45	243			+					
322	Oakhurst Complex : 12000-8300 BP					-8	9,95	243			+					
323	OxA-4070	375	65	1443	1662	1,55	0,4	213	B	Bone point				+		
323	OxA-4069	715	65	1228	1403	1,32	0,63	213	B	Bone point				+		
324	Robberg Industry : 40000 - 12000 BP					-25	26,95	243			+					
Libya																
325	20 000-10 000 BP					-15	16,95	176			+					
326	Graeco-Roman and Pharaonic Age					-1	2,95	176			+					
327	Iberomaurusian, about 18 000 - 10 000 BP					-15	16,95	176			+					
328	Grn-3541	6950	250	-6380	-5471	-5,93	7,88	95	Ch	Layer VIII-X				+		
329	NPL40	5800	100	-4928	-4407	-4,67	6,62	327	Ch					+		
329	Grn-3541	6950	250	-6380	-5471	-5,93	7,88	95	Ch	Layer VIII-X						
329	NPL41	4860	100	-3938	-3375	-3,66	5,61	327	Ch					+		
330	1800-1700 y BP					0,3	1,65	2			+					

N site	Date N / Other chronological attribution	Age BP	±	cal BC		ky Mid BC/AD	ky Mid BP	Citation	Material	Note (taxon, level)	ND	1D	Env	Arch	Fau	Unk
331	4300 y BP					-3	4,95	2			+					
332		8110	70	-7330	-6826	-7,08	9,03	223								
333	R-1404b	8400	90	-7591	-7189	-7,39	9,34	113	Ch	III						
333	R-1402	9350	110	-9119	-8298	-8,71	10,66	113	Ch	III						
333		8630	50	-7754	-7575	-7,66	9,61	113		III						
333	R-1408	8450	60	-7592	-7359	-7,48	9,43	113	Ch	II						
333		8840	60	-8220	-7750	-7,99	9,94	113		I						
333		8650	105	-8179	-7518	-7,85	9,8	113		II						
334	R-1031a	5970	50	-4983	-4727	-4,86	6,81	113	Ch	II						
334	R-1029	5260	130	-4350	-3791	-4,07	6,02	113	Co	Bovidae, I				+		
334	R-1030	5360	120	-4449	-3961	-4,21	6,16	113	Co	Bovidae, I				+		
334	R-1028	5600	50	-4529	-4348	-4,44	6,39	113	Co	Bovidae, I						
335		8440	50	-7586	-7367	-7,48	9,43	113		CI						
335		8460	50	-7591	-7383	-7,49	9,44	113		CI						
335		7990	70	-7072	-6683	-6,88	8,83	113		CII						
335		8460	60	-7595	-7365	-7,48	9,43	113		CII						
335		8520	60	-7649	-7478	-7,56	9,51	113		I						
335		8540	140	-8169	-7187	-7,68	9,63	113		Rsup						
335		8640	70	-7937	-7545	-7,74	9,69	113		Rinf						
335		8670	60	-7937	-7583	-7,76	9,71	113		Rinf						
335		9080	70	-8543	-8003	-8,27	10,22	113		Rinf						
335		7730	80	-6752	-6429	-6,59	8,54	113		Rsup						
336	Gx-20754	8935	100	-8299	-7749	-8,02	9,97	69	Ch	Unit1						
336	Gx-20346	8330	100	-7573	-7085	-7,33	9,28	69	Ch	Organic sand, Unit1						
336	Gx-20010	8765	105	-8207	-7598	-7,9	9,85	69	W	Unit1						
337	Gx-19118	9260	290	-9295	-7684	-8,49	10,44	69	Ch	Unit2						
337	Gx-20750	9765	105	-9649	-8798	-9,22	11,17	69	Ch	Unit2						
338	Gx-17816AMS	7823	95	-7029	-6476	-6,75	8,7	94	Ch	layer 3				+		
338	Gx-17815AMS	7550	120	-6645	-6106	-6,38	8,33	68	Ch	layer 4				+		
339	Gd-4290	2220	220	-807	225	-0,29	2,24	68		*Citrullus,* B - layer 1				+		
339	Gd-2854	3810	80	-2472	-2032	-2,25	4,2	68	V	*Balanites* , B - layer 1						
339	Gd-2962	3720	90	-2457	-1891	-2,17	4,12	68	V	*Balanites* , A - l. 1-1a						
339	Gd-4363	3800	140	-2832	-1784	-2,31	4,26	68	V	*Balanites* , A - l. 1-1a						
340	Ud-225	6035	100	-5215	-4718	-4,97	6,92	68	Ch	A - layer 2-2c						
340	Pi-125b	5405	180	-4669	-3802	-4,24	6,19	68	L	Trench 1a						
340	Gd-2958	4980	70	-3945	-3651	-3,8	5,75	68	V	*Balanites* , B - l. 1a-2d						
340	Gd-4362	5290	100	-4346	-3848	-4,1	6,05	68	V	*Balanites* , A - l. 2-2c						
340	Gd-4361	5480	120	-4581	-3996	-4,29	6,24	68	V	*Balanites* , B - l. 2a-2d						
341	Mesolithic					-10	11,95	68			+					

N site	Date N / Other chronological attribution	Age BP	±	cal BC		ky Mid BC/AD	ky Mid BP	Citation	Material	Note (taxon, level)	ND	1D	Env	Arch	Fau	Unk
342	Rome-292	8800	100	-8215	-7611	-7,91	9,86	69	Ch	TLII-13						
342	Gx-17818-AMS	8690	110	-8201	-7548	-7,87	9,82	69	Ch	TL-III-20						
342	Rome-293	8880	100	-8280	-7686	-7,98	9,93	69	Ch	TLIII-14						
343	Geo	6745	175	-5991	-5371	-5,68	7,63	68		Layer IV-V						
344	Gx-18101	4675	75	-3641	-3131	-3,39	5,34	68	OM	Layer I						
345	Bo-187	5900	80	-4981	-4554	-4,77	6,72	68		Layer II-IIIa						
345	Bo-186	5320	75	-4328	-3989	-4,16	6,11	68	OM	Layer II-IIIa						
346	GX-28483-AMS	3030	40	-1408	-1132	-1,27	3,22	80						+		
346	GX-28482-AMS	3530	40	-1963	-1745	-1,85	3,8	80						+		
347	GX-28441-AMS	3370	40	-1750	-1531	-1,64	3,59	80								
347	GX-28477-AMS	2930	40	-1266	-1009	-1,14	3,09	80								
347	GX-28479-AMS	2720	40	-971	-804	-0,89	2,84	80								
347	GX-28478-AMS	2550	190	-1192	-202	-0,7	2,65	80								
347	GX-28476-AMS	3850	40	-2461	-2205	-2,33	4,28	80								
Madagascar																
348	Holocene					-5	6,95	89			+					
349	Holocene					-5	6,95	89			+					
Malawi																
350	7-14 th century AD					1	0,95	331			+					
351	Contemporaneous to level 2					1,7	0,25	158			+					
352	Pta-2806	240	40	1629	1953	1,79	0,16	158	Ch					+		
353	Pta-2805	720	40	1273	1392	1,33	0,62	158	Ch					+		
353	Har-3801	590	50	1305	1448	1,38	0,57	158	Ch					+		
353	Pta-2810	470	45	1407	1622	1,51	0,44	158	Ch					+		
353	R-4094	390	80	1417	1796	1,61	0,34	158	Ch					+		
353	Har-4793	660	50	1287	1411	1,35	0,6	158	Ch					+		
354	I-11845	910	75	1030	1275	1,15	0,8	158	Ch					+		
354	Har-3802	630	50	1296	1427	1,36	0,59	158	Ch					+		
355	R-4109	8560	240	-8274	-7036	-7,66	9,61	158	Ch					+		
356	7-14 th century AD for the whole site					1	0,95	331			+					
357	7-14 th century AD for the whole site					1	0,95	331			+					
358	7-14 th century AD for the whole site					1	0,95	331			+					
359	7-14 th century AD					1	0,95	331			+					
Mali																
360	Beta-40446	1630	60	257	560	0,41	1,54	285	Ch	AK3, lv 36-38						
361	Beta-40444	1430	70	433	765	0,6	1,35	285		AK1-lev41?						
361	Beta-49720	1130	100	669	1148	0,91	1,04	285	Ch	AK3-lev12	+					
362	Beta-40445	1390	80	434	852	0,64	1,31	285		Base AK1 lev67						
362	Beta-40443	980	70	898	1214	1,06	0,89	285	Ch	AK1-lev24						

N site	Date N	Age BP	±	cal BC		ky Mid BC/AD	ky Mid BP	Citation	Material	Note (taxon, level)	ND	1D	Env	Arch	Fau	Unk
	Other chronological attribution															
362	Beta-40442	750	70	1052	1395	1,22	0,73	285	Ch	AK1-lev10						
362	Beta-40447	910	50	1025	1217	1,12	0,83	285		AK4, lev1						
362	Beta-40441	640	70	1261	1426	1,34	0,61	285	Ch	AK1-lev6						
363		6390				-5,4	7,35	273	B	Human		+		+		
364	Gx-27055	2470	70	-775	-408	-0,59	2,54	16		F081						
364	Gx-27051	2220	100	-522	3	-0,26	2,21	16		C105						
364	Gx-27054	1990	50	-152	125	-0,01	1,96	16		A115						
364	Gx-27050-AMS	2450	30	-753	-410	-0,58	2,53	16		C092						
364	Gx-25785-AMS	2470	40	-765	-414	-0,59	2,54	16		B095						
364	Gx-25786-AMS	2550	50	-811	-518	-0,66	2,61	16		B122						
364	Gx-27053	2070	100	-372	123	-0,12	2,07	16		C077						
365	Ca 300BC-300AD					0	1,95	16			+					
366	Gx-25634-LS	310	90	1425	1954	1,69	0,26	16		B050						
366	Gx-25637-AMS	1200	40	690	946	0,82	1,13	16	Ch	Grave, B071						
367	Gx-25633-LS	less than 100 years - 15th to 17th C AD				1,6	0,35	16		Floor 2, unit B	+					
368	Gif5813	4480	110	-3506	-2896	-3,2	5,15	154	Ch	MK42						
368	Gif5232	6540	130	-5716	-5229	-5,47	7,42	154	Ch	AZ22						
368	Gif5227	3750	100	-2466	-1921	-2,19	4,14	154	Ch	AZ21						
368	Gif5228	6340	130	-5551	-4996	-5,27	7,22	154	Ch	AZ22						
368	Gif5441	3600	180	-2466	-1530	-2	3,95	154	B	Human, AZ21						
368	Gif5442	3990	130	-2883	-2146	-2,51	4,46	154	B	Bovidae, AZ21						
369	Pa374/GrA22579	3940	45	-2570	-2296	-2,43	4,38	154	En	Mammal, MN24						
370	Pa-269	2680	120	-1190	-417	-0,8	2,75	327	Ch			+		+		
371	Late 13th - early 14 th century, period I					1,4	0,55	196			+					
372	Pa-412	1000	70	893	1206	1,05	0,9	327	Ch			+		+		
373	19th century					1,85	0,1	57			+					
374	Pa1058	6795	200	-6071	-5361	-5,72	7,67	154	B	Human, MK40						
374	Gif5495	6970	130	-6072	-5630	-5,85	7,8	154	B	Fish, AR7						
374	Gif6459	6235	300	-5721	-4501	-5,11	7,06	154	Ce	MN27						
374	Pa373/GrA22576	5065	45	-3966	-3716	-3,84	5,79	154	En	Mammal, MN27						
374	Pa376/GrA22581	4910	45	-3785	-3637	-3,71	5,66	154	En	Mammal, MN35						
374	Pa1024	4885	200	-4227	-3104	-3,67	5,62	154	Ce	MN10						
374	Pa375/GrA22580	4255	45	-3010	-2677	-2,84	4,79	154	En	Mammal, MN6						
375						-1	2,95	273			+					
376	RL1574	1860	120	-162	422	0,13	1,82	293	Ch					+		
376	RL1573	1750	100	66	535	0,3	1,65	293						+		
376	RL1580	1750	100	66	535	0,3	1,65	293						+		
376	RL1622	2090	120	-397	132	-0,13	2,08	293	Ch					+		
376	RL1581	1800	120	-44	534	0,25	1,7	293	Ch					+		

N site	Date N	Age BP	±	cal BC		ky Mid BC/AD	ky Mid BP	Citation	Material	Note (taxon, level)	ND	1D	Env	Arch	Fau	Unk
Other chronological attribution																
376	RL1621	1910	110	-175	381	0,1	1,85	293	Ch					+		
376	RL1619	2050	120	-383	211	-0,09	2,04	293	Ch					+		
376	RL1620	2060	110	-387	135	-0,13	2,08	293	Ch					+		
376	RL1576	1790	120	-39	534	0,25	1,7	293	Ch					+		
377	RL1618	1300	110	552	977	0,76	1,19	293	Ch							
377	RL1578	1310	110	542	975	0,76	1,19	293	Ch							
378	RL1571	1060	110	710	1210	0,96	0,99	293	Ch							
378	RL1808	970	110	784	1271	1,03	0,92	293								
378	RL1616	550	100	1262	1627	1,44	0,51	293								
378	RL1617	790	100	1026	1391	1,21	0,74	293								
378	RL1807	940	110	885	1280	1,08	0,87	293								
379	OxA-16974	3853	32	-2461	-2206	-2,33	4,28	91	Ch	Context 77 at 0,4m				+		
379	OxA-16895	3894	32	-2471	-2288	-2,38	4,33	91	Ch	Context 32 at 0,6m				+		
379	OxA-16896	3882	32	-2469	-2235	-2,35	4,3	91	V	Euphorbiacea, Context 50 at 1,3 m				+		
379	OxA-16897	3856	32	-2461	-2208	-2,33	4,28	91	Ch	Context 81 at 1,6m				+		
379	OxA-16919	4011	33	-2620	-2467	-2,54	4,49	91	V	*Pennisetum* , Context 31 at 0,4m				+		
379	OxA-16973	3988	32	-2580	-2459	-2,52	4,47	91		Leguminosae/*Vitis* , Context 166 at 2,4m				+		
379	OxA-16975	3877	33	-2468	-2212	-2,34	4,29	91	Ch	Context 77 at 0,5m				+		
379	OxA-16976	3913	33	-2479	-2292	-2,39	4,34	91	V	*Sterculia* , Context 147 at 0,1m				+		
379	OxA-16977	3986	31	-2578	-2462	-2,52	4,47	91	Ch	Context 87 at 4,4m				+		
379	OxA-16898	3889	32	-2471	-2243	-2,36	4,31	91	V	Combretacea, Context 85 at 4m				+		
379	OxA-16920	3922	33	-2489	-2296	-2,39	4,34	91	Ch	Context 37 at 1m				+		
380	N-1395	3960	160	-2890	-2036	-2,46	4,41	271, 272	Ch	KS2, 140 cm				+		
380	N-1397	3710	85	-2435	-1886	-2,16	4,11	271, 272	Ch	KN1, 120 cm				+		
380	N-1396	3950	90	-2853	-2151	-2,5	4,45	271, 272	Ch	KN1, 210 cm				+		
380	N-1398	3620	80	-2201	-1756	-1,98	3,93	271, 272	Ch	KN2, 40 cm				+		
380	Gif851	3310	110	-1890	-1387	-1,64	3,59	271, 272	B	KS-surface				+		
380	N-1394	3640	100	-2299	-1740	-2,02	3,97	271, 272	Ch	KS2, 10-20 cm				+		
381	Pa-221	3335	100	-1882	-1430	-1,66	3,61	156	Bb	Hippo, surface, 84-P2						
381	Pa-1341	3450	80	-1961	-1532	-1,75	3,7	156	B	Human, surface, H4						
381	Pa-1340	3320	100	-1879	-1416	-1,65	3,6	156	B	Human, surface, H90						
381	Pa-916	3305	80	-1859	-1418	-1,64	3,59	156	Ch	C/150-168 cm; 89-61						
382	Gx-21192-AMS	3365	70	-1877	-1499	-1,69	3,64	189	En	Hippo, phase I						
382	Gx-19814-AMS	3084	73	-1500	-1129	-1,31	3,26	189	En	*Bos taurus* , Interface II-I						
383	Gx-19814-AMS	3084	73	-1500	-1129	-1,31	3,26	189	En	*Bos taurus* , Interface II-I						
384	Gx-18516	1020	105	774	1224	1	0,95	192	W	Transition II-I						
385	Gx-18376	5185	95	-4258	-3773	-4,02	5,97	192	W	Transition III-II						
386	Before 10-12th centuries AD					0,8	1,15	149			+					
387	Pa172	3920	100	-2850	-2059	-2,45	4,4	154	Ch	MN51						

N site	Date N	Age BP	±	cal BC		ky Mid BC/AD	ky Mid BP	Citation	Material	Note (taxon, level)	ND	1D	Env	Arch	Fau	Unk
Other chronological attribution																
387	Pa181	4320	100	-3339	-2638	-2,99	4,94	154	Ch	MN51						
388		3980	60	-2836	-2292	-2,56	4,51	183	MoS							
389	GX-21728	1105	80	693	1148	0,92	1,03	183	Ch							
389	Gx-19233	1605	130	136	655	0,4	1,55	189	Ch	Lev36, hearth						
390	Gx-19990	3635	90	-2281	-1751	-2,02	3,97	191	Ch	WK-I, level 17						
390	Gx-19234	3115	195	-1877	-899	-1,39	3,34	191	Ch	WK-I, level 7 (midden)						
391	Gx-19254-AMS	3319	57	-1742	-1461	-1,6	3,55	191	En	*Phaco. aethiopicus* , WKSO unit 3, lev 1						
392	Gx-19232-AMS	2607	86	-928	-417	-0,67	2,62	191	Ch	Tumulus 7						
Mauritania																
393	Pa-1576	6700	200	-6014	-5234	-5,62	7,57	157	Bb	Mammal, square 2						
394	Gif-2492	3500	120	-2141	-1524	-1,83	3,78	327	Ch					+		
394	Gif-2163	2100	180	-746	325	-0,21	2,16	327	B	Human				+		
395	Gx-1890	3700	80	-2389	-1883	-2,14	4,09	215	MoS							
396	I-3566	2330	105	-767	-175	-0,47	2,42	215	Ch							
396	GX-1329	2600	110	-975	-408	-0,69	2,64	215	Ch							
397	GX-1324	2700	115	-1209	-522	-0,87	2,82	215	Ch							
397	GX-1888	2885	140	-1413	-808	-1,11	3,06	215	Ch							
397	I-3564	3205	105	-1743	-1215	-1,48	3,43	215	Ch							
398	GX-1323	3205	95	-1736	-1262	-1,5	3,45	215	Ch							
398	I-3563	2950	100	-1413	-917	-1,17	3,12	215	Ch							
398	I-3561	3350	110	-1917	-1418	-1,67	3,62	215	Ch							
398	GX-1889	3425	130	-2120	-1436	-1,78	3,73	215	Ch							
398	I-3562	3190	100	-1736	-1213	-1,47	3,42	215	Ch							
398	GX-1421	3465	160	-2271	-1422	-1,85	3,8	215	BoA	Fish						
399	I-3565	3100	108	-1612	-1052	-1,33	3,28	215	Ch							
399	GX-1325	2780	140	-1387	-564	-0,98	2,93	215	Ch							
399	GX-1326	2600	110	-975	-408	-0,69	2,64	215	Ch							
400	Gif-6083	2430	80	-774	-393	-0,58	2,53	143								
401	Gif-2884	3830	120	-2620	-1937	-2,28	4,23	143								
402	Pa2153	3765	35	-2292	-2041	-2,17	4,12	225	Ch	S5-midden (bottom)						
403	Gif-2550	2030	100	-359	210	-0,07	2,02	40	Ce					+		
403	Gif-2552	4850	130	-3945	-3368	-3,66	5,61	40	Ce					+		
403	Gif-2551	3120	110	-1630	-1055	-1,34	3,29	40	B	Human				+		
404	Pa-1641	4255	30	-2918	-2761	-2,84	4,79	328	MoS	*Dentalium*			+			
404	Pa-	3100	70	-1516	-1132	-1,32	3,27	328	Ce					+		
404	Pa-1650	3080	30	-1421	-1268	-1,34	3,29	328	OES					+		
404	Pa-1651	3055	30	-1411	-1221	-1,32	3,27	328	B	Human				+		
405	Ly-2974	3040	160	-1660	-851	-1,26	3,21	14	Ce	Occupation phase III						
405	Gif-6891	2830	60	-1193	-836	-1,01	2,96	14	Ce	Occupation phase IV						

N site	Date N	Age BP	±	cal BC		ky Mid BC/AD	ky Mid BP	Citation	Material	Note (taxon, level)	ND	1D	Env	Arch	Fau	Unk
	Other chronological attribution															
405	Ly-2502	3310	200	-2189	-1058	-1,62	3,57	14	Ce	Occupation phase II						
405	Ly2501	3350	130	-2014	-1386	-1,7	3,65	14	Ce	Occupation phase I						
406	Pa1536	3930	80	-2832	-2147	-2,49	4,44	4	B	Human				+		
406	Ly-4059	2760	40	-1002	-826	-0,91	2,86	327	Ce					+		
407	Dak-193	1851	100	-51	410	0,18	1,77	327	Ch					+		
407	Dak-193	1233	100	640	1012	0,83	1,12	327	Ch					+		
408	Dak-95	831	117	982	1394	1,19	0,76	327	Ch					+		
408	Dak-93	881	118	895	1382	1,14	0,81	327	Ch					+		
409	12th to 13th century AD					1,3	0,65	26			+					
410	Posterior to 14th century AD					1,5	0,45	26			+					
411	I-3566	2330	105	-767	-175	-0,47	2,42	216	Ch	Taidart II						
411	GX-1326	2600	110	-975	-408	-0,69	2,64	216	Ch	Bledd Initi						
412	I-3819	2170	105	-409	57	-0,18	2,13	216	Ch	Niemilane IV						
413	I-3565	3100	105	-1609	-1055	-1,33	3,28	216	Ch	Seyyid Ouinquil						
413	GX-1325	2780	140	-1387	-564	-0,98	2,93	216	Ch	Le Baidha I						
414	I-3562	3190	105	-1739	-1209	-1,47	3,42	216	Ch	Goungou B, 4						
414	I-3563	2950	100	-1413	-917	-1,17	3,12	216	Ch	Goungou B, 2						
414	GX-1889	3425	130	-2120	-1436	-1,78	3,73	216	Ch	Goungou B, 3						
415	I-3561	3350	110	-1917	-1418	-1,67	3,62	216	Ch	Goungou A						
415	GX-1421	3465	160	-2271	-1422	-1,85	3,8	216	BoA	Fish, Goungou A						
415	GX-1323	3205	95	-1736	-1262	-1,5	3,45	216	Ch	Goungou A						
416	GX-1888	2885	140	-1413	-808	-1,11	3,06	216	Ch	Naghez						
416	I-3564	3205	105	-1743	-1215	-1,48	3,43	216	Ch	Naghez						
416	GX-1324	2700	115	-1209	-522	-0,87	2,82	216	Ch	Naghez						
417	Ca 1025 bc					-1	2,95	216			+					
418	C	5020	160	-4235	-3385	-3,81	5,76	327	MoS	*Cymbium*				+		
418	Gif-1884	2460	100	-803	-389	-0,6	2,55	327	Ce					+		
418	Ly-459	3930	120	-2864	-2045	-2,45	4,4	327	Ce					+		
418	C	5520	100	-4584	-4061	-4,32	6,27	327	MoS	*Anadara*				+		
418	Ly-503	4270	100	-3324	-2576	-2,95	4,9	327	MoS	*Cymbium*				+		
418	Ly-460	3530	130	-2267	-1526	-1,9	3,85	327	MoS	*Anadara*				+		
418	Gif-2484	3240	100	-1755	-1271	-1,51	3,46	327	B					+		
Morocco																
419	IA, Protohistory					-0,5	2,45	220			+					
420	Proto-campaniform (recent Neolithic)					-3,2	5,15	220			+					
421	Cardial					-4,5	6,45	220			+					
422	TO-2049	25580	130	-23895	-23375	-23,64	25,59	5	MoS	*Helix*		+	+	+		
423	Ly-2149	5980	210	-5368	-4402	-4,89	6,84	327	B	Human				+		
423	UQ-1601	5800	150	-5017	-4347	-4,68	6,63	327	MoS	*Cardium*				+	+	

N site	Date N / Other chronological attribution	Age BP	±	cal BC		ky Mid BC/AD	ky Mid BP	Citation	Material	Note (taxon, level)	ND	1D	Env	Arch	Fau	Unk
424	5450-4400 BP					-3,8	5,75	217			+					
425	Modern					1,9	0,05	217			+					
426	Ancient Neolithic					-4,5	6,45	220			+					
427	Recent Neolithic					-3,2	5,15	220			+					
428	OxA-11321	9470	55	-9122	-8624	-8,87	10,82	29	Ch	*Phillyrea/Rhamnus*				+		
428	OxA-11322	11180	65	-11260	-10992	-11,13	13,08	29	Ch	*Phillyrea/Rhamnus*				+		
428	OxA-11323	11125	65	-11210	-10960	-11,09	13,04	29	Ch	*Phillyrea/Rhamnus*				+		
429	Protohistory					-2	3,95	220			+					
430	Protohistory, Campaniforme					-2,5	4,45	220			+					
431	Recent Neolithic					-3,5	5,45	220			+					
432	Ly-3821	6050	120	-5297	-4707	-5	6,95	73	Ch							
432	Ly-7288	6520	80	-5619	-5341	-5,48	7,43	73	Ch							
433	~ 7000 BP					-5	6,95	220			+					
434	Ly-7695	9865	75	-9661	-9221	-9,44	11,39	73	Ch							
434	Ly-7287	9910	50	-9653	-9267	-9,46	11,41	73	Ch							
435	Ly-7289	13300	180	-14426	-13264	-13,85	15,8	73	Ch							
436	OxA-11929	14110	60	-15245	-14442	-14,84	16,79	13	Ch	*Junip./Tetraclinus* , l. 5				+		
436	OxA-11926	13345	50	-14283	-13531	-13,91	15,86	29	Ch	*Pinus,* layer 3				+		
436	OxA-11417	15940	80	-17386	-17005	-17,2	19,15	13	Ch	Leguminosae, l. 6 base				+		
436	OxA-11927	13805	55	-14883	-14132	-14,51	16,46	13	Ch	*Pinus,* layer 4a				+		
436	OxA-11928	14005	55	-15105	-14349	-14,73	16,68	13	Ch	*Pinus,* layer 6				+		
437	Gif-2888	4750	110	-3786	-3123	-3,45	5,4	220			+					
437		5254	521			-5,3	7,25	220	Ce							
437		5035	343			-5	6,95	220	Ce							
437	Gif-2889	5160	110	-4243	-3708	-3,98	5,93	220								
438	IA-bronze					-1,8	3,75	220			+					
439	Ly-3087	4881	190	-4224	-3104	-3,66	5,61	72	B	*Ceratotherium simum*						
440	OxA-13516	11065	45	-11134	-10950	-11,04	12,99	29	Ch	*Pinus,* G89-203				+		
440	OxA-13480	10950	45	-11011	-10890	-10,95	12,9	29	Ch	*Pinus,* G89-202				+		
440	OxA-13479	10935	40	-10991	-10891	-10,94	12,89	29	Ch	*Pinus,* G88-200				+		
440	OxA-13478	12495	50	-12996	-12291	-12,64	14,59	29	Ch	*Juniperus/Tetraclinus* , G99-90				+		
440	OxA-13517	10990	45	-11085	-10908	-11	12,95	29	Ch	Dicotyled., G89-204				+		
440	OxA-13477	12675	50	-13241	-12755	-13	14,95	29	Ch	Conifer, G97-36				+		
Mozambique																
441	Phase 1 : AD1150-1350					1,25	0,7	12			+					
442	Phase 2 : AD1350 - late 18th centurymidden C: 1350/1450middei					1,58	0,37	12			+					
443	Phase 3 : ~ AD 1600					1,6	0,35	12			+					
444	Phase 4 : early 17th century					1,62	0,33	12			+					
445	Phase 5 : 17th century					1,65	0,3	12			+					

N site	Date N / Other chronological attribution	Age BP	±	cal BC		ky Mid BC/AD	ky Mid BP	Citation	Material	Note (taxon, level)	ND	1D	Env	Arch	Fau	Unk
Namibia																
446	LSA					-6	7,95	282			+					
447	Pta-1918	1960	45	1	235	0,12	1,83	172								+
447	KN-1870	1670	55	260	574	0,42	1,53	172								+
447	KN-1846	1460	55	540	768	0,65	1,3	172			+					+
448	LSA Wilton ~ 10000 to 6000 BP					-7	8,95	282			+					
449	Early LSA: 20000 to 12500 BP					-15	16,95	282			+					
450	500 BP					1,4	0,55	303		Umuab 21-BBC	+					
450	460 BP					1,4	0,55	303		Umuab A-BBC	+					
450	350 BP					1,6	0,35	303		Katarakt	+					
451	700-800 BP					1,3	0,65	303		Umuab 28-LSA	+					
451	500 BP					1,4	0,55	303		Umuab 21-LSA	+					
452	5300-2700 BP					-2,5	4,45	303		Amis 10-LSA	+					
452	3000-2500 BP					-0,3	2,25	303		Amis 11-LSA	+					
452	2400-2700 BP					-0,7	2,65	303		Amis 17-LSA	+					
452	3300 BP					-1,6	3,55	303		Umuab B2-LSA	+					
453		>44000				-20	21,95	303	Ch							
454	KN-1732	2190	40	-358	-45	-0,2	2,15	267	Ch			+				
455	Pta-4416	800	50	1182	1381	1,28	0,67	278	Ch	C7: LIA , 0,04 m		+		+		
456	Pta-5875	1970	40	19	224	0,12	1,83	278	Ch	C7:L8 0,46 m				+		
456	Pta-4413	1980	50	-41	220	0,09	1,86	278	Ch	C6:L7 0,36 m				+		
456	Pta-5871	2090	45	-172	65	-0,05	2	278	Ch	C7:L9- 0,51 m				+		
456	Pta-5873	2040	50	-157	129	-0,01	1,96	278	Ch	C7/B7:L10- 0,55 m				+		
456	Pta-4414	2110	60	-347	75	-0,14	2,09	278	Ch	C7:L11- 0,63 m				+		
456	Pta-4419	1790	80	86	532	0,31	1,64	278	OM	Dung, B6:L5 0,26 m				+		
456	Pta-5872	2300	50	-400	-186	-0,29	2,24	278	Ch	C7/L13- 0,72m				+		
456	Pta-2720	1790	50	139	418	0,28	1,67	278	Ch	B5 : 0,3/0,4 m				+		
457	LSA					-6	7,95	282			+					
458	LSA					-6	7,95	282			+					
459	LSA					-6	7,95	282			+					
460	Pa652	12091	59	-12136	-11851	-11,99	13,94	142	EnA	Bovidae					+	
460	Pa650	10778	54	-10927	-10774	-10,85	12,8	142	BoA	Bovidae					+	
460	Pa651	11489	56	-11493	-11281	-11,39	13,34	142	BoA	Bovidae					+	
461	LSA					-6	7,95	282			+					
462						-6	7,95	282			+					
463	LSA					-6	7,95	282			+					
Niger																
464	UCLA-1658	5760	500	-5723	-3637	-4,68	6,63	41	BoC	*Bos taurus*					+	
464	N-870	4910	135	-3977	-3373	-3,68	5,63	41	B	*Bos taurus* , midden					+	

N site	Date N	Age BP	±	cal BC		ky Mid BC/AD	ky Mid BP	Citation	Material	Note (taxon, level)	ND	1D	Env	Arch	Fau	Unk
Other chronological attribution																
464		6325	300	-5838	-4591	-5,21	7,16	264	B	*Bos taurus*					+	
465	Washington	9130	65	-8542	-8247	-8,39	10,34	265	Ch							
465	Washington	9030	190	-8701	-7613	-8,16	10,11	265	Ch							
465	Pa-	9100	150	-8715	-7824	-8,27	10,22	265	Ch							
466	Contemp. with site 465, 473, 495					-8	9,95	125			+					
467	CAMS-103520	5315	40	-4312	-4003	-4,16	6,11	276	En	*Bos taurus*				+	+	
467	CAMS-103518	4145	45	-2879	-2581	-2,73	4,68	276	En	*Bos taurus*				+	+	
467	CAMS-103528	3880	40	-2471	-2209	-2,34	4,29	276	En	*Bos taurus* , Greboun Wadi				+	+	
467	CAMS-103519	4310	40	-3023	-2879	-2,95	4,9	276	En	*Gazella,* Greboun Wadi				+	+	
467	CAMS-77911	3570	40	-2030	-1774	-1,9	3,85	276		Agorass in-Tchissikan				+		
467	CAMS-103523	4690	40	-3630	-3368	-3,5	5,45	276	En	*Alcelaphus buselaphus* area 1				+	+	
467	CAMS-103529	4730	50	-3636	-3375	-3,51	5,46	276	En	*Bos taurus* , area 5				+	+	
467	CAMS-103521	4755	40	-3640	-3379	-3,51	5,46	276	En	*Bos taurus* , area 2				+	+	
467	CAMS-103531	4795	40	-3655	-3384	-3,52	5,47	276	En	*Bos taurus* , area 2				+	+	
467	CAMS-103526	4800	40	-3658	-3384	-3,52	5,47	276	En	*Bos taurus* , area 5				+	+	
467	CAMS-103525	4660	45	-3627	-3357	-3,49	5,44	276	En	*Bos taurus* , area 2				+	+	
467	CAMS-103527	5355	40	-4326	-4052	-4,19	6,14	276	En	*Bos taurus* , area 4				+	+	
467	CAMS-103524	5430	40	-4355	-4176	-4,27	6,22	276	En	*Bos taurus* , area 5				+	+	
467	CAMS-103532	4855	40	-3711	-3527	-3,62	5,57	276	En	*Bos taurus* , area 2				+	+	
468	Gif-3057	5380	130	-4488	-3956	-4,22	6,17	108	Ch			+				+
469	Gif-1797	2640	100	-1015	-417	-0,72	2,67	108	B					+		
470	Pa-363	4585	120	-3635	-2941	-3,29	5,24	82	MoS							
471	Pa-458	3830	130	-2831	-1907	-2,37	4,32	224	Sed	Palaeosoil						
471	Pa-1054	3910	150	-2872	-2024	-2,45	4,4	224	B	Human						
471	Pa-663	4390	80	-3338	-2891	-3,11	5,06	224	MoS	Pit base						
471	Pa-660	4555	130	-3631	-2920	-3,28	5,23	224	MoS	Pit base						
471	Pa-1043	3775	150	-2620	-1771	-2,2	4,15	224	B	Human						
471	Pa-295	3730	200	-2849	-1626	-2,24	4,19	224	B	Human						
471	Pa-251	3725	120	-2473	-1777	-2,13	4,08	224	Ch	Hearth 2						
471	Orsay	3385	65	-1879	-1522	-1,7	3,65	224	MoS	Hearth 1						
471	Pa-438	3365	100	-1901	-1436	-1,67	3,62	224	MoS	*Mutela*						
471	Pa-667	3810	200	-2868	-1752	-2,31	4,26	224	Ce							
471	Pa-285	3325	260	-2341	-943	-1,64	3,59	224	Bb	Bovinae B4						
472	Pa-367	4100	70	-2877	-2491	-2,68	4,63	82	MoS	*Caelatura*						
473	P-683	7480	40	-6430	-6249	-6,34	8,29	270	BoA	*Lates niloticus* , area 5				+	+	
473	P-682	7700	40	-6629	-6461	-6,55	8,5	270	BoA	*Redunca redunca* , area 5				+	+	
474	UCIAMS-35944	5065	20	-3951	-3797	-3,87	5,82	270	MoS	Midden 1				+	+	
474	UCIAMS-35599	4445	25	-3331	-2945	-3,14	5,09	270	Bb	Midden 2				+	+	
474	UCIAMS-35598	4445	25	-3331	-2945	-3,14	5,09	270	Ch	Midden 2				+		

N site	Date N	Age BP	±	cal BC		ky Mid BC/AD	ky Mid BP	Citation	Material	Note (taxon, level)	ND	1D	Env	Arch	Fau	Unk
Other chronological attribution																
474	P-549	4210	40	-2905	-2668	-2,79	4,74	270	BoA	*Bos taurus* , Midd 1-2				+	+	
474	UCIAMS-35940	5665	20	-4541	-4457	-4,5	6,45	270	MoS	Midden 4				+	+	
474	UCIAMS-35946	5040	15	-3943	-3781	-3,86	5,81	270	MoS	Midden 1				+	+	
474	UCIAMS-35941	5535	20	-4448	-4340	-4,39	6,34	270	MoS	Midden 4				+	+	
474	UCIAMS-35945	5005	15	-3909	-3711	-3,81	5,76	270	MoS	Midden 1				+	+	
475	Pa-309	5065	100	-4052	-3644	-3,85	5,8	82	OES	Bead						
475	Pa-1614	5010	120	-4052	-3526	-3,79	5,74	82	BoA	(purified) Human						
476	Gif-374	5520	250	-4946	-3796	-4,37	6,32	55				+				
477	Pa-1624	2740	100	-1256	-669	-0,96	2,91	82	BoA	(purified) Human						
478	Pa-415	4590	130	-3637	-2936	-3,29	5,24	82	Ce	OM, MNTS 13						
479	Pa-1575	7160	80	-6220	-5890	-6,06	8,01	82	MoS	*Caelatura lacoini*						
479	Pa-1572	7725	80	-6750	-6427	-6,59	8,54	82	MoS	*Mutela*						
479	Pa-1573	4765	80	-3695	-3368	-3,53	5,48	82	MoS	*Melanoides tuberculata*						
480	Or-2552	4440	220	-3651	-2500	-3,08	5,03	82	MoS							
480	Or-1764	8320	240	-7946	-6656	-7,3	9,25	82	MoS	*Melanoides tuberculata*						
480	Or-2638	6760	170	-6000	-5379	-5,69	7,64	82	MoS	*Melanoides tuberculata*						
481	Pa-1610	6120	120	-5322	-4747	-5,03	6,98	82	MoS	*Cael. lacoini* , MNTS 4						
482	Pa-1434	5225	80	-4314	-3805	-4,06	6,01	82	MoS	*Caelatura,* MNTS 5						
483	Pa-492	4610	180	-3760	-2892	-3,33	5,28	82	Ce	Organic matter, MNTS 6						
484	Pa-576	4080	250	-3356	-1976	-2,67	4,62	82	Ce	Organic matter, MNTS 7						
485	Pa-1625	5390	60	-4344	-4052	-4,2	6,15	82	MoS	*Melanoides tuberculata*						
486	Or-1660	5220	180	-4443	-3653	-4,05	6	82	MoS	*Caelatura*						
487	Pa-571	6300	200	-5622	-4798	-5,21	7,16	82	Ce							
488	Pa-311	6310	110	-5481	-5006	-5,24	7,19	82	MoS	*Limicolaria*						
488	Pa-1599	6410	120	-5616	-5074	-5,35	7,3	82	Ce							
488	Pa-301	5945	300	-5510	-4240	-4,88	6,83	82	Ce							
488	Pa-308	6220	300	-5715	-4496	-5,11	7,06	82	BoC	Human						
488	Pa-1041	5960	200	-5322	-4375	-4,85	6,8	82	BoA	(purified) Human						
488	Pa-300	6065	90	-5219	-4746	-4,98	6,93	82	MoS	*Limicolaria*						
488	Pa-304	7775	110	-7029	-6437	-6,73	8,68	82	MoS	*Melanoides tuberculata*						
488	Pa-302	5000	150	-4228	-3382	-3,81	5,76	82	Ca							
489	Pa-444	2010	200	-716	503	-0,11	2,06	82	Ce							
490	Pa-454	5880	120	-5046	-4461	-4,75	6,7	82	Ce							
491	Or-1751	3950	120	-2872	-2137	-2,5	4,45	82	MoS	*Melanoides tuberculata*						
492	Or-1750	9025	150	-8611	-7741	-8,18	10,13	82	MoS	*Mutela*						
493	Pa-368	4510	80	-3495	-2926	-3,21	5,16	82	MoS	*Caelatura*						
494	Pa-890	4630	120	-3649	-3022	-3,34	5,29	82	MoS	*Melanoides*						
494	Pa-620	8300	200	-7731	-6698	-7,21	9,16	82	MoS	*Melanoides*						
495	Pa-	9260	100	-8739	-8287	-8,51	10,46	265	Ch							

N site	Date N	Age BP	±	cal BC		ky Mid BC/AD	ky Mid BP	Citation	Material	Note (taxon, level)	ND	1D	Env	Arch	Fau	Unk
	Other chronological attribution															
495	Orsay	9220	140	-9114	-7998	-8,56	10,51	265	Ch							
496	Pa-411	3060	140	-1622	-927	-1,27	3,22	82	Ce	Human burial 2						
496	Pa-738	3640	150	-2465	-1643	-2,05	4	82	Ch	Hearth						
496	Pa-735	3735	150	-2571	-1746	-2,16	4,11	82	Ce	Internal deposit						
496	Pa-410	3060	90	-1504	-1048	-1,28	3,23	82	Ch	Human burial 2						
496	Pa-1052	3860	160	-2864	-1901	-2,38	4,33	82	Bb	Fauna in human burial 1						
497	Pa-1068	5230	100	-4325	-3800	-4,06	6,01	82	BoA	Human						
497	Pa-1047	4870	110	-3944	-3376	-3,66	5,61	82	BoA	Human						
497	Pa-1042	5245	150	-4356	-3712	-4,03	5,98	82	BoA	Human						
497	Pa-373	5060	100	-4047	-3646	-3,85	5,8	82	MoS	*Limicolaria*						
497	Pa-436	4210	110	-3091	-2487	-2,79	4,74	82	Ca							
497	Pa-306	5895	90	-4991	-4546	-4,77	6,72	82	MoS	*Caelatura*						
Nigeria																
498	UtC-6779	3205	43	-1607	-1406	-1,51	3,46	183	Ch							
498	KI-4206	2940	125	-1432	-844	-1,14	3,09	183	Ch							
498	UtC-6841	2991	46	-1388	-1057	-1,22	3,17	183	Ch							
499	Gajiganna II, Bama-Konduga Group					-0,5	2,45	183			+					
500	I-2945	2520	10	-777	-556	-0,67	2,62	327	Ch					+		
500	I-2372	2400	95	-792	-234	-0,51	2,46	327	Ch					+		
501	Erl-2971	1797	38	127	338	0,23	1,72	183	Ch							
501	Erl-3103	1546	35	426	592	0,51	1,44	183	V	*Sorghum*						
501	Erl-3104	1559	35	420	575	0,5	1,45	183	Ch							
501	KI-4730	1610	30	393	539	0,47	1,48	183	Ch							
502	KI-4743	1770	60	93	407	0,25	1,7	183	Ch							
502	KI-4742	1660	35	257	532	0,39	1,56	183	V	*Sorghum*						
503	UtC-2795	2960	50	-1371	-1019	-1,2	3,15	30	Ch	Upper layer, trench, mound top						
503	KIA-603	2180	80	-395	-46	-0,22	2,17	30	B	Human						
503	KN-4674	2141	66	-379	-39	-0,21	2,16	30	B	*Bos taurus* , basal layer, trench, mound top						
503	UtC-2329	2930	60	-1370	-939	-1,15	3,1	30	Ch	Basal layer, trench, mound top						
504	UtC-2330	3150	70	-1608	-1261	-1,43	3,38	30	Ch	Lower layer, trench, mound top						
504	UtC-2332	3140	110	-1686	-1116	-1,4	3,35	30	Ch	Lower layer, trench, mound top						
504	KI-3605	3040	120	-1600	-932	-1,27	3,22	30	Ch	Basal filling of pit, mound top						
505	UtC-2796	2730	50	-996	-804	-0,9	2,85	30	Ch	Upper layer, trench, mound foot						
505	UtC-2797	2750	70	-1111	-797	-0,95	2,9	30	Ch	Upper layer, trench, mound foot						
505	UtC-2331	2740	50	-999	-808	-0,9	2,85	30	Ch	Upper layer, trench, mound foot						
505	KN-4675	2880	165	-1500	-771	-1,14	3,09	30	B	*Bos taurus* , lower layer, trench, mound top						
506	UtC-2799	3070	70	-1494	-1127	-1,31	3,26	183	Ch							
507	UtC-2802	230	40	1521	1955	1,74	0,21	183	Ch							
507	KI-3756	180	36	1649	1955	1,8	0,15	183	Ch							

N site	Date N	Age BP	±	cal BC		ky Mid BC/AD	ky Mid BP	Citation	Material	Note (taxon, level)	ND	1D	Env	Arch	Fau	Unk
	Other chronological attribution															
507	KI-3757	180	45	1648	1954	1,8	0,15	183	Ch							
508	KI-4745	2910	70	-1367	-916	-1,14	3,09	183	Ch							
508	Erl-3105	2859	40	-1191	-912	-1,05	3	183	Ch							
509	Gajiganna IIb					-0,5	2,45	183			+					
510	Gajiganna IIc					-1	2,95	183			+					
511	KI-4073	2780	95	-1250	-795	-1,02	2,97	183	Ch							
511	UtC-5158	3146	38	-1501	-1316	-1,41	3,36	183	Ch							
511	UtC-5159	3076	43	-1436	-1215	-1,33	3,28	183	Ch							
512	N-796	2590	170	-1209	-261	-0,74	2,69	183	Ch							
512	N-794	3960	160	-2890	-2036	-2,46	4,41	183	Ch							
512	N-794?	2960	160	-1521	-813	-1,17	3,12	183	Ch							
512	KI-4206	2940	125	-1432	-844	-1,14	3,09	183	Ch							
512	UtC-5746	2920	35	-1259	-1011	-1,14	3,09	183	Ch							
512	N-795	2880	140	-1411	-807	-1,11	3,06	183	Ch							
512	UtC-5747	2773	38	-1008	-832	-0,92	2,87	183	Ch							
512	KI-4204	2780	95	-1250	-795	-1,02	2,97	183	Ch							
513	UtC-3517	2860	60	-1258	-896	-1,08	3,03	130	Ch	~-4,80 m						
514	Iron Age					0,5	1,45	183			+					
515	UtC-5452	2430	70	-766	-397	-0,58	2,53	183	V	*Pennisetum*						
515	UtC-5453	2290	70	-726	-171	-0,45	2,4	183	V	*Pennisetum*						
515	UtC-3516	2250	80	-510	-58	-0,28	2,23	183	Ch							
515	UtC-3517	2860	60	-1258	-896	-1,08	3,03	183	Ch							
515	UtC-3518	1920	50	-38	219	0,09	1,86	183	Ch							
515	N-480	2880	140	-1411	-807	-1,11	3,06	183	Ch							
515	UtC-6476	2615	36	-891	-668	-0,78	2,73	183	Ch							
516	Erl-5141	1614	44	340	550	0,45	1,5	183	Ch							
516	Erl-3106	1944	37	-40	130	0,05	1,9	183	Ch							
517	UtC-6964	3275	33	-1631	-1456	-1,54	3,49	183	Ch							
517	UtC-6781	3223	33	-1606	-1423	-1,51	3,46	183	Ch							
518	UtC-6472	2481	42	-771	-415	-0,59	2,54	183	Ch							
518	UtC-4933	2524	34	-795	-539	-0,67	2,62	183	Ch							
518	UtC-4204	2659	36	-897	-792	-0,84	2,79	183	Ch							
518	UtC-4934	2570	50	-827	-538	-0,68	2,63	183	Ch							
519	UtC-8507	1382	41	575	765	0,67	1,28	183	Ch							
519	UtC-6471	1310	60	635	876	0,76	1,19	183	Ch							
520	Iron Age, Daima III					1	0,95	183			+					
521	KN-4812	365	66	1434	1650	1,54	0,41	183	Ch							
521	UtC-8508	399	33	1436	1631	1,53	0,42	183	Ch							
521	UtC-4935	411	36	1428	1629	1,53	0,42	183	Ch							

N site	Date N	Age BP	±	cal BC		ky Mid BC/AD	ky Mid BP	Citation	Material	Note (taxon, level)	ND	1D	Env	Arch	Fau	Unk
	Other chronological attribution															
522	UtC-3513	2470	70	-775	-408	-0,59	2,54	183	Ch							
523	UtC-3514	3120	70	-1599	-1134	-1,37	3,32	183	Ch							
524	UtC-8297	3489	34	-1901	-1695	-1,8	3,75	183	Ch							
524	UtC-3151	3690	120	-2461	-1772	-2,12	4,07	183	Ch							
525	UtC-5147	1490	33	443	646	0,54	1,41	183	Ch							
525	UtC-5148	2565	39	-811	-544	-0,68	2,63	183	Ch							
525	UtC-5149	1630	39	268	540	0,4	1,55	183	Ch							
525	UtC-5150	2584	34	-819	-566	-0,69	2,64	183	Ch							
525	UtC-5151	1467	33	545	647	0,6	1,35	183	Ch							
525	UtC-5146	2099	40	-347	-1	-0,17	2,12	183	Ch							
526	UtC-5152	1165	33	776	971	0,87	1,08	183	Ch							
527	10-12th C. AD					1,1	0,85	183			+					
528	14/16-18th C. AD					1,6	0,35	183			+					
529	19-20th C. AD					1,9	0,05	183			+					
530	Erl-3312	2359	42	-735	-365	-0,55	2,5	183	V	*Pennisetum*						
530	Erl-3130	2483	41	-772	-416	-0,59	2,54	183	Ch							
530	UtC-13071	2508	33	-789	-521	-0,66	2,61	183	Ch							
530	Kn-5457	2510	40	-795	-421	-0,61	2,56	183	Ch							
530	Erl-4835	2537	46	-804	-519	-0,66	2,61	183	Ch							
530	Erl-4836	2548	41	-806	-539	-0,67	2,62	183	Ch							
530	Ki-4959	2480	25	-768	-419	-0,59	2,54	183	Ch							
Rhodesia																
531	Pta-605	250	50	1510	1953	1,73	0,22	147		Hut 1		+		+		
532	SR-25	650	120	1161	1616	1,39	0,56	261		Floor B				+		
532	SR-71	440	90	1329	1667	1,5	0,45	261	Ch	Midden in layer 5				+		
532	SR-70	610	90	1237	1496	1,37	0,58	261	Ch	Layer above floor B				+		
533	IA, Inyanga Lowland culture					0,5	1,45	261			+					
534	SR-43	180	120	1505	1950	1,73	0,22	146								+
534	SR-79	570	110	1272	1627	1,45	0,5	146								+
Rwanda																
535	GrN-7671	1075	95	780	1210	1	0,95	291								
535	GrN-7672	845	75	1043	1380	1,21	0,74	291								
536		2300 bp				-0,4	2,35	291		Fauna maybe more recent						
537	GrN-7589	185	45	1646	1954	1,8	0,15	291		Mid 17-begin. 19th C AD						
537	GrN-7588	155	45	1664	1952	1,81	0,14	291		Mid 17-begin. 19th C AD						
Senegal																
538	Occupation mound dated to 0-250AD					0,13	1,82	193			+					
539	Occupation mound dated to 250-400 AD					0,33	1,62	193			+					
540	Occupation mound dated to 400-600 AD					0,5	1,45	193			+					

N site	Date N	Age BP	±	cal BC		ky Mid BC/AD	ky Mid BP	Citation	Material	Note (taxon, level)	ND	1D	Env	Arch	Fau	Unk
	Other chronological attribution															
541	Occupation mound dated to 600-950 AD					0,78	1,17	193			+					
542	Beta-85886	2370	50	-751	-366	-0,56	2,51	11	MoS	*Anadara senilis* , -200 cm						
542	Pa-?	1555	80	336	650	0,49	1,46	11	MoS	*Anadara senilis* , -150 cm						
543	Gif-4529	910	90	977	1275	1,13	0,82	327	Ch					+		
543	Gif-4530	1020	90	782	1214	1	0,95	327	Ch					+		
543	Ly-2159	1910	210	-396	553	0,08	1,87	327	Ch					+		
543	Ly-2034	790	100	1026	1391	1,21	0,74	327	Ch					+		
544	Ly-1741	1470	260	-44	1119	0,54	1,41	283	Sed	Ash, M16 - 205 cm	+					
544	Dak-155	1363	120	425	945	0,69	1,26	283	Ch	M18 - 270 cm	+					
544	Dak-192	900	110	897	1288	1,09	0,86	283	Ch	L14 - 225 cm	+					
544	Ly-1745	1550	140	137	766	0,45	1,5	283	Ch	K16-K17 - 165 cm	+					
544	Ly-1742	970	150	712	1288	1	0,95	283	Ch	O16 - 160 cm	+					
544	Ly-1744	1090	160	652	1225	0,94	1,01	283	Sed	Ash, K17 - 134 cm	+					
544	Ly-1743	1460	220	85	1012	0,55	1,4	283	Ch	K16 - 96 cm	+					
544	Gif-4522	920	80	988	1262	1,13	0,82	283	Ch	X16 - 255 cm	+					
545	Dated to 0-250AD					0,13	1,82	193			+					
546	Dated to 250-400 AD					0,33	1,62	193			+					
547	Dated to 400-600 AD					0,5	1,45	193			+					
548	Dated to 950-1400 AD					0,78	1,17	193			+					
549	Ly-3875	1490	100	337	766	0,55	1,4	327	Ch					+		
549	Ly-3874	1390	100	431	869	0,65	1,3	327	Ch					+		
549	Ly-3873	1230	90	653	982	0,82	1,13	327	Ch					+		
550	AA-41040	2201	39	-381	-177	-0,28	2,23	77		W1-5						
550	AA-41041	2218	38	-386	-198	-0,29	2,24	77		W1-12						
550	AA-41042	2364	43	-744	-367	-0,56	2,51	77		W1-17						
551	AA-41043	2497	43	-789	-417	-0,6	2,55	77		W1-20						
551	AA-41044	2592	43	-834	-550	-0,69	2,64	77		W1-22						
551	AA-41045	2470	53	-767	-412	-0,59	2,54	77		W1-24						
551	AA-41046	2496	43	-788	-417	-0,6	2,55	77		W1-25						
552	AA-41047	2394	43	-749	-390	-0,57	2,52	77		W2-4a						
552	AA-41048	2412	43	-752	-397	-0,57	2,52	77		W2-5a						
552	AA-41049	2543	43	-804	-538	-0,67	2,62	77		W2-6						
Sinai																
553	AA33714	6195	65	-5310	-4991	-5,15	7,1	64	Ch	Structure 3				+		
553	AA34154	6230	65	-5323	-5008	-5,17	7,12	64	Ch	Structure 4				+		
553	AA33713	6225	75	-5358	-4993	-5,18	7,13	64	Ch	Structure 3				+		
553	AA34150	6045	60	-5207	-4789	-5	6,95	64	Ch	Structure 2				+		
553	AA33711	5965	65	-5016	-4708	-4,86	6,81	64	Ch	Structure 3				+		
553	AA34151	6065	65	-5208	-4806	-5,01	6,96	64	Ch	Structure 2				+		

N site	Date N	Age BP	±	cal BC		ky Mid BC/AD	ky Mid BP	Citation	Material	Note (taxon, level)	ND	1D	Env	Arch	Fau	Unk
	Other chronological attribution															
553	AA33710	6595	80	-5664	-5378	-5,52	7,47	64	Ch	Structure 2				+		
553	AA34152	6185	75	-5311	-4947	-5,13	7,08	64	Ch	Structure 3				+		
553	AA33712	6175	65	-5301	-4962	-5,13	7,08	64	Ch	Structure 3				+		
554	AA33716	6980	65	-5988	-5737	-5,86	7,81	64	Ch	Tomb 1				+		
554	AA3317	7410	75	-6428	-6097	-6,26	8,21	64	Ch	Tomb 1				+		
554	AA33715	6255	60	-5357	-5052	-5,2	7,15	64	Ch	Tomb 1				+		
South Africa																
555	Pta-5530	1980	120	-358	322	-0,02	1,97	140				+				+
556	16th Century AD					1,6	0,35	184			+					
557	Marateng phase : AD 1650 to 1820					1,8	0,15	66			+					
558	Pta-6175	1900	40	68	317	0,19	1,76	140, 141	MoS	Marine				+		
558	OxA-4543	1960	50	34	237	0,14	1,81	140, 141	B	*Ovis aries*				+		
558	OxA-4544	1880	55	62	344	0,2	1,75	140, 141	B	*Ovis aries*				+		
558	Pta-6185	1840	50	90	388	0,24	1,71	140, 141	Ch					+		
558	Pta-6246	1880	50	71	336	0,2	1,75	140, 141	MoS	Marine				+		
558	Pta-6247	1940	50	4	246	0,13	1,82	140, 141	MoS	Marine				+		
559	SR-132	1980	120	-341	389	0,02	1,93	169		lower limit		+				+
560	SR-132	1980	120	-341	389	0,02	1,93	169		upper limit		+				+
561	SR-142	3090	100	-1506	-997	-1,25	3,2	169		lower limit		+				+
562	SR-142	3090	100	-1506	-997	-1,25	3,2	169		upper limit						+
562	SR-152	3650	120	-2296	-1626	-1,96	3,91	169		lower limit						+
563	Robberg Tradition: 11 to 21 ka and older (>40 ka)					-15	16,95	167			+					
564	Albany Tradition : 10400 to 9100 (12 to 8ka)					-10	11,95	167			+					
565	Wilton Tradition 6400 to ~ 2000 BP					-3	4,95	167			+					
566	Pta-6247	1940	50	4	246	0,13	1,82	275		Herder Unit				+		
566	OxA-4544	1880	55	62	344	0,2	1,75	275						+		
566	OxA-4543	1960	50	34	237	0,14	1,81	275						+		
566	Pta-6185	1840	50	90	388	0,24	1,71	275						+		
566	Pta-6246	1880	50	71	336	0,2	1,75	275						+		
566	Pta-6175	1900	40	68	317	0,19	1,76	275						+		
567	Iron Age ~ 800 to 200 BP					1,5	0,45	166								
568	Pta-1807	22575	270	-21186	-20102	-20,64	22,59	168, 218	Ch	HE2 +-190 cm				+		
568	UW-339	22800	850	-22781	-19296	-21,04	22,99	168, 218	Ch	HE2 +-190 cm				+		
568	Pta-1484	8960	80	-8276	-7760	-8,02	9,97	168, 218	Ch	HE1 +-145 cm				+		
568	Pta-1805	11875	115	-12029	-11502	-11,77	13,72	168, 218	Ch	HE2 +-185 cm				+		
569	UW-340	7300	120	-6389	-5896	-6,14	8,09	168, 218	Ch	CHE 3 +-125 cm				+		
569	Pta-1841	7460	65	-6423	-6097	-6,26	8,21	168, 218	Ch	CHE 3 +-125 cm				+		
570	Dated between 7300+-120 BP 5120 +- 65 BP					-5	6,95	168, 218			+					
571	Pta-1481	5120	65	-3986	-3701	-3,84	5,79	168, 218	Ch	BOL2 +-65 cm				+		

N site	Date N	Age BP	±	cal BC		ky Mid BC/AD	ky Mid BP	Citation	Material	Note (taxon, level)	ND	1D	Env	Arch	Fau	Unk
	Other chronological attribution															
572	UW-341	3840	80	-2467	-1981	-2,22	4,17	168, 218	Ch	BOL1 +-15 cm				+		
573	Pta-196	170	50	1670	1953	1,81	0,14	169				+		+		
574	Pta-1175	12310	120	-12917	-11982	-12,45	14,4	238	Ch	Base level 14				+		
574	GrN-4854	9510	55	-9140	-8645	-8,89	10,84	238	Ch	Level 5				+		
574	GrN-4813	9940	80	-9796	-9263	-9,53	11,48	238	Ch	Level 7				+		
574	GrN-4814	12090	95	-12206	-11806	-12,01	13,96	238	Ch	Level 14				+		
574	GrN-5874	9570	55	-9184	-8767	-8,98	10,93	238	Ch	Level 2				+		
575	GrN-4815	12160	95	-12371	-11824	-12,1	14,05	238	Ch	Level 15/16				+		
575	Pta-1176	12470				-12,5	14,45	238	Ch	Level 16				+		
575	GrN-4816	12510	105	-13086	-12224	-12,66	14,61	238	Ch	Level16-18				+		
575	Pta-1177	12800	75	-13496	-12909	-13,2	15,15	238	Ch	Level 17				+		
575	Pta-1178	12950	40	-13628	-13090	-13,36	15,31	238	Ch	On top of level 18				+		
576	~ 12500 BP					-12,5	14,45	170		Level 19	+	+				
577	~ 9700 BP					-9,2	11,15	170		Levels 18-14	+					
578	~ 6400 BP					-5,4	7,35	170		Levels 13-10	+					
579	~ 4000 (Lev 9-6) ~ 3500 (Lev 5-2) ~ 3200 and younger (Lev1: 1880+-50 BP)															
579	Pta-1865	1880	50	71	336	0,2	1,75	170		Level 1		+				+
580	Weighmean	1821	23	140	379	0,26	1,69	267	Ch			+		+		
581	Pta-2550	6270	40	-5304	-5048	-5,18	7,13	219		Layer 4		+		+		
582	Pta-2547	70	40	1697	1950	1,82	0,13	219		Layer 1				+		
582	Pta-2608	920	50	1037	1265	1,15	0,8	219		Layer 2				+		
582	Pta-2549	1890	45	69	325	0,2	1,75	219		Layer 3				+		
583	Contemp. to Eland´s Bay Open-b (~ 1600to 600 BP)					0,9	1,05	173			+					
584	Pta-6785	3620	60	-2130	-1740	-1,94	3,89	350	Ch					+		
584	Pta-6787	2850	60	-1121	-816	-0,97	2,92	350	Ch					+		
585	Late Holocene ~ 4000 to 1000 y BP (non calibrated?)					-0,6	2,55	269			+					
586	GX-1688	1960	85	-109	336	0,11	1,84	172		Layer 1	+					+
586	Gak-3877	1650	90	255	637	0,45	1,5	268	Ch	Layer 2				+		
586	GX-1687	1960	95	161	341	0,25	1,7	172		Layer 1						+
586	GX-1686	2020	95	-342	314	-0,01	1,96	172								+
586	GaK-3956	1590	80	347	656	0,5	1,45	172								+
586	GaK-3955	1600	120	223	763	0,49	1,46	172	Ch	Layer 2						+
586	GX-1685	1465	100	420	862	0,64	1,31	172	Ch	Layer 2						+
586	Gak-3878	2620	100	-896	-409	-0,65	2,6	268	Ch	Layer 2				+		
587	Pta-1065	3060	60	-1410	-1053	-1,23	3,18	169				+		+		
588	Pta-1507	570	45	1319	1451	1,39	0,56	169		III		+		+		
589	Pta-186	1120	40	890	1029	0,96	0,99	169		I				+		
589	Pta-187	1120	40	890	1029	0,96	0,99	169		I				+		
590	IA					0,5	1,45	277			+					

N site	Date N Other chronological attribution	Age BP	±	cal BC		ky Mid BC/AD	ky Mid BP	Citation	Material	Note (taxon, level)	ND	1D	Env	Arch	Fau	Unk
591	Pta-5478	1050	60	898	1177	1,04	0,91	277	Ch	Bottom of square C6		+				
592	IA					0,5	1,45	277			+					
593	Pta-1524	1680	50	260	559	0,41	1,54	244						+		
593	Pta-1607	1560	50	432	643	0,54	1,41	244						+		
594	Pta-1668	1000	50	933	1185	1,06	0,89	244						+		
594	Pta-1745	825	50	1156	1372	1,26	0,69	244						+		
595	Pta-686	9600	90	-9251	-8749	-9	10,95	173	Ch	Level 12						+
595	Pta-737	10700	100	-10930	-10448	-10,69	12,64	173	Ch	Level 13						+
595	UW-192	11070	140	-11275	-10884	-11,08	13,03	173	Ch	Level 15						+
595		12450	280	-13404	-11838	-12,62	14,57	173		Level 16						+
596	Levels 18-20, dated from 18 to 14-13 ka, Robberg Industry					-15	16,95	173			+					
597		7910	80	-7036	-6509	-6,77	8,72	173		Level 11						+
598		3780	85	-2456	-1894	-2,18	4,13	173		Level 9						+
598		3510	45	-1895	-1634	-1,76	3,71	173		Level 8						+
599	GaK-4337	1520	80	409	761	0,59	1,36	173		Level 7-5						+
599	GaK-4335	1120	85	772	1158	0,97	0,98	173		Level 7-5						+
600		315	50	1477	1798	1,64	0,31	173								+
601	Level 4, older than 1500 BP					0,5	1,45	173			+					
602	Pta-2465	705	45	1277	1395	1,34	0,61	172								+
602	Pta-2460	590	50	1305	1448	1,38	0,57	172								+
602	Pta-2569	1470	50	539	760	0,65	1,3	172	Ch					+		
603	Late Holocene					-3	4,95	171								
604	Pta-2452	2390	60	-749	-204	-0,48	2,43	175	Ch	Middle 1A				+		
605	6th century AD					0,55	1,4	253			+					
606	9th century AD					0,85	1,1	253			+					
607	14th century AD					1,35	0,6	253			+					
608	Pta-4549	10170	80	-10171	-9455	-9,81	11,76	356	MoS	*Donax serra*		+			+	
609	End of occupation: AD 1780-1800					1,8	0,15	88			+					
610	SR-121	90	90	1669	1950	1,81	0,14	169								+
610	GX-1295	235	80	1508	1953	1,73	0,22	169		Lower limit						+
611	GX-1295	235	80	1508	1953	1,73	0,22	169		Upper limit		+				+
612	AD1450 to AD 1900 (date, ceramic, items)					1,5	0,45	97			+					
613	Contemp. to Eland´s Bay Open-b (~ 1600to 600 BP)					0,9	1,05	173			+					
614	LIA, 1720 AD					1,72	0,23	240			+					
615	AD : Weighmean	1540	35	440	649	0,54	1,41	339	Ch			+		+		
616	Pta-100	13100	100	-13961	-13176	-13,57	15,52	330		Hnk 2,27 mid stratum 1				+		
616	Pta-114	10430	150	-10848	-9875	-10,36	12,31	330		Hnk 1,56 top stratum 1				+		
617	Pta-100	13100	100	-13961	-13176	-13,57	15,52	330		Hnk 2,27 mid stratum 1				+		
617	Pta-101	24630	300	-23306	-22102	-22,7	24,65	330		Hnk 4,39 mid stratum 1	+			+		

N site	Date N	Age BP	±	cal BC		ky Mid BC/AD	ky Mid BP	Citation	Material	Note (taxon, level)	ND	1D	Env	Arch	Fau	Unk
	Other chronological attribution															
618	Group 2: end 17th-18th century AD					1,7	0,25	250			+					
619	Pta-5049	4870	70	-3760	3375	-0,19	2,14	177	Ch	Gravel l., Sq.4 - 200-300 mm		+		+		
620	Pta-5056	640	50	1292	1421	1,36	0,59	177	Ch	Humus l. outside in sq 2/3/5/6/9, 0-380 mm				+		
620	(Pta-5058	710	50	1269	1400	1,33	0,62	177	Ch	Humus l. outside in Sq. 6, 0-80				+		
620	Pta-5057	300	40	1500	1799	1,65	0,3	177	Ch	Humus l. outside in Sq. 12Al1618A, 0-250 mm				+		
621	~ 4000 and 5000 BP.					-3,2	5,15	177			+					
622	Pta-5048	210	45	1645	1953	1,8	0,15	177	Ch	Humus l. from hearth, q. H3/G3, 50-100 mm		+		+		
623	~ AD 1330 14 th century AD					1,33	0,62	332			+					
624	Wits-1436	8500	240	-8227	-6831	-7,53	9,48	286		R D2/D3				+		
625	Wits-1398	1840	50	90	388	0,24	1,71	286		B C4				+		
625	Wits-1399	1350	120	535	994	0,76	1,19	286		RR D2				+		
625	Wits-1381	1550	40	435	647	0,54	1,41	286		MS C2				+		
626	Wits-1382	4090	80	-2874	-2346	-2,61	4,56	286		BM B1				+		
626	Wits-1449	3250	60	-1623	-1316	-1,47	3,42	286		LYN 4 C2				+		
626	Pta-3563	3730	60	-2275	-1888	-2,08	4,03	286		RL2 D1				+		
626	Wits-1450	5250	129	-4326	-3712	-4,02	5,97	286		K11 D3				+		
626	Wits-1462	6490	200	-5736	-4911	-5,32	7,27	286		BM/R C2				+		
626	Wits-1214	3100	150	-1631	-901	-1,27	3,22	286		RL C1/D1				+		
627	Pta-3461	1790	40	175	416	0,3	1,65	172								+
627	Pta-3711	1860	60	77	382	0,23	1,72	172								+
627	OxA-3864	1630	60	343	613	0,48	1,47	140								+
627	OxA-3865	1430	55	568	772	0,67	1,28	275								+
628	Pta-3737	190	40	1657	1953	1,81	0,14	172								+
628	Pta-3742	920	40	1042	1261	1,15	0,8	172								+
628	Pta-3747	880	50	1046	1282	1,16	0,79	172								+
629	19th-20th century AD					1,9	0,05	253			+					
630	LSA : 5000 to 0 BP					-2,5	4,45	165			+					
631	GX-970	4755	95	-3694	-3111	-3,4	5,35	164	Ch							
631	GX-973	4695	180	-3771	-2906	-3,34	5,29	164	Ch							
632	GX-971	2795	85	-1189	-769	-0,98	2,93	164	MoS	Marine						
632	GX-969	2525	85	-789	-404	-0,6	2,55	164	Ch							
633	GX-1397	2285	105	-731	4	-0,36	2,31	164	MoS	Marine				+		
633	GX-1378	4110	160	-3022	-2137	-2,58	4,53	164	Ch					+		
634	Pta-4671	1990	50	-45	216	0,09	1,86	7, 317	Ch	Unit C 180 mm				+		
634	Pta-4672	1230	50	708	984	0,85	1,1	7, 317	Ch	Bedding unit 70 mm				+		
635	Pta-1747	970	40	1027	1202	1,11	0,84	86	Ch	Pit				+		
635	Pta-1663	1160	50	782	1022	0,9	1,05	86	Ch	Feature d				+		
635	Pta-2160	1120	50	830	1131	0,98	0,97	86	Ch	Feature g				+		

N site	Date N / Other chronological attribution	Age BP	±	cal BC		ky Mid BC/AD	ky Mid BP	Citation	Material	Note (taxon, level)	ND	1D	Env	Arch	Fau	Unk
636	Wits-1219	1380	50	610	855	0,73	1,22	244				+		+		
637	Pta-5517	970	50	1025	1210	1,12	0,83	277	Ch	Square E6 20-25 cm		+		+		
638	Wits-1237	1230	70	687	989	0,84	1,11	244						+		
638	Wits-1280	1190	70	695	1024	0,86	1,09	244						+		
639	GaK-1540	2270	100	-520	17	-0,25	2,2	75	W	Hearth in 16/19-18/21 l. 2B				+		
639	I-2565	4860	115	-3939	-3348	-3,64	5,59	75	W	Hearth in 16/17-16/18 l. 3F				+		
640	LIA, 1830 AD					1,83	0,12	240			+					
641	AD 550-750					0,65	1,3	239			+					
642	300-600 AD, EIA					0,45	1,5	239			+					
643	+-AD 830					0,83	1,12	239			+					
644	Wits-2009	4860	100	-3797	-3361	-3,58	5,53	350	OES					+		
644	Wits-2010	5210	100	-4233	-3712	-3,97	5,92	350	OES					+		
645	Pta-1621	1620	50	393	606	0,5	1,45	169		III		+		+		
646	Pta-1759	1430	50	572	771	0,67	1,28	169	Ch	IV-V				+		
646	Weighmean	1681	31	345	538	0,44	1,51	267	Ch					+		
647	Pta-1634	1410	50	595	776	0,69	1,26	87		Pit 1, 450-650 mm				+		
647	Pta-328	1460	50	543	764	0,65	1,3	87	Ch	Pit 2				+		
648	+- 470 AD, EIA					0,47	1,48	239			+					
649	AD 800-900					0,85	1,1	239			+					
650	6-8 th century AD					0,7	1,25	335			+					
651	Pta-2874	1360	50	644	866	0,76	1,19	267	Ch			+		+		
652	AD 760					0,76	1,19	335			+					
653	Early Holocene					-7	8,95	171			+					
654				670		0,7	1,25	337	B	Area II, ashy soil				+		
654				774		0,8	1,15	337	Ch	Area II, ashy soil				+		
654				905		0,9	1,05	337	IR	Furnace 1				+		
655	Group 1. 19th century AD					1,85	0,1	250			+					
656	950-1150 AD					1,15	0,8	336			+					
657	LKA~950 to 1130 AD					1	0,95	336			+					
658	1070-1150 AD					1,1	0,85	336			+					
659	Group 3: 19th century AD					1,85	0,1	250			+					
660	Pta-5929	200	50	1648	1953	1,8	0,15	350				+				+
661	17-18th Century					1,7	0,25	246			+					
662	Pta-2878	1400	50	597	779	0,69	1,26	267	Ch			+		+		
663	pCW: 2700-100 BP					-1	2,95	243			+					
664	300-600 AD, EIA					0,45	1,5	239			+					
665	Weighmean	1348	24	665	773	0,72	1,23	267	Ch			+		+		
666	Group 2: end 17th-19th century AD					1,7	0,25	250			+					
667	Weighmean	1205	35	779	985	0,88	1,07	267	Ch			+				

N site	Date N / Other chronological attribution	Age BP	±	cal BC		ky Mid BC/AD	ky Mid BP	Citation	Material	Note (taxon, level)	ND	1D	Env	Arch	Fau	Unk
668	UW-164	10285	85	-10621	-9767	-10,19	12,14	164	Ch					+		
668	UW-177	11950	150	-12183	-11498	-11,84	13,79	164	Ch					+		
668	Pta-391	8990	80	-8288	-7796	-8,04	9,99	164	Ch					+		
668	Pta-392	10150	90	-10170	-9406	-9,79	11,74	164	Ch					+		
668	UW-178	10540	110	-10857	-10193	-10,53	12,48	164	Ch					+		
668	UW-162	11540	110	-11709	-11256	-11,48	13,43	164	Ch					+		
668	Q-1085	10256	210	-10733	-9339	-10,04	11,99	164	Ch					+		
669	UW-218	11250	270	-11772	-10846	-11,31	13,26	164	Ca					+		
669	I-6516	16700	240	-18407	-17477	-17,94	19,89	164	Ch					+		
669	GrN-5889	18660	110	-20495	-20086	-20,29	22,24	164	OES					+		
669	UW-185	18100	550	-20938	-18185	-19,56	21,51	164	Ch					+		
669	I-6515	11080	260	-11631	-10454	-11,04	12,99	164	Ch					+		
670	UW-222	6060	125	-5286	-4587	-4,94	6,89	164	Ch					+		
670	UW-179	9080	180	-8761	-7719	-8,24	10,19	164	MoS	Marine				+		
670	UW-184	8570	170	-8182	-7084	-7,63	9,58	164	MoS	Marine				+		
670	UW-181	8120	240	-7545	-6477	-7,01	8,96	164	Ch					+		
670	UW-186	6050	80	-5206	-4691	-4,95	6,9	164	Ch					+		
670	UW-176	6020	160	-5294	-4493	-4,89	6,84	164	Ch					+		
670	UW-217	5350	65	-4321	-3976	-4,15	6,1	164	MoS	Marine				+		
670	UW-187	5825	150	-4996	-4336	-4,67	6,62	164	Ch					+		
671	End of occupation: AD 1780-1800					1,8	0,15	88			+					
672	LSA					-2	3,95	169			+					
673	+-AD 830					0,83	1,12	239			+					
674	Ending 19th Century, burned on 1879					1,9	0,05	351			+					
675	Late Holocene					-3	4,95	171			+					
676	AD1669 to 1732					1,7	0,25	71			+					
677	AD 800-850					0,85	1,1	239			+					
678	Pta-5921	880	50	1046	1282	1,16	0,79	151	Ch	Misterio - K5				+		
678	Pta-5605	570	20	1396	1436	1,42	0,53	151	Ch	Latina - N7				+		
679	Pta-5602	2640	60	-896	-515	-0,71	2,66	151	Ch	Temprano - N7				+		
679	Pta-5990	2940	20	-1193	-942	-1,07	3,02	151	Ch	Sh. Gringo - N7				+		
680	Pta-5743	3570	60	-2022	-1688	-1,86	3,81	151	Ch	Sh. Pancho II - L7				+		
680	Pta-5923	3060	60	-1410	-1053	-1,23	3,18	151	Ch	Penco - N7				+		
681	Pta-1615	870	50	1048	1284	1,17	0,78	262		Layer 1				+		
681	Pta-1616	855	45	1151	1286	1,22	0,73	262		Layer 1				+		
682	Between 1-2 and 5-6					-0,1	2,05	262			+					
683	Pta-1717	3510	60	-1938	-1616	-1,78	3,73	262		Layer 5				+		
684	LIA, 1200 AD					1,2	0,75	240			+					
685	LIA, 1200-1400 AD					1,3	0,65	240			+					

N site	Date N / Other chronological attribution	Age BP	±	cal BC		ky Mid BC/AD	ky Mid BP	Citation	Material	Note (taxon, level)	ND	1D	Env	Arch	Fau	Unk
686	AD 800-900					0,85	1,1	239			+					
687	Group 3: 19th century AD					1,85	0,1	250			+					
688	Pta-1818	800	50	1182	1381	1,28	0,67	247		Unit 1		+				+
689	Pta-1959	1140	50	782	1031	0,91	1,04	247		Unit 4		+				+
690	Pta-1763	2250	65	-393	-60	-0,23	2,18	169		Surface				+		
690	Pta-1508	3680	60	-2197	-1776	-1,99	3,94	169		III				+		
691	Middle to Late Holocene					-6	7,95	240			+					
692	LIA					1,75	0,2	32			+					
693	Late Iron Age, 17-18 th century AD					1,7	0,25	242			+					
694	pCW: 2700-100 BP					-1	2,95	243			+					
695	Robberg Industry : 40000 - 12000 BP					-25	26,95	243			+					
696	17th century AD					1,65	0,3	253			+					
697	GX-723	2645	95	-914	-412	-0,66	2,61	169				+				+
698	GGX-865	2285	115	-746	23	-0,36	2,31	169		Lower limit		+				+
699	GX-724	2530	105	-824	-381	-0,6	2,55	169				+				+
700	Late Holocene > 3000 BP					-3	4,95	274			+					
701	Wits-2005	3740	280	-2878	-1449	-2,16	4,11	284		Square DS1, level 3				+		
701	Wits-2003	6150	100	-5298	-4787	-5,04	6,99	284		Square DS1, level 4				+		
702	Pta-5762	100	60	1672	1950	1,81	0,14	284		Level 2 R				+		
702	OxA-3386	415	65	1435	1641	-1,54	3,49	284		Level 3				+		
702	Pta-5761	100	45	1684	1953	1,82	0,13	284		Level 2 Sad				+		
703	Pta-5931	1290	50	666	934	0,8	1,15	162	Ch	Ash level at a depth of 40 mm				+		
703	Pta-5932	1920	60	4	325	0,16	1,79	162	Ch	Ashy level, -200 mm				+		
704	gRn-5298	1100	30	898	1031	0,96	0,99	350	Ch					+		
704	Pta-5622	680	50	1281	1404	1,34	0,61	350	Ch					+		
704	Pta-6788	500	50	1393	1618	1,51	0,44	350	Ce					+		
704	Pta-350	610	50	1300	1439	1,37	0,58	350	Ce					+		
705	Pta-5601	13360	150	-14453	-13390	-13,92	15,87	350	Ch					+		
705	Pta-5593	12690	120	-13399	-12366	-12,88	14,83	350	Ch					+		
706	Pta-5600	8380	70	-7542	-7144	-7,34	9,29	350	Ch					+		
706	Pta-5560	8614	38	-7651	-7523	-7,59	9,54	350	Ch					+		
706	Pta-5599	9250	70	-8608	-8282	-8,45	10,4	350	Ch					+		
707	Pta-5934	5970	70	-4988	-1607	-3,3	5,25	350	Ch					+		
707	Pta-6783	7630	80	-6593	-6250	-6,42	8,37	350	Ch					+		
708	AD 800-850					0,8	1,15	336			+					
709	Group 1. 19th century AD					1,85	0,1	250								
710	Group 2: second half of 19th century AD					1,88	0,07	250			+					
711	Group 1. 17-19th century AD					1,7	0,25	250			+					
712	AD 800-900					0,85	1,1	239			+					

N site	Date N	Age BP	±	cal BC		ky Mid BC/AD	ky Mid BP	Citation	Material	Note (taxon, level)	ND	1D	Env	Arch	Fau	Unk
	Other chronological attribution															
713	LIA, 1200 AD - 18th-19th century AD					1,2	0,75	240			+					
714	LIA, 1890 AD					1,89	0,06	240			+					
715	Group 1. Early occupation: 10th century AD					0,95	1	250			+					
716	Group 1. Late occupation: 18-19th century AD					1,8	0,15	250			+					
717	+- 740 AD					0,75	1,2	239			+					
718	~ 7000-2000 BP					-5	6,95	240			+					
719	Pta-2200	1175	35	782	994	0,89	1,06	198		Layer 4				+		+
719	Pta-2198	1420	35	604	768	0,69	1,26	198		Layer 9				+		+
720	Group 2: end 17th-18th century AD					1,7	0,25	250			+					
721	Pta-6749	1930	50	6	313	0,16	1,79	329		C6-Twiggy 6/l. 6 (45 cm)				+		
721	Pta-6750	1450	50	554	766	0,66	1,29	329		C6-CST/Layer (14cm)				+		
721	Pta-4745	1920	40	54	248	0,15	1,8	140	Ch	-91 cm, 1987 squares				+		
721	OxA-3862	2105	65	-350	115	-0,12	2,07	140	B	*Ovis aries*					+	
721	Pta-6334	2020	60	-157	213	0,03	1,92	329	OM	Crayfish mantle, C9 Hearth 3 (20 cm)				+		
721	Pta-4753	1390	50	603	806	0,7	1,25	140	Ch	-20 cm, squares 1987				+		
722	Pta-7200	2400	25	-518	-366	-0,44	2,39	329	Ch	C6 SAS/l.10 (64) cm				+		
723	Pta-6987	3820	60	-2457	-1980	-2,22	4,17	329	Ch	D5 BSB2/l.12 (75) cm				+		
723	Pta-6754	3520	50	-1921	-1636	-1,78	3,73	329	Ch	D5 Health 34/l.13 (50) cm				+		
724	Estimated to~ 2000 to 1500 BP (on artefact)					0,3	1,65	173			+					
725	Pta-6424	2200	50	-362	-47	-0,2	2,15	9		Layer 1				+		
725	Pta-6136	2200	60	-377	-42	-0,21	2,16	9		Layer 1				+		
726	Pta-6498	2360	45	-511	-204	-0,36	2,31	9		Layer 2		+		+		
727	Pta-6505	2490	50	-758	-402	-0,58	2,53	9		Layer 3				+		
727	Pta-6134	2690	60	-969	-546	-0,76	2,71	9		Layer 3				+		
728	Pta-6805	3640	60	-2133	-1751	-1,94	3,89	9		Layer 4a				+		
728	Pta-6794	3510	50	-1907	-1629	-1,77	3,72	9		Layer 4a				+		
729	Pta-6420	3990	60	-2580	-2206	-2,39	4,34	9		Layer 4b				+		
729	Pta-6808	6070	80	-5207	-4717	-4,96	6,91	9		Layer 4b				+		
730	Wits-2004	860	60	1046	1294	1,17	0,78	350	Ch			+		+		
731	LIA, 1750-1850 AD					1,8	0,15	240			+					
732	LIA, 1500-1800 AD					1,65	0,3	240			+					
733	Pta-1586	95	45	1679	1940	1,81	0,14	169		II				+		
733	Pta-1583	80	45	1680	1939	1,81	0,14	169		I				+		
734	Pta-1585	150	45	1665	1952	1,81	0,14	169				+		+		
735		7700	70	-6640	-6394	-6,52	8,47	173		Level 14		+				+
736		4190	60	-2887	-2500	-2,69	4,64	173		Level 10						+
736		4330	50	-3082	-2678	-2,88	4,83	173		Level 13-12						+
736		3520	60	-1944	-1622	-1,78	3,73	173		Levels 8-4						+
736		4020	60	-2833	-2213	-2,52	4,47	173		Level 8-4						+

N site	Date N / Other chronological attribution	Age BP	±	cal BC		ky Mid BC/AD	ky Mid BP	Citation	Material	Note (taxon, level)	ND	1D	Env	Arch	Fau	Unk
737	Pta-3310	1620	50	393	606	0,5	1,45	173		Level 3						+
737	Pta-3512	1680	50	260	559	0,41	1,54	173		Level 3						+
738	Pta-6300	760	50	1216	1390	1,3	0,65	173		Level 1A		+				+
739	+- 510 AD					0,51	1,44	239			+					
740		1880	50	71	336	0,2	1,75	350				+				+
741	1829-1838					1,83	0,12	251			+					
742	LSA phase 6 = Smithfield					-1	2,95	169			+					
743	Pta-5520	520	45	1391	1495	1,44	0,51	277		VI-3				+		
743	Pta-5473	520	45	1391	1495	1,44	0,51	277		VI-1				+		
744	Pta-5551	1920	50	24	318	0,17	1,78	277	Ch	Gran. Orange l. 0,3-0,4 m				+		
744	Pta-5547	373	21	1481	1631	1,56	0,39	277	Ch	Loose Grey, 0,03-0,05 m				+		
744	Pta-5527	420	40	1445	1626	1,54	0,41	277	Ch	Charcoal l., 0,15-0,02 m				+		
745	Pta-5764	120	45	1677	1954	1,82	0,13	284		Heart B, Level 2		+				
746	19th century AD					1,85	0,1	253			+					
747	Pta-5467	330	45	1464	1667	1,57	0,38	277	Ch	0,15-0,3 m - sq I1				+		
747	Pta-5469	500	50	1393	1618	1,51	0,44	277	Ch	0,12-0,15 m - sq I1				+		
748	Between 500 and 1400 BP					1	0,95	277			+					
749	Pta-4609	1850	50	86	382	0,23	1,72	277	Ch	0,8m - sq F1		+		+		
749	Pta-4608	1380	50	610	855	0,73	1,22	277	Ch	0,4m - sq F1		+		+		
750	~ 3000 BP					-2,5	4,45	277			+					
751	Pta-4104	1430	60	556	776	0,67	1,28	318	Ch	GII, layer 4				+		
751	Pta-4100	1460	50	543	764	0,65	1,3	318	Ch	GII, layer 3				+		
752	Pta-4094	1290	50	666	934	0,8	1,15	318	Ch	GI, 30 cm				+		
752	Pta-4095	1270	60	671	965	0,82	1,13	318	Ch	GII, layer 2				+		
753	Terminal EIA ~ 800 AD					0,8	1,15	318			+					
Sudan																
754	T-8525	8390	70	-7584	-7200	-7,39	9,34	133	MoS	101x/98y, level 4				+		
754	T-8626	8330	100	-7573	-7085	-7,33	9,28	133	MoS	102x/103y, level 1				+		
754	T-5728	7700	140	-7028	-6256	-6,64	8,59	133	MoS	101x/90y, level 1				+		
754	T5727	7860	190	-7310	-6394	-6,85	8,8	133	MoS	101x/90y, level 4				+		
754	T-8624	8640	120	-8199	-7491	-7,85	9,8	133	MoS	101x/99v, level 5				+		
754	T-6381	8500	100	-7784	-7310	-7,55	9,5	133	MoS	108x-96y, level 1				+		
755	45 000 to 20 000 BP					-30	31,95	104			+					
756	Ca 3000-2000 cal bc					-2,5	4,45	255			+					
757	T-8646	7790	90	-7025	-6456	-6,74	8,69	133	MoS	Area 4, 48x/146y, level 2				+		
757	T-6384	7450	100	-6467	-6087	-6,28	8,23	133	MoS	Area 2, 46x/98y				+		
757	T-8643	8230	120	-7568	-6841	-7,2	9,15	133	MoS	Area 2, 58x/98y, level 3				+		
757	T-6383	7290	150	-6446	-5890	-6,17	8,12	133	MoS	Area 1, 100x/98y				+		

N site	Date N	Age BP	±	cal BC		ky Mid BC/AD	ky Mid BP	Citation	Material	Note (taxon, level)	ND	1D	Env	Arch	Fau	Unk
	Other chronological attribution															
757	T-6382	7310	110	-6408	-6001	-6,2	8,15	133	MoS	Area 1,100 x/98y, level 1				+		
757	T-7480	7560	130	-6655	-6098	-6,38	8,33	133	MoS	Area 1, 103x/96y, level 1				+		
757	T-8642	7470	110	-6506	-6074	-6,29	8,24	133	MoS	Area 4,46 x146/y, level 4				+		
757	T-8636	7570	60	-6568	-6256	-6,41	8,36	133	MoS	Area 4, 45x/148y, level 3				+		
757	T-6385	7730	110	-7025	-6394	-6,71	8,66	133	MoS	Area 4, 43x/147y, level 1				+		
757	T-8645	7660	50	-6599	-6433	-6,52	8,47	133	MoS	Area 4, 48x/146y, level 3				+		
757	T-8647	7830	100	-7031	-6478	-6,75	8,7	133	MoS	Area 4, 48x/146y, level 5				+		
757	T-8641	7890	100	-7058	-6529	-6,79	8,74	133	MoS	Area 4, 46x/146y, level 5				+		
757	T-8648	8090	60	-7303	-6824	-7,06	9,01	133	MoS	Area 4, 48x/146y, level 6				+		
757	T-8644	7950	100	-7128	-6595	-6,86	8,81	133	MoS	Area 2, 59x/89y, level 3				+		
757	T-7481	6820	170	-6048	-5473	-5,76	7,71	133	MoS	Area 2, 56x/93y, level 1				+		
757	T-7483	7780	110	-7029	-6440	-6,73	8,68	133	MoS	Area 4, 49x/147y				+		
757	T-7482	7910	140	-7173	-6471	-6,82	8,77	133	MoS	Area 3, 64x/107y, level 5				+		
758	T-6655	7750	90	-6983	-6428	-6,71	8,66	39	MoS							
759	Postmeroitic, 4C AD					0,35	1,6	179			+					
760	Early part of middle kingdom, when burnt in 1675 BC					-1,8	3,75	65			+					
761	T-5064	1250	70	652	950	0,8	1,15	263								+
761	T-5065	1600	140	91	671	0,38	1,57	263								+
761	T-5066	340	110	1399	1954	1,68	0,27	263								+
761	T-5067	CAL BC 1100+		-1180	-1020	-1,1	3,05	263								+
761	T-5068	CAL BC 1340+		-1470	-1210	-1,34	3,29	263								+
761	T-5063	1200	80	670	983	0,83	1,12	263								+
762	T-4811	3300	60	-1737	-1447	-1,59	3,54	263								+
762	T-5032	3240	80	-1736	-1321	-1,53	3,48	263								+
762	T-4562	2760	70	-1112	-801	-0,96	2,91	263								+
763	KN-3417	5030	260	-4451	-3120	-3,79	5,74	327	Ch					+		
763	Kn-3405	4000	200	-3085	-1947	-2,52	4,47	327	Ch					+		
763	Kn-2940	3530	120	-2200	-1536	-1,87	3,82	327	Ch					+		
763	Kn-3416	3360	110	-1926	-1425	-1,68	3,63	327	Ch					+		
763	Kn-3554	4200	120	-3265	-2467	-2,87	4,82	327	Ch					+		
763	Kn-3521	4210	130	-3323	-2467	-2,9	4,85	327	Ch					+		
764	T-8632	7860	120	-7051	-6479	-6,77	8,72	234	MoS	22x/98y, level 3				+		
764	T-8633	7920	100	-7075	-6534	-6,8	8,75	234	MoS	21x/54y, level 8				+		
764	T-8640	8010	120	-7309	-6635	-6,97	8,92	234	MoS	22x/99y, level 3				+		
764	T-8634	7850	170	-7248	-6396	-6,82	8,77	234	MoS	22x/99y, level 4				+		
764	T-8638	7780	110	-7029	-6440	-6,73	8,68	234	MoS	21x/54-55y, level 5				+		
764	T-7484	7690	110	-6901	-6260	-6,58	8,53	234	MoS	54x/140y, level 2-3				+		
764	T-8631	7260	110	-6388	-5921	-6,15	8,1	234	MoS	137x/54y				+		

N site	Date N	Age BP	±	cal BC		ky Mid BC/AD	ky Mid BP	Citation	Material	Note (taxon, level)	ND	1D	Env	Arch	Fau	Unk
	Other chronological attribution															
764	T-9698	8040	120	-7329	-6646	-6,99	8,94	234	MoS	125x/109y, level 4				+		
764	T-7485	8390	50	-7570	-7342	-7,46	9,41	133	MoS	105x/87y, level4				+		
764	T-8637	7970	90	-7128	-6639	-6,88	8,83	234	MoS	21x/55y, level 7-8				+		
764	T-9689	7910	115	-7078	-6502	-6,79	8,74	234	MoS	125x/107y, level 4				+		
764	T-8635	7580	110	-6641	-6231	-6,44	8,39	234	MoS	57x/138y, level 2				+		
764	T-8639	7820	110	-7031	-6470	-6,75	8,7	234	MoS	21x/54-55y, level 4				+		
765	Gif-6809	4840	70	-3776	-3381	-3,58	5,53	107	MoS	*Aspatharia,* KDD 98 - tomb 3						
765	Gif-5771	4730	80	-3655	-3359	-3,51	5,46	107	MoS	*Aspatharia,* KDD 76 - tomb 3						
765	Gif-5213	4830	50	-3710	-3385	-3,55	5,5	107	MoS	*Aspatharia,* KDD 22 - tomb 100						
765	Gif-5508	4790	110	-3911	-3345	-3,63	5,58	107	MoS	*Pila,* KDD 21 - habitat						
765	Gif-4675	4630	80	-3634	-3104	-3,37	5,32	107	MoS	*Aspatharia,* KDD 22 - tomb 5						
766	T-6647	5290	80	-4326	-3968	-4,15	6,1	38	MoS	*Limicolaria,* El Kenger East, S2, level 1						
766	Gif-6812	5080	70	-4036	-3707	-3,87	5,82	38	MoS	*Limicolaria,* El Kenger middle, surface						
766	T-6648	5620	80	-4681	-4334	-4,51	6,46	38	MoS	*Limicolaria,* El Kenger middle, S1, level 3						
767	T-10946	7705	145	-7030	-6257	-6,64	8,59	51	MoS	*Pila*				+		
767		6940	85	-5989	-5673	-5,83	7,78	51								+
768	T-3700	5290	100	-4346	-3848	-4,1	6,05	84	MoS	*Etheria elliptica*						
768	T-3701	5520	130	-4681	-4047	-4,36	6,31	84	MoS	*Etheria elliptica*						
769	Rom-257	6220	80	-5362	-4964	-5,16	7,11	39	MoS							
769	T-6656	6620	90	-5716	-5382	-5,55	7,5	39	MoS					+		
770	Rom-263	6780	90	-5876	-5529	-5,7	7,65	39	MoS							
771	T-2818	5350	90	-4346	-3986	-4,17	6,12	84	MoS	*Etheria elliptica*				+		
771	T-3050	5660	80	-4686	-4354	-4,52	6,47	84	MoS	*Etheria elliptica*				+		
772	Early Holocene					-6	7,95	221			+					
773	T-3222	5240	80	-4321	-3815	-4,07	6,02	84	MoS	*Etheria elliptica*				+		
773	T-3699	5550	90	-4612	-4178	-4,4	6,35	84	MoS	*Etheria elliptica*				+		
773	T-3223	5370	80	-4352	-3996	-4,17	6,12	84	MoS	*Etheria elliptica*				+		
774	Amm Adam Group					-5	6,95	119			+					
775	Kassala group					-3,5	5,45	119			+					
776	Hv-15568	3665	250	-2859	-1458	-2,16	4,11	221	BoA	*Bos taurus*				+		
776	Hv-15564	4325	100	-3340	-2671	-3,01	4,96	221	BoA	Crocodyle				+		
776	Hv-15566	4460	115	-3503	-2886	-3,19	5,14	221	BoA	Hippo				+		
777	T-5022	5570	100	-4684	-4236	-4,46	6,41	327	MoS	*Pila*				+		
778	Here : 2050 to 1750 bc mostly					-1,9	3,85	52								
779	SUA-211	5480	90	-4498	-4054	-4,28	6,23	1	MoS	*Pila wernei*			+			
780	UCLA-1864	1600	80	255	614	0,43	1,52	62, 128	Ch	TT1-40-50 (A1)				+		
781	T-5301	8240	120	-7573	-6859	-7,22	9,17	83	MoS	Trench 65 cm below						
781	T-5300	8920	180	-8531	-7596	-8,06	10,01	83	MoS	Trench 40 cm below						
782	T-6645	6150	80	-5303	-4856	-5,08	7,03	39	MoS					+		

N site	Date N / Other chronological attribution	Age BP	±	cal BC		ky Mid BC/AD	ky Mid BP	Citation	Material	Note (taxon, level)	ND	1D	Env	Arch	Fau	Unk
782	Rom-262	6280	85	-5469	-5041	-5,26	7,21	39	MoS							
783	T-2188	5260	90	-4331	-3821	-4,08	6,03	113	MoS					+		
783	T-2189	5030	70	-3965	-3665	-3,82	5,77	113	MoS					+		
783	T-3260	5360	80	-4346	-3995	-4,17	6,12	113	MoS					+		
783	SMU-482	5630	70	-4653	-4341	-4,5	6,45	113	MoS		+					
784	Attributed to 3062+-70 and to Kerma phase					-1,3	3,25	46			+					
785	?	3650	60	-2201	-1883	-2,04	3,99	48		Dog inhumation						
786	T-6654	7470	90	-6476	-6098	-6,29	8,24	39	MoS							
786	Rom-259	6630	90	-5721	-5385	-5,55	7,5	39	MoS							
787	8000-7000 BP					-6,4	8,35	226			+					
788	4700-2700 BP uncalibrated					-2	3,95	292			+					
789		6215	75	-5339	-4963	-5,15	7,1	292				+		+		
790	SMU1149	10230	270	-10826	-9281	-10,05	12	203								
791	SMU1607	7700	90	-6754	-6393	-6,57	8,52	203	MoS							
792	Hv-15545	9320	120	-9119	-8283	-8,7	10,65	221					+			
792	Hv-15543	9290	120	-9113	-8272	-8,69	10,64	221					+			
792	Hv-15544	9450	120	-9219	-8457	-8,84	10,79	221					+			
792	Hv-15547	9375	120	-9130	-8304	-8,72	10,67	221					+			
792	Hv-15546	9060	120	-8609	-7841	-8,23	10,18	221					+			
792	Hv-15549	8815	120	-8232	-7610	-7,92	9,87	221					+			
793	3700 BP					-2,1	4,05	314								
794	3700 BP					-2,1	4,05	314								
795	KN-3088	7490	300	-7076	-5750	-6,41	8,36	138								
795	KN-3034	6550	65	-5622	-5376	-5,5	7,45	138	MoS							
796	Hv-14433	3825	115	-2577	-1949	-2,26	4,21	222	B	Giraffa, lacustrine sed					+	
797	Cal 1900-1700 bc					-1,8	3,75	117			+					
798	Gif-7652	4010	90	-2872	-2292	-2,58	4,53	117	Ch		+					
799	Cal 1400-1000 bc					-1,2	3,15	117			+					
800	Cal 1700-1500/1400 bc					-1,6	3,55	117			+					
801	Gif-7651	3780	90	-2469	-1973	-2,22	4,17	117	Ch		+					
802	Before 3rd millenium bc					-3,5	5,45	117			+					
803	Dated ca 5000-4000 cal bc					-4,5	6,45	233			+					
804	KN2939	5220	50	-4230	-3955	-4,09	6,04	327	B					+		
804	KN-2938	4850	60	-3776	-3386	-3,58	5,53	327	OES		+					
804	KN2693	5580	80	-4608	-4262	-4,44	6,39	327	MoS					+		
805	T-5024	7230	100	-6361	-5903	-6,13	8,08	37	MoS	*Pila,* 2nd occup. -65 cm						
805	T-5025	7410	100	-6443	-6072	-6,26	8,21	37	MoS	*Pila,* 1st occup. -145 cm						
805	T-5026	7250	110	-6381	-5913	-6,15	8,1	37	MoS	*Pila,* 2nd occup. -30 cm						
805	T-5027	7320	110	-6413	-6008	-6,21	8,16	37	MoS	*Pila,* 1st occup. -110 cm						

N site	Date N / Other chronological attribution	Age BP	±	cal BC		ky Mid BC/AD	ky Mid BP	Citation	Material	Note (taxon, level)	ND	1D	Env	Arch	Fau	Unk
806	Beginning of New Kingdom (post Kerma)					-1,5	3,45	53			+					
807	HAR-3475	9370	110	-9121	-8309	-8,72	10,67	160	Ch	Level 4				+		
807	HAR-3476	9330	110	-9116	-8291	-8,7	10,65	160	Ch	Level 3 (-30 cm)				+		
808	SMU-955	5940	100	-5196	-4549	-4,87	6,82	138	Ch					+		
809	KN-4033	6030	160	-5313	-4554	-4,93	6,88	178				+				+
809	SMU-954	6380	130	-5613	-5046	-5,33	7,28	138	Ch							
810	SUA-440	7470	240	-7027	-5886	-6,46	8,41	327	B	Human				+		
810	SUA-298	7050	120	-6209	-5714	-5,96	7,91	138	MoS					+		
811	SMU-1208	4046	101	-2884	-2309	-2,6	4,55	231	Ch					+		
811	SMU-1133	3615	88	-2272	-1742	-2,01	3,96	231	Ch					+		
812		7056	321	-6595	-5376	-5,99	7,94	231								
812	SMU-1134	5584	74	-4592	-4268	-4,43	6,38	231	Ch							
813		7417	67	-6430	-6100	-6,27	8,22	231								
814		4590	45	-3515	-3104	-3,31	5,26	51								+
814	T-10950	5555	60	-4520	-4269	-4,39	6,34	51	MoS	*Limicolaria*				+		
815		7930	50	-7035	-6661	-6,85	8,8	51								+
815		7600	80	-6607	-6256	-6,43	8,38	51								+
816	45 000 to 20 000 BP					-30	31,95	104			+					
817	6th to 9th century AD : transitional and early medieval period					0,65	1,3	49			+					
818	I- 1485	8370	350	-8286	-6532	-7,41	9,36	1	MoS	1,6 to 1,9 m			+			
818	SUA-68	8130	225	-7587	-6534	-7,06	9,01	1	MoS	1,4 to 1,2 m			+			
819	T-3697	6010	90	-5207	-4714	-4,96	6,91	84	MoS	*Etheria elliptica*						
819	T-4045	5600	110	-4707	-4247	-4,48	6,43	84	MoS	*Etheria elliptica*				+		
819	T-3261	4950	80	-3953	-3545	-3,75	5,7	84	MoS	*Etheria elliptica*				+		
820	Rom-258	7300	90	-6373	-6014	-6,19	8,14	39	MoS							
820	T-6644	7240	90	-6354	-5921	-6,14	8,09	39	MoS							
821	Ca 4800 to 3500 BP					-2,7	4,65	228			+					
822	Ca 4800 to 3500 BP					-2,7	4,65	228			+					
823	Pleistocene					-15	16,95	182			+					
824	KIA-17543	3385	25	-1743	-1623	-1,68	3,63	153	OM	Faeces, fireplace						
824	KIA-17510	3355	25	-1736	-1536	-1,64	3,59	153	Ch	Fireplace						
824	KIA-17544	4415	35	-3323	-2917	-3,12	5,07	153	Ch	Fireplace						
825	Attributed to ~ 2200-1100 BC					-1,5	3,45	153			+					
826	KN-5318	3560	150	-2340	-1522	-1,93	3,88	153	B	*Equus asinus*						
826	KN-5317	3875	35	-2468	-2210	-2,34	4,29	153	Ch							
826	KN-5316	4075	45	-2862	-2481	-2,67	4,62	153	Ch							
827	KN-5327	3675	40	-2196	-1943	-2,07	4,02	153	Ch	Fireplace						
827	KN-5447	3785	40	-2397	-2043	-2,22	4,17	153	Ch	Fireplace						
828	5200 BP					-4	5,95	314								

N site	Date N / Other chronological attribution	Age BP	±	cal BC		ky Mid BC/AD	ky Mid BP	Citation	Material	Note (taxon, level)	ND	1D	Env	Arch	Fau	Unk
829	Hv-15584	3140	80	-1613	-1211	-1,41	3,36	327	Ch					+		
829	Hv-15558	3195	210	-2009	-925	-1,47	3,42	327						+		
830	4200 BP					-2,8	4,75	314								
831	Hv-14439	3540	180	-2456	-1456	-1,96	3,91	327	B	*Bos taurus*				+		
832	Hv-	4720	110	-3711	-3105	-3,41	5,36	327						+		
833	Ca 1700-1500 cal bc					-1,6	3,55	233			+					
834	ZU-2168	4720	110	-3711	-3105	-3,41	5,36	222	B	Hippo		+			+	
Tanzania																
835	UCLA-1913	4860	100	-3797	-3361	-3,58	5,53	209	Ch	Bed III, 1m depth				+		
835	FRA-1	4890	70	-3780	-3377	-3,58	5,53	209	B	Human, bed III, 1m depth				+		
Western Sahara																
836	Gif-2909	6100	120	-5310	-4729	-5,02	6,97	27	B	Human						
836	Mc-670	7000	500	-7082	-4851	-5,97	7,92	72	Ch							
837	GIF-3466	1780	100	18	532	0,28	1,67	236	B	Human				+		
838	Gif-3465	3100	110	-1617	-1048	-1,33	3,28	236	B	Human				+		
839	6180-5250 BP					-4,5	6,45	236	MoS		+			+		
840	Neolithic, 5000 BP					-3,8	5,75	27			+					
841	Gif-291	4450	110	-3497	-2886	-3,19	5,14	27	Ch							
Zaire																
842	Hv-6257	2035	130	-381	229	-0,08	2,03	291	Ch	Square B, 10-16 cm						
843	SI-7065	19870	240	-22466	-20971	-21,72	23,67	230	MoS	10,5-10,45 m above lake						+
843	SI-7064	23760	385	-22621	-21076	-21,85	23,8	230	MoS	*Melanoides tuberculata,* 10,6-10,55 m above lake						+
843	W-283	21000	500	...	-22246	-22,2	24,15	33	MoS	10,5 m above lake						+
844	Early Holocene wet period ~ 12000-9000 BP					-9,5	11,45	230			+					
845	Post-date ? Early Holocene period ~ 9000-7000 BP					-7	8,95	230			+					
846	Recent Pleistocene					-15	16,95	291			+					
847	1200 BP					0,8	1,15	288			+					
848	Not older than 200 years					1,9	0,05	291			+					
849	Classic Kasilien, 10th AD ; Kabambien A: 13-14 AD					1	0,95	288			+					
850	Classic Kasilien, 10th Century AD					1	0,95	288			+					
851	Kabambien A : 13-14th century AD					1,4	0,55	288			+					
852	Kabambien B : 16-18th century AD					1,7	0,25	288			+					
853	Hv-13453	260	55	1470	1954	1,71	0,24	291	Ch							
853	Hv-13452	205	55	1525	1955	1,74	0,21	291								
853	Hv-13454	140	55	1665	1952	1,81	0,14	291								
853	Hv-13451	265	55	1470	1954	1,71	0,24	291	Ch							
854	GrN-7244	720	45	1217	1389	1,3	0,65	315		Level I				+		
855	GrN-7347	21630	410	-20556	-18911	-19,73	21,68	289		12				+		

N site	Date N / Other chronological attribution	Age BP	±	cal BC		ky Mid BC/AD	ky Mid BP	Citation	Material	Note (taxon, level)	ND	1D	Env	Arch	Fau	Unk
855	GrN-7348	22100	240	-20645	-19690	-20,17	22,12	289		13				+		
855	GrN-7344	16470	150	-17992	-17392	-17,69	19,64	289		9				+		
855	GrN-6918	21210	400	...	-22621	-22,6	24,55	289		4				+		
855	GrN-7343	12050	250	-12907	-11454	-12,18	14,13	289		7+8				+		
855	GrN-6917	14230	220	-15919	-14405	-15,16	17,11	289		1				+		
856	Hv-5258	2145	45	-350	24	-0,16	2,11	291	Ch							
856	Hv-6258	2035	65	-171	210	0,02	1,93	291	Ch							
857	18th c. AD?? - 19th Century AD. Abandoned in 1915					1,9	0,05	291			+					
858	10th century AD, classic Kissalien					1	0,95	288			+					
Zambia																
859	8th-10th century AD					0,85	1,1	253			+					
860	Nachikufian I - Spits 17-24 : 16500 to ~10 000 BP					-12	13,95	171			+					
861	Nachikufian II - Spits 9-16 : 10 000 to ~5 000 BP					-7	8,95	171			+					
862	Nachikufian III - Spits 1-8 : 5 000 to ~2 000 BP					-3	4,95	171			+					
863	Spits 25-40 : 23500 to 16500 BP Pre-Nachikufian or Proto-LSA					-18	19,95	171			+					
864	Weighmean	1376	62	610	869	0,74	1,21	267	Ch			+		+		
865	Birm-836	1240	100	660	1020	0,84	1,11	211	Ch					+		
865	Birm-835	1190	100	667	1129	0,9	1,05	211	Ch					+		
865	Birm-834	980	120	868	1295	1,08	0,87	211	Ch					+		
866	N-3429	1420	95	437	889	0,66	1,29	211	Ch					+		
866	N-3428	1300	75	653	965	0,81	1,14	211	Ch					+		
866	N-3427	1110	80	779	1156	0,97	0,98	211	Ch					+		
Zanzibar																
867	Just before 1107 AD, medieval					1,1	0,85	298			+					
Zimbabwe																
868	Pta-944	720	40	1279	1392	1,34	0,61	148	Ch			+				
869	Pta-7060	385	15	1463	1625	1,54	0,41	253	Ch	0,45-,055 mm		+		+		
870	I-3726	25650	1800	-28869	-20582	-24,73	26,68	70				+				
871	LIA, later Zimbabwe and Muzengezi wares					1,7	0,25	96			+					
872	Wits-1210	1110	70	666	1268	0,97	0,98	148	Ch			+		+		

Chapter 5 Species Distribution

Order Afrosoricida

Afrosoricidae

Country	Site N	SITE
Kenya	295	Ndabibi
Lesotho	304	Ha Makotoko
	312	Liphofung-CW
	314	Muela
	315	Ntloana Tsoana
	316	Sehonghong-a
	321	Tloutle-CWI
	322	Tloutle-O
Morocco	428	Ghar-Khal-Ib
Namibia	455	Geduld-LIA
	456	Geduld-LSA
South Africa	580	Collingham Shelter
	619	HKLP Ia
	620	HKLP Ib
	639	Large Rock Shelter
	663	Mhlwazini Cave
	701	Rooikrans-CW

Chrysochloris (Chrysochloris) asiatica

Country	Site N	SITE
South Africa	725	Steenbokfontein-1

Eremitalpa granti

Country	Site N	SITE
South Africa	721	Spoegrivier-a
	725	Steenbokfontein-1
	726	Steenbokfontein-2
	727	Steenbokfontein-3
	728	Steenbokfontein-4a
	729	Steenbokfontein-4b

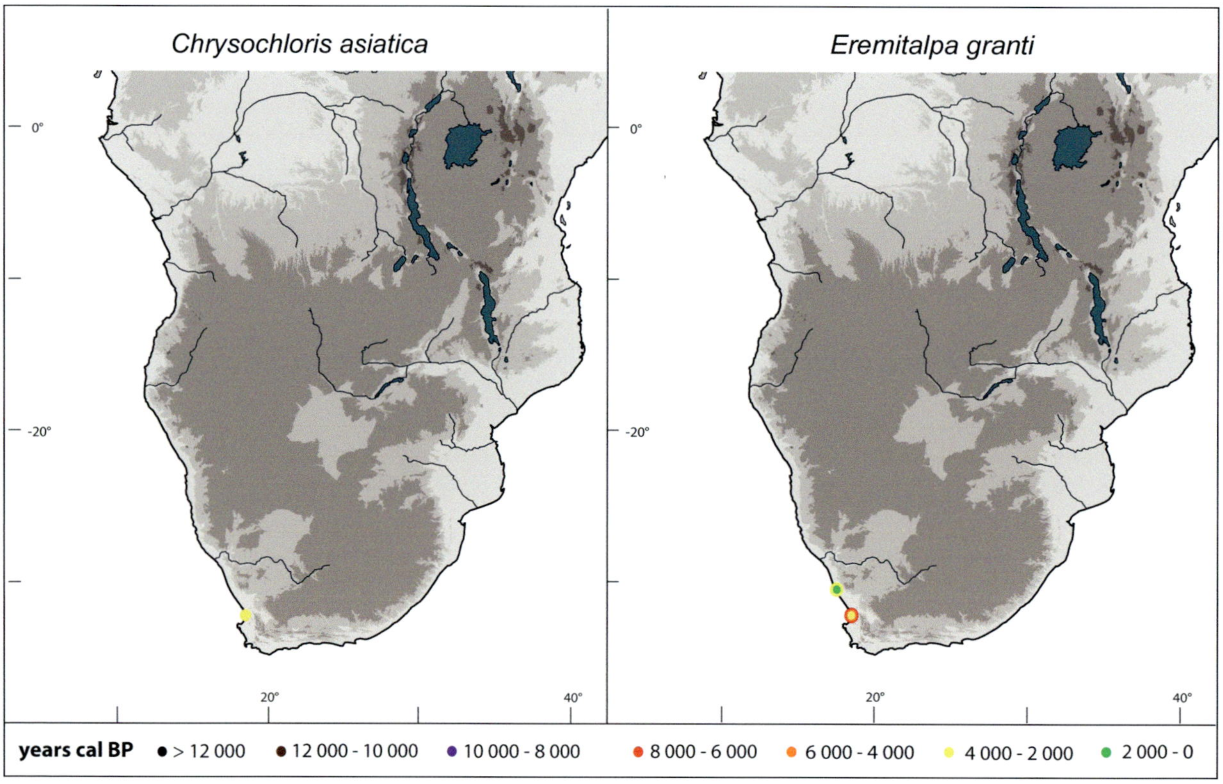

Order Macroscelidea

Macroscelidae

Country	Site N	SITE
Botswana	26	Bosutwe-Mapungubwe
	28	Bosutwe-Toutswe
Namibia	446	Aar I and II
	449	Apollo 11-c
	454	Fackelträger
	455	Geduld-LIA
	456	Geduld-LSA

Elephantulus sp.

Country	Site N	SITE
Namibia	450	Brandberg-BBC
	452	Brandberg-LSA
South Africa	589	Doornfontein
	744	Voëlvlei
Zaire	846	Kakontwe

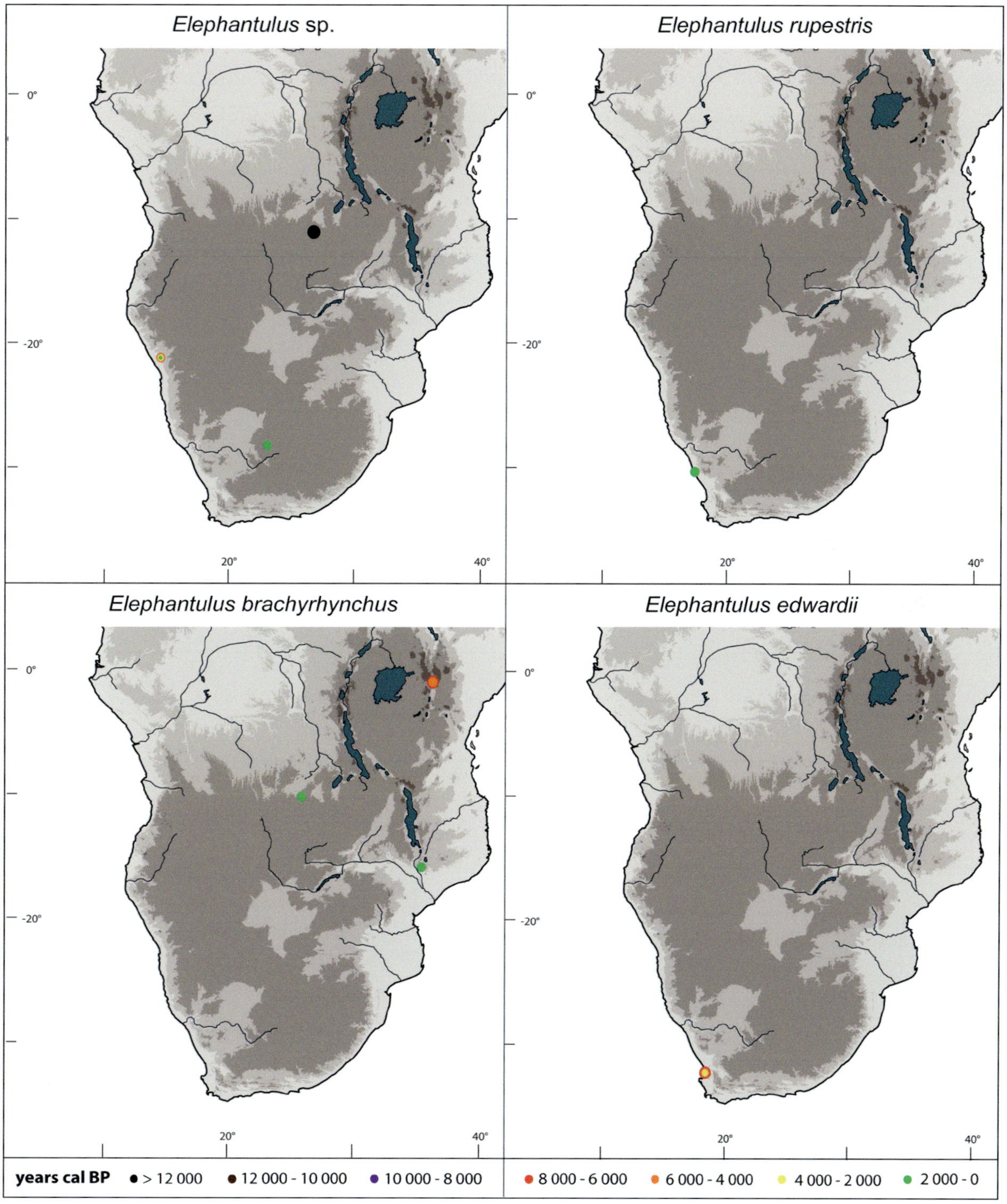

Elephantulus rupestris

Country	Site N	SITE
South Africa	721	Spoegrivier-a

Elephantulus brachyrhynchus

Country	Site N	SITE
Kenya	259	Enkapune Ya Muto-Eburran 4b
Malawi	351	Malowa-1
	352	Malowa-2
	353	Malowa-3
Zaire	848	Kiantapo

Petrodromus tetradactylus

Country	Site N	SITE
Zaire	848	Kiantapo
Zanzibar	867	Kizimkazi Dimbani

Macroscelides proboscideus

Country	Site N	SITE
South Africa	721	Spoegrivier-a

Elephantulus edwardii

Country	Site N	SITE
South Africa	725	Steenbokfontein-1
	726	Steenbokfontein-2
	727	Steenbokfontein-3
	728	Steenbokfontein-4a
	729	Steenbokfontein-4b

Rhynchocyon cirnei

Country	Site N	SITE
Zaire	854	Matupi-I

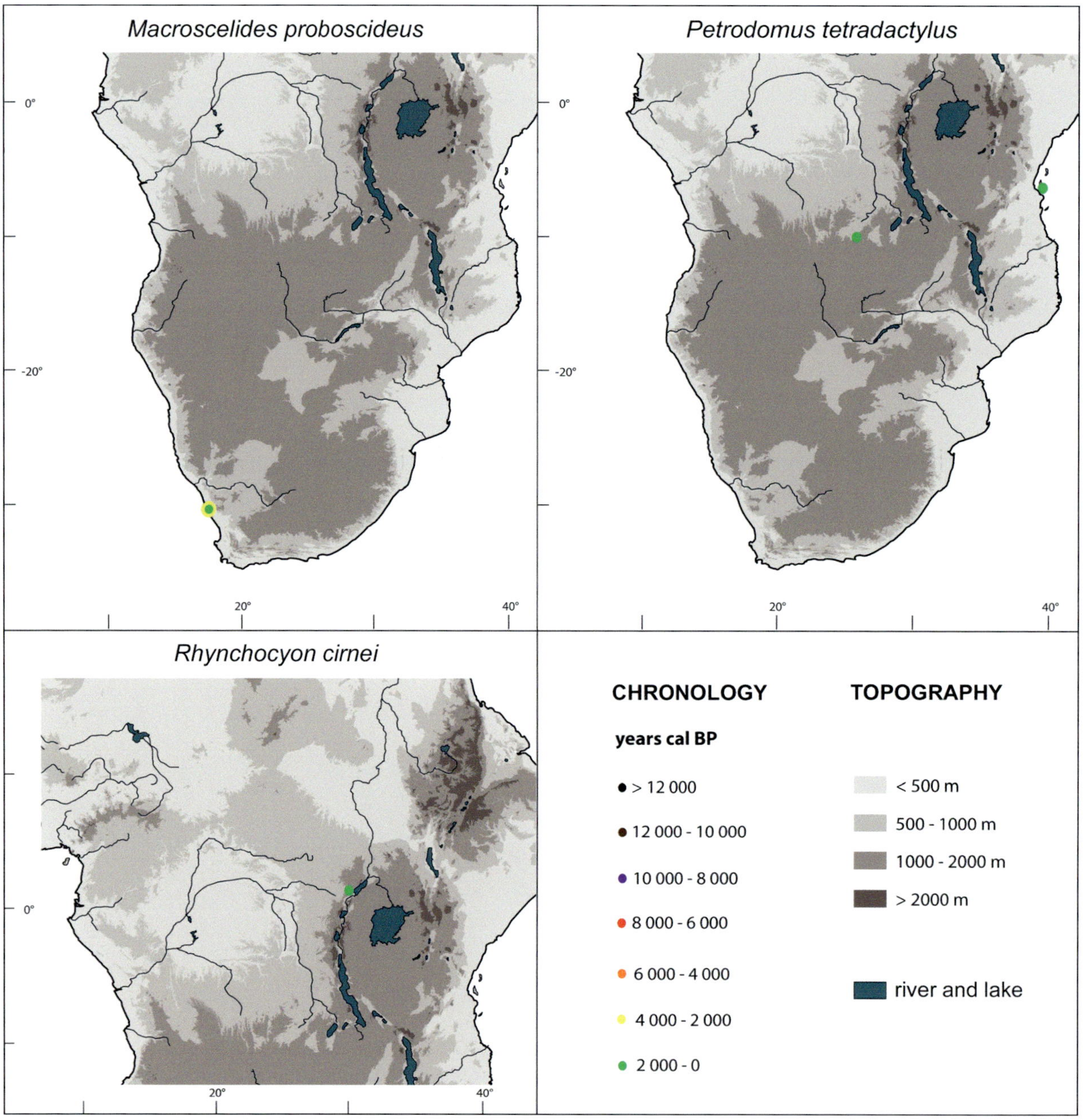

ORDER TUBULIDENTATA

Orycteropus afer

Country	Site N	SITE
Botswana	28	Bosutwe-Toutswe
	45	Toteng3-SU3
	46	Toteng3-SU4
	48	White Paintings RS-2ab
Burkina Faso	54	Oursi hu-beero
Ghana	241	Mandara
Kenya	254	Ele Bor A-B
	255	Ele Bor A-C
	275	GvJm 19-HA
	278	GvJm 22-LHC
Libya	347	Wadi Tannezzuft-96/129 (cf.)
Mali	369	Est-Hassi el Abiod
	374	Hassi el-Abiod
	376	Jenné-Jeno - phase I/II
	380	Karkarichinkat N-S
Mauritania	402	DN4-BKH-S5
Nigeria	511	Kariari
South Africa	574	Bushman Rock Shelter-LSA
	575	Bushman Rock Shelter-MSA
	595	Eland´s Bay Cave-AI
	596	Eland´s Bay Cave-RI
	600	Eland´s Bay Cave-Wid
	604	Equus Cave 1A
	616	Heuningneskrans-I/O
	617	Heuningneskrans-R
	619	HKLP Ia
	626	Jubilee Shelter-W
	627	Kasteelberg A
	628	Kasteelberg B
	640	Le2
	641	Le6
	648	Ma38
	668	Nelson Bay Cave- PW
	676	Oudepost I
	697	Riversmead Shelter I-II
	698	Riversmead Shelter III-IV
	699	Riversmead Shelter VI-IX
	711	Serotwe Hill
	729	Steenbokfontein-4b
	733	Tienfontein 2
	736	Tortoise cave-b
	752	Wosi-middle-late EIA
	753	Wosi-terminal EIA
Sudan	756	Abu Tabari 02/1
	771	El Zakiab
	777	Geili
	788	Khasm el Girba - Kassala
	789	Khasm el Girba - KG14
	798	Mahal Teglinos K1-EG
	800	Mahal Teglinos K1-LG
	802	Mahal Teglinos K1-PG
	811	Shaqaduq-S1-A
	821	Upper Atbara hinterland
	822	Upper Atbara Valley
Tanzania	835	Mumba
Zaire	843	Ishango-AB
	848	Kiantapo
	854	Matupi-I
	855	Matupi-II
Zambia	862	Leopard's Hill-NIII

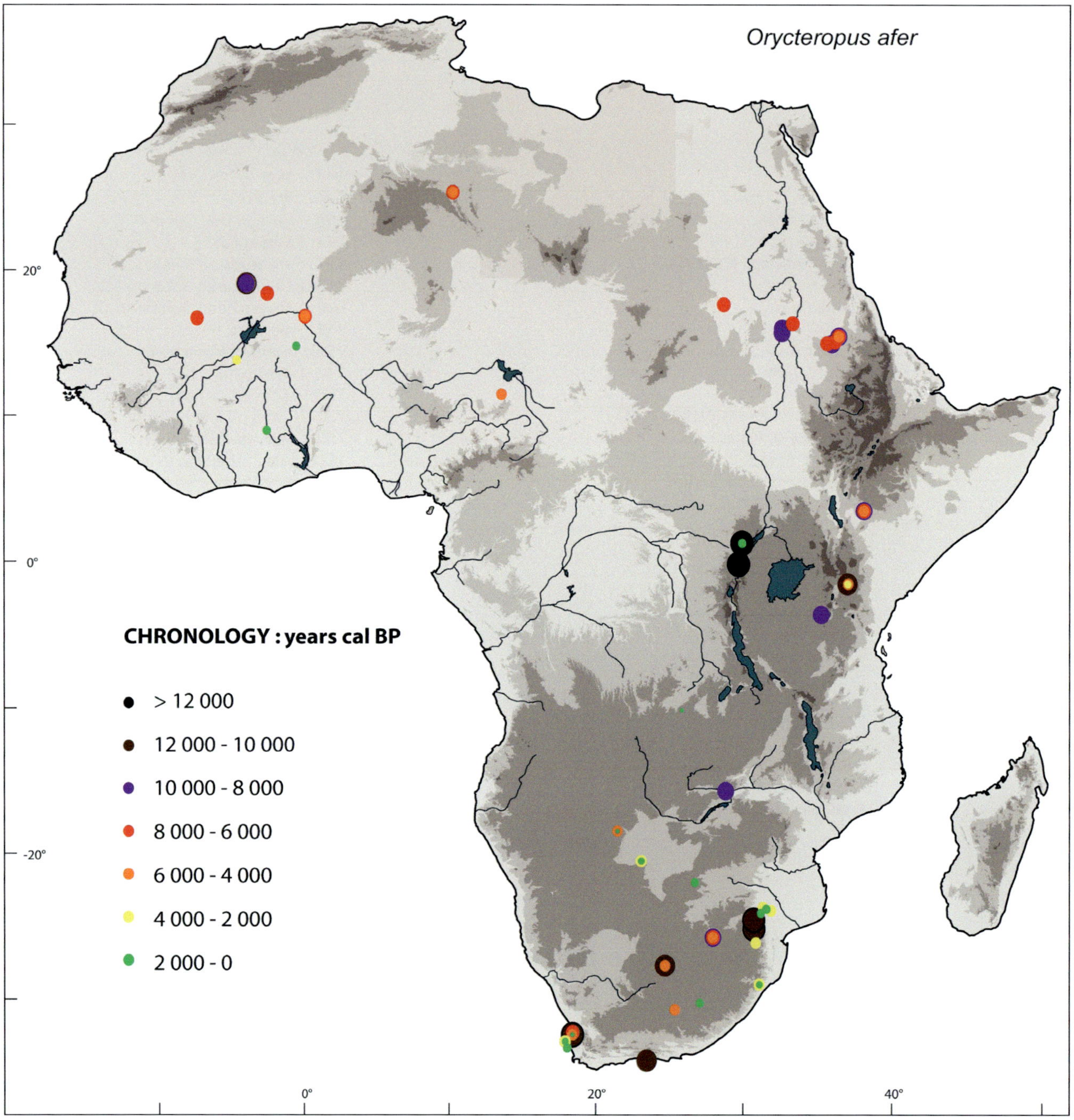
Orycteropus afer
CHRONOLOGY : years cal BP
> 12 000
12 000 - 10 000
10 000 - 8 000
8 000 - 6 000
6 000 - 4 000
4 000 - 2 000
2 000 - 0
20°
0°
-20°
0°
20°
40°

ORDER HYRACOIDEA

HYRACOIDAE

Country	Site N	SITE
Ethiopia	216	Moche Borago-1b
	217	Moche Borago-1c
	218	Moche Borago-2
Kenya	252	Dongodien
	255	Ele Bor A-C
	258	Enkapune Ya Muto - Eburran 4a
	259	Enkapune Ya Muto - Eburran 4b
	260	Enkapune Ya Muto - Eburran 5a
	275	GvJm 19-HA
	276	GvJm 19-HC
	281	GvJm 62-H
	282	GvJm 62-LPG
	283	GvJm 62-LPNG
Malawi	352	Malowa-2
South Africa	567	Border Cave
	606	Ficus B
	696	Rhenosterkloof
	753	Wosi-terminal EIA
Sudan	775	Gash Delta-K
Tanzania	835	Mumba
Zaire	854	Matupi-I
	855	Matupi-II
Zambia	860	Leopard's Hill-NI
	861	Leopard's Hill-NII
	862	Leopard's Hill-NIII
	863	Leopard's Hill-PN

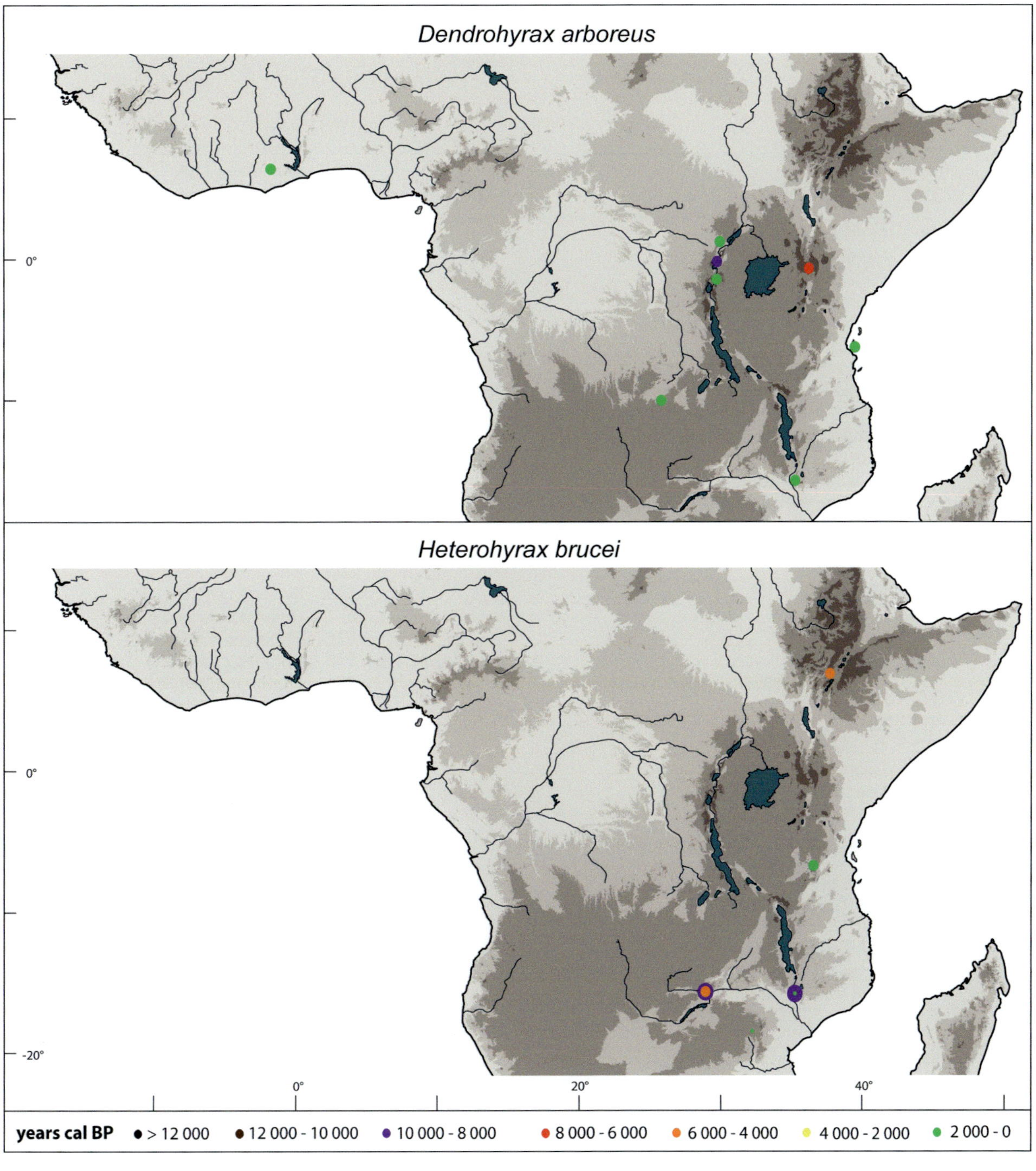

Dendrohyrax sp.

Country	Site N	SITE
Zaire	854	Matupi-I

Dendrohyrax arboreus

Country	Site N	SITE	
Ghana	223	Anyinam-Area A-B-C	
	225	Asantamanso-ASII/III	
	227	Asantamanso-ASV	
	233	Esiease - A-C-D	
Kenya	292	Marula Rock Shelter	
Malawi	352	Malowa-2	
	353	Malowa-3	
Rwanda	535	Akameru-Cyinkomane	*intr
Zaire	845	Ishango-Z-Post-Em	
	848	Kiantapo	
Zanzibar	867	Kizimkazi Dimbani	

Heterohyrax brucei

Country	Site N	SITE
Ethiopia	216	Moche Borago-1b
Kenya	293	Naivasha Railway Shelter
Malawi	354	Malowa-4
	355	Malowa-5
Rhodesia	533	Harleigh Farm-2
Zambia	861	Leopard's Hill-NII
	862	Leopard's Hill-NIII

Procavia capensis

Country	Site N	SITE	
Algeria	21	Ti-n-Hanakaten	
	5	Amekni	**P. ruficeps*?
Burkina Faso	60	Péntènga 89/1-spit 13	
	61	Péntènga 89/1-spits 15-21	
Egypt	153	Gilf Kebir	
	184	Sodmein Cave - level 5 (MP5)	
	186	Sodmein Cave younger backfill	
	195	Tree Shelter-5	
	198	Wadi el Akhdar	
Ethiopia	216	Moche Borago-1b	
Kenya	255	Ele Bor A-C	
	292	Marula Rock Shelter	
Lesotho	303	Bolahla	
	304	Ha Makotoko	
	305	Lequetsoana	
	307	Likoaeng-III-V	
	308	Likoaeng-VII-IX	
	309	Likoaeng-XI	
	310	Likoaeng-XIII	
	312	Liphofung-CW	
	313	Liphofung-pCW	
	314	Muela	
	315	Ntloana Tsoana	
	316	Sehonghong-a	
	321	Tloutle-CWI	
	322	Tloutle-O	
	323	Tloutle-pCW	
	324	Tloutle-R	
Libya	333	Ti-n-Thora East	
	335	Ti-n-Thora Two Caves	
	340	Uan Muhuggiag (MP)	
	342	Uan Tabu-14-20	
	343	Uan Telocat (EP)	
	344	Uan Telocat (LP)	
	345	Uan Telocat (MP)	
Malawi	351	Malowa-1	
	352	Malowa-2	
	353	Malowa-3	
	354	Malowa-4	
	358	Matope Court-LIA	
Namibia	446	Aar I and II	
	447	Apollo 11-a	
	448	Apollo 11-b	
	449	Apollo 11-c	
	452	Brandberg-LSA	
	454	Fackelträger	
	455	Geduld-LIA	
	455	Geduld-LIA	* cf.
	456	Geduld-LSA	
	457	Haalenberg	
	458	Namtib	
	459	Nos	
	461	Pokkenbank	
	463	Tiras	
South Africa	559	Blydefontein I-II	
	560	Blydefontein III-IV	
	561	Blydefontein V	
	562	Blydefontein VI-VII	
	563	Boomplass Cave-a	
	564	Boomplass Cave-b	
	565	Boomplass Cave-c	
	566	Boomplass Cave-d	
	568	Buffelskloof RS-a	
	570	Buffelskloof RS-c	
	571	Buffelskloof RS-d	
	572	Buffelskloof RS-e	
	573	Burchell´s Shelter	
	574	Bushman Rock Shelter-LSA	
	576	Byeneskranskop Cave 1a	
	577	Byeneskranskop Cave 1b	
	578	Byeneskranskop Cave 1c	
	579	Byeneskranskop Cave 1d	
	580	Collingham Shelter	
	581	Colwinton Shelter-CW	
	582	Colwinton Shelter-pCW	
	583	Connies Limpet Bar	
	586	Die Kelders	
	587	Dikbosch 1	
	588	Dikbosch 2	
	590	Drie Susters	
	591	Drie Susters Main	
	592	Driebos	
	595	Eland´s Bay Cave-AI	
	596	Eland´s Bay Cave-RI	
	597	Eland´s Bay Cave-Wia	
	598	Eland´s Bay Cave-Wib	
	599	Eland´s Bay Cave-Wic	
	600	Eland´s Bay Cave-Wid	

Country	**Site N**	**SITE**	
	601	Eland´s Bay Open-a	
	602	Eland´s Bay Open-b	
	603	Emambeni	
	604	Equus Cave 1A	
	610	Green Elliott Shelter I-III	
	611	Green Elliott Shelter IV-VI	
	613	Hail Stone Midden	
	616	Heuningneskrans-I/O	
	617	Heuningneskrans-R	
	619	HKLP Ia	
	620	HKLP Ib	
	622	HKLP Vb	
	624	Jubilee Shelter-O	
	625	Jubilee Shelter-PW	
	626	Jubilee Shelter-W	
	631	Klasies River Mouth-1I	
	634	Klein Kliphuis	
	639	Large Rock Shelter	*cf.
	644	Leliehoek	
	645	Limerock 1	
	646	Limerock 2	
	647	Lydenburg Heads site	
	656	Mapungubwe-Hill	
	657	Mapungubwe-K2/TS3	
	659	Marupale Hill	
	660	Mauermanshoek	
	663	Mhlwazini Cave	
	668	Nelson Bay Cave- PW	
	669	Nelson Bay Cave- R	
	670	Nelson Bay Cave- W	
	671	Oakhurst	
	672	OFD1	
	674	Ondini	
	675	Onink	
	678	Pancho's Kitchen Midden 1-2	
	680	Pancho's Kitchen Midden 6-7	
	688	Pont-Drift-LKA	
	689	Pont-Drift-Zhizo	
	690	Powerhouse Cave	
	692	QwaQwa Museum	
	694	Ravenscraig Shelter-pCW	
	697	Riversmead Shelter I-II	
	698	Riversmead Shelter III-IV	
	699	Riversmead Shelter VI-IX	
	700	Rooiels	
	701	Rooikrans-CW	
	702	Rooikrans-pCW	
	704	Rose Cottage Cave-A-Mn	
	706	Rose Cottage Cave-OIC	
	707	Rose Cottage Cave-WI	
	708	Schroda	
	711	Serotwe Hill	
	718	Sk4	
	719	Smitswinkelbaai Cave	
	720	Sonkoanini Hill	
	721	Spoegrivier-a	
	722	Spoegrivier-b	
	723	Spoegrivier-c	
	724	Spring Cave	
	725	Steenbokfontein-1	
	726	Steenbokfontein-2	
	727	Steenbokfontein-3	
	728	Steenbokfontein-4a	
	729	Steenbokfontein-4b	
	730	Tandjesberg	
	733	Tienfontein 2	
	735	Tortoise cave-a	
	736	Tortoise cave-b	
	737	Tortoise cave-c	
	738	Tortoise cave-d	
	740	Twyfelpoort	
	742	Ventershoek	
	743	Vlaeberg	*cf.
	744	Voëlvlei	
Sudan	800	Mahal Teglinos K1-LG	
Zaire	854	Matupi-I	
	855	Matupi-II	
Zambia	861	Leopard's Hill-NII	
Zimbabwe	870	Redcliff Cave	

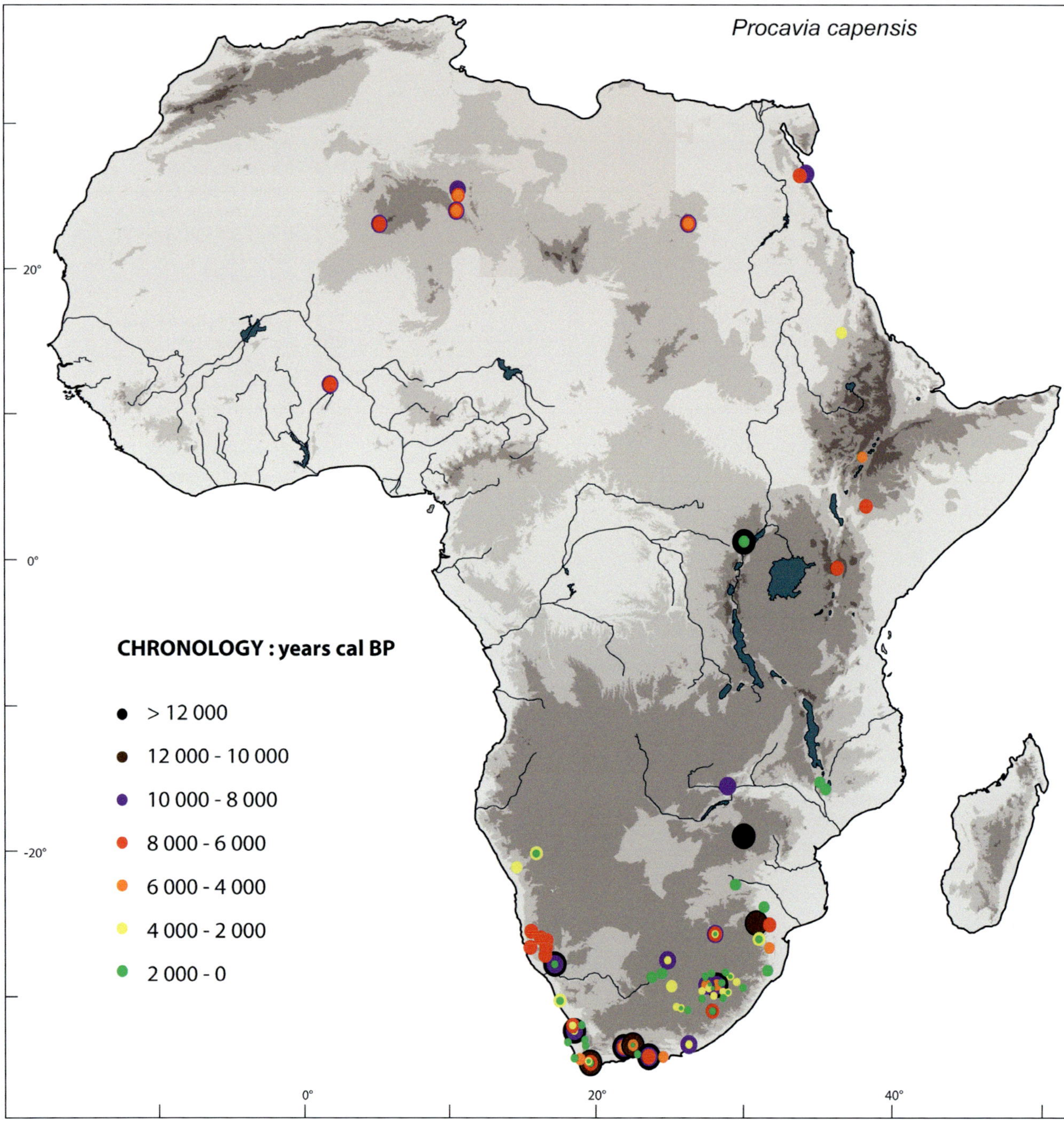
Procavia capensis
CHRONOLOGY : years cal BP
> 12 000
12 000 - 10 000
10 000 - 8 000
8 000 - 6 000
6 000 - 4 000
4 000 - 2 000
2 000 - 0
20°
0°
-20°
0°
20°
40°

ORDER PROBOSCIDEA

Loxodonta africana

Country	Site N	SITE	
Algeria	11	Gueldaman Akbou	
Botswana	26	Bosutwe-Mapungubwe	
	27	Bosutwe-Taukome	
	28	Bosutwe-Toutswe	
	29	Bosutwe-Zimbabwe	
	40	Taukome	* ivory
Burkina Faso	54	Oursi hu-beero	
Cameroon	73	Blé E	
	74	Djaba Hosséré	
C.A.R.	80	Lingbangbo	
Chad	82	Enneri Bardagué	
Congo	87	Ntadi Yom ba-level B	
Egypt	107	Berenike 95/97- Late Roman	
	109	Bir Kiseiba-EN	*intr?
	122	Dakhleh Ouasis, Ba	
	154	Hierakonpolis HK6-Naqada IC-IIA	
	155	Hierakonpolis HK6-Naqada IIA-B	
	156	Hierakonpolis HK6-Naqada III	
	157	Hierakonpolis HK6-Naqada I-II	
	174	Qantir	
	184	Sodmein Cave - level 5 (MP5)	
	197	Umm el-Qaab, Friedhof U	
	94	Abu Ballas - 85/56	
Ethiopia	207	Axum-D	
Ghana	225	Asantamanso-ASII/III	
	244	Ntereso	
	245	Nyanawase/Akwamu Amanfoso	
Kenya	250	Causeway Site	
	270	GaJi2	
Libya	327	Hauah Fteah-Iberomaurusian	
Malawi	356	Matope Court-EIA	
	357	Matope Court-IA	
Mali	361	Akumbu-EIA	
	370	Fanfannyégèné I	
	374	Hassi el-Abiod	
	375	In Begaouaen	
	391	Windé Koroji SO	
Mauritania	395	Dhar Tichitt- highest beach	*?
	405	Khatt Lemaiteg	
Morocco	422	El Harhoura 1	
	437	Kehf-el-Baroud	
	438	Kehf-el-Baroud-aven	
Mozambique	441	Manekweni-1	
	442	Manekweni-2-ABC	
	444	Manekweni-4	
	445	Manekweni-5	
Niger	467	Adrar Bous-T	
	470	Awzerkat-1	
	471	Chin Tafidet	
	474	Gobero-T	
	484	Mentes 7	
	490	Takene Bawat 6	
	495	Tin Ouaffadene	
	497	TMY3	
Nigeria	528	Ngala IIIb	
Rhodesia	532	Harleigh Farm-1	*ivory
Rwanda	535	Akameru-Cyinkomane	
Senegal	540	Cubalel-II	
	544	Sintiou-Bara	
South Africa	566	Boomplass Cave-d	
	571	Buffelskloof RS-d	cf.
	577	Byeneskranskop Cave 1b	
	579	Byeneskranskop Cave 1d	
	595	Eland´s Bay Cave-Al	
	596	Eland´s Bay Cave-Rl	
	597	Eland´s Bay Cave-Wia	
	607	Ficus C	*ivory
	628	Kasteelberg B	
	636	Klipspring	
	654	Mamba 1	
	666	Muhululu Hill	
	688	Pont-Drift-LKA	
	689	Pont-Drift-Zhizo	
	708	Schroda	
	713	SH16	
	739	Tsh1	
	751	Wosi-early EIA	
	752	Wosi-middle-late EIA	
	753	Wosi-terminal EIA	
Sudan	754	Abu Darbein	
	756	Abu Tabari 02/1	
	757	Aneibis	
	764	El Damer	
	765	El Kadada	
	769	El Qala´a	
	773	Esh Shaheinab	
	776	Gebel Tageru	
	777	Geili	
	782	Kabbashi	
	784	Kadruka	*?(ivory)
	785	Kerma	*imported?
	787	Khartoum Hopital	
	788	Khasm el Girba - Kassala	
	789	Khasm el Girba - KG14	
	791	Khasm el Girba - KG68	
	795	Laqiya area - 82/lake	
	804	Rahib-80/87	
	805	Saggai 1	
	810	Shabona	
	813	Shaqaduq-S21	
	815	Sheikh Mustafa	
	822	Upper Atbara Valley	
Zaire	855	Matupi-II	
Zanzibar	867	Kizimkazi Dimbani	
Zimbabwe	870	Redcliff Cave	

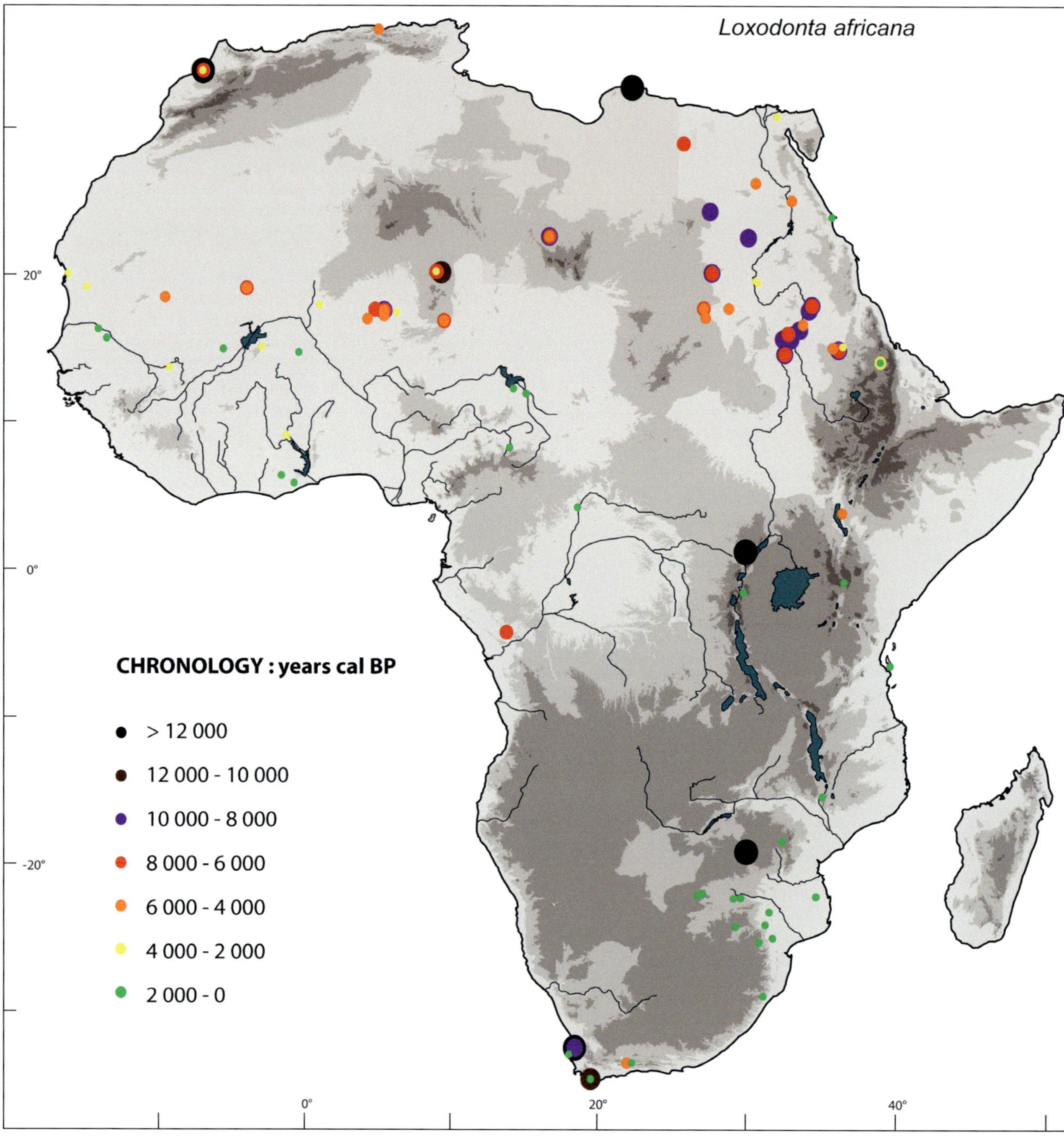
Loxodonta africana
20°
0°
-20°
CHRONOLOGY : years cal BP
> 12 000
12 000 - 10 000
10 000 - 8 000
8 000 - 6 000
6 000 - 4 000
4 000 - 2 000
2 000 - 0
0°
20°
40°

ORDER SIRENIA

Dugong dugon

Country	Site N	SITE
Egypt	106	Berenike 95/97- Early Roman
	96	Abu Sha´ar-5C

Trichechus senegalensis

Country	Site N	SITE
Mali	381	Kobadi
Senegal	542	Ndiamon-Badat

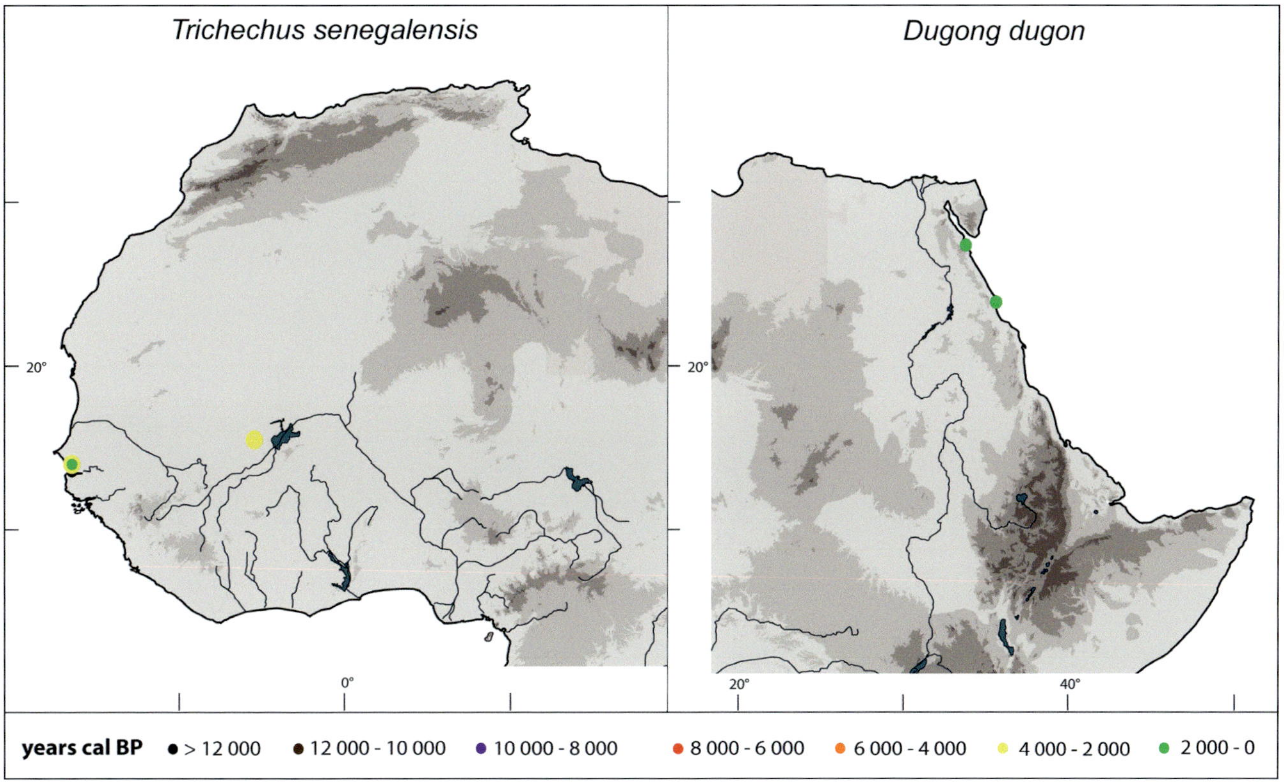

Order Primates

Galagidae

Galago senegalensis

Country	Site N	SITE
Botswana	28	Bosutwe-Toutswe
Ethiopia	216	Moche Borago-1b
Ghana	225	Asantamanso-ASII/III
Zaire	848	Kiantapo

Galago demidoff

Country	Site N	SITE
Zaire	854	Matupi-I
	855	Matupi-II

Otolemur crassicaudatus

Country	Site N	SITE
Malawi	357	Matope Court-IA
Zaire	848	Kiantapo

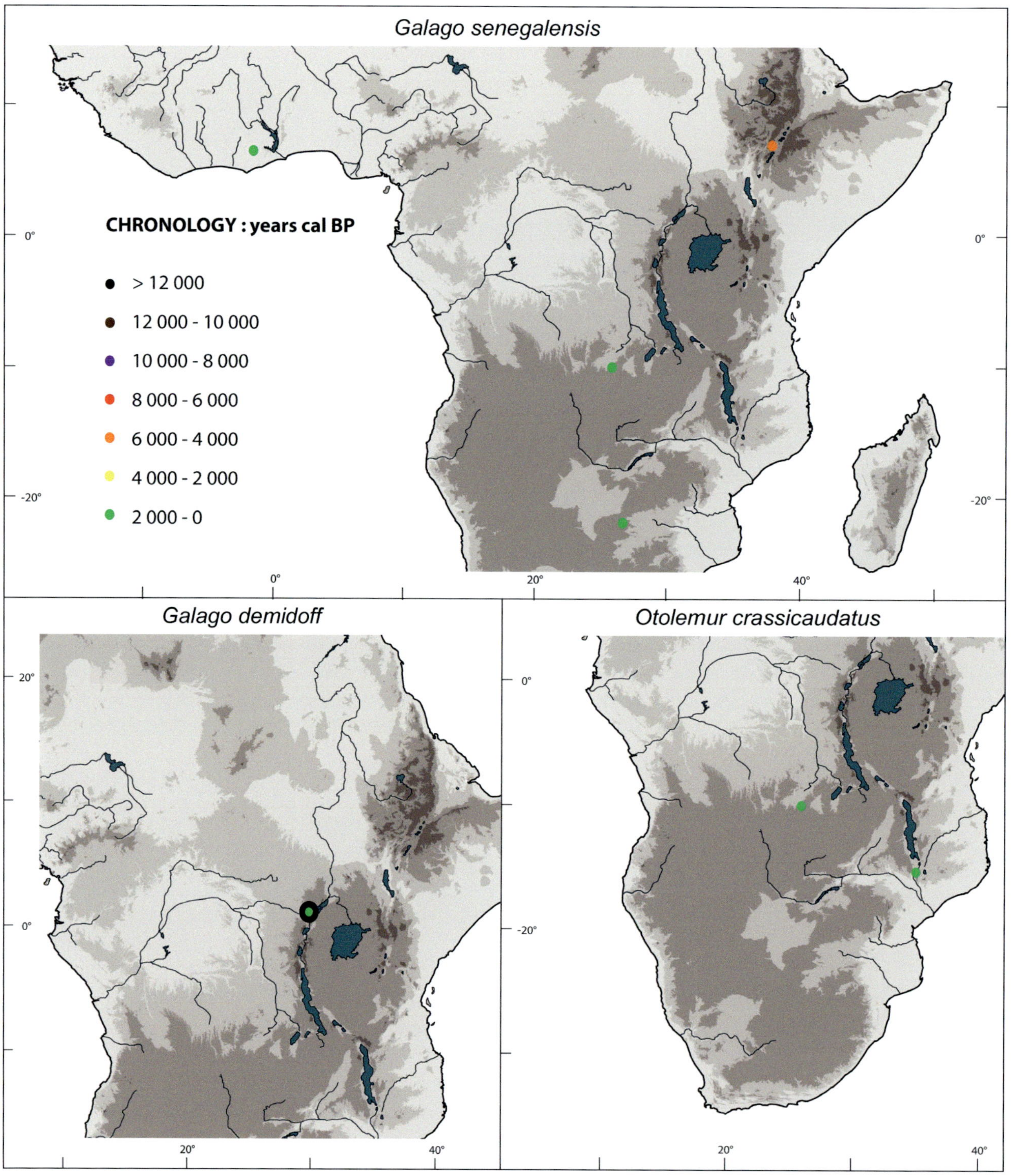

LORISIDAE

Perodicticus potto

Country	Site N	SITE
Ghana	230	Bono Manso

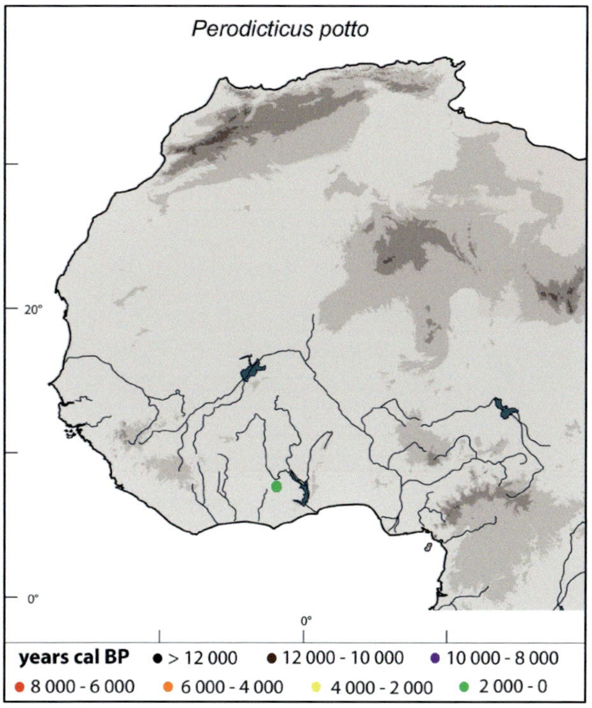

CERCOPITHECIDAE

Macaca sylvanus

Country	Site N	SITE
Algeria	20	Tamar Hat
	8	Djurdjura
Egypt	104	Baboon Catacomb
	196	Tuna el-Gebel

Cercocebus torquatus

Country	Site N	SITE
Ghana	225	Asantamanso-ASII/III
	227	Asantamanso-ASV
	232	Dawu

Cercopithecus sp.

Country	Site N	SITE	
Cameroon	78	Shum Laka IIa+b	
C.A.R.	80	Lingbangbo	
Congo	87	Ntadi Yomba-level B	
	88	Ntadi Yomba-level A	*?
Egypt	185	Sodmein Cave older backfill	
Ghana	244	Ntereso	
Mauritania	402	DN4-BKH-S5	
Nigeria	503	Gajiganna A	
Senegal	549	Tulel-Fobo	
South Africa	741	Umgungundlovu	
Sudan	789	Khasm el Girba - KG14	
	812	Shaqaduq-S1-B	
	821	Upper Atbara hinterland	
Zaire	848	Kiantapo	
	854	Matupi-I	
	855	Matupi-II	
	857	Nkile	
Zanzibar	867	Kizimkazi Dimbani	

Cercopithecus mona

Country	Site N	SITE
Ghana	223	Anyinam-Area A-B-C
	224	Anyinam-Area D
	225	Asantamanso-ASII/III
	227	Asantamanso-ASV
	233	Esiease - A-C-D

Cercopithecus mitis

Country	Site N	SITE
Kenya	293	Naivasha Railway Shelter
Mozambique	442	Manekweni-2-ABC
	444	Manekweni-4
	445	Manekweni-5

Chlorocebus aethiops

Country	Site N	SITE	
Botswana	49	White Paintings RS-2b-3ab	
Egypt	104	Baboon Catacomb	
	196	Tuna el-Gebel	
Ethiopia	216	Moche Borago-1b	
Ghana	238	Kintampo Rockshelter K6-K	*?
	239	Kintampo Rockshelter K6-K/P	
	241	Mandara	
Kenya	258	Enkapune Ya Muto-Eburran 4a	
	259	Enkapune Ya Muto-Eburran 4b	
	260	Enkapune Ya Muto-Eburran 5a	
	261	Enkapune Ya Muto-Eburran 5b	
Lesotho	308	Likoaeng-VII-IX	
	314	Muela	
South Africa	603	Emambeni	
	642	Le7a	
	675	Onink	
	707	Rose Cottage Cave-WI	
	720	Sonkoanini Hill	
	751	Wosi-early EIA	
	752	Wosi-middle-late EIA	
	753	Wosi-terminal EIA	
Sudan	765	El Kadada	
	773	Esh Shaheinab	
	777	Geili	
	785	Kerma	
	805	Saggai 1	
Zimbabwe	869	Muozi	

Chlorocebus pygerythrus

Country	Site N	SITE
Botswana	41	Toteng1-SU1

Erythrocebus patas

Country	Site N	Site	
Egypt	196	Tuna el-Gebel	
Mali	361	Akumbu-EIA	
Niger	467	Adrar Bous-T	* cf, intr

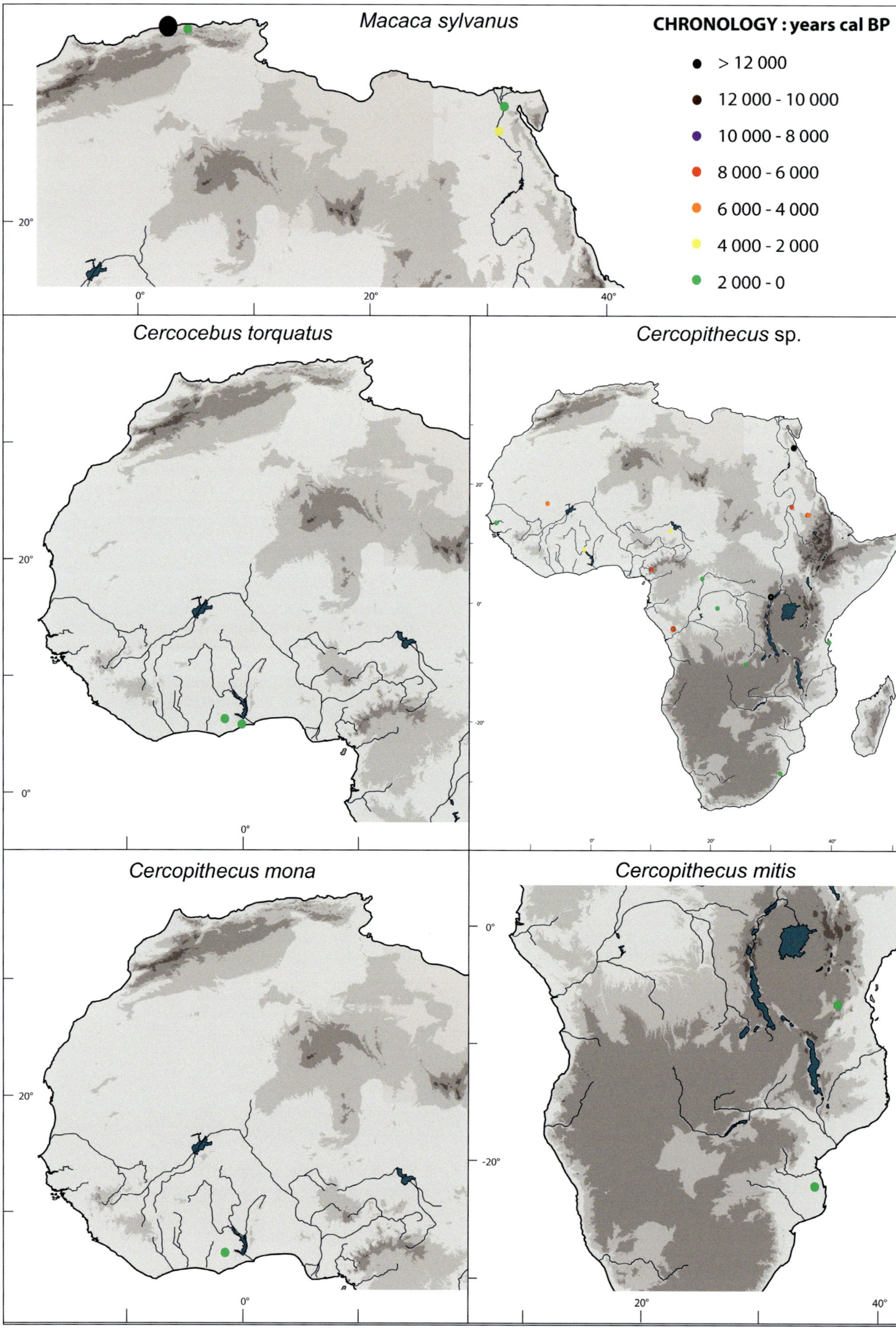
Macaca sylvanus
CHRONOLOGY : years cal BP
> 12 000
12 000 - 10 000
10 000 - 8 000
8 000 - 6 000
6 000 - 4 000
4 000 - 2 000
2 000 - 0
20°
0°
20°
40°
Cercocebus torquatus
20°
0°
0°
Cercopithecus sp.
Cercopithecus mona
20°
0°
Cercopithecus mitis
0°
-20°
20°
40°

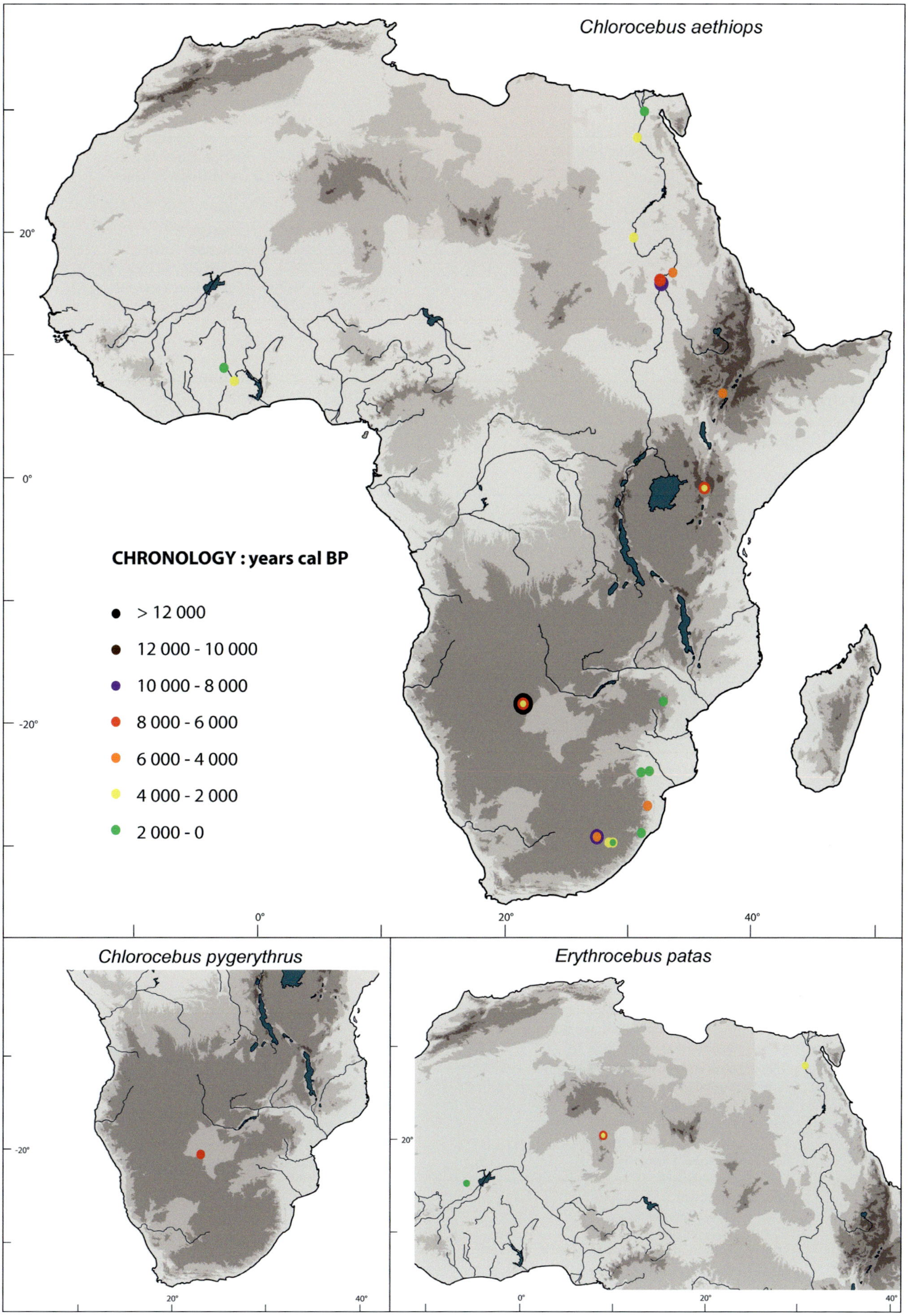
Chlorocebus aethiops
CHRONOLOGY : years cal BP
> 12 000
12 000 - 10 000
10 000 - 8 000
8 000 - 6 000
6 000 - 4 000
4 000 - 2 000
2 000 - 0
20°
0°
-20°
0°
20°
40°
Chlorocebus pygerythrus
-20°
20°
40°
Erythrocebus patas
20°
0°
20°
40°

Papio sp.

Country	Site N	Site
Cameroon	67	Abeke
	76	Shum Laka II c
	78	Shum Laka IIa+b
Ethiopia	216	Moche Borago-1b
Ghana	238	Kintampo Rockshelter K6-K
	239	Kintampo Rockshelter K6-K/P
Kenya	254	Ele Bor A-B
	255	Ele Bor A-C
	256	Ele Bor A-D
	258	Enkapune Ya Muto-Eburran 4-5
	259	Enkapune Ya Muto-Eburran 4b
	260	Enkapune Ya Muto-Eburran 5a
	261	Enkapune Ya Muto-Eburran 5b
Malawi	357	Matope Court-IA
Mozambique	441	Manekweni-1
Namibia	448	Apollo 11-b
Tanzania	835	Mumba

Papio cynocephalus

Country	Site N	Site
Kenya	276	GvJm 19-HC
Libya	333	Ti-n-Thora East
Sudan	781	Jebel Umm Marrahi
	791	Khasm el Girba - KG68
Zaire	848	Kiantapo

Papio ursinus

Country	Site N	Site
Botswana	26	Bosutwe-Mapungubwe
	28	Bosutwe-Toutswe
Lesotho	303	Bolahla
	304	Ha Makotoko
	312	Liphofung-CW
	314	Muela
	315	Ntloana Tsoana
	321	Tloutle-CWI
	322	Tloutle-O
	323	Tloutle-pCW
Rhodesia	532	Harleigh Farm-1
South Africa	560	Blydefontein III-IV
	563	Boomplass Cave-a
	564	Boomplass Cave-b
	565	Boomplass Cave-c
	566	Boomplass Cave-d
	567	Border Cave
	568	Buffelskloof RS-a
	570	Buffelskloof RS-c
	571	Buffelskloof RS-d
	574	Bushman Rock Shelter-LSA
	575	Bushman Rock Shelter-MSA
	577	Byeneskranskop Cave 1b
	578	Byeneskranskop Cave 1c
	579	Byeneskranskop Cave 1d
	580	Collingham Shelter
	581	Colwinton Shelter-CW
	582	Colwinton Shelter-pCW
	583	Connies Limpet Bar
	586	Die Kelders
	588	Dikbosch 2
	589	Doornfontein
	595	Eland´s Bay Cave-AI
	598	Eland´s Bay Cave-Wib
	599	Eland´s Bay Cave-Wic
	600	Eland´s Bay Cave-Wid
	601	Eland´s Bay Open-a
	602	Eland´s Bay Open-b
	603	Emambeni
	604	Equus Cave 1A
	609	Glentyre
	611	Green Elliott Shelter IV-VI
	617	Heuningneskrans-R
	624	Jubilee Shelter-O
	625	Jubilee Shelter-PW
	626	Jubilee Shelter-W
	632	Klasies River Mouth-1II
	633	Klasies River Mouth-5II
	639	Large Rock Shelter
	642	Le7a
	643	Le7b
	646	Limerock 2
	663	Mhlwazini Cave
	668	Nelson Bay Cave- PW
	669	Nelson Bay Cave- R
	670	Nelson Bay Cave- W
	673	Ol20
	675	Onink
	680	Pancho's Kitchen Midden 6-7
	690	Powerhouse Cave
	691	Pr34
	694	Ravenscraig Shelter-pCW
	697	Riversmead Shelter I-II
	699	Riversmead Shelter VI-IX
	700	Rooiels
	702	Rooikrans-pCW
	704	Rose Cottage Cave-A-Mn
	705	Rose Cottage Cave-DB2
	706	Rose Cottage Cave-OIC
	707	Rose Cottage Cave-WI
	717	SK17
	718	Sk4
	719	Smitswinkelbaai Cave
	751	Wosi-early EIA
	752	Wosi-middle-late EIA
	753	Wosi-terminal EIA
Zimbabwe	870	Redcliff Cave

Papio hamadryas

Country	Site N	Site
Egypt	196	Tuna el-Gebel
Lesotho	306	Likoaeng-I
	308	Likoaeng-VII-IX
	309	Likoaeng-XI
	310	Likoaeng-XIII
	313	Liphofung-pCW
	316	Sehonghong-a

Papio anubis

Country	Site N	Site	
Egypt	104	Baboon Catacomb	
	154	Hierakonpolis HK6-Naqada IC-IIA	
	155	Hierakonpolis HK6-Naqada IIA-B	
	156	Hierakonpolis HK6-Naqada III	
	196	Tuna el-Gebel	
Ethiopia	216	Moche Borago-1b	
Ghana	223	Anyinam-Area A-B-C	
	225	Asantamanso-ASII/III	
	227	Asantamanso-ASV	
	241	Mandara	
Kenya	292	Marula Rock Shelter	
Niger	467	Adrar Bous-T	*cf.
Nigeria	510	Gilgila-Iic	
Zaire	846	Kakontwe	
	854	Matupi-I	
	855	Matupi-II	

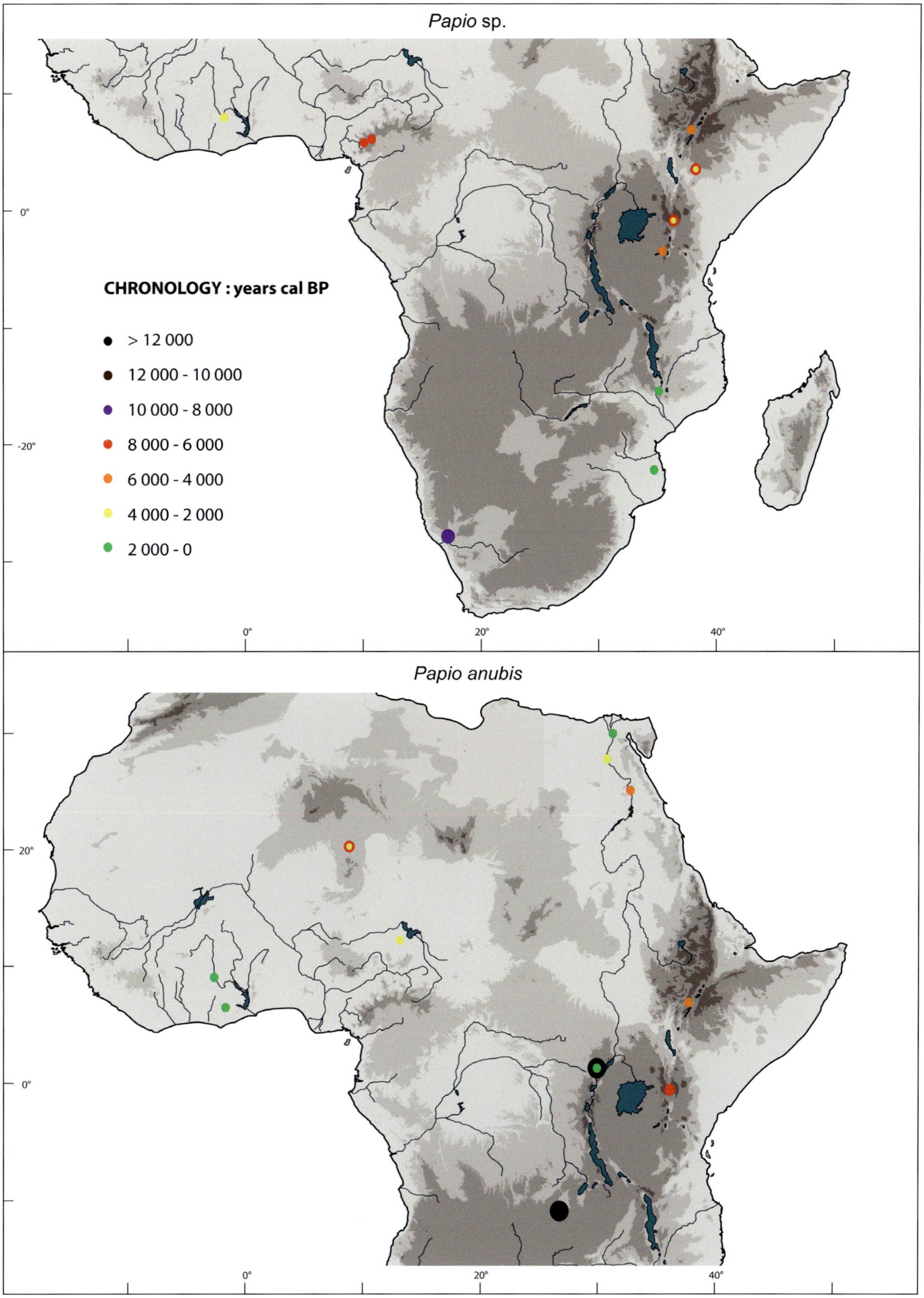
Papio sp.
CHRONOLOGY : years cal BP
> 12 000
12 000 - 10 000
10 000 - 8 000
8 000 - 6 000
6 000 - 4 000
4 000 - 2 000
2 000 - 0
0°
-20°
0°
20°
40°
Papio anubis
20°
0°
0°
20°
40°

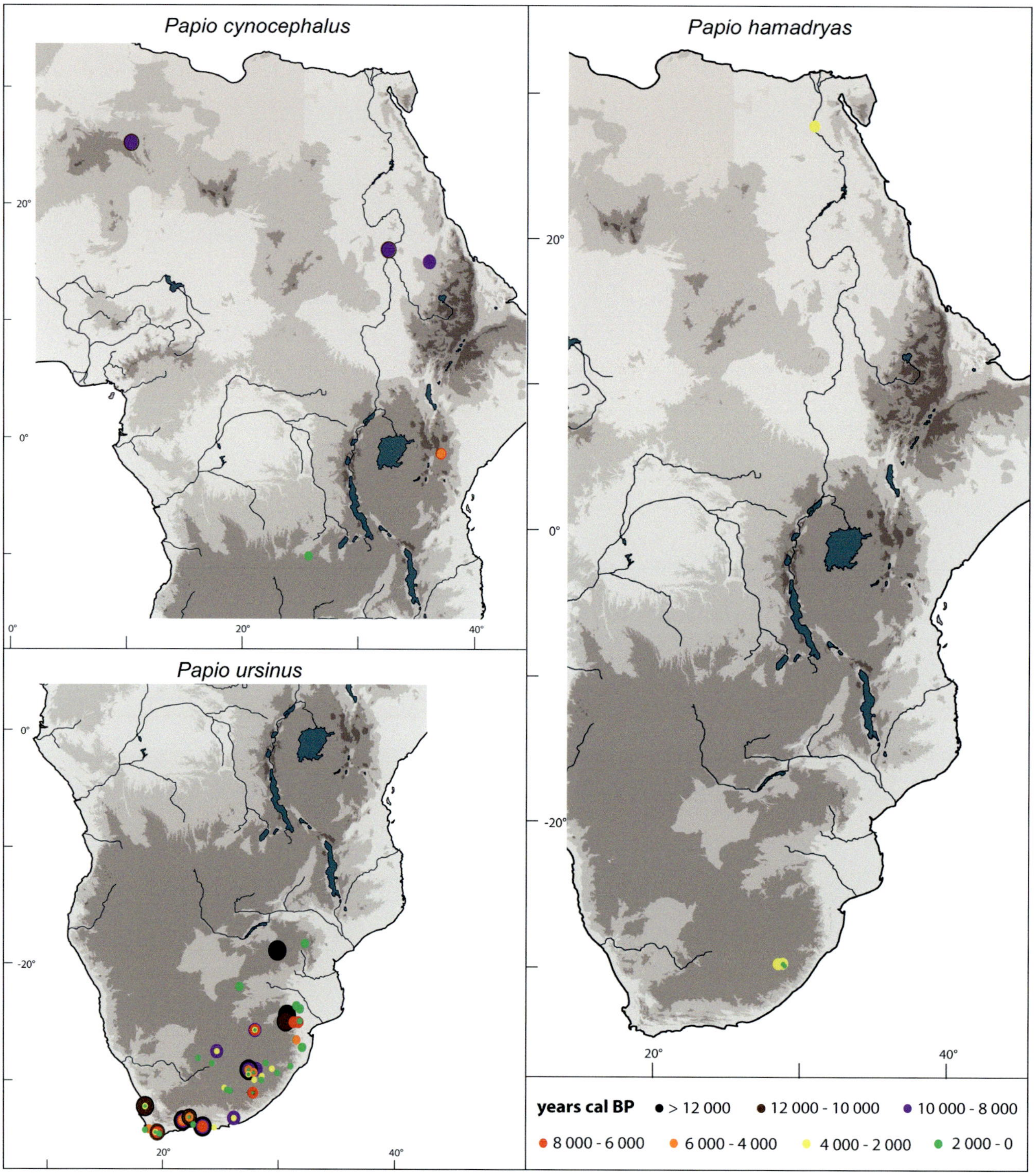
Papio cynocephalus
20°
0°
0°
20°
40°
Papio hamadryas
20°
0°
-20°
20°
40°
Papio ursinus
0°
-20°
20°
40°
years cal BP
> 12 000
12 000 - 10 000
10 000 - 8 000
8 000 - 6 000
6 000 - 4 000
4 000 - 2 000
2 000 - 0

Colobinae

Colobus sp.

Country	Site N	Site
Congo	89	Tongo
Kenya	297	Ngenyn
Zaire	855	Matupi-II

Colobus guereza

Country	Site N	Site
Ethiopia	216	Moche Borago-1b
Ethiopia	218	Moche Borago-2

Procolobus verus

Country	Site N	Site
Ghana	232	Dawu

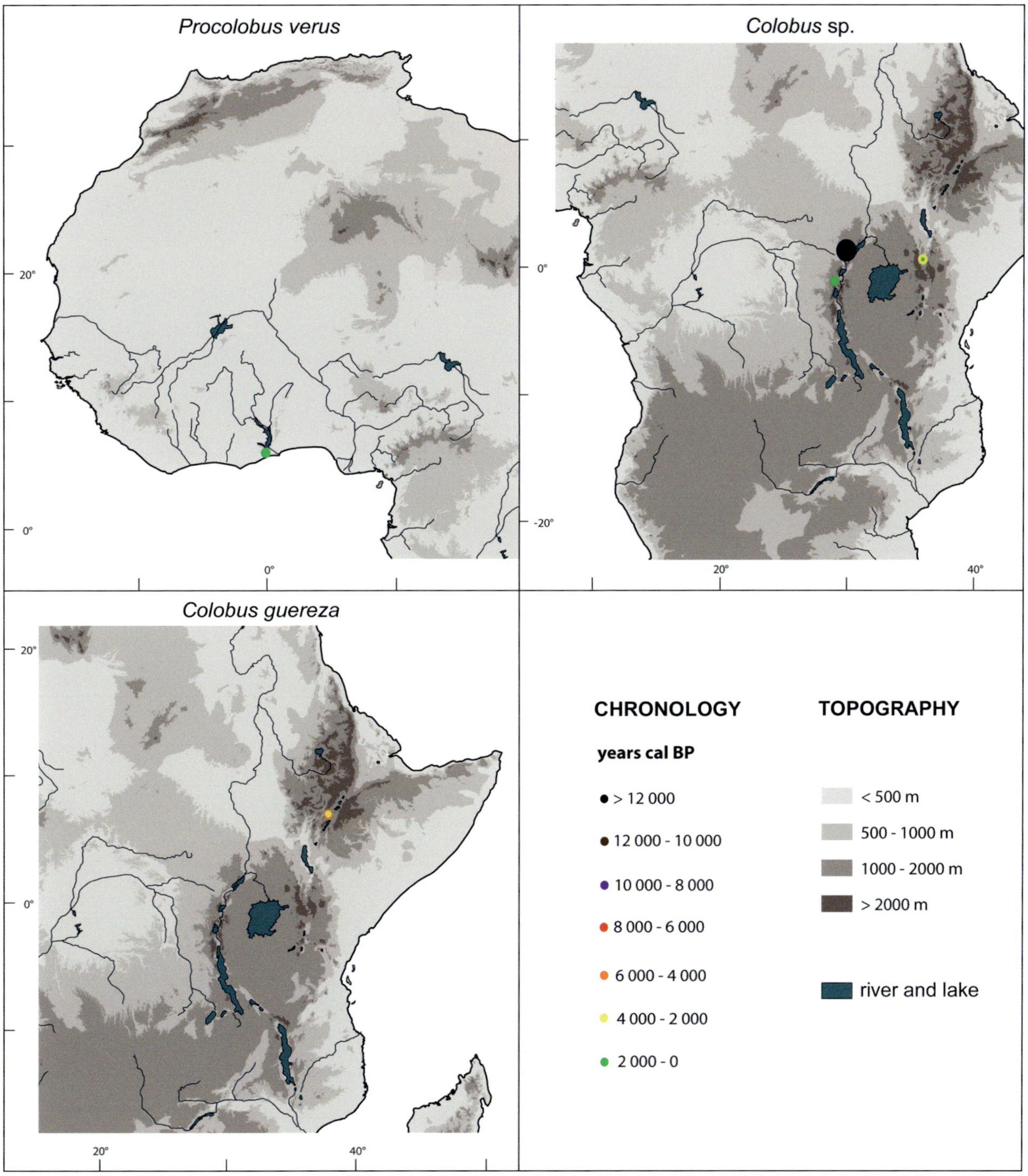

Hominidae

Gorilla gorilla

Country	Site N	Site
Cameroon	77	Shum Laka II d

Pan troglodytes

Country	Site N	Site	
Cameroon	78	Shum Laka IIa+b	
Tanzania	835	Mumba	*

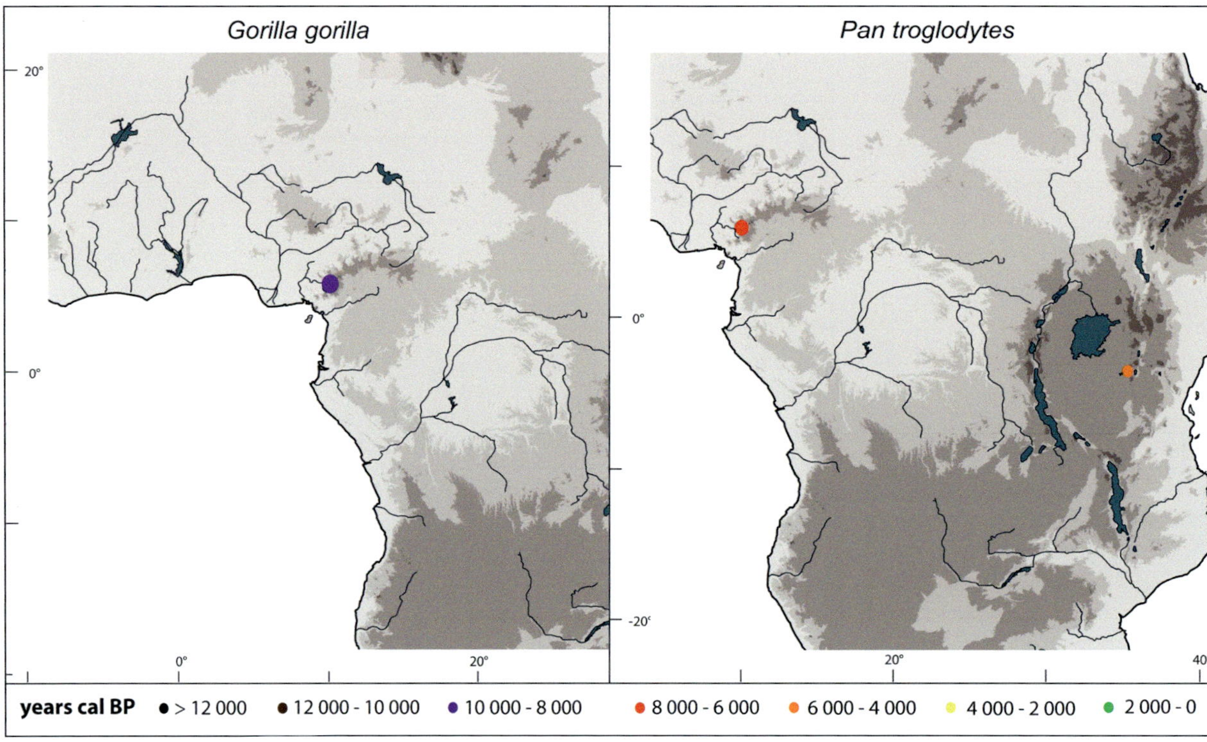

ORDER RODENTIA

SCIURIDAE

Country	Site N	Site
Botswana	26	Bosutwe-Mapungubwe
Congo	87	Ntadi Yom ba-level B
	87	Ntadi Yom ba-level B
Ghana	223	Anyinam-Area A-B-C
	227	Asantamanso-ASV
	232	Dawu

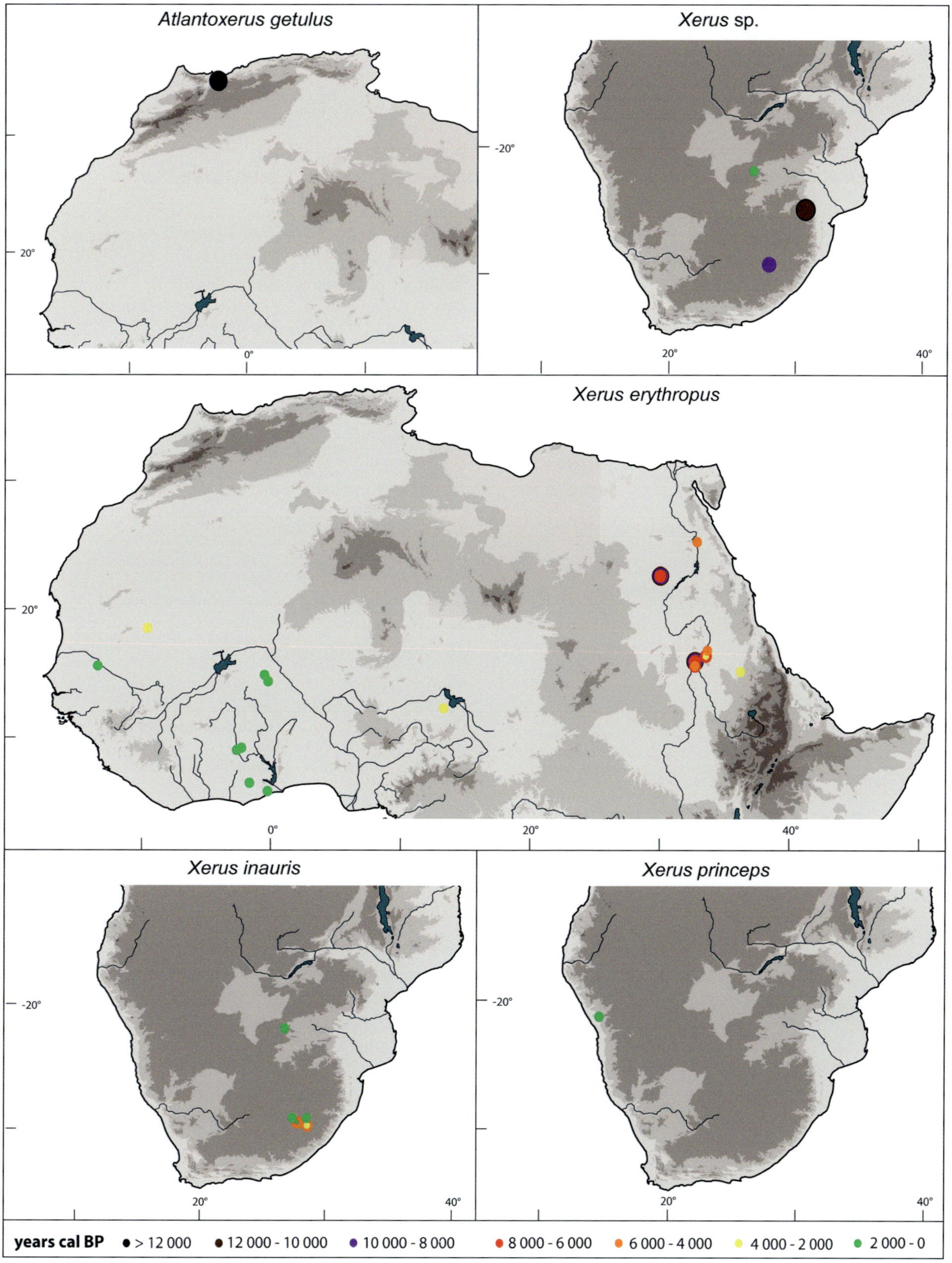

Atlantoxerus getulus

Country	Site N	Site
Morocco	440	Taforalt

Xerus sp.

Country	Site N	Site
Botswana	26	Bosutwe-Mapungubwe
	28	Bosutwe-Toutswe
Lesotho	304	Ha Makotoko
South Africa	574	Bushman Rock Shelter-LSA

Xerus (Euxerus) erythropus

Country	Site N	Site	
Burkina Faso	54	Oursi hu-beero	
	63	Saouga 94/120	
Egypt	100	Adaima-habitat	*intr
	109	Bir Kiseiba-EN	
	111	Bir Kiseiba-MN	
Ghana	223	Anyinam-Area A-B-C	
	227	Asantamanso-ASV	
	236	Frederiksgave plantation	
	241	Mandara	
	242	Mankuma	
	246	Nyanga	
Mauritania	411	Tichitt - Akjinjeir phase	
Nigeria	505	Gajiganna B2	
Senegal	543	Ogo region	
Sudan	765	El Kadada	
	770	El Temeyim	
	773	Esh Shaheinab	
	781	Jebel Umm Marrahi	
	782	Kabbashi	
	783	Kadero	
	788	Khasm el Girba - Kassala	
	805	Saggai 1	
	811	Shaqaduq-S1-A	
	812	Shaqaduq-S1-B	
	820	Umm Singid	

Xerus (Geosciurus) inauris

Country	Site N	Site	
Botswana	26	Bosutwe-Mapungubwe	
	28	Bosutwe-Toutswe	*cf.
	29	Bosutwe-Zimbabwe	
Lesotho	312	Liphofung-CW	
	313	Liphofung-pCW	
	321	Tloutle-CWI	
	323	Tloutle-pCW	
South Africa	704	Rose Cottage Cave-A-Mn	
	730	Tandjesberg	

Xerus (Geosciurus) princeps

Country	Site N	Site
Namibia	450	Brandberg-BBC

Funisciurus sp.

Country	Site N	Site
Congo	87	Ntadi Yom ba-level B
Zaire	854	Matupi-I
	855	Matupi-II

Funisciurus leucogenys

Country	Site N	Site
Ghana	223	Anyinam-Area A-B-C
	227	Asantamanso-ASV

Heliosciurus sp.

Country	Site N	Site
Cameroon	76	Shum Laka II c

Heliosciurus rufobrachium

Country	Site N	Site
Ghana	223	Anyinam-Area A-B-C
	227	Asantamanso-ASV
Zaire	854	Matupi-I

Paraxerus sp.

Country	Site N	Site
Zaire	854	Matupi-I

Paraxerus poensis

Country	Site N	Site
Congo	87	Ntadi Yom ba-level B

Protoxerus stangeri

Country	Site N	Site
Congo	87	Ntadi Yom ba-level B
Zaire	854	Matupi-I
	855	Matupi-II

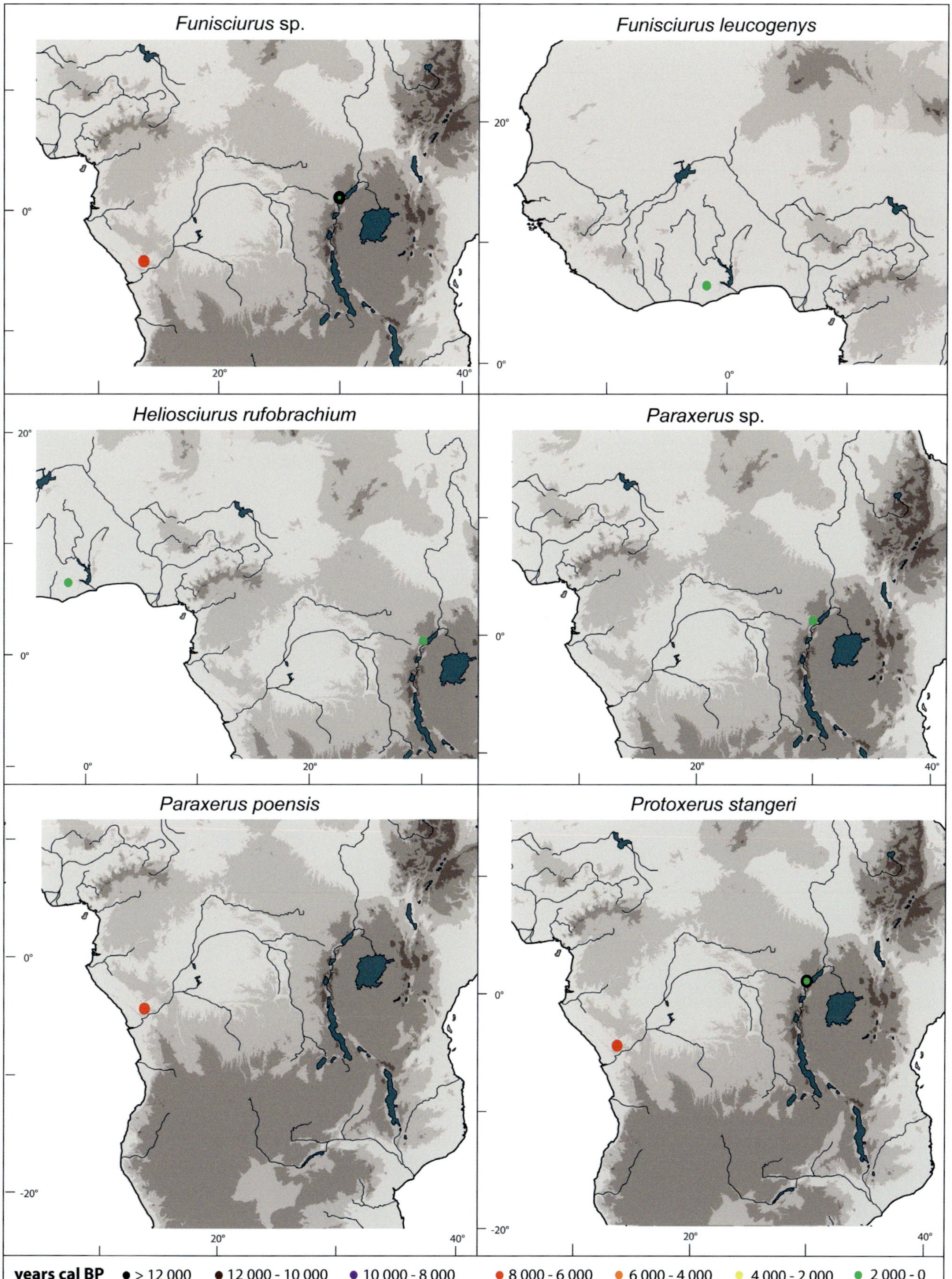
Funisciurus sp.
0°
20°
40°
Funisciurus leucogenys
20°
0°
0°
Heliosciurus rufobrachium
20°
0°
0°
20°
Paraxerus sp.
0°
20°
40°
Paraxerus poensis
0°
-20°
20°
40°
Protoxerus stangeri
0°
-20°
20°
40°
years cal BP
> 12 000
12 000 - 10 000
10 000 - 8 000
8 000 - 6 000
6 000 - 4 000
4 000 - 2 000
2 000 - 0

Gliridae

Graphiurus platyops

Country	Site N	Site
South Africa	721	Spoegrivier-a

Graphiurus ocularis

Country	Site N	Site
South Africa	725	Steenbokfontein-1

Eliomys quercinus

Country	Site N	Site
Morocco	428	Ghar-Khal-Ib

Dipodidae

Jaculus sp.

Country	Site N	Site
Sudan	783	Kadero

Jaculus jaculus

Country	Site N	Site	
Egypt	138	Elkab-Epipaleo	*intr
	168	Merimde Benisâlame	
Libya	335	Ti-n-Thora Two Caves	
Mali	380	Karkarichinkat N-S	
Mauritania	396	Dhar Tichitt- latest phase	
	411	Tichitt - Akjinjeir phase	

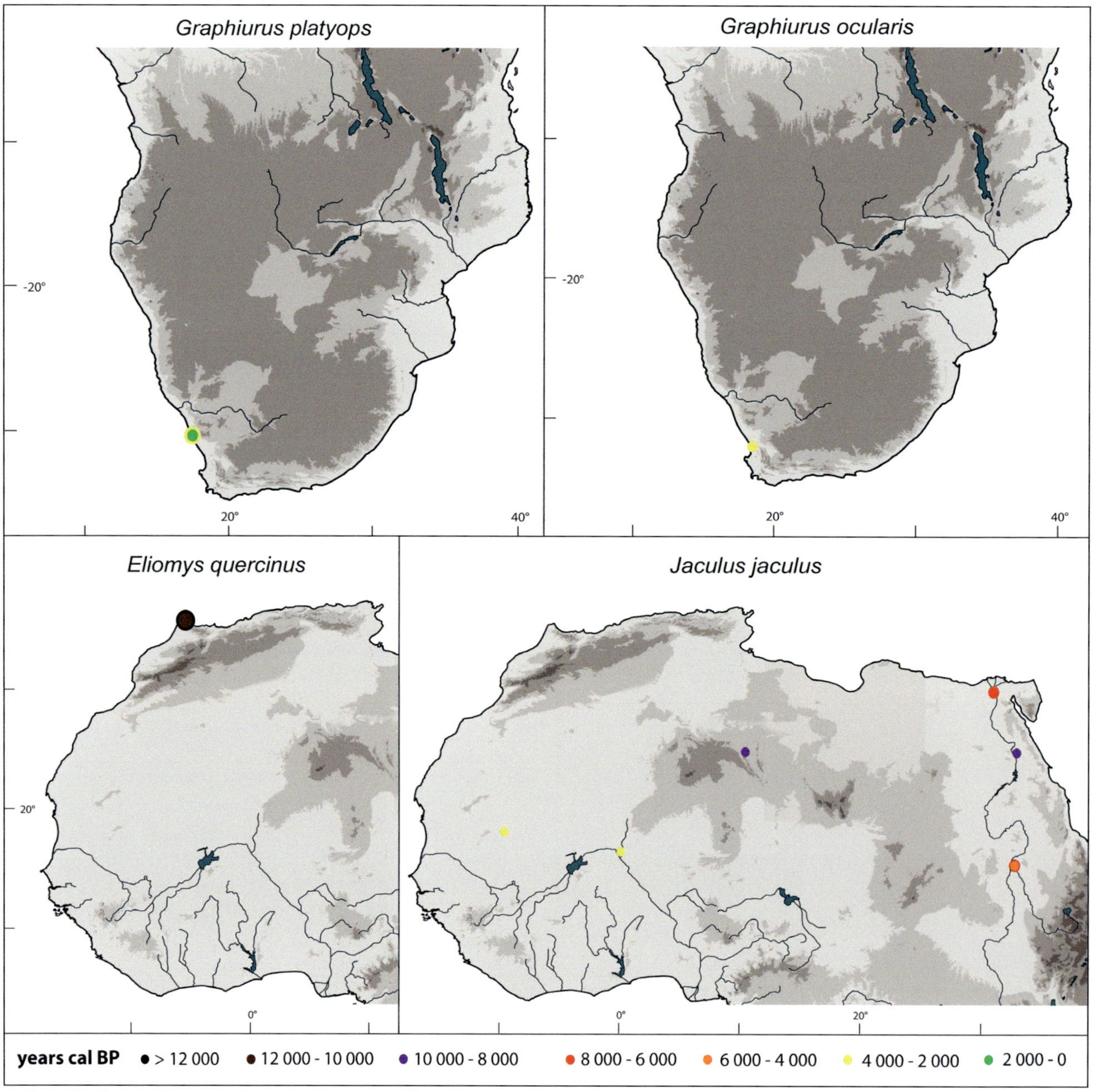

NESOMYIDAE

Petromyscus collinus

Country	Site N	Site	
Namibia	457	Haalenberg	*?

Cricetomys sp.

Country	Site N	Site
Cameroon	76	Shum Laka II c
	77	Shum Laka II d
	78	Shum Laka IIa+b
C.A.R.	80	Lingbangbo
Congo	87	Ntadi Yom ba-level B
	89	Tongo
Ghana	238	Kintampo Rockshelter K6-K
Mauritania	400	Dhar Tichitt Site 38
Rwanda	535	Akameru-Cyinkomane
Zaire	842	Dimba
	854	Matupi-I
	855	Matupi-II

Cricetomys gambianus

Country	Site N	Site
Burkina Faso	53	Kissi 40
	54	Oursi hu-beero
	58	Oursi-IA
	61	Péntènga 89/1-spits 15-21
	63	Saouga 94/120
Ghana	223	Anyinam-Area A-B-C
	224	Anyinam-Area D
	225	Asantamanso-ASII/III
	227	Asantamanso-ASV
	229	Bibease Plantation
	232	Dawu
	233	Esiease - A-C-D
	234	Esiease - H
	235	Esiease -recent
	241	Mandara
	242	Mankuma
	246	Nyanga
Malawi	351	Malowa-1
	352	Malowa-2
	353	Malowa-3
Mali	361	Akumbu-EIA
	362	Akumbu-MIA
Mauritania	411	Tichitt - Akjinjeir phase
	417	Tichitt - Nkhal phase
Nigeria	498	Bukarkurari-I
	502	Elkido North
	503	Gajiganna A
	507	Galaga
Zaire	848	Kiantapo
Zanzibar	867	Kizimkazi Dimbani

Saccostomus sp.

Country	Site N	Site
Namibia	455	Geduld-LIA
	456	Geduld-LSA
South Africa	589	Doornfontein
	674	Ondini

Saccostomus campestris

Country	Site N	Site
Botswana	42	Toteng1-SU2
	48	White Paintings RS-2ab
	49	White Paintings RS-2b-3ab
Zaire	848	Kiantapo

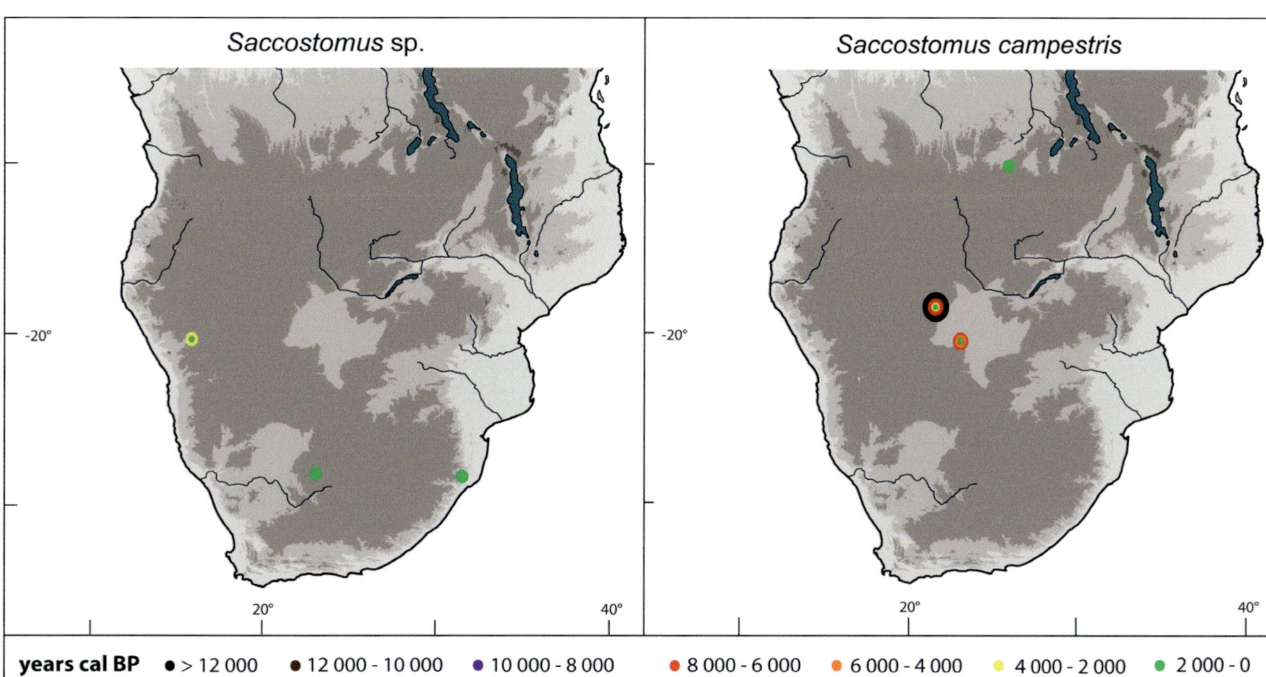

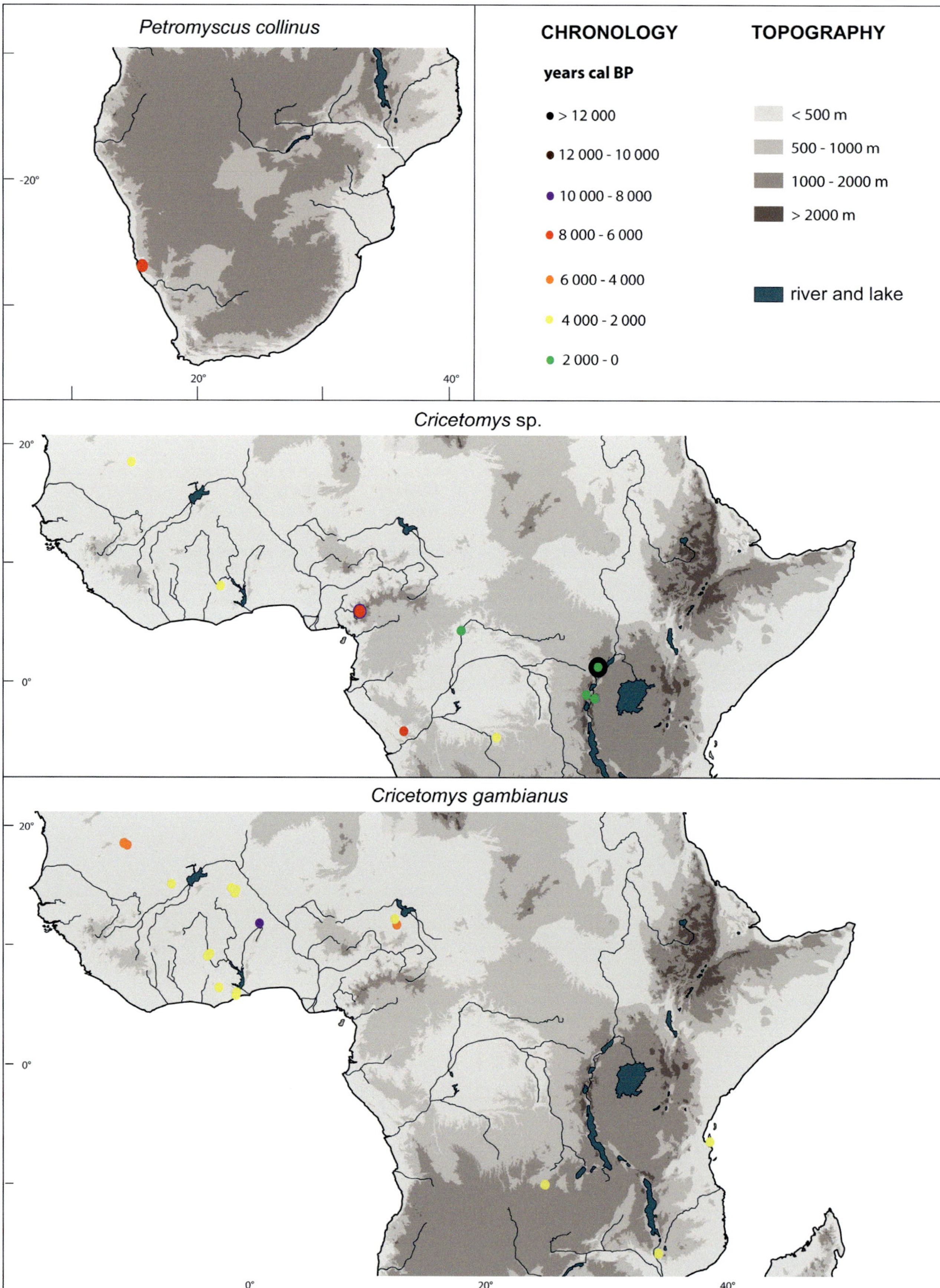
Petromyscus collinus
-20°
20°
40°
CHRONOLOGY
years cal BP
> 12 000
12 000 - 10 000
10 000 - 8 000
8 000 - 6 000
6 000 - 4 000
4 000 - 2 000
2 000 - 0
TOPOGRAPHY
< 500 m
500 - 1000 m
1000 - 2000 m
> 2000 m
river and lake
Cricetomys sp.
20°
0°
Cricetomys gambianus
20°
0°
0°
20°
40°

Dendromus sp.

Country	Site N	Site
Botswana	35	Drotsky's Cave-P
Kenya	259	Enkapune Ya Muto-Eburran 4b
South Africa	589	Doornfontein

Dendromus mesomelas

Country	Site N	Site
South Africa	725	Steenbokfontein-1
	726	Steenbokfontein-2
	727	Steenbokfontein-3

Dendromus melanotis

Country	Site N	Site
South Africa	683	Paternoster5-6
	721	Spoegrivier-a
	725	Steenbokfontein-1
	726	Steenbokfontein-2
	727	Steenbokfontein-3
	728	Steenbokfontein-4a
	729	Steenbokfontein-4b

Malacothrix typica

Country	Site N	Site	
Botswana	26	Bosutwe-Mapungubwe	
	29	Bosutwe-Zimbabwe	*cf.
South Africa	729	Steenbokfontein-4b	

CRICETIDAE

Country	Site N	Site
Kenya	254	Ele Bor A-B
Mali	376	Jenné-Jeno - phase I/II
	377	Jenné-Jeno - phase III

Steatomys sp.

Country	Site N	Site
Namibia	455	Geduld-LIA

Steatomys krebsii

Country	Site N	Site
South Africa	721	Spoegrivier-a
	725	Steenbokfontein-1
	726	Steenbokfontein-2
	727	Steenbokfontein-3
	728	Steenbokfontein-4a
	729	Steenbokfontein-4b

Mystromys albicaudatus

Country	Site N	Site
Botswana	35	Drotsky's Cave-P
South Africa	589	Doornfontein
	721	Spoegrivier-a
	725	Steenbokfontein-1
	726	Steenbokfontein-2
	727	Steenbokfontein-3
	728	Steenbokfontein-4a
	729	Steenbokfontein-4b

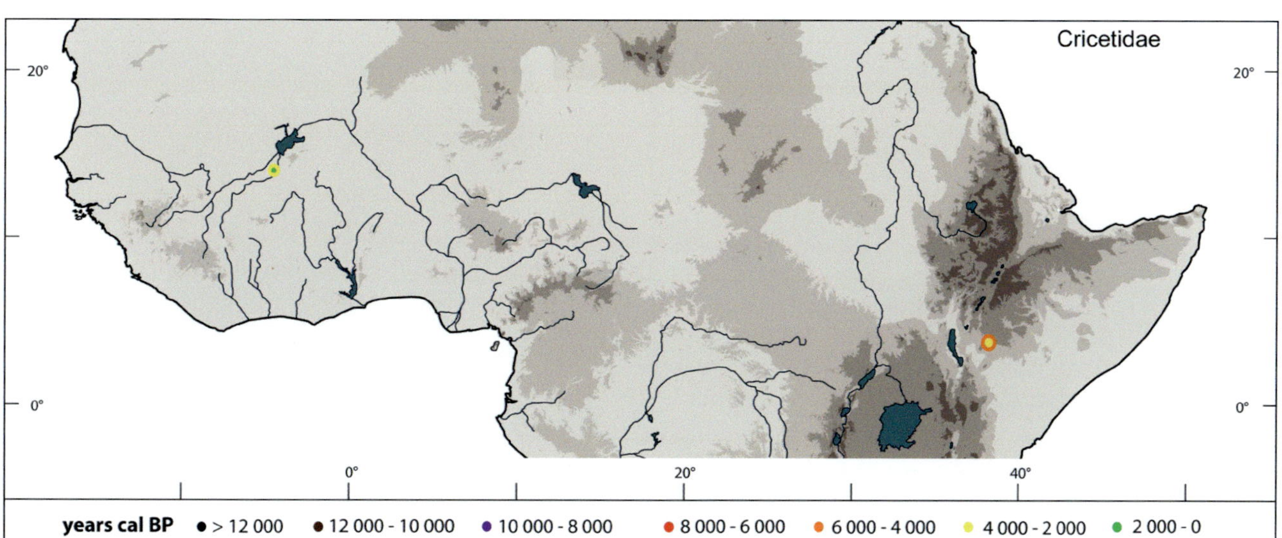

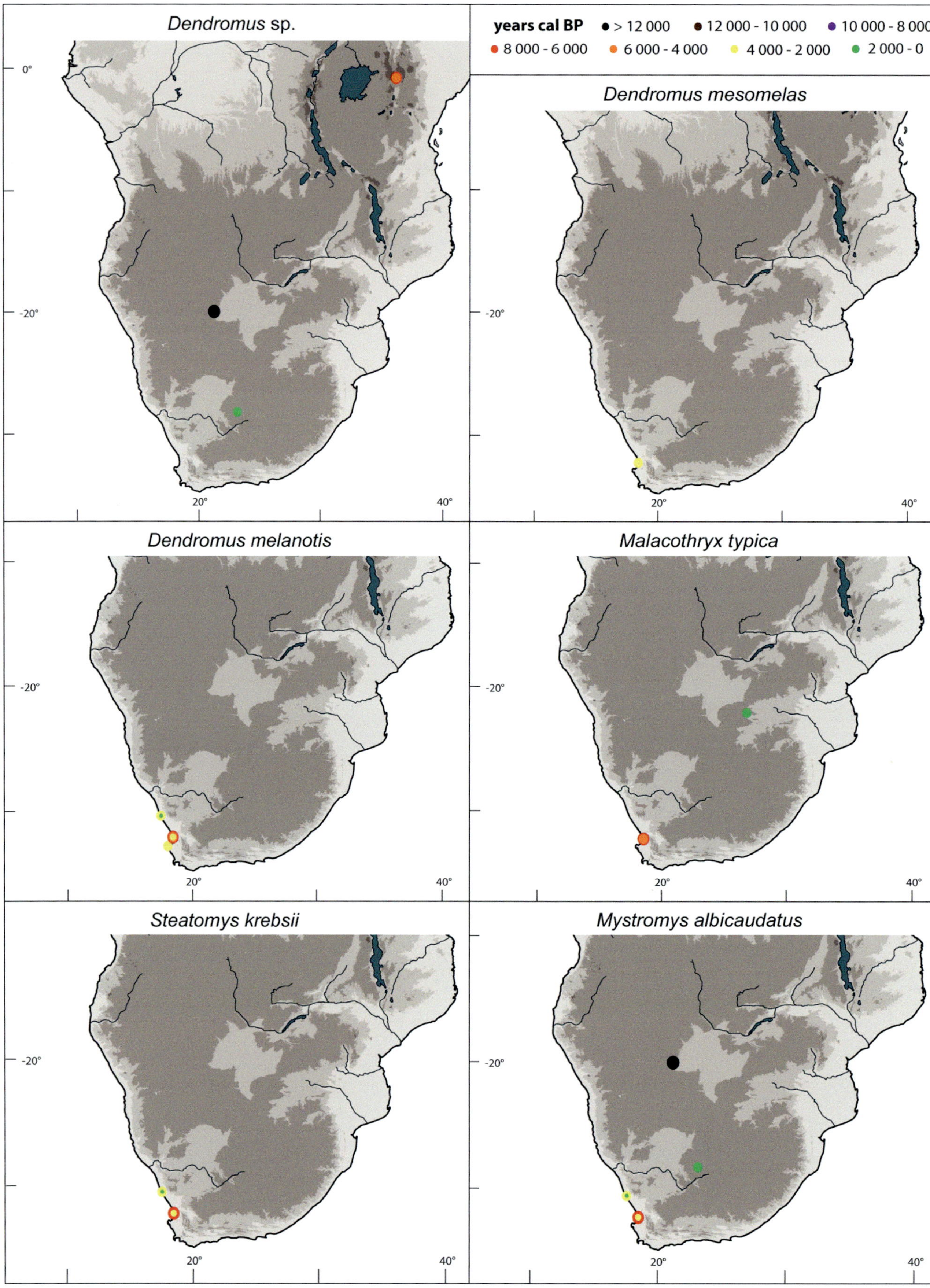
Dendromus sp.
years cal BP
> 12 000
12 000 - 10 000
10 000 - 8 000
8 000 - 6 000
6 000 - 4 000
4 000 - 2 000
2 000 - 0
Dendromus mesomelas
Dendromus melanotis
Malacothryx typica
Steatomys krebsii
Mystromys albicaudatus
0°
-20°
20°
40°

Muridae

Deomyinae

Acomys sp.

Country	Site N	Site
Egypt	135	Elephantine-necropolis
Kenya	251	Crescent Island main
	254	Ele Bor A-B
	298	Prolonged Drift

Acomys cahirinus

Country	Site N	Site
Burkina Faso	62	Péntènga 89/1-spits 4-10 *intr
Egypt	114	Buto-9-12
Mali	361	Akumbu-EIA *cf.

Lophuromys flavopunctatus/sikapusi

Country	Site N	Site
Kenya	258	Enkapune Ya Muto-Eburran 4a
	259	Enkapune Ya Muto-Eburran 4b
	260	Enkapune Ya Muto-Eburran 5a

Nesokia indica

Country	Site N	Site
Egypt	112	Buto/1-3
	113	Buto-6
	162	Kom Ombo
	168	Merimde Benisâlame
	201	Wadi Kubbaniya - E-78-3
	203	Wadi Kubbaniya - E-81-1

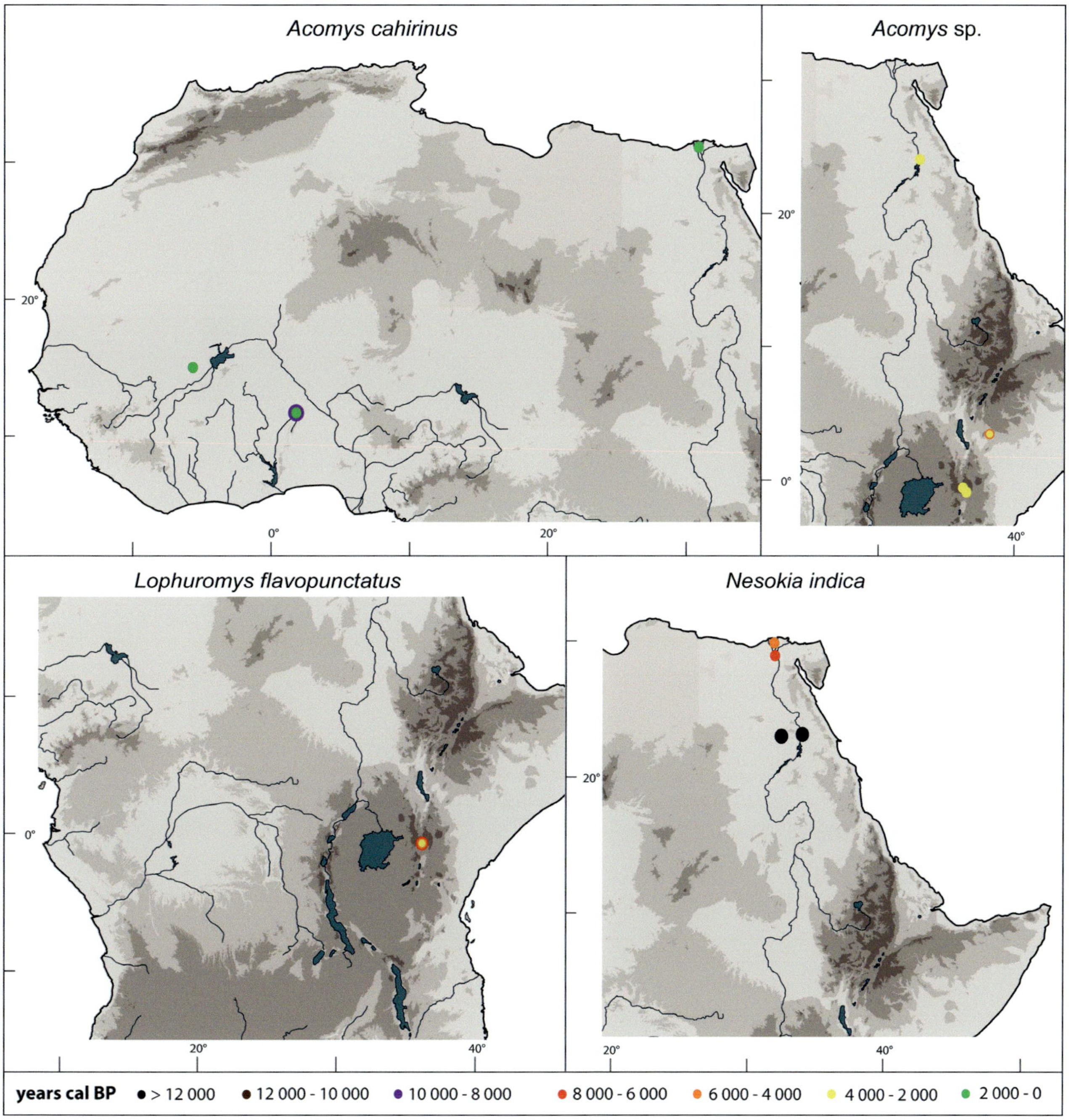

Gerbillinae

Tatera sp.

Country	Site N	Site
Algeria	21	Ti-n-Hanakaten
	5	Amekni
Botswana	33	Drotsky's Cave-H
	34	Drotsky's Cave-LP
	35	Drotsky's Cave-P
Burkina Faso	52	Kissi 22
	62	Péntènga 89/1-spits 4-10 *intr
Egypt	193	Tree Shelter-3
Ghana	227	Asantamanso-ASV
Mauritania	397	Dhar Tichitt- lowest beach
	398	Dhar Tichitt- middle beach
	414	Tichitt - Goungou phase
	417	Tichitt - Nkhal phase
Namibia	455	Geduld-LIA
	456	Geduld-LSA
South Africa	589	Doornfontein
Sudan	783	Kadero
	788	Khasm el Girba - Kassala
	789	Khasm el Girba - KG14
	805	Saggai 1
	812	Shaqaduq-S1-B
Zaire	848	Kiantapo

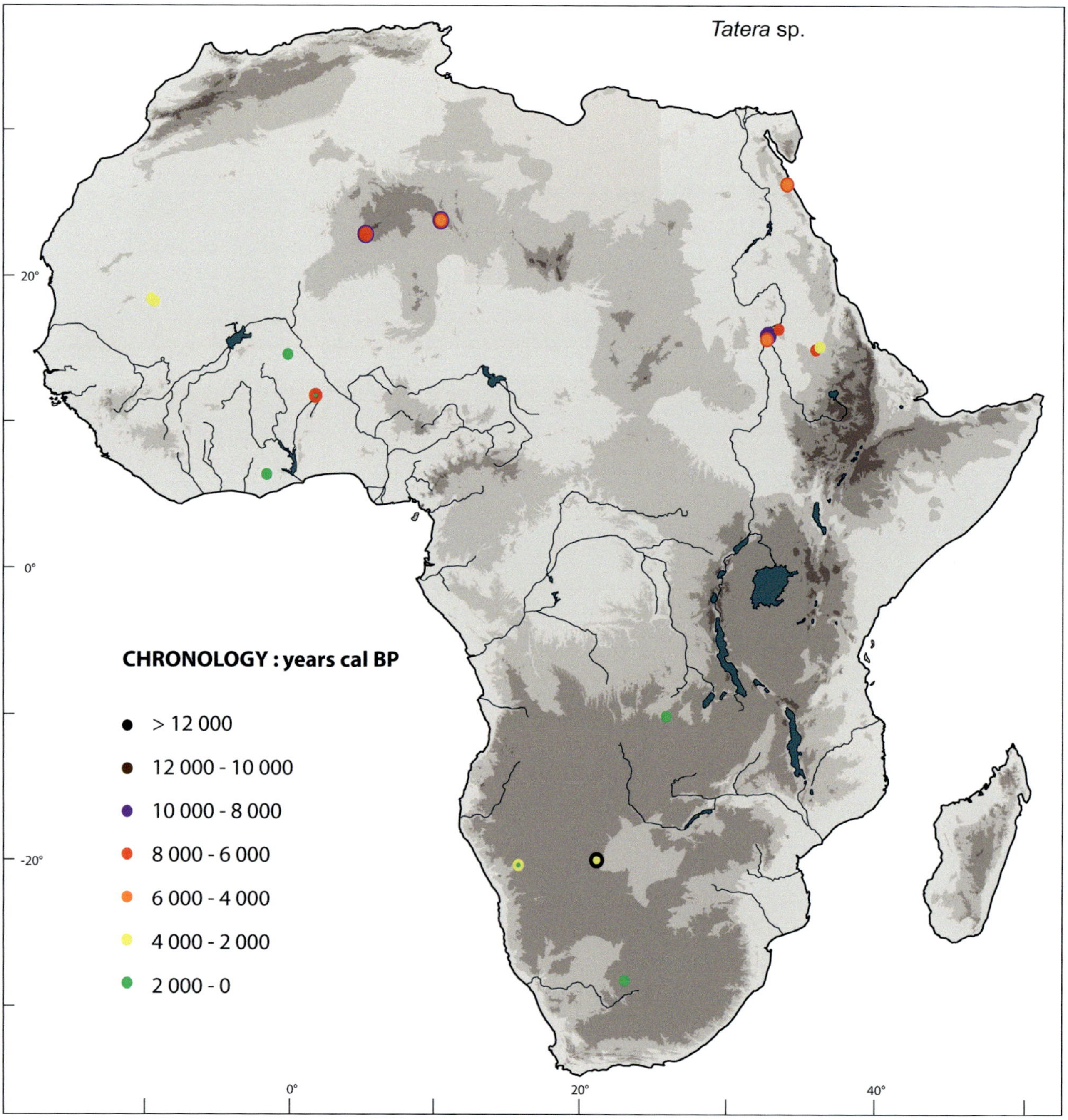

Tatera leucogaster

Country	Site N	Site	
Botswana	34	Drotsky's Cave-LP	*cf.
	48	White Paintings RS-2ab	*cf.
	48	White Paintings RS-1	*cf.
	49	White Paintings RS-2b-3ab	*
South Africa	589	Doornfontein	
	720	Sonkoanini Hill	

Tatera afra

Country	Site N	Site
South Africa	725	Steenbokfontein-1
	726	Steenbokfontein-2
	727	Steenbokfontein-3
	728	Steenbokfontein-4a
	729	Steenbokfontein-4b

Tatera nigricaudata

Country	Site N	Site
Kenya	293	Naivasha Railway Shelter

Tatera brantsii

Country	Site N	Site	
Botswana	33	Drotsky's Cave-H	*cf.
	34	Drotsky's Cave-LP	*cf.
	35	Drotsky's Cave-P	*cf.
South Africa	589	Doornfontein	

Taterillus sp.

Country	Site N	Site
Kenya	253	Ele Bor A-A
	254	Ele Bor A-B
	255	Ele Bor A-C
	256	Ele Bor A-D

Desmodilliscus braueri

Country	Site N	Site
Burkina Faso	63	Saouga 94/120

Dipodillus campestris

Country	Site N	Site
Morocco	435	Kaf-taht-el-Ghar-G
	437	Kehf-el-Baroud

Gerbilliscus validus

Country	Site N	Site
Malawi	351	Malowa-1
	352	Malowa-2
	353	Malowa-3
	354	Malowa-4

Gerbilliscus robustus

Country	Site N	Site
Sudan	765	El Kadada

Gerbillurus paeba

Country	Site N	Site	
Botswana	33	Drotsky's Cave-H	
	34	Drotsky's Cave-LP	
	35	Drotsky's Cave-P	
	48	White Paintings RS-1	
	49	White Paintings RS-2b-3ab	
Namibia	457	Haalenberg	*cf.
South Africa	589	Doornfontein	
	641	Le6	
	721	Spoegrivier-a	
	725	Steenbokfontein-1	
	726	Steenbokfontein-2	
	727	Steenbokfontein-3	
	728	Steenbokfontein-4a	
	729	Steenbokfontein-4b	

Gerbillus sp.

Country	Site N	Site	
Algeria	21	Ti-n-Hanakaten	
	22	Zmeilet el-Barka	
	5	Amekni	
	9	Dra-Mta-el-Ma-el-Abiod	
Burkina Faso	58	Oursi-IA	
	62	Péntènga 89/1-spits 4-10	*
Egypt	100	Adaima-habitat	*intr
	129	El Abadiya 2 North sector	
	130	El Abadiya 2 South sector	
	168	Merimde Benisâlame	
	98	Adaima-burials	
	99	Adaima-habitat	
Mauritania	396	Dhar Tichitt- latest phase	
	399	Dhar Tichitt- post beach	
Morocco	428	Ghar-Khal-Ib	
	436	Kehf el Hammar	
Namibia	455	Geduld-LIA	
	456	Geduld-LSA	

Gerbillus mauritaniae

Country	Site N	Site
Algeria	9	Dra-Mta-el-Ma-el-Abiod

Gerbillus gerbillus

Country	Site N	Site	
Egypt	109	Bir Kiseiba-EN	*intr

Gerbillus pyramidum

Country	Site N	Site	
Egypt	128	Ein el Gazzareen	**G.p.gedeedus*
Sudan	765	El Kadada	

Gerbillus nanus

Country	Site N	Site
Libya	335	Ti-n-Thora Two Caves

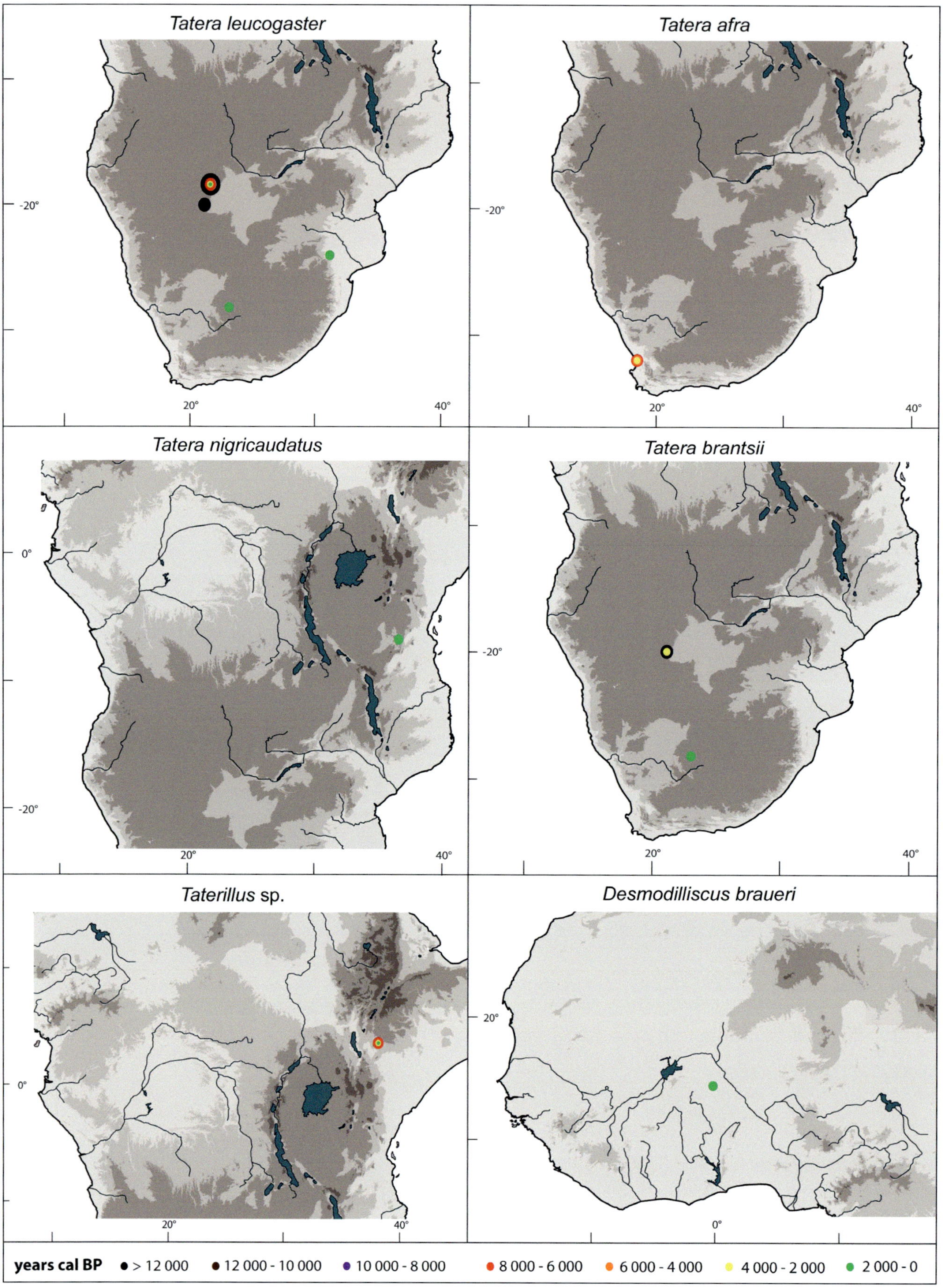
Tatera leucogaster
-20°
20°
40°
Tatera afra
-20°
20°
40°
Tatera nigricaudatus
0°
-20°
20°
40°
Tatera brantsii
-20°
20°
40°
Taterillus sp.
0°
20°
40°
Desmodilliscus braueri
20°
0°
years cal BP
> 12 000
12 000 - 10 000
10 000 - 8 000
8 000 - 6 000
6 000 - 4 000
4 000 - 2 000
2 000 - 0

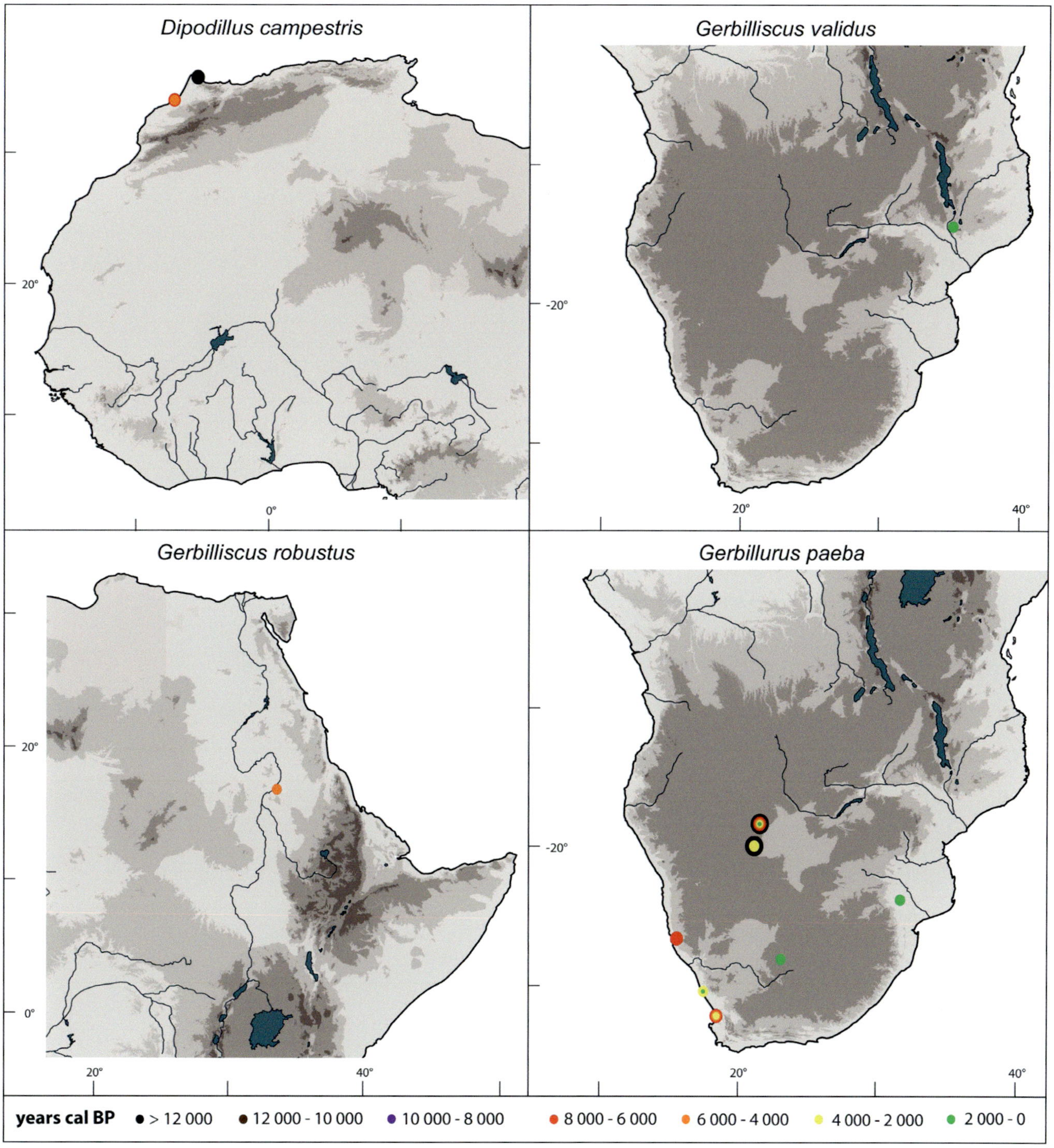

Dipodillus campestris
Gerbilliscus validus
Gerbilliscus robustus
Gerbillurus paeba
20°
0°
-20°
40°
years cal BP
> 12 000
12 000 - 10 000
10 000 - 8 000
8 000 - 6 000
6 000 - 4 000
4 000 - 2 000
2 000 - 0

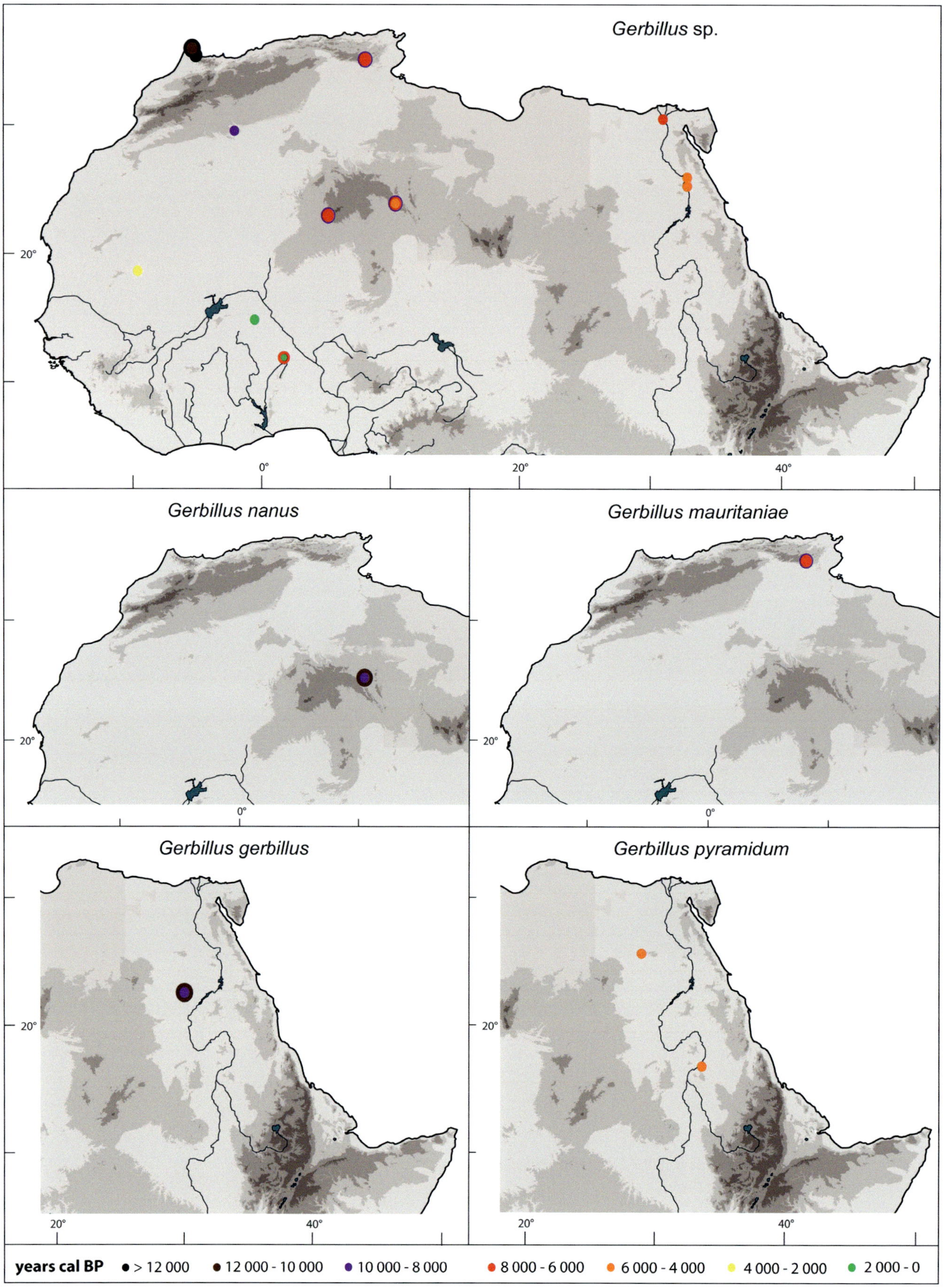
Gerbillus sp.
20°
0°
20°
40°
Gerbillus nanus
20°
0°
Gerbillus mauritaniae
20°
0°
Gerbillus gerbillus
20°
20°
40°
Gerbillus pyramidum
20°
20°
40°
years cal BP
> 12 000
12 000 - 10 000
10 000 - 8 000
8 000 - 6 000
6 000 - 4 000
4 000 - 2 000
2 000 - 0

Meriones sp.

Country	Site N	Site
Algeria	22	Zmeilet el-Barka
Egypt	112	Buto/1-3
	168	Merimde Benisâlame
Morocco	436	Kehf el Hammar

Meriones crassus

Country	Site N	Site	
Sudan	825	Wadi Hariq 01/4	*cf.

Meriones shawi

Country	Site N	Site
Morocco	435	Kaf-taht-el-Ghar-G
	437	Kehf-el-Baroud
	438	Kehf-el-Baroud-aven

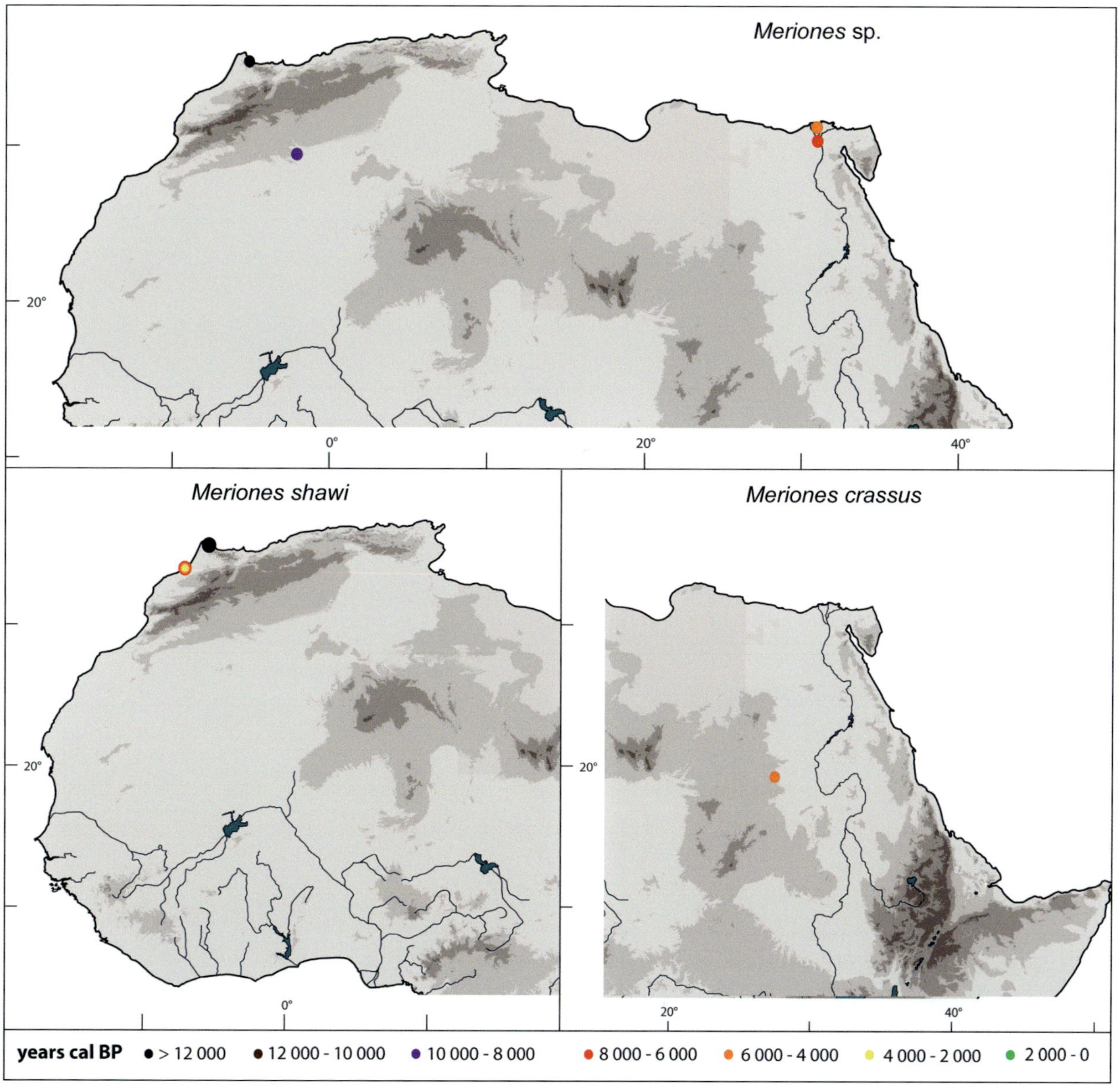

Murinae

Praomys sp.

Country	Site N	Site
Sudan	775	Gash Delta-K
	787	Khartoum Hopital
Zaire	858	Sanga

Praomys natalensis

Country	Site N	Site
Lesotho	308	Likoaeng-VII-IX
South Africa	650	Magogo-1
	651	Magogo-2

Praomys jacksoni/natalensis

Country	Site N	Site
Kenya	259	Enkapune Ya Muto-Eburran 4b
	260	Enkapune Ya Muto Eburran 5a

Apodemus sylvaticus

Country	Site N	Site
Morocco	420	Bou Saria B
	431	Kaf-taht-el-Ghar-C
	437	Kehf-el-Baroud

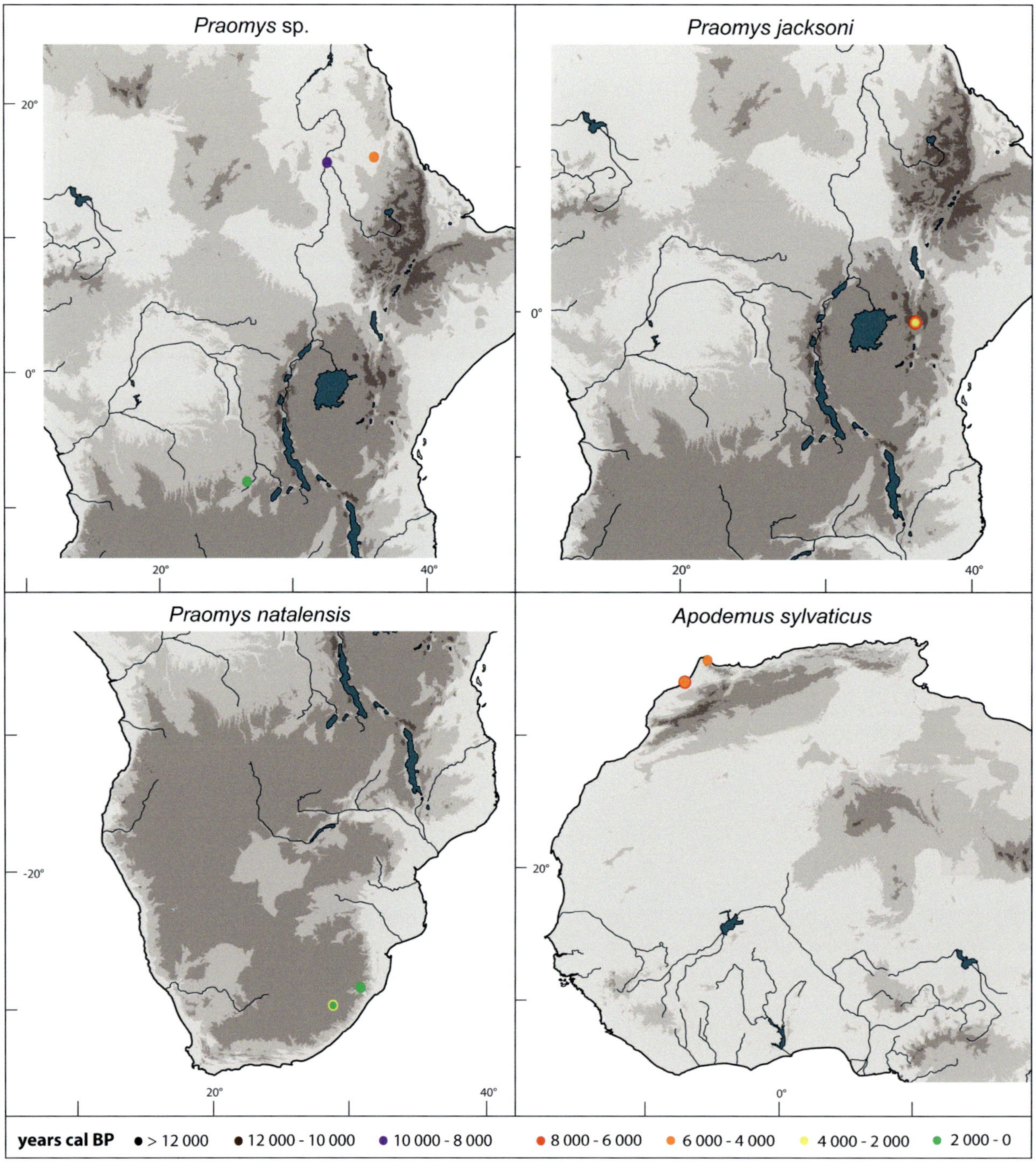

Aethomys sp.

Country	Site N	Site
Botswana	28	Bosutwe-Toutswe
Namibia	455	Geduld-LIA
Zaire	847	Kamilamba

Aethomys kaiseri

Country	Site N	Site
Zaire	848	Kiantapo

Aethomys chrysophilus

Country	Site N	Site
South Africa	620	HKLP Ib
	651	Magogo-2

Aethomys namaquensis

Country	Site N	Site	
Namibia	452	Brandberg-LSA	
	453	Brandberg-MSA	
South Africa	589	Doornfontein	*?
	721	Spoegrivier-a	
	725	Steenbokfontein-1	
	726	Steenbokfontein-2	
	727	Steenbokfontein-3	
	729	Steenbokfontein-4b	

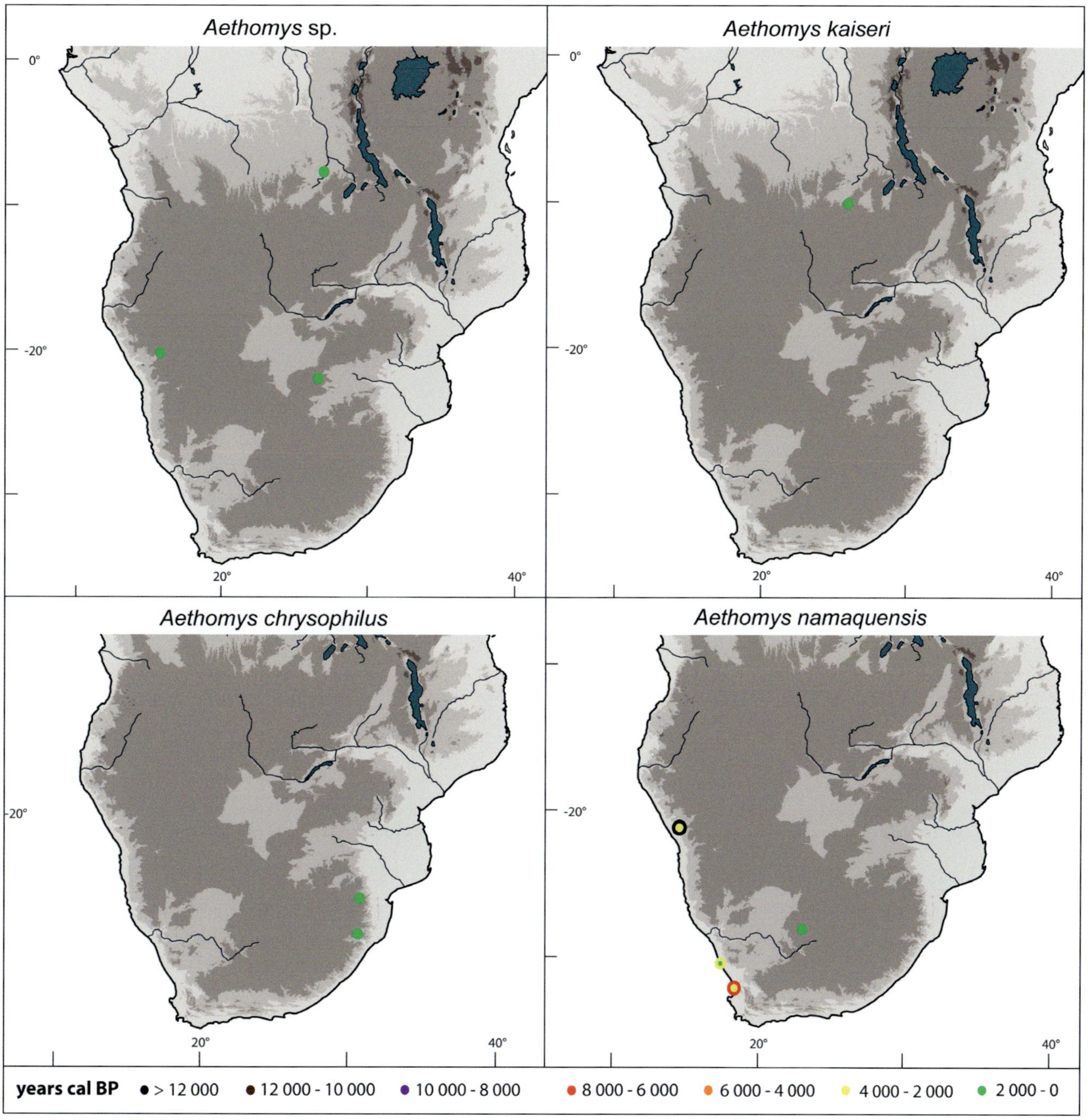

Arvicanthis niloticus

Country	Site N	Site	
Algeria	21	Ti-n-Hanakaten	
	5	Amekni	
Burkina Faso	53	Kissi 40	
	59	Oursi-LSA	
	63	Saouga 94/120	
	64	Saouga 95/7	
Egypt	100	Adaima-habitat	*intr.
	109	Bir Kiseiba-EN	
	111	Bir Kiseiba-MN	
	114	Buto-9-12	
	129	El Abadiya 2 North sector	
	130	El Abadiya 2 South sector	
	135	Elephantine-necropolis	
	137	Elephantine-Satet temple	
	164	Maadi	
	165	Mahgar Dendera 2	*intr.
	168	Merimde Benisâlame	
	196	Tuna el-Gebel	
	197	Umm el-Qaab, Friedhof U	
	98	Adaima-burials	
	99	Adaima-habitat	
Kenya	259	Enkapune Ya Muto-Eburran 4b	
	260	Enkapune Ya Muto-Eburran 5a	
Libya	333	Ti-n-Thora East	
Nigeria	502	Elkido North	
	503	Gajiganna A	
	509	Gilgila-lib	
	514	Kursakata-IA	
	516	Labe Kanuri	
	525	Ngala I	
	526	Ngala II	
	527	Ngala IIIa	
Sudan	762	Debbat El Eheima	
	765	El Kadada	
	787	Khartoum Hopital	
	812	Shaqaduq-S1-B	
	821	Upper Atbara hinterland	

Arvicanthis abyssinicus

Country	Site N	Site
Kenya	293	Naivasha Railway Shelter

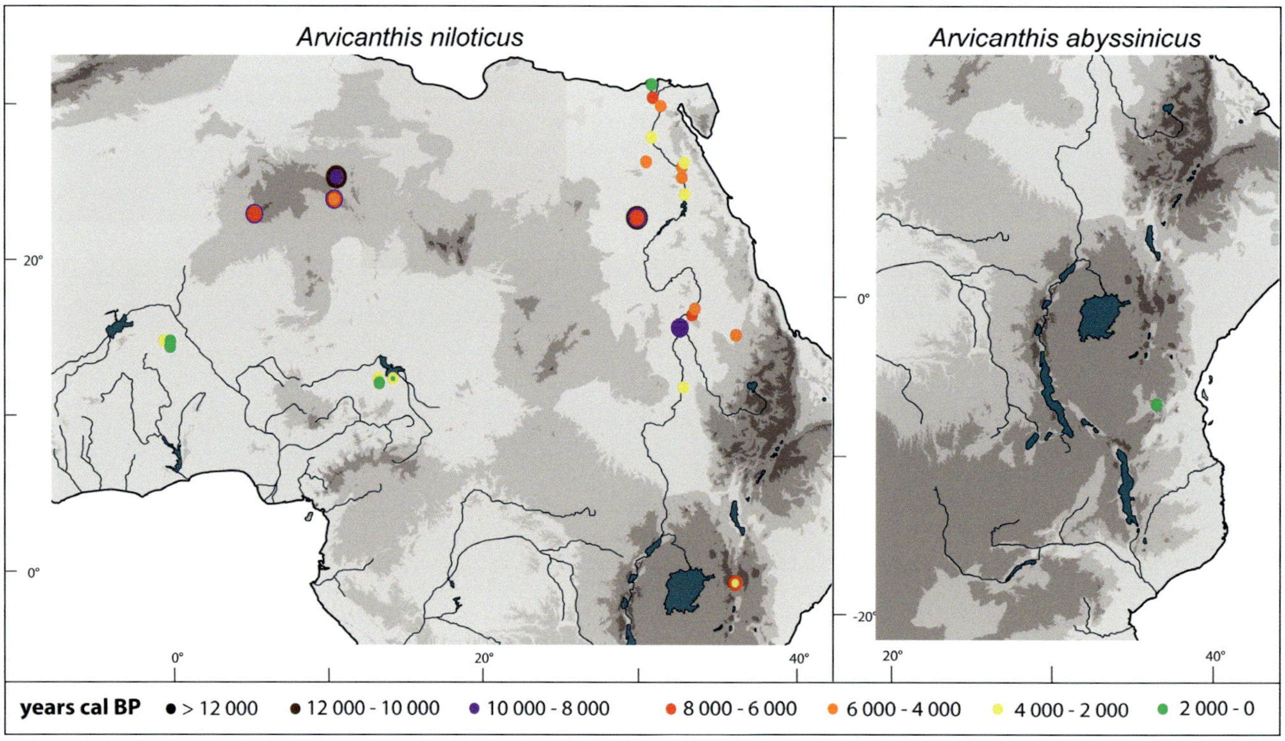

Dasymys incomtus

Country	Site N	Site	
Algeria	21	Ti-n-Hanakaten	
Kenya	258	Enkapune Ya Muto-Eburran 4a	
	259	Enkapune Ya Muto-Eburran 4b	
	260	Enkapune Ya Muto-Eburran 5a	
	261	Enkapune Ya Muto-Eburran 5b	
Rwanda	535	Akameru-Cyinkomane	*intr.
Zaire	848	Kiantapo	
	855	Matupi-II	

Lemniscomys sp.

Country	Site N	Site
Kenya	258	Enkapune Ya Muto-Eburran 4a

Lemniscomys barbarus

Country	Site N	Site
Morocco	437	Kehf-el-Baroud

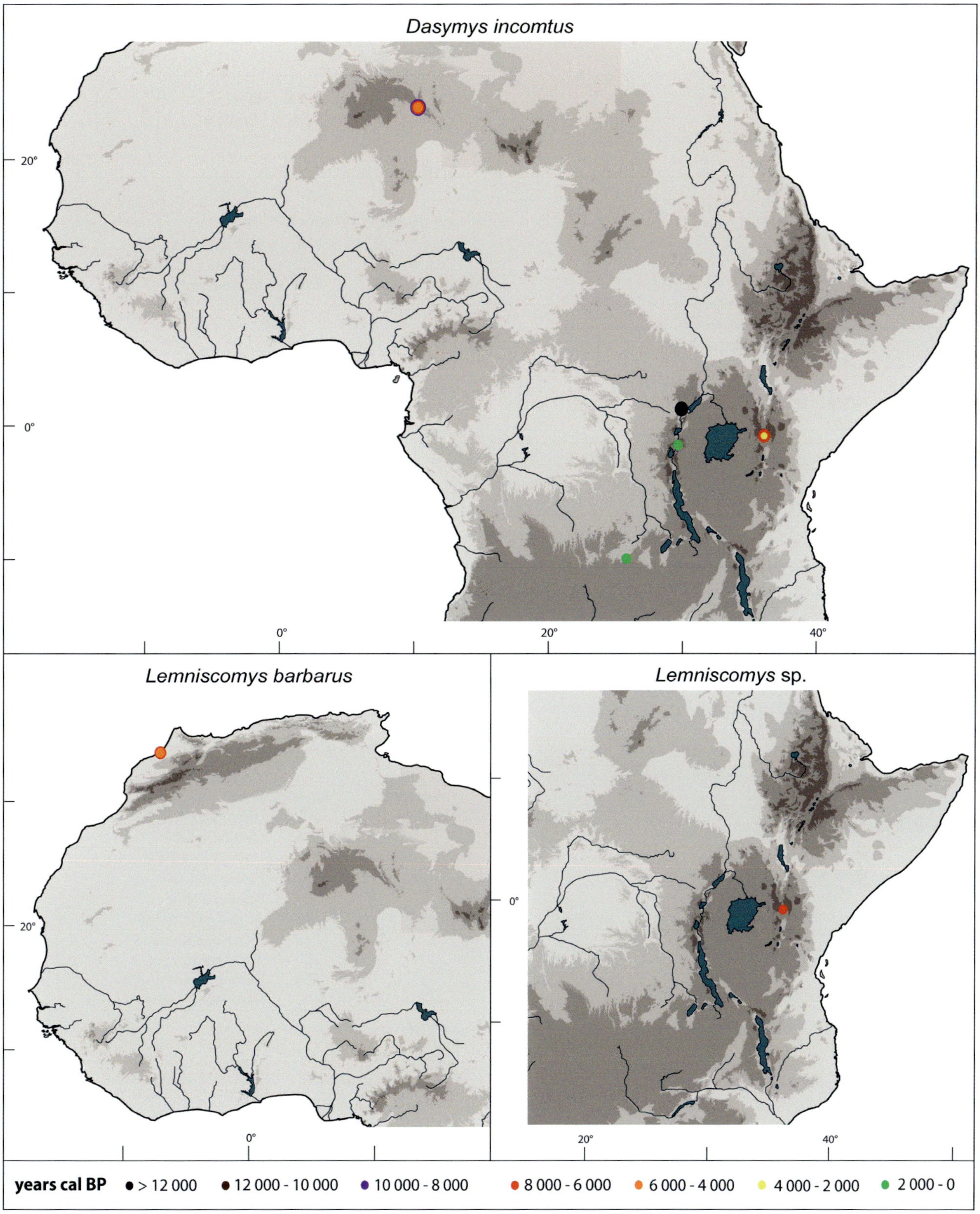
Dasymys incomtus
20°
0°
0°
20°
40°
Lemniscomys barbarus
20°
0°
Lemniscomys sp.
0°
20°
40°
years cal BP
> 12 000
12 000 - 10 000
10 000 - 8 000
8 000 - 6 000
6 000 - 4 000
4 000 - 2 000
2 000 - 0

Mastomys sp.

Country	Site N	Site
South Africa	717	SK17
Sudan	761	Debbat Bangdit

Mastomys natalensis

Country	Site N	Site
Burkina Faso	52	Kissi 22
	53	Kissi 40
	54	Oursi hu-beero
	55	Oursi Village-EIA
	57	Oursi Village-MIA
	58	Oursi-IA
	63	Saouga 94/120
Nigeria	501	Dorota
	507	Galaga
	514	Kursakata-IA
	516	Labe Kanuri
	519	Mege II
	521	Mege IV
	525	Ngala I
	526	Ngala II
	527	Ngala IIIa

Mus spretus

Country	Site N	Site
Morocco	428	Ghar-Khal-Ib
	431	Kaf-taht-el-Ghar-C
	436	Kehf el Hammar
	437	Kehf-el-Baroud
	438	Kehf-el-Baroud-aven

Mus minutoides

Country	Site N	Site
South Africa	721	Spoegrivier-a
	726	Steenbokfontein-2
	727	Steenbokfontein-3

Mus musculus

Country	Site N	Site
Egypt	128	Ein el Gazzareen **M.m.praetextus*
	168	Merimde Benisâlame

Mylomys dybowskii

Country	Site N	Site
Kenya	258	Enkapune Ya Muto-Eburran 4a
	259	Enkapune Ya Muto-Eburran 4b

Oenomys hypoxanthus

Country	Site N	Site
Kenya	258	Enkapune Ya Muto-Eburran 4a
	259	Enkapune Ya Muto-Eburran 4b

Pelomys fallax

Country	Site N	Site
Zaire	848	Kiantapo

Rattus sp.

Country	Site N	Site
Algeria	16	Meniet
Kenya	251	Crescent Island main
South Africa	661	Mgoduyanuka

Rattus rattus

Country	Site N	Site	
Botswana	26	Bosutwe-Mapungubwe	
	27	Bosutwe-Taukome	
	28	Bosutwe-Toutswe	*cf.
	29	Bosutwe-Zimbabwe	
Burundi	66	Ruzizi Plain	
Egypt	106	Berenike 95/97-Early Roman	
	107	Berenike 95/97-Late Roman	
	108	Berenike 96/97-Ptolemaic	
	179	Quseir al-Qadim	
	188	Tell el-Dab´a - b	
Kenya	293	Naivasha Railway Shelter	
South Africa	667	Ndondondwane	
	688	Pont-Drift-LKA	
	689	Pont-Drift-Zhizo	
Zimbabwe	869	Muozi	*cf.

Rhabdomys pumilio

Country	Site N	Site
Namibia	452	Brandberg-LSA
	456	Geduld-LSA
South Africa	682	Paternoster3-4
	683	Paternoster5-6
	721	Spoegrivier-a
	725	Steenbokfontein-1
	726	Steenbokfontein-2
	727	Steenbokfontein-3
	728	Steenbokfontein-4a
	729	Steenbokfontein-4b

Thallomys sp.

Country	Site N	Site
Namibia	455	Geduld-LIA
	456	Geduld-LSA

Thamnomys venustus/rutilans

Country	Site N	Site
Kenya	258	Enkapune Ya Muto-Eburran 4a
	259	Enkapune Ya Muto-Eburran 4b

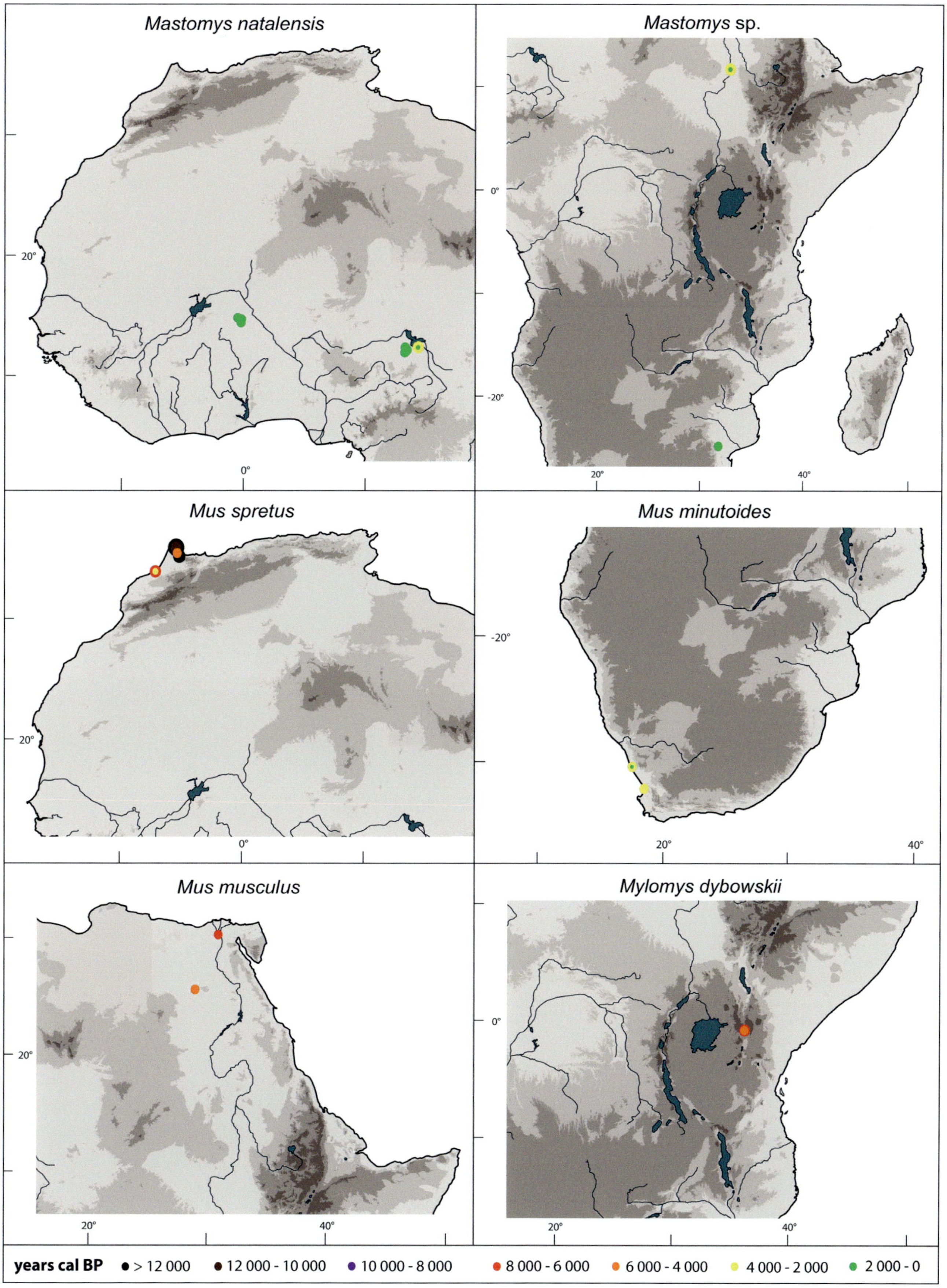

Mastomys natalensis
20°
0°
Mastomys sp.
0°
-20°
20°
40°
Mus spretus
20°
0°
Mus minutoides
-20°
20°
40°
Mus musculus
20°
20°
40°
Mylomys dybowskii
0°
20°
40°
years cal BP
> 12 000
12 000 - 10 000
10 000 - 8 000
8 000 - 6 000
6 000 - 4 000
4 000 - 2 000
2 000 - 0

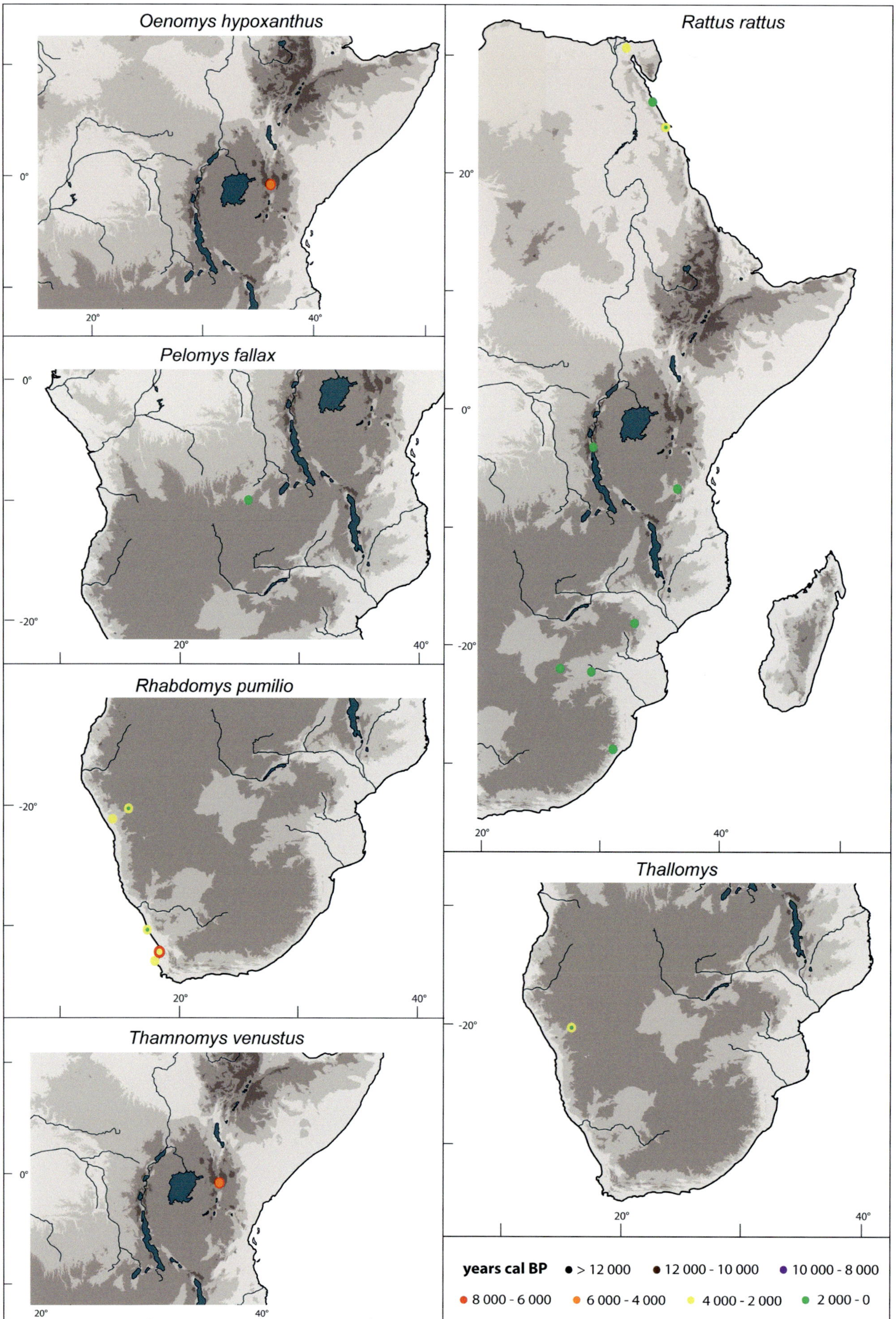
Oenomys hypoxanthus
0°
20°
40°
Pelomys fallax
0°
-20°
20°
40°
Rhabdomys pumilio
-20°
20°
40°
Thamnomys venustus
0°
20°
40°
Rattus rattus
20°
0°
-20°
20°
40°
Thallomys
-20°
20°
40°
years cal BP
> 12 000
12 000 - 10 000
10 000 - 8 000
8 000 - 6 000
6 000 - 4 000
4 000 - 2 000
2 000 - 0

Otomyinae

Otomys sp.

Country	Site N	Site
Botswana	26	Bosutwe-Mapungubwe
	28	Bosutwe-Toutswe
Kenya	253	Ele Bor A-A
	255	Ele Bor A-C
	284	Kulchurdo
	293	Naivasha Railway Shelter
Lesotho	303	Bolahla
	305	Lequetsoana
	306	Likoaeng-I
	307	Likoaeng-III-V
	308	Likoaeng-VII-IX
	309	Likoaeng-XI
Namibia	455	Geduld-LIA
	457	Haalenberg
South Africa	574	Bushman Rock Shelter-LSA
	589	Doornfontein
	660	Mauermanshoek
	691	Pr34
	710	Selongwe Hill
	713	SH16
	740	Twyfelpoort

Otomys irroratus

Country	Site N	Site	
Kenya	258	Enkapune Ya Muto-Eburran 4a	
	259	Enkapune Ya Muto-Eburran 4b	
	260	Enkapune Ya Muto-Eburran 5a	
	261	Enkapune Ya Muto-Eburran 5b	
	263	Enkapune Ya Muto BSS-HTL	
Lesotho	308	Likoaeng-VII-IX	
	310	Likoaeng-XIII	
South Africa	589	Doornfontein	
	619	HKLP Ia	*cf.
	620	HKLP Ib	*cf.
	644	Leliehoek	
	721	Spoegrivier-a	
	725	Steenbokfontein-1	
	726	Steenbokfontein-2	
	727	Steenbokfontein-3	
	728	Steenbokfontein-4a	
	729	Steenbokfontein-4b	

Otomys tropicalis

Country	Site N	Site
Malawi	351	Malowa-1
	352	Malowa-2
	353	Malowa-3

Otomys saundersiae

Country	Site N	Site
South Africa	721	Spoegrivier-a
	725	Steenbokfontein-1
	726	Steenbokfontein-2
	727	Steenbokfontein-3
	728	Steenbokfontein-4a
	729	Steenbokfontein-4b

Otomys angoniensis

Country	Site N	Site	
Botswana	34	Drotsky's Cave-LP	*cf.
	35	Drotsky's Cave-P	*cf.
	41	Toteng1-SU1	*cf.
	48	White Paintings RS-1	*cf.
	48	White Paintings RS-2ab	*cf.
	49	White Paintings RS-2b-3ab	*
South Africa	713	SH16	

Otomys unisulcatus

Country	Site N	Site
South Africa	634	Klein Kliphuis
	682	Paternoster3-4
	683	Paternoster5-6

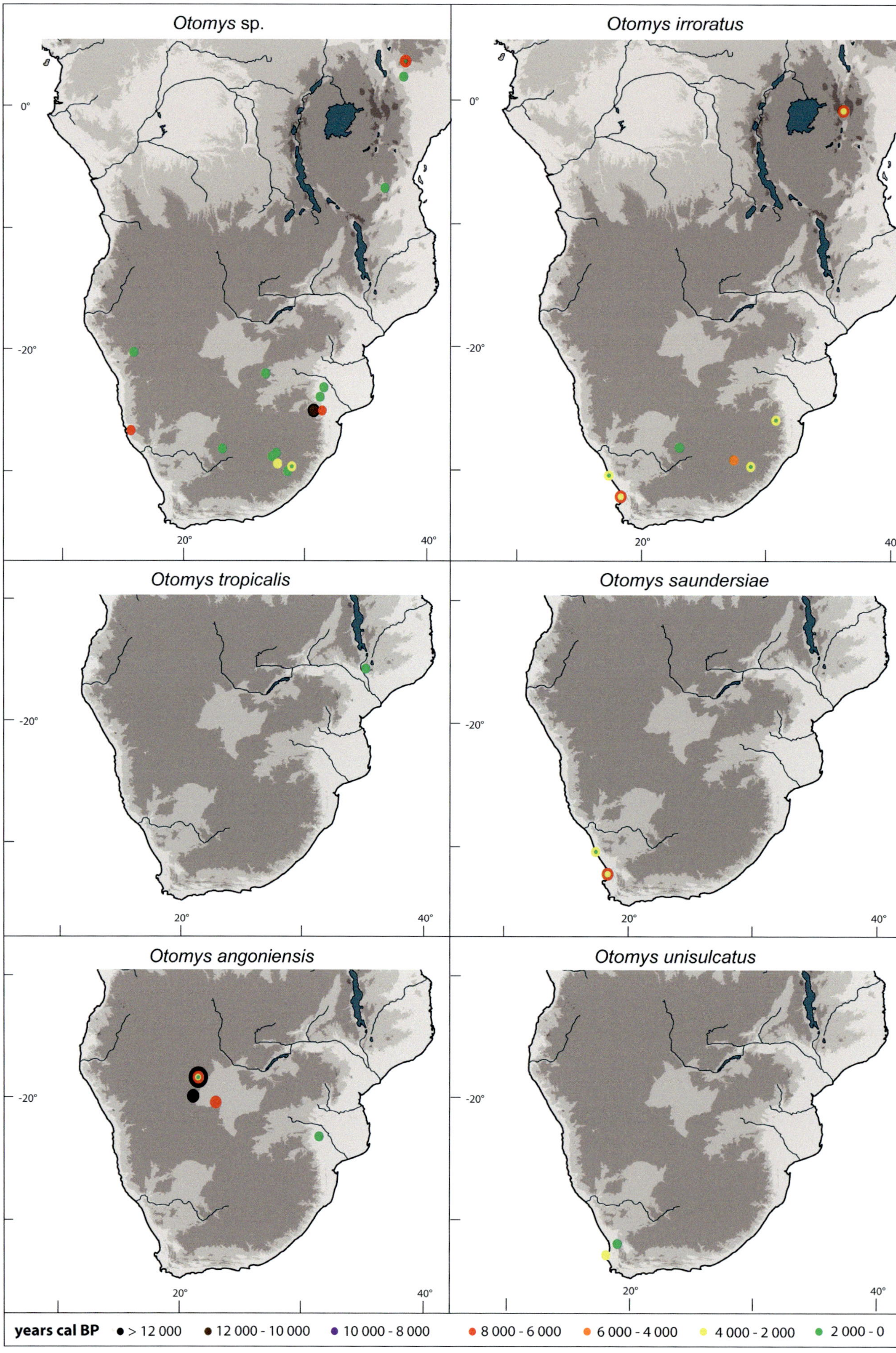
Otomys sp.
Otomys irroratus
Otomys tropicalis
Otomys saundersiae
Otomys angoniensis
Otomys unisulcatus
0°
-20°
20°
40°
years cal BP
> 12 000
12 000 - 10 000
10 000 - 8 000
8 000 - 6 000
6 000 - 4 000
4 000 - 2 000
2 000 - 0

Myotomys unisulcatus

Country	Site N	Site
South Africa	721	Spoegrivier-a
	725	Steenbokfontein-1
	726	Steenbokfontein-2
	727	Steenbokfontein-3
	728	Steenbokfontein-4a
	729	Steenbokfontein-4b

Parotomys brantsii

Country	Site N	Site
South Africa	721	Spoegrivier-a
	726	Steenbokfontein-2

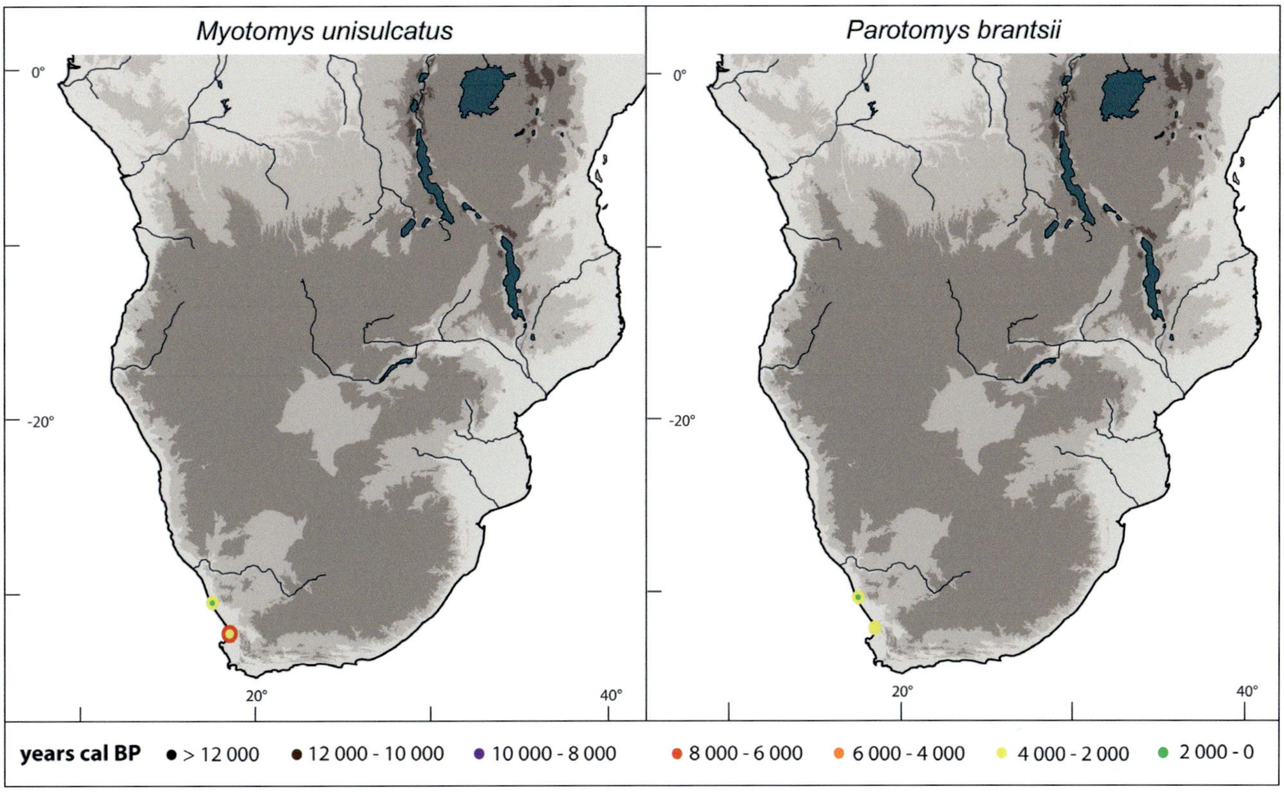

Spalacidae

Tachyoryctes sp.

Country	Site N	Site	
Kenya	257	Engaruani	*cf.
	301	Salasun-PN	

Tachyoryctes splendens

Country	Site N	Site
Kenya	250	Causeway Site
	251	Crescent Island main
	254	Ele Bor A-B
	258	Enkapune Ya Muto-Eburran 4a
	259	Enkapune Ya Muto-Eburran 4b
	260	Enkapune Ya Muto-Eburran 5a
	262	Enkapune Ya Muto-Eburran 5c
	282	GvJm 62-LPG
	292	Marula Rock Shelter
	293	Naivasha Railway Shelter
	295	Ndabibi
	297	Ngenyn
	298	Prolonged Drift
	299	Salasun-LSA
	300	Salasun-PIA
Malawi	352	Malowa-2

Tachyoryctes ruandae

Country	Site N	Site	
Rwanda	535	Akameru-Cyinkomane	*intr.

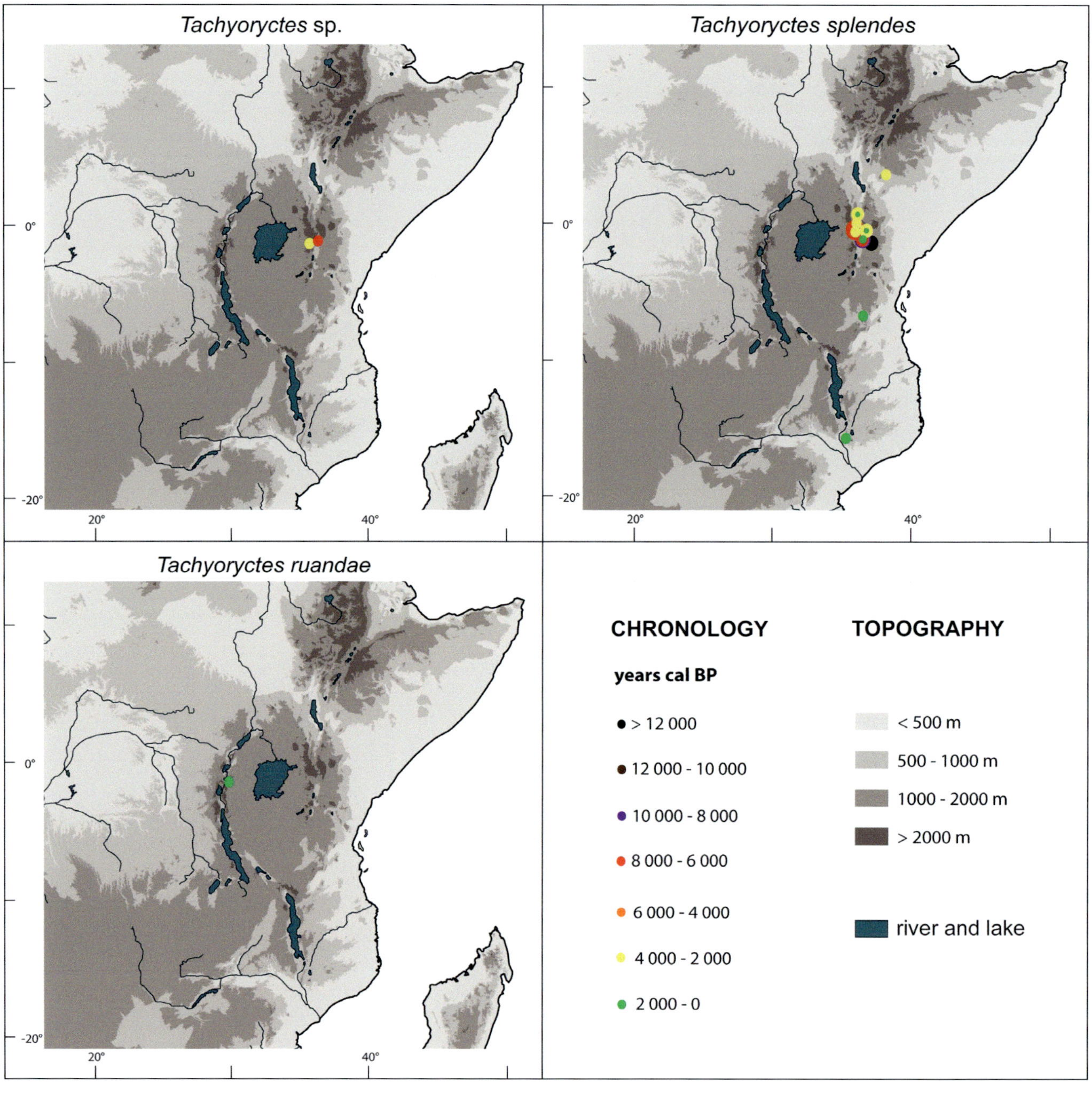

Pedetidae

Pedetes capensis

Country	Site N	Site	
Botswana	26	Bosutwe-Mapungubwe	
	27	Bosutwe-Taukome	
	28	Bosutwe-Toutswe	
	29	Bosutwe-Zimbabwe	
	31	Commando Kop	
	33	Drotsky's Cave-H	
	34	Drotsky's Cave-LP	
	35	Drotsky's Cave-P	
	41	Toteng1-SU1	
	42	Toteng1-SU2	
	43	Toteng1-SU3	
	48	White Paintings RS-2ab	
	48	White Paintings RS-1	
	49	White Paintings RS-2b-3ab	
Kenya	257	Engaruani	* cf., intr.
	275	GvJm 19-HA	
	276	GvJm 19-HC	
	277	GvJm 19-TLPG	
	281	GvJm 62-H	
	282	GvJm 62-LPG	
	283	GvJm 62-LPNG	
Lesotho	304	Ha Makotoko	
	305	Lequetsoana	
	310	Likoaeng-XIII	
	316	Sehonghong-a	
	321	Tloutle-CWI	
South Africa	585	Deelpan	
	604	Equus Cave 1A	
	610	Green Elliott Shelter I-III	
	611	Green Elliott Shelter IV-VI	
	619	HKLP Ia	
	620	HKLP Ib	
	622	HKLP Vb	
	640	Le2	
	641	Le6	
	644	Leliehoek	
	645	Limerock 1	
	646	Limerock 2	
	663	Mhlwazini Cave	
	673	Ol20	
	674	Ondini	
	690	Powerhouse Cave	
	691	Pr34	
	697	Riversmead Shelter I-II	
	698	Riversmead Shelter III-IV	
	699	Riversmead Shelter VI-IX	
	701	Rooikrans-CW	
	702	Rooikrans-pCW	
	704	Rose Cottage Cave-A-Mn	
	705	Rose Cottage Cave-DB2	
	718	Sk4	
	740	Twyfelpoort	
Tanzania	835	Mumba	
Zimbabwe	870	Redcliff Cave	

Pedetes surdaster

Country	Site N	Site
Kenya	295	Ndabibi

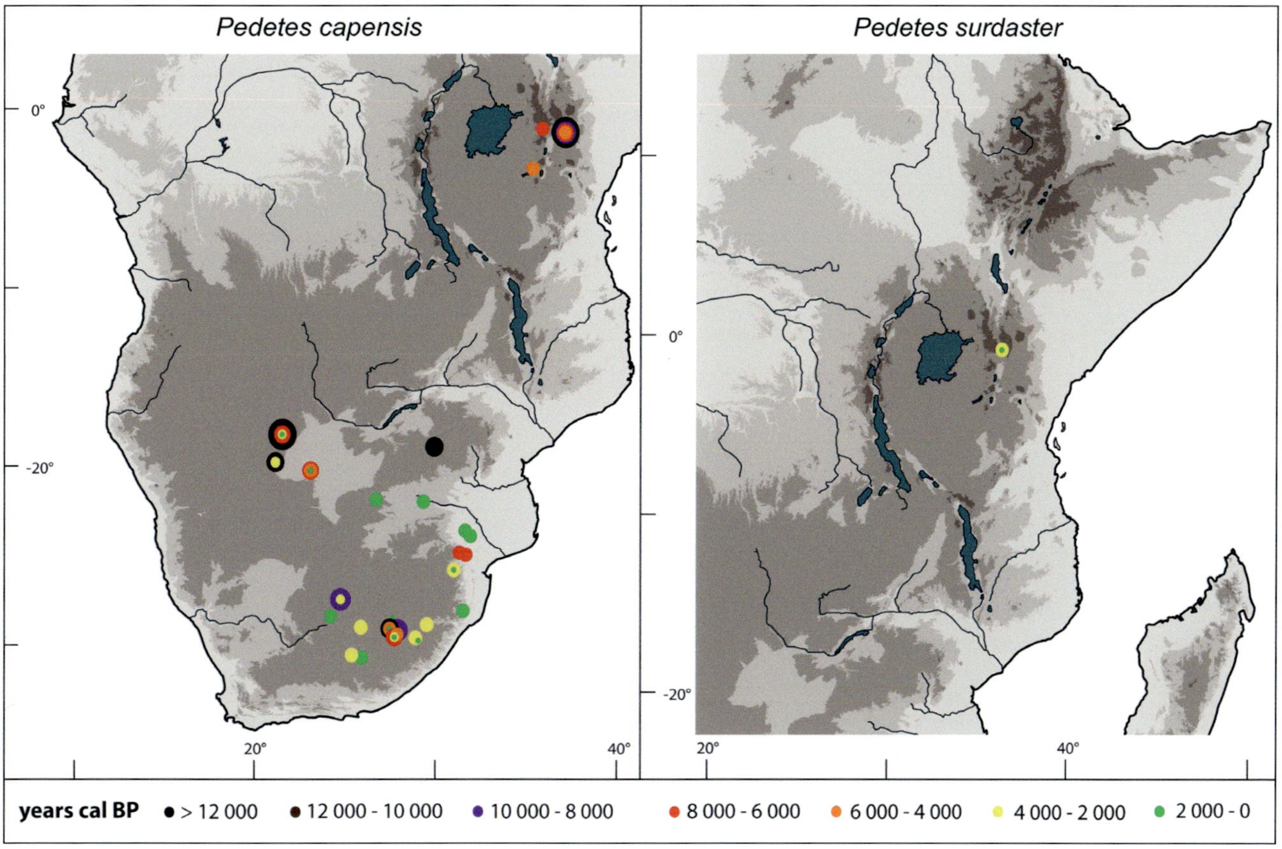

Ctenodactylidae

Ctenodactylus gundi

Country	Site N	Site
Algeria	21	Ti-n-Hanakaten

Massoutiera mzabi

Country	Site N	Site
Libya	333	Ti-n-Thora East
	334	Ti-n-Thora North
	335	Ti-n-Thora Two Caves

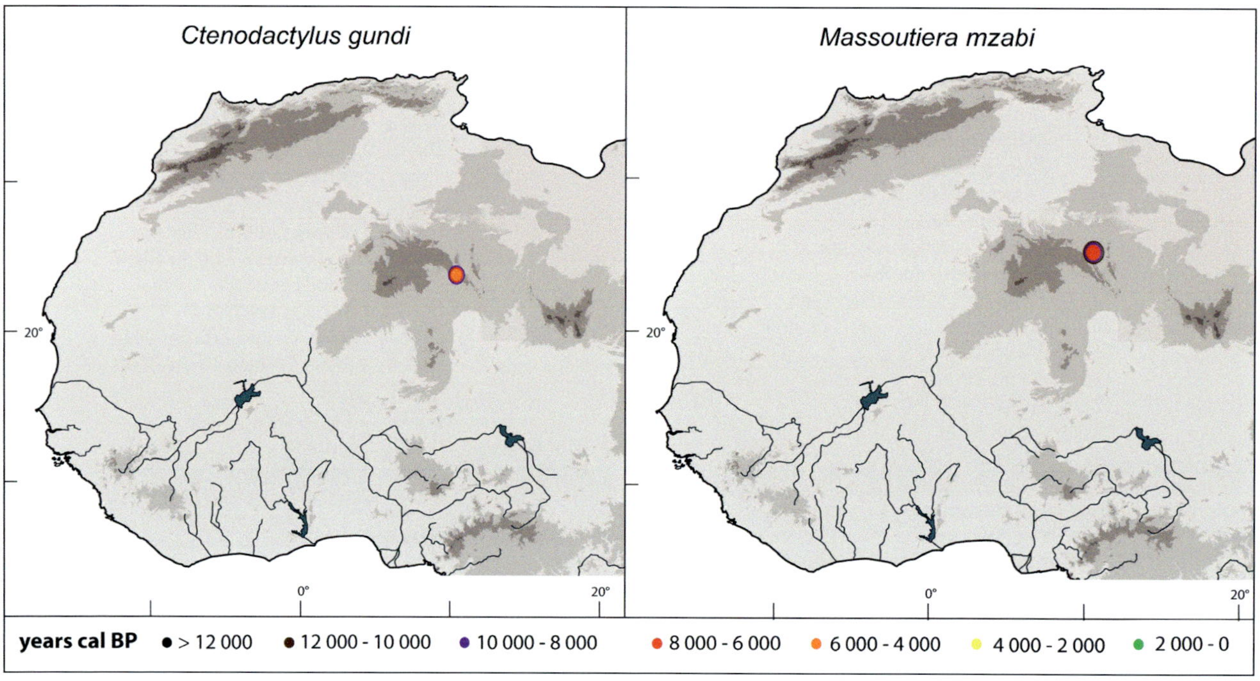

Bathyergidae

Bathyergus janetta

Country	Site N	Site
South Africa	721	Spoegrivier-a

Bathyergus suillus

Country	Site N	Site
South Africa	558	Blombos Cave
	576	Byeneskranskop Cave 1a
	577	Byeneskranskop Cave 1b
	578	Byeneskranskop Cave 1c
	579	Byeneskranskop Cave 1d
	586	Die Kelders
	591	Drie Susters Main
	595	Eland´s Bay Cave-AI
	596	Eland´s Bay Cave-RI
	597	Eland´s Bay Cave-Wia
	598	Eland´s Bay Cave-Wib
	599	Eland´s Bay Cave-Wic
	600	Eland´s Bay Cave-Wid
	601	Eland´s Bay Open-a
	602	Eland´s Bay Open-b
	627	Kasteelberg A
	628	Kasteelberg B
	634	Klein Kliphuis
	637	Kreeftebaai
	676	Oudepost I
	679	Pancho's Kitchen Midden 3-4
	680	Pancho's Kitchen Midden 6-7
	682	Paternoster3-4
	683	Paternoster5-6
	719	Smitswinkelbaai Cave
	724	Spring Cave
	725	Steenbokfontein-1
	726	Steenbokfontein-2
	727	Steenbokfontein-3
	729	Steenbokfontein-4b
	735	Tortoise cave-a
	736	Tortoise cave-b
	737	Tortoise cave-c
	738	Tortoise cave-d
	744	Voëlvlei
	747	Witklip-1
	748	Witklip-2
	749	Witklip-3
	750	Witklip-4

Cryptomys sp.

Country	Site N	Site	
Botswana	28	Bosutwe-Toutswe	*cf.
Malawi	357	Matope Court-IA	
South Africa	589	Doornfontein	
Zaire	846	Kakontwe	

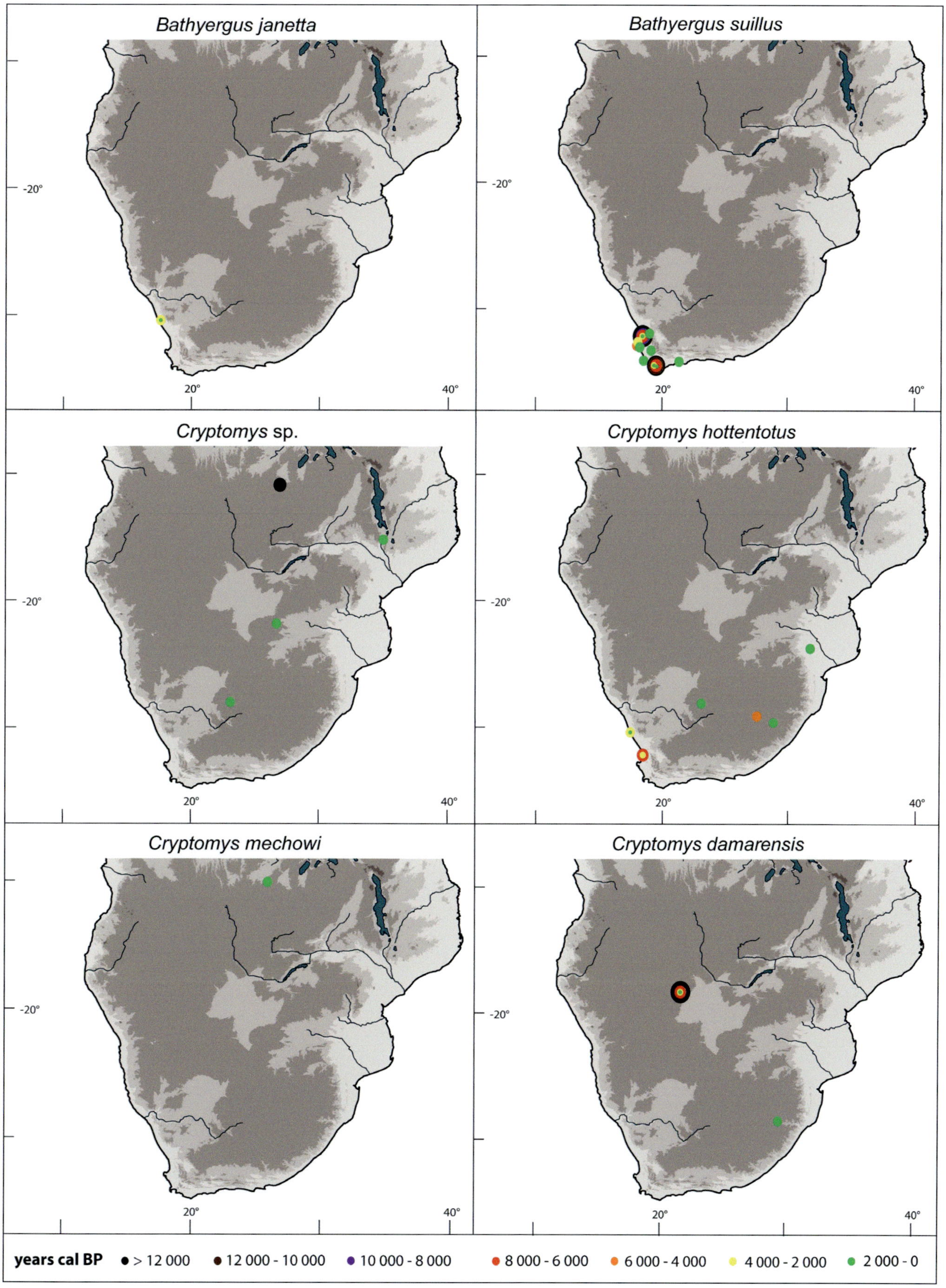
Bathyergus janetta
Bathyergus suillus
Cryptomys sp.
Cryptomys hottentotus
Cryptomys mechowi
Cryptomys damarensis
-20°
20°
40°
years cal BP
> 12 000
12 000 - 10 000
10 000 - 8 000
8 000 - 6 000
6 000 - 4 000
4 000 - 2 000
2 000 - 0

Cryptomys hottentotus

Country	Site N	Site	
Lesotho	306	Likoaeng-I	*cf.
South Africa	589	Doornfontein	
	642	Le7a	
	644	Leliehoek	
	721	Spoegrivier-a	
	725	Steenbokfontein-1	
	726	Steenbokfontein-2	
	727	Steenbokfontein-3	
	729	Steenbokfontein-4b	

Cryptomys mechowi

Country	Site N	Site	
Zaire	848	Kiantapo	*?
	848	Kiantapo	

Cryptomys damarensis

Country	Site N	Site
Botswana	48	White Paintings RS-1
	48	White Paintings RS-2ab
	49	White Paintings RS-2b-3ab

Cryptomys natalensis

Country	Site N	Site
South Africa	661	Mgoduyanuka

Hystricidae

Atherurus africanus

Country	Site N	Site
C.A.R.	79	Mondongué
Congo	87	Ntadi Yom ba-level B
Ghana	223	Anyinam-Area A-B-C
	225	Asantamanso-ASII/III
	227	Asantamanso-ASV
	232	Dawu
	233	Esiease - A-C-D
Zaire	842	Dimba
	854	Matupi-I
	855	Matupi-II

Hystrix sp.

Country	Site N	Site
Algeria	11	Gueldaman Akbou
	15	Medjez II
	9	Dra-Mta-el-Ma-el-Abiod
Kenya	259	Enkapune Ya Muto-Eburran 4b
South Africa	610	Green Elliott Shelter I-III
	611	Green Elliott Shelter IV-VI
Sudan	755	Abu Hugar
	783	Kadero
Tanzania	835	Mumba
Zaire	845	Ishango-Z-Post-Em
	854	Matupi-I
	855	Matupi-II

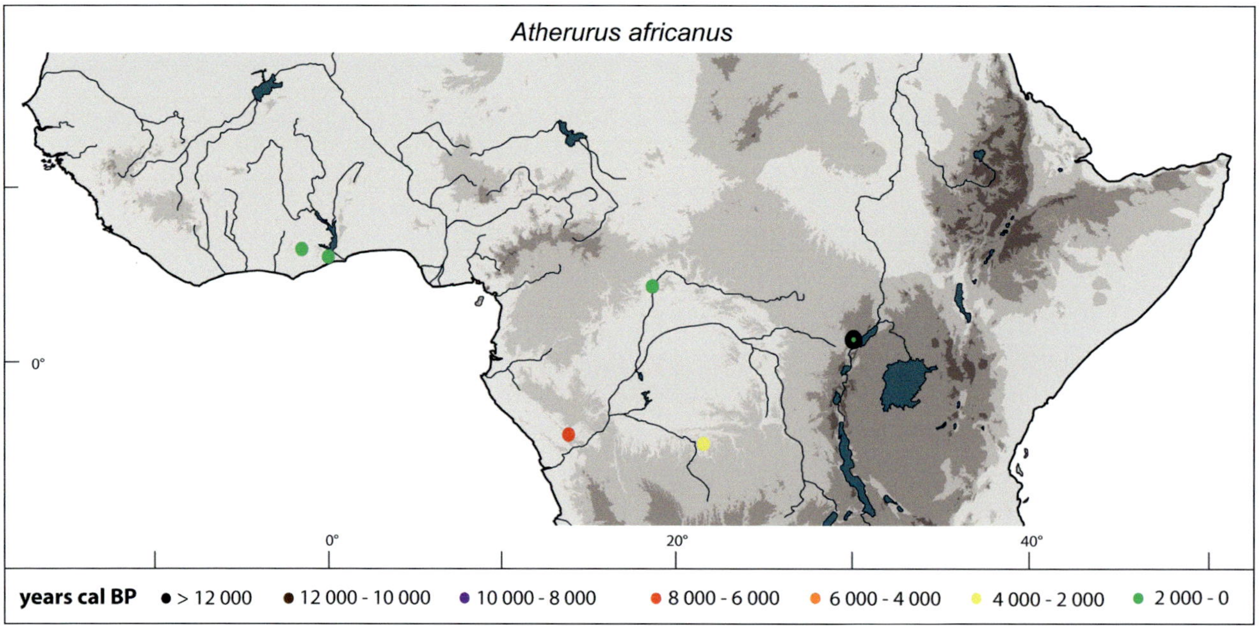

Hystrix africaeaustralis

Country	Site N	Site	
Botswana	42	Toteng1-SU2	
	43	Toteng1-SU3	
	48	White Paintings RS-1	
	48	White Paintings RS-2ab	
	49	White Paintings RS-2b-3ab	
Kenya	276	GvJm 19-HC	
Lesotho	303	Bolahla	
	304	Ha Makotoko	
	305	Lequetsoana	
	308	Likoaeng-VII-IX	
	310	Likoaeng-XIII	
	313	Liphofung-pCW	
	314	Muela	
	316	Sehonghong-a	
	321	Tloutle-CWI	
	322	Tloutle-O	
	323	Tloutle-pCW	
	324	Tloutle-R	
Malawi	357	Matope Court-IA	
Namibia	449	Apollo 11-c	
	461	Pokkenbank	
South Africa	556	2627 CD7	
	563	Boomplass Cave-a	
	564	Boomplass Cave-b	
	565	Boomplass Cave-c	
	566	Boomplass Cave-d	
	568	Buffelskloof RS-a	
	571	Buffelskloof RS-d	
	573	Burchell´s Shelter	
	574	Bushman Rock Shelter-LSA	
	577	Byeneskranskop Cave 1b	
	579	Byeneskranskop Cave 1d	
	580	Collingham Shelter	
	581	Colwinton Shelter-CW	
	584	De Hoop	
	586	Die Kelders	
	588	Dikbosch 2	
	595	Eland´s Bay Cave-AI	
	596	Eland´s Bay Cave-RI	
	598	Eland´s Bay Cave-Wib	
	600	Eland´s Bay Cave-Wid	
	602	Eland´s Bay Open-b	
	603	Emambeni	
	604	Equus Cave 1A	
	613	Hail Stone Midden	
	616	Heuningneskrans-I/O	
	617	Heuningneskrans-R	
	619	HKLP Ia	
	620	HKLP Ib	*cf.
	621	HKLP Va	
	625	Jubilee Shelter-PW	
	626	Jubilee Shelter-W	
	627	Kasteelberg A	
	628	Kasteelberg B	
	631	Klasies River Mouth-1I	
	632	Klasies River Mouth-1II	
	633	Klasies River Mouth-5II	
	644	Leliehoek	
	648	Ma38	
	659	Marupale Hill	
	663	Mhlwazini Cave	
	668	Nelson Bay Cave- PW	
	669	Nelson Bay Cave- R	
	670	Nelson Bay Cave- W	
	674	Ondini	
	675	Onink	
	676	Oudepost I	
	681	Paternoster1-2	
	683	Paternoster5-6	
	690	Powerhouse Cave	
	697	Riversmead Shelter I-II	
	698	Riversmead Shelter III-IV	
	699	Riversmead Shelter VI-IX	
	702	Rooikrans-pCW	
	704	Rose Cottage Cave-A-Mn	
	706	Rose Cottage Cave-OIC	
	707	Rose Cottage Cave-WI	
	717	SK17	
	718	Sk4	
	719	Smitswinkelbaai Cave	
	720	Sonkoanini Hill	
	721	Spoegrivier-a	
	722	Spoegrivier-b	
	723	Spoegrivier-c	
	724	Spring Cave	
	729	Steenbokfontein-4b	
	733	Tienfontein 2	
	736	Tortoise cave-b	
	737	Tortoise cave-c	
	738	Tortoise cave-d	
	739	Tsh1	
	744	Voëlvlei	
	749	Witklip-3	
	750	Witklip-4	
	752	Wosi-middle-late EIA	
Zaire	846	Kakontwe	
	848	Kiantapo	
Zambia	860	Leopard's Hill-NI	
	861	Leopard's Hill-NII	
	862	Leopard's Hill-NIII	
	863	Leopard's Hill-PN	
Zimbabwe	870	Redcliff Cave	

Hystrix cristata

Country	Site N	Site	
Algeria	21	Ti-n-Hanakaten	
	22	Zmeilet el-Barka	
	5	Amekni	
	7	Damous el Ahmar	
Chad	85	Koyom	
Egypt	109	Bir Kiseiba-EN	
	111	Bir Kiseiba-MN	
	138	Elkab-Epipaleo	
	171	Napta Playa 77-1/4/5/6	*?
	172	Napta Playa 77-8	*?
	173	Napta Playa E77-3/7	*?
Ghana	223	Anyinam-Area A-B-C	
Ghana	230	Bono Manso	
	245	Nyanawase/Akwamu Amanfoso	
Kenya	253	Ele Bor A-A	
	255	Ele Bor A-C	*cf.
	297	Ngenyn	
Libya	328	Hauah Fteah-Libyco-capsian	
	329	Hauah Fteah-Neolithic	
	333	Ti-n-Thora East	
	335	Ti-n-Thora Two Caves	
	336	Uan Afuda-US1-3	
	338	Uan Muhuggiag (LA)	
	345	Uan Telocat (MP)	
Morocco	419	Bou Saria A	
	421	Bou Saria C-D	
	422	El Harhoura 1	
	423	El Harhoura 2	
	425	Ghar-Khal I	
	426	Ghar-Khal IIIb-V	
	429	Kaf-taht-el-Ghar-A	
	431	Kaf-taht-el-Ghar-C	
	432	Kaf-taht-el-Ghar-D	

	435	Kaf-taht-el-Ghar-G
	436	Kehf el Hammar
	437	Kehf-el-Baroud
	438	Kehf-el-Baroud-aven
Sudan	762	Debbat El Eheima
	764	El Damer
	765	El Kadada
	773	Esh Shaheinab
	775	Gash Delta-K
	780	Jebel et Tomat
	781	Jebel Umm Marrahi
	787	Khartoum Hopital
	788	Khasm el Girba - Kassala
	790	Khasm el Girba - KG15-16
	791	Khasm el Girba - KG68
	801	Mahal Teglinos K1-MG
	805	Saggai 1
	811	Shaqaduq-S1-A
	812	Shaqaduq-S1-B
	813	Shaqaduq-S21
	814	Sheikh el Amin
	815	Sheikh Mustafa
	821	Upper Atbara hinterland

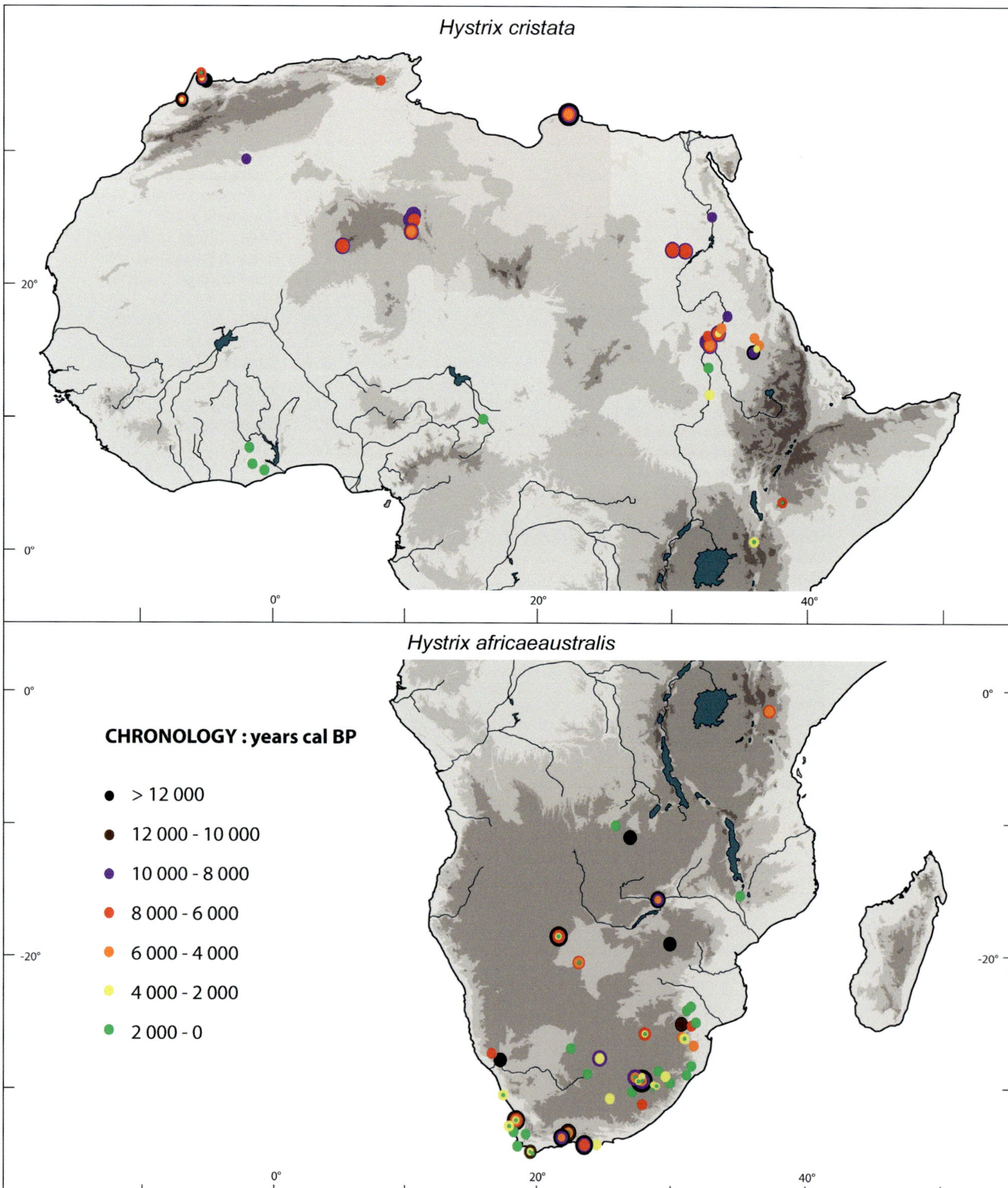

PETROMURIDAE

Petromus sp.

Country	Site N	Site
Namibia	454	Fackelträger

Petromus typicus

Country	Site N	Site	
Namibia	450	Brandberg-BBC	
	451	Brandberg-late LSA	
	452	Brandberg-LSA	
	453	Brandberg-MSA	
	457	Haalenberg	
South Africa	584	De Hoop	
Sudan	810	Shabona	*cf.

Petromus cunealis

Country	Site N	Site
Namibia	446	Aar I and II
	447	Apollo 11-a
	449	Apollo 11-c
	458	Namtib
	461	Pokkenbank

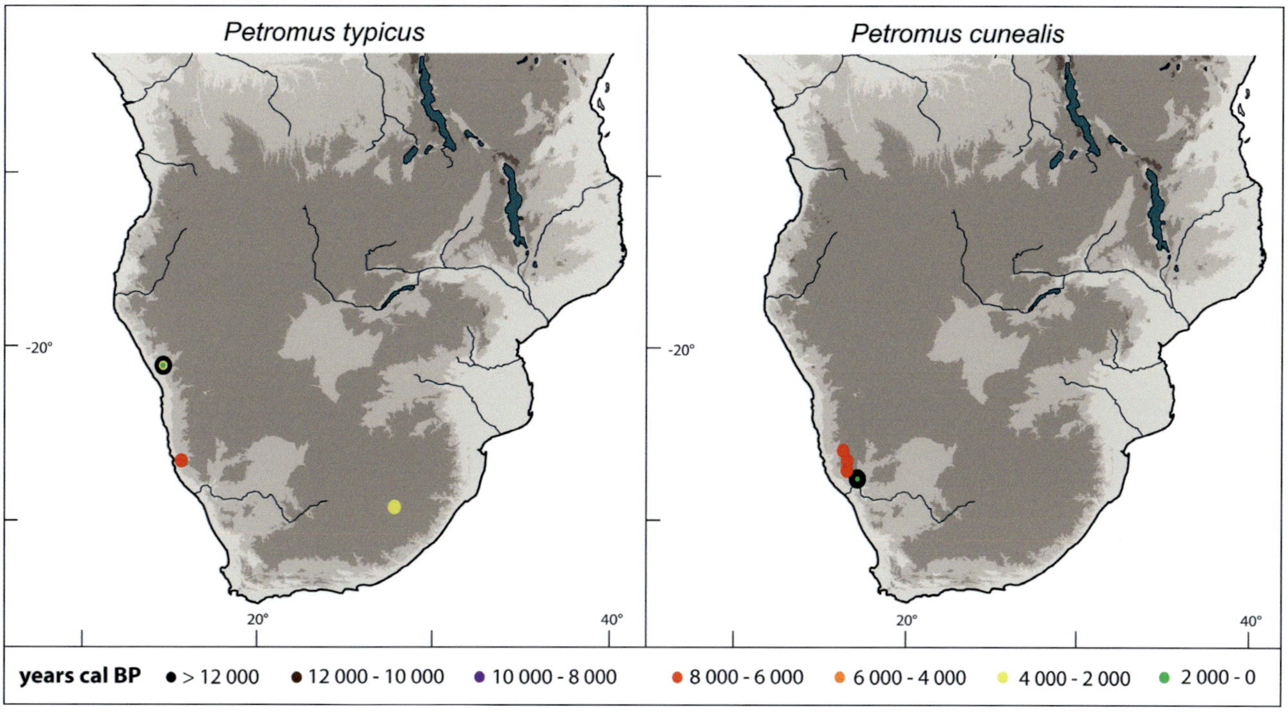

THRYONOMYIDAE

Thryonomys sp.

Country	Site N	Site	
Algeria	16	Meniet	
	5	Amekni	
Cameroon	75	Nkang	
Chad	85	Koyom	
Ghana	231	Daboya	
Kenya	272	Gogo Falls-I	
	273	Gogo Falls-II-III	
Mauritania	396	Dhar Tichitt- latest phase	
	399	Dhar Tichitt- post beach	
Mozambique	442	Manekweni-2-ABC	
	443	Manekweni-3	
	444	Manekweni-4	
	445	Manekweni-5	
Rwanda	535	Akameru-Cyinkomane	
South Africa	619	HKLP Ia	
	620	HKLP Ib	*cf.
Sudan	772	Erg of Ennedi	
	813	Shaqaduq-S21	
	828	Wadi Howar - 80/73	
Zaire	848	Kiantapo	
	854	Matupi-I	
	855	Matupi-II	

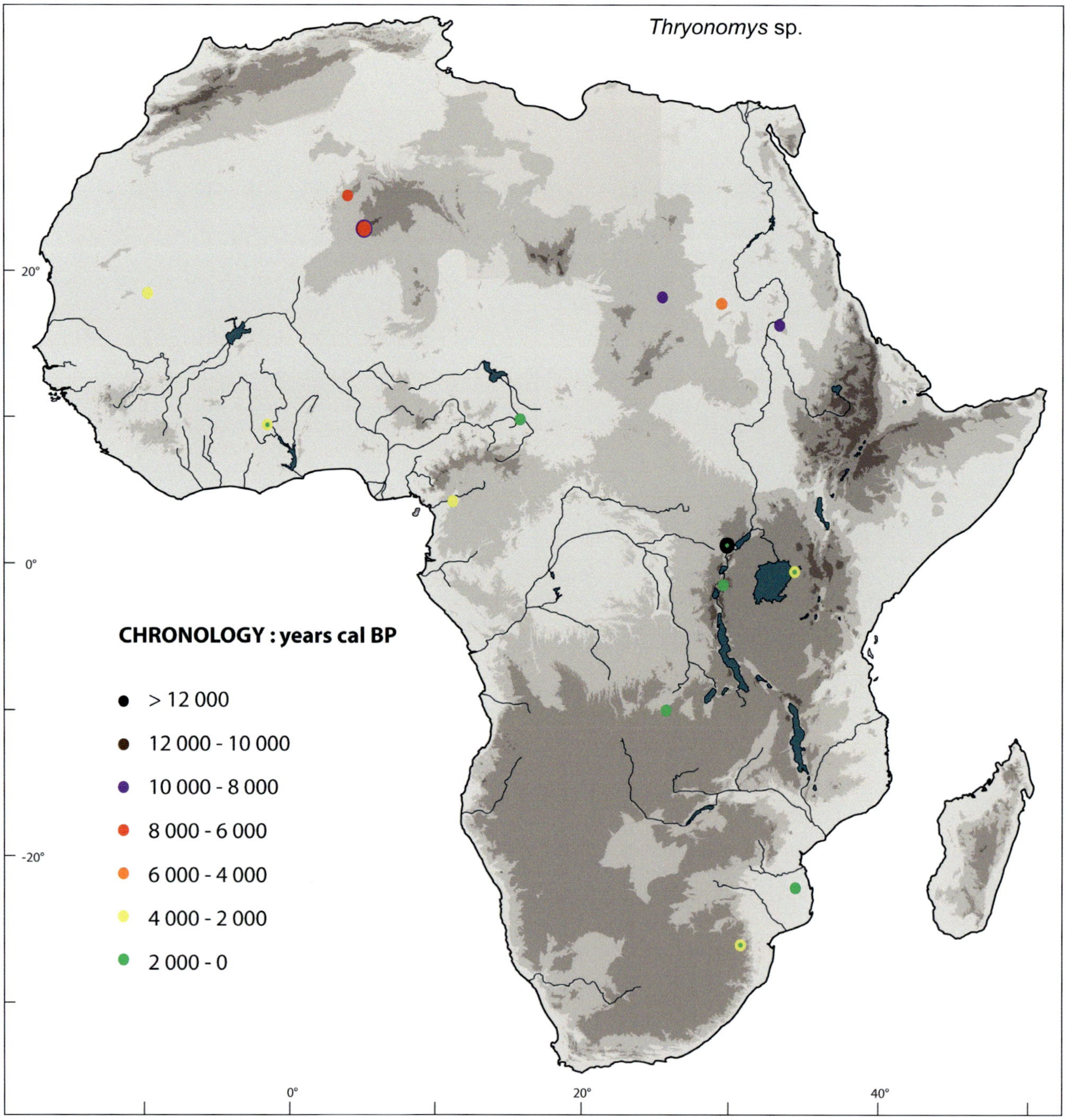

Thryonomys swinderianus

Country	Site N	Site	
Algeria	17	Oued Adjemet	
Botswana	31	Commando Kop	
Burkina Faso	51	Cocorba-LSA	
	61	Péntènga 89/1-spits 15-21	
Cameroon	76	Shum Laka II c	
	77	Shum Laka II d	
	78	Shum Laka IIa+b	
Ghana	223	Anyinam-Area A-B-C	
	227	Asantamanso-ASV	
	229	Bibease Plantation	
	230	Bono Manso	
	232	Dawu	
	233	Esiease - A-C-D	
	238	Kintampo Rockshelter K6-K	
	241	Mandara	
	242	Mankuma	
	244	Ntereso	
	246	Nyanga	
Malawi	352	Malowa-2	
	356	Matope Court-EIA	
	357	Matope Court-IA	
Mauritania	397	Dhar Tichitt- lowest beach	*cf.
	398	Dhar Tichitt- middle beach	*cf.
	402	DN4-BKH-S5	
	417	Tichitt - Nkhal phase	*cf.
Niger	473	Gobero-K	
South Africa	642	Le7a	
	659	Marupale Hill	
	667	Ndondondwane	
	674	Ondini	
	675	Onink	
	713	SH16	
	714	Sh27	
	739	Tsh1	
	751	Wosi-early EIA	
	753	Wosi-terminal EIA	
Sudan	780	Jebel et Tomat	
	787	Khartoum Hopital	
	789	Khasm el Girba - KG14	
	805	Saggai 1	
Zaire	843	Ishango-AB	
	845	Ishango-Z-Post-Em	
	848	Kiantapo	
	853	Mashita Mbanza	
	854	Matupi-I	
	855	Matupi-II	
	856	Ngovo	*cf.
Zambia	863	Leopard's Hill-PN	

Thryonomys gregorianus

Country	Site N	Site
Kenya	258	Enkapune Ya Muto-Eburran 4a
	259	Enkapune Ya Muto-Eburran 4b
	260	Enkapune Ya Muto-Eburran 5a
Zaire	848	Kiantapo
	854	Matupi-I
	855	Matupi-II

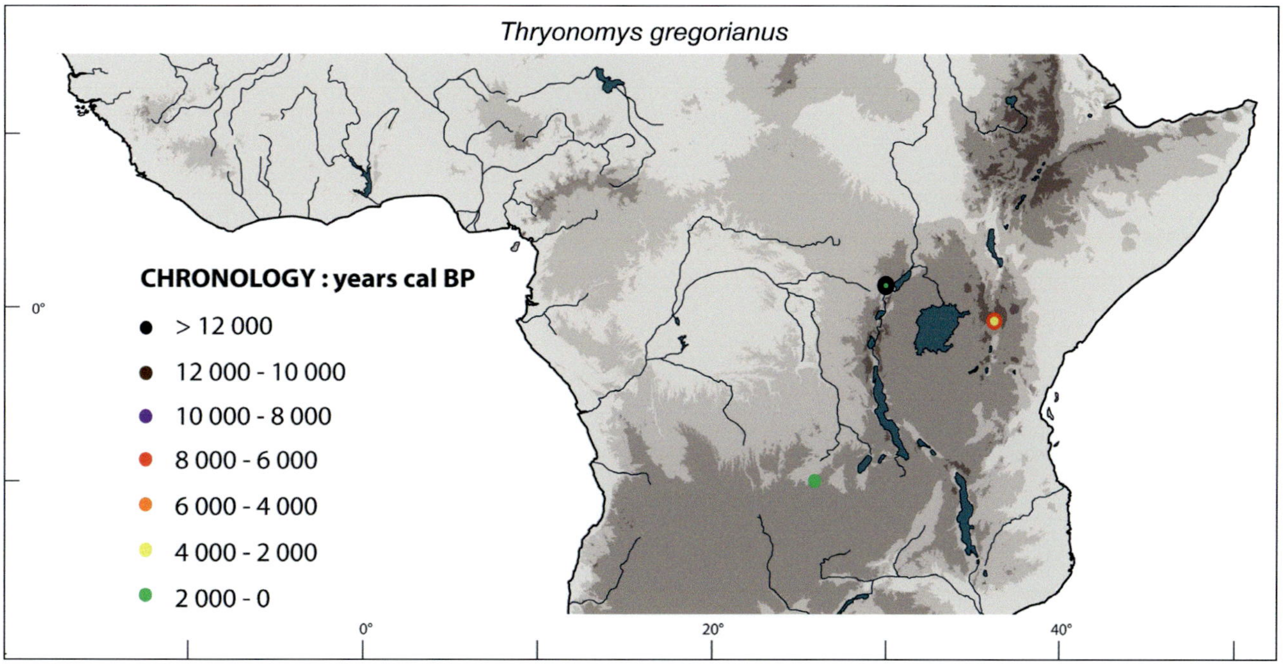

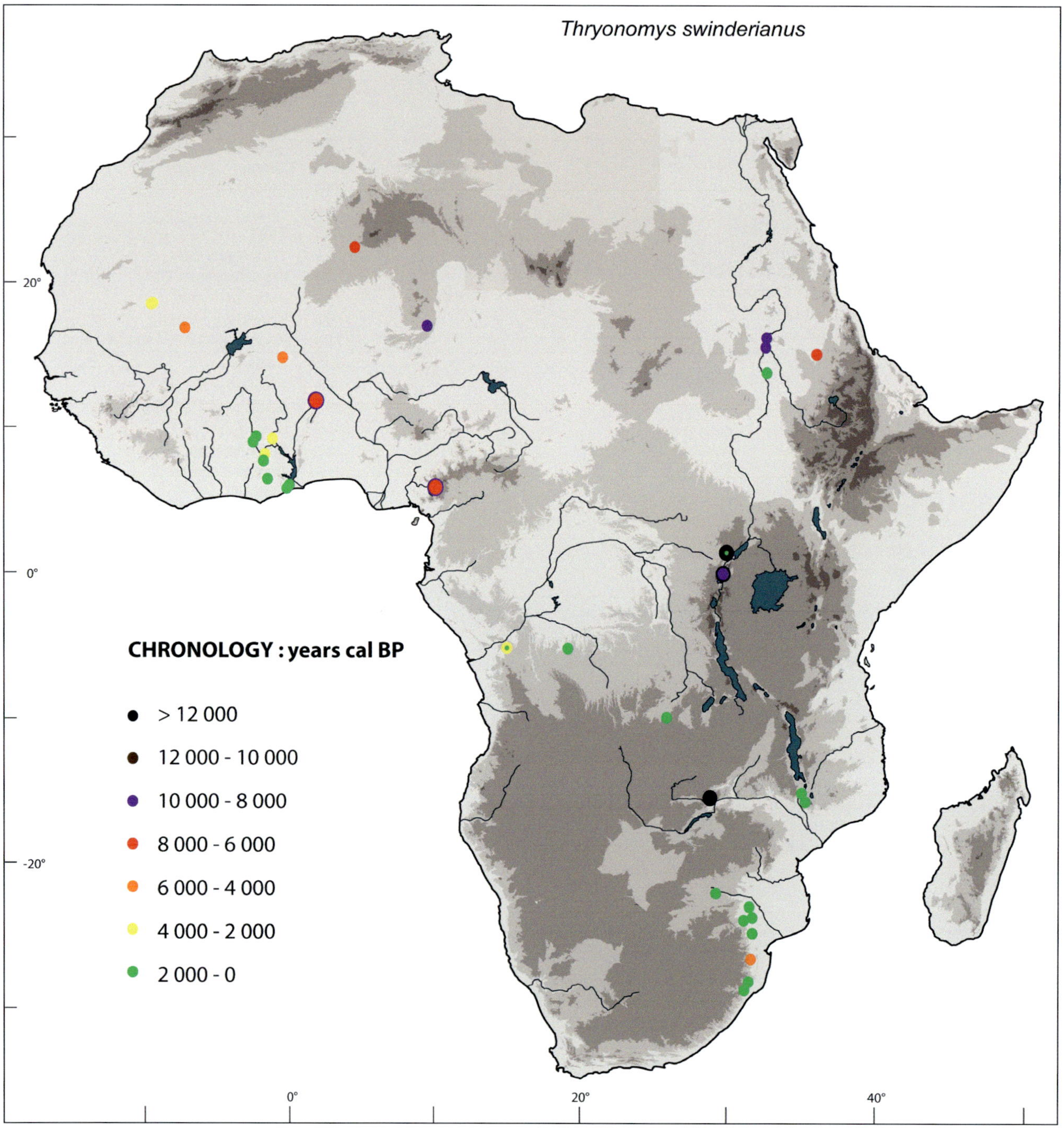
Thryonomys swinderianus
CHRONOLOGY : years cal BP
> 12 000
12 000 - 10 000
10 000 - 8 000
8 000 - 6 000
6 000 - 4 000
4 000 - 2 000
2 000 - 0
20°
0°
-20°
0°
20°
40°

ORDER LAGOMORPHA

LEPORIDAE

Lepus sp.

Country	Site N	Site	
Algeria	13	Kef Zoura D	
	15	Medjez II	*L. kabylicus
	22	Zmeilet el-Barka	
	9	Dra-Mta-el-Ma-el-Abiod	*L. kab.
Botswana	26	Bosutwe-Mapungubwe	
	27	Bosutwe-Taukome	
	28	Bosutwe-Toutswe	
	48	White Paintings RS-1	
	48	White Paintings RS-2ab	
	49	White Paintings RS-2b-3ab	
Burkina Faso	52	Kissi 22	
	53	Kissi 40	
	54	Oursi hu-beero	
	56	Oursi Village-LIA	
	57	Oursi Village-MIA	
	58	Oursi-IA	
	63	Saouga 94/120	
	64	Saouga 95/7	
Cameroon	74	Djaba Hosséré	
Egypt	115	Catfish Cave-L1	
	117	Catfish Cave-L3	
	168	Merimde Benisâlame	
Ghana	223	Anyinam-Area A-B-C	
	241	Mandara	
	242	Mankuma	
	246	Nyanga	
Lesotho	314	Muela	
	316	Sehonghong-a	
	321	Tloutle-CWI	
Mali	361	Akumbu-EIA	
	362	Akumbu-MIA	
	378	Jenné-Jeno - phase IV	
	380	Karkarichinkat N-S	
	383	Kolima Sud-II	
Mauritania	396	Dhar Tichitt- latest phase	
	399	Dhar Tichitt- post beach	
	411	Tichitt - Akjinjeir phase	
	416	Tichitt - Naguez phase	
Morocco	423	El Harhoura 2	
Mozambique	441	Manekweni-1	
	442	Manekweni-2-ABC	
	444	Manekweni-4	
Namibia	446	Aar I and II	
	447	Apollo 11-a	
	448	Apollo 11-b	
	449	Apollo 11-c	
	455	Geduld-LIA	*cf.
	455	Geduld-LIA	
	456	Geduld-LSA	
	456	Geduld-LSA	
	458	Namtib	
	461	Pokkenbank	
	463	Tiras	
Niger	467	Adrar Bous-T	
	489	Takene Bawat 5	
	495	Tin Ouaffadene	
Nigeria	501	Dorota	
	507	Galaga	
	516	Labe Kanuri	
Rhodesia	533	Harleigh Farm-2	
Senegal	540	Cubalel-II	
South Africa	556	2627 CD7	
	576	Byeneskranskop Cave 1a	
	577	Byeneskranskop Cave 1b	
	578	Byeneskranskop Cave 1c	
	579	Byeneskranskop Cave 1d	
	586	Die Kelders	
	589	Doornfontein	
	615	Happy Rest	
	624	Jubilee Shelter-O	
	625	Jubilee Shelter-PW	
	626	Jubilee Shelter-W	
	627	Kasteelberg A	
	628	Kasteelberg B	
	639	Large Rock Shelter	
	644	Leliehoek	
	651	Magogo-2	
	652	Magogo-3	
	657	Mapungubwe-K2/TS3	
	663	Mhlwazini Cave	
	701	Rooikrans-CW	
	702	Rooikrans-pCW	
	704	Rose Cottage Cave-A-Mn	
	707	Rose Cottage Cave-WI	
	719	Smitswinkelbaai Cave	
	730	Tandjesberg	
	740	Twyfelpoort	
	741	Umgungundlovu	
Sudan	773	Esh Shaheinab	
	775	Gash Delta-K	
	778	Gism el-Arba	
	780	Jebel et Tomat	
	785	Kerma	
	811	Shaqaduq-S1-A	
	812	Shaqaduq-S1-B	
	817	Soba	
	821	Upper Atbara hinterland	

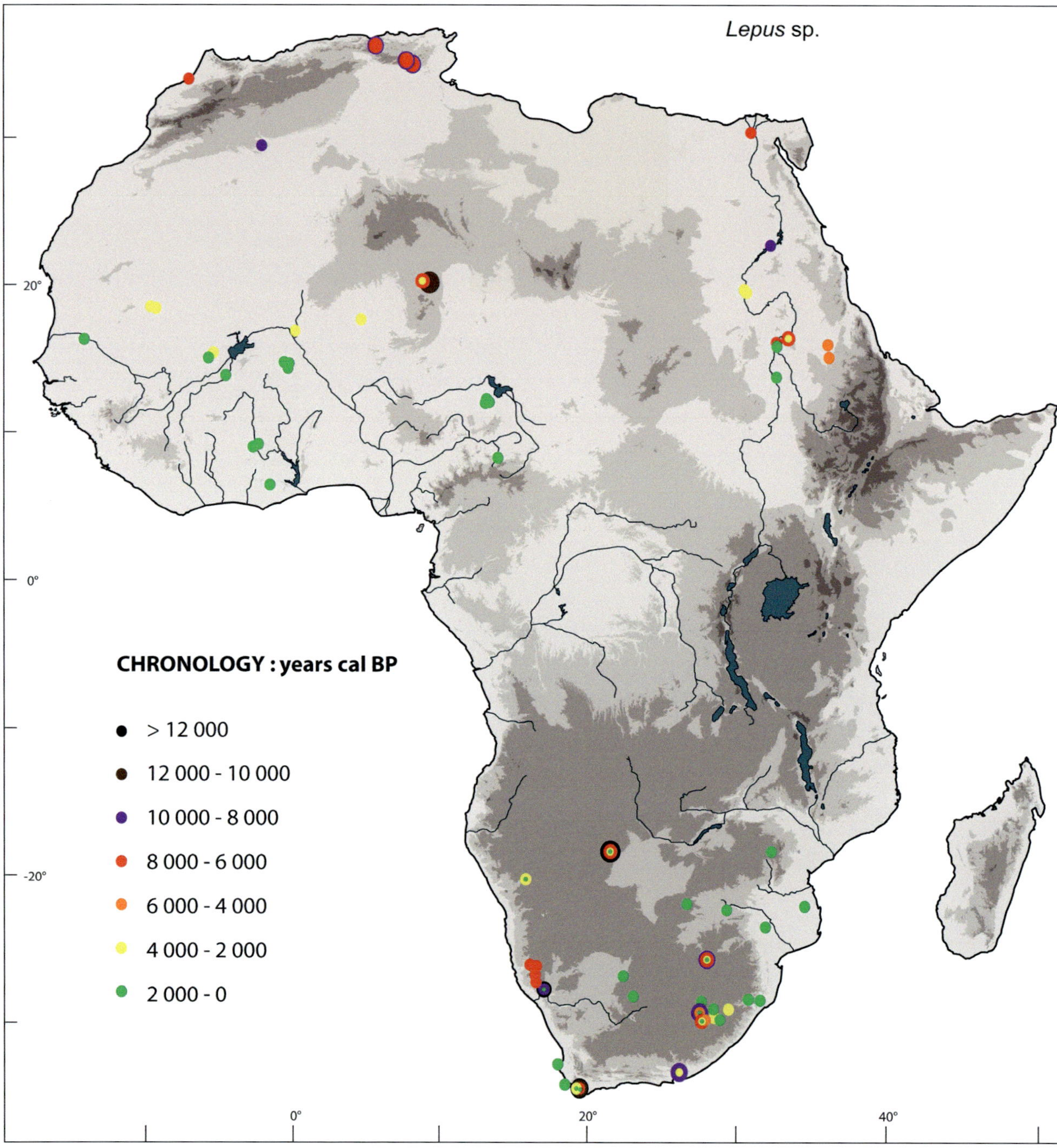

Lepus capensis

Country	Site N	Site
Algeria	10	Grotte Capeletti
	21	Ti-n-Hanakaten
	22	Zmeilet el-Barka
Djibouti	90	Asa Koma
Egypt	100	Adaima-habitat
	106	Berenike 95/9 Early Roman
	107	Berenike 95/97Late Roman
	108	Berenike 96/97 Ptolem.
	109	Bir Kiseiba-EN
	111	Bir Kiseiba-MN
	121	Dakhleh Oasis, SMU
	124	Dakhleh Ouasis,OK
	126	Eastpan 95-1/1
	127	Eastpan 95-2
	137	Elephantine-Satet temple
	139	Elkab-necropolis
	149	Fayoum Site 2 *cf.
	153	Gilf Kebir
	156	Hierakonpolis HK6-Naq III
	162	Kom Ombo
	165	Mahgar Dendera 2
	166	Makhadma 2
	167	Makhadma 4
	171	Napta Playa 77-1/4/5/6
	172	Napta Playa 77-8
	173	Napta Playa E77-3/7
	188	Tell el-Dab´a - b
	201	Wadi Kubbaniya - E-78-3
	203	Wadi Kubbaniya - E-81-1
	205	Willmann´s camp - 81/61
	91	Abu Ballas - 83/39
	92	Abu Ballas - 85/50-1
	93	Abu Ballas - 85/51-3
	94	Abu Ballas - 85/56
	99	Adaima-habitat
Ethiopia	215	Moche Borago-1a
	216	Moche Borago-1b
	217	Moche Borago-1c

	218	Moche Borago-2	
	219	Moche Borago-3	
Kenya	254	Ele Bor A-B	*cf.
	255	Ele Bor A-C	*cf.
	256	Ele Bor A-D	*cf.
	259	Enkapune Ya Muto - Eb 4b	
	293	Naivasha Railway Shelter	
	295	Ndabibi	
	298	Prolonged Drift	
	299	Salasun-LSA	
Libya	327	Hauah Fteah-Iberom	
	328	Hauah Fteah-Lib-caps	
	331	In Aghelachem-T1	
	333	Ti-n-Thora East	
	335	Ti-n-Thora Two Caves	
	338	Uan Muhuggiag (LA)	
	340	Uan Muhuggiag (MP)	
	345	Uan Telocat (MP)	
Mauritania	394	Chami-A2	*cf.
Morocco	436	Kehf el Hammar	
South Africa	568	Buffelskloof RS-a	*cf.
	570	Buffelskloof RS-c	*cf.
	571	Buffelskloof RS-d	*cf.
	572	Buffelskloof RS-e	*cf.
	591	Drie Susters Main	
	618	Hill X	
	659	Marupale Hill	
	669	Nelson Bay Cave- R	*cf.
	680	Pancho's Kitchen Midden 6-7	
	692	QwaQwa Museum	*cf.
	720	Sonkoanini Hill	
Sudan	756	Abu Tabari 02/1	
	761	Debbat Bangdit	
	763	Djarabona 84/13	
	765	El Kadada	
	766	El Kenger Middle/East	
	783	Kadero	
	797	Mahal Teglinos K1-CG	
	798	Mahal Teglinos K1-EG	
	800	Mahal Teglinos K1-LG	
	801	Mahal Teglinos K1-MG	
	802	Mahal Teglinos K1-PG	
	825	Wadi Hariq 01/4	
	831	Wadi Howar - 84/13	
	832	Wadi Howar - 84/19	

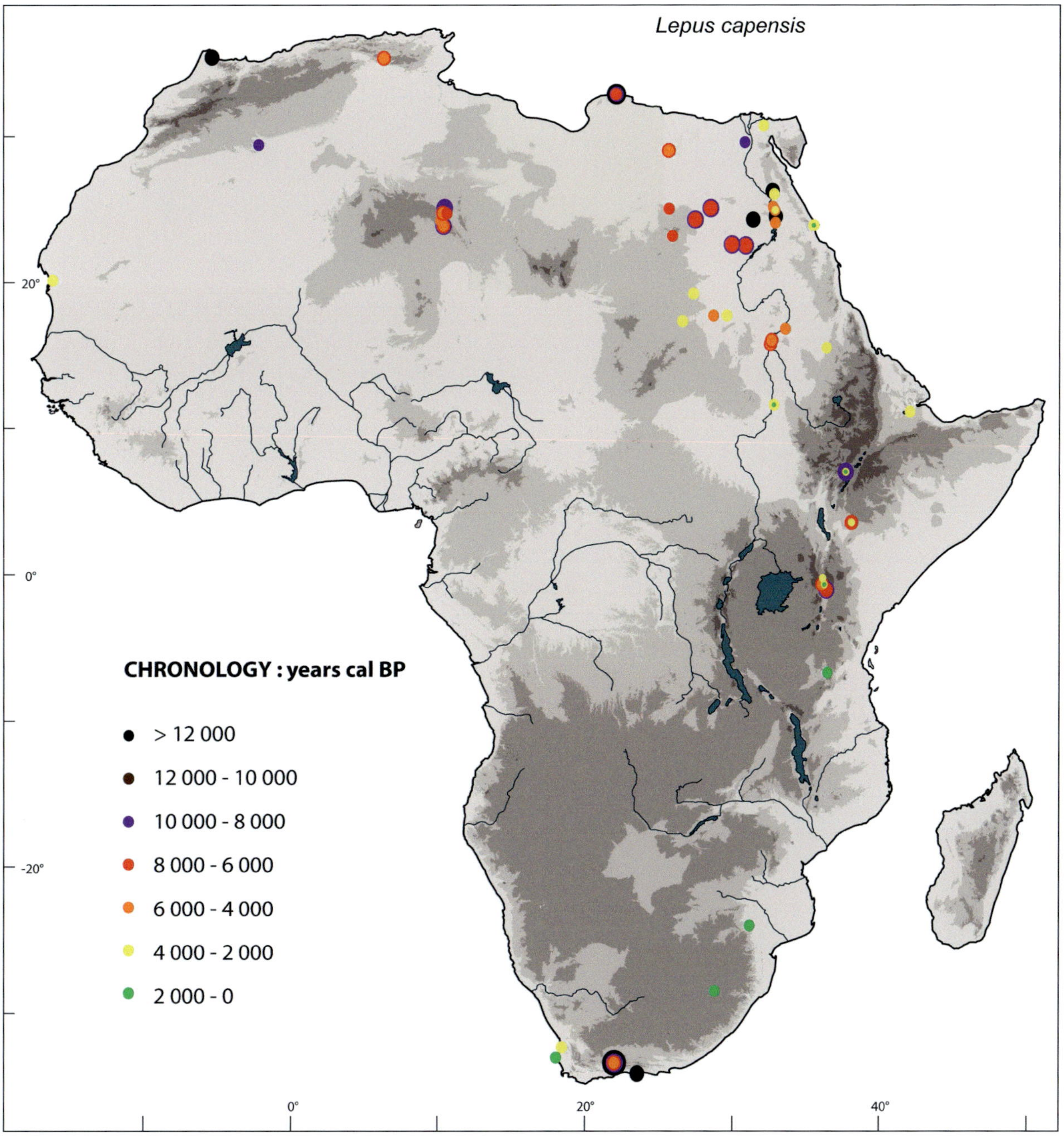

Lepus saxatilis

Country	Site N	Site	
Botswana	26	Bosutwe-Mapungubwe	*cf.
	27	Bosutwe-Taukome	*cf.
	28	Bosutwe-Toutswe	*cf.
Lesotho	307	Likoaeng-III-V	
	308	Likoaeng-VII-IX	
	309	Likoaeng-XI	
	310	Likoaeng-XIII	
	316	Sehonghong-a	
South Africa	574	Bushman RS-LSA	
	594	Eiland Salt Works-b	
	620	HKLP Ib	*cf.
	634	Klein Kliphuis	
	640	Le2	
	664	Mo8	
	673	Ol20	
	704	Rose Cottage Cave-A-Mn	
	709	Sebatini Hill	
	710	Selongwe Hill	*cf.
	711	Serotwe Hill	

Bunolagus monticularis

Country	Site N	Site	
South Africa	568	Buffelskloof RS-a	*cf.
	569	Buffelskloof RS-b	*cf.
	570	Buffelskloof RS-c	*cf.
	571	Buffelskloof RS-d	*cf.
	572	Buffelskloof RS-e	*cf.

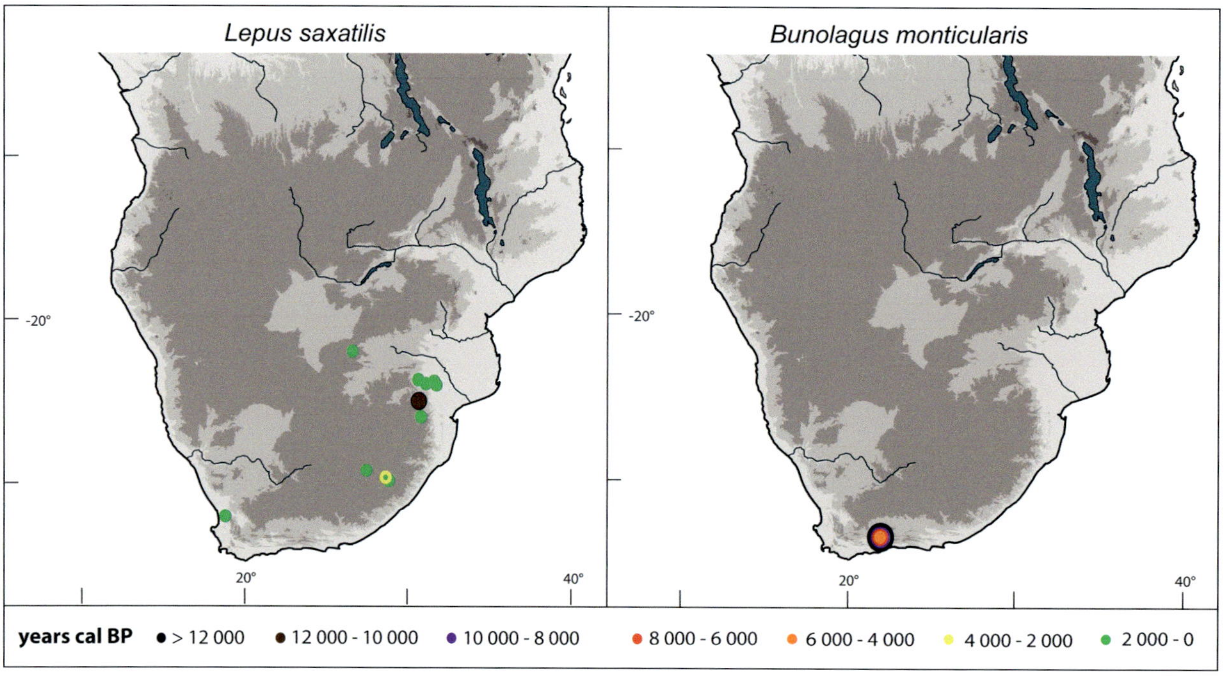

Pronolagus sp.

Country	Site N	Site
South Africa	574	Bushman RS-LSA

Pronolagus crassicaudatus

Country	Site N	Site	
South Africa	663	Mhlwazini Cave	
	700	Rooiels	*cf.

Pronolagus randensis

Country	Site N	Site
Namibia	450	Brandberg-BBC
	451	Brandberg-late LSA
	452	Brandberg-LSA
	453	Brandberg-MSA
	458	Namtib

Pronolagus rupestris

Country	Site N	Site
Lesotho	314	Muela
South Africa	680	Pancho's Kitchen Midden 6-7

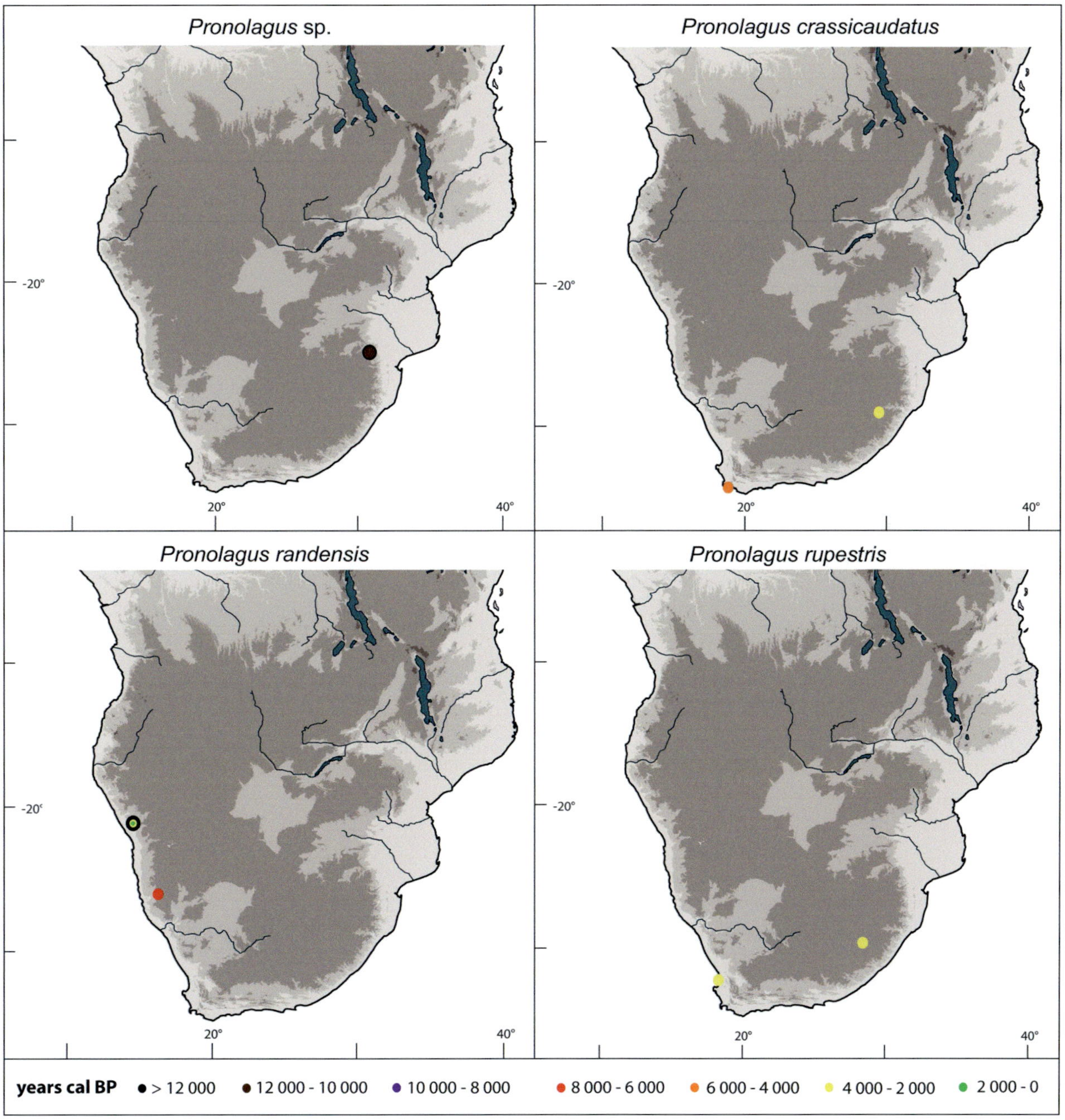

ORDER ERINACEOMORPHA

ERINACEOMORPHA

Country	Site N	Site
Egypt	171	Napta Playa 77-1/4/5/6
	172	Napta Playa 77-8
Sudan	756	Abu Tabari 02/1
	780	Jebel et Tomat

Atelerix albiventris

Country	Site N	Site
Burkina Faso	52	Kissi 22
	58	Oursi-IA
	63	Saouga 94/120
	64	Saouga 95/7

Atelerix algirus

Country	Site N	Site
Algeria	10	Grotte Capeletti
	15	Medjez II
	4	Ain Mistehayia
	9	Dra-Mta-el-Ma-el-Abiod
Morocco	426	Ghar-Khal IIIb-V
	431	Kaf-taht-el-Ghar-C
	432	Kaf-taht-el-Ghar-D
	433	Kaf-taht-el-Ghar-E
	436	Kehf el Hammar
	437	Kehf-el-Baroud
	438	Kehf-el-Baroud-aven

Atelerix frontalis

Country	Site N	Site
Botswana	26	Bosutwe-Mapungubwe
Malawi	352	Malowa-2
South Africa	595	Eland´s Bay Cave-AI
	596	Eland´s Bay Cave-RI
	597	Eland´s Bay Cave-Wia
	599	Eland´s Bay Cave-Wic
	736	Tortoise cave-b
	739	Tsh1
Zimbabwe	870	Redcliff Cave

Hemiechinus auritus

Country	Site N	Site
Algeria	21	Ti-n-Hanakaten

Paraechinus aethiopicus

Country	Site N	Site
Algeria	22	Zmeilet el-Barka **P.a. deserti*
Egypt	109	Bir Kiseiba-EN
	111	Bir Kiseiba-MN
	168	Merimde Benisâlame
Libya	333	Ti-n-Thora East
	335	Ti-n-Thora Two Caves
	336	Uan Afuda-US1-3

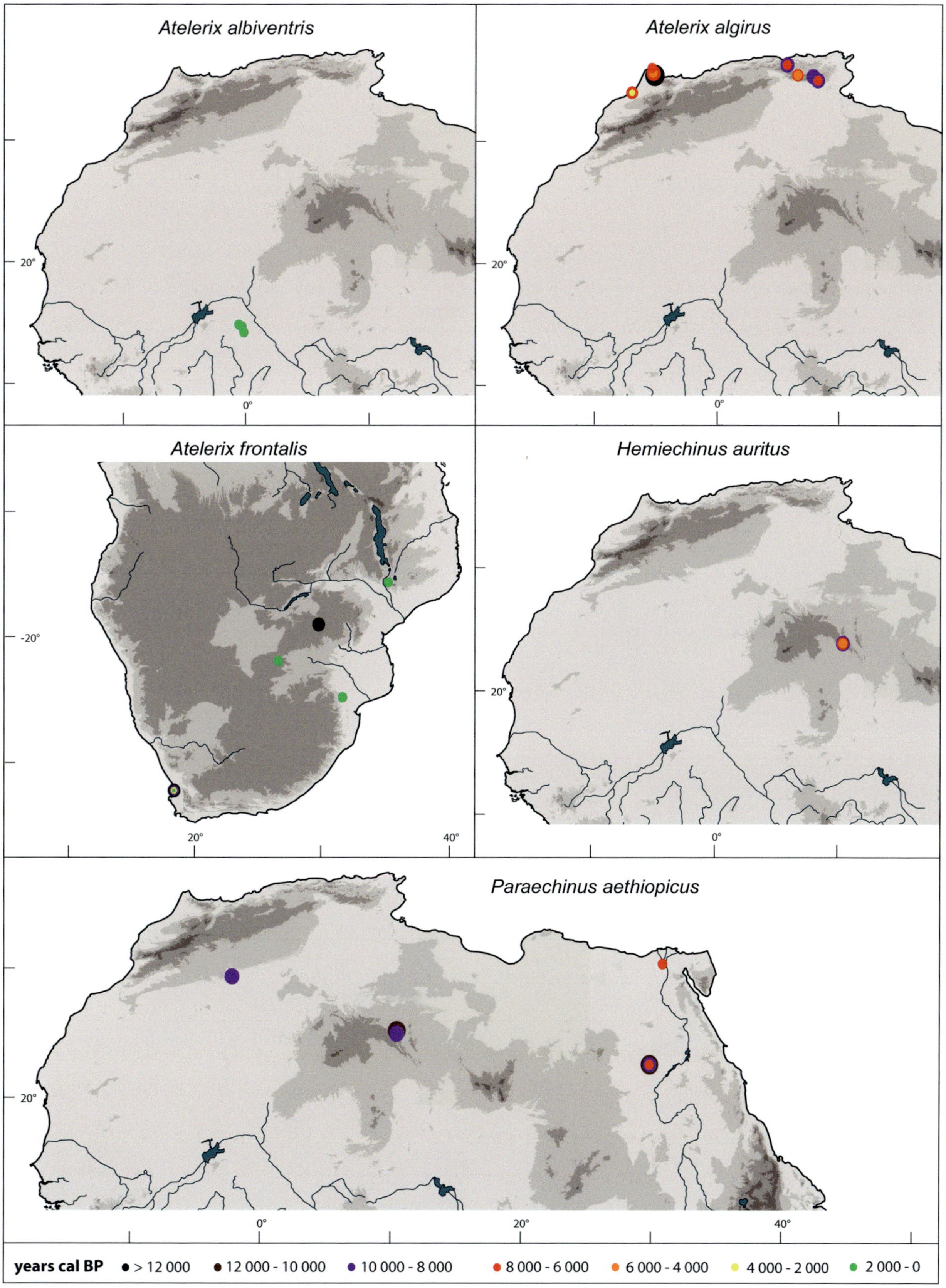
Atelerix albiventris
20°
0°
Atelerix algirus
20°
0°
Atelerix frontalis
-20°
20°
40°
Hemiechinus auritus
20°
0°
Paraechinus aethiopicus
20°
0°
20°
40°
years cal BP
> 12 000
12 000 - 10 000
10 000 - 8 000
8 000 - 6 000
6 000 - 4 000
4 000 - 2 000
2 000 - 0

Order Soricomorpha

Soricidae

Country	Site N	Site
Lesotho	303	Bolahla
Egypt	124	Dakhleh Ouasis,OK

Crocidura sp.

Country	Site N	Site
Burkina Faso	54	Oursi hu-beero
	59	Oursi-LSA
	64	Saouga 95/7
Ghana	237	Kintampo RS K6-IA
Kenya	259	Enkapune Ya Muto - Eb 4b
Libya	333	Ti-n-Thora East
	335	Ti-n-Thora Two Caves
Malawi	351	Malowa-1
	352	Malowa-2
	353	Malowa-3
	354	Malowa-4
Morocco	435	Kaf-taht-el-Ghar-G
	436	Kehf el Hammar
	437	Kehf-el-Baroud
Namibia	450	Brandberg-BBC
	452	Brandberg-LSA
Nigeria	514	Kursakata-IA
	525	Ngala I
South Africa	589	Doornfontein

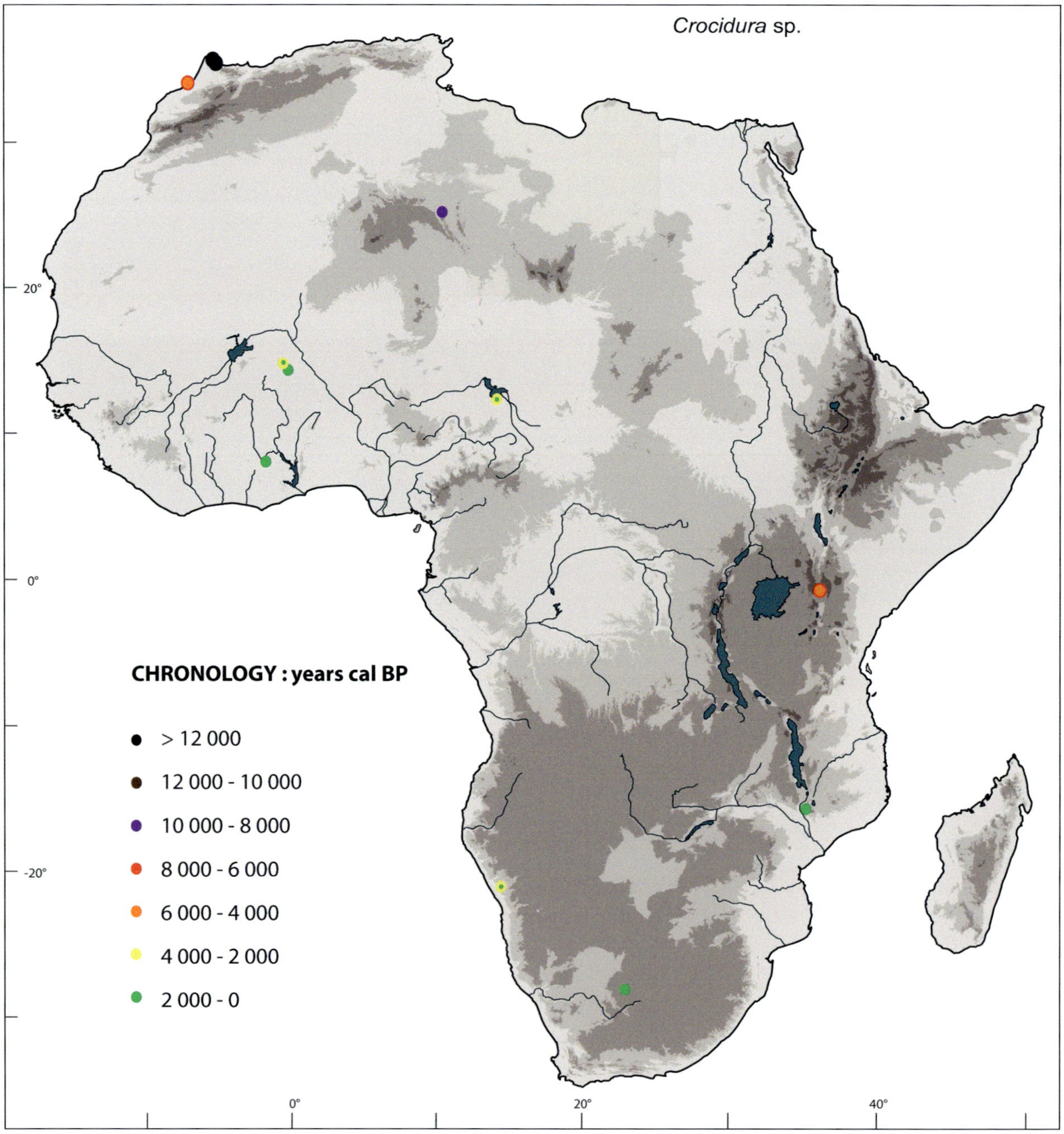

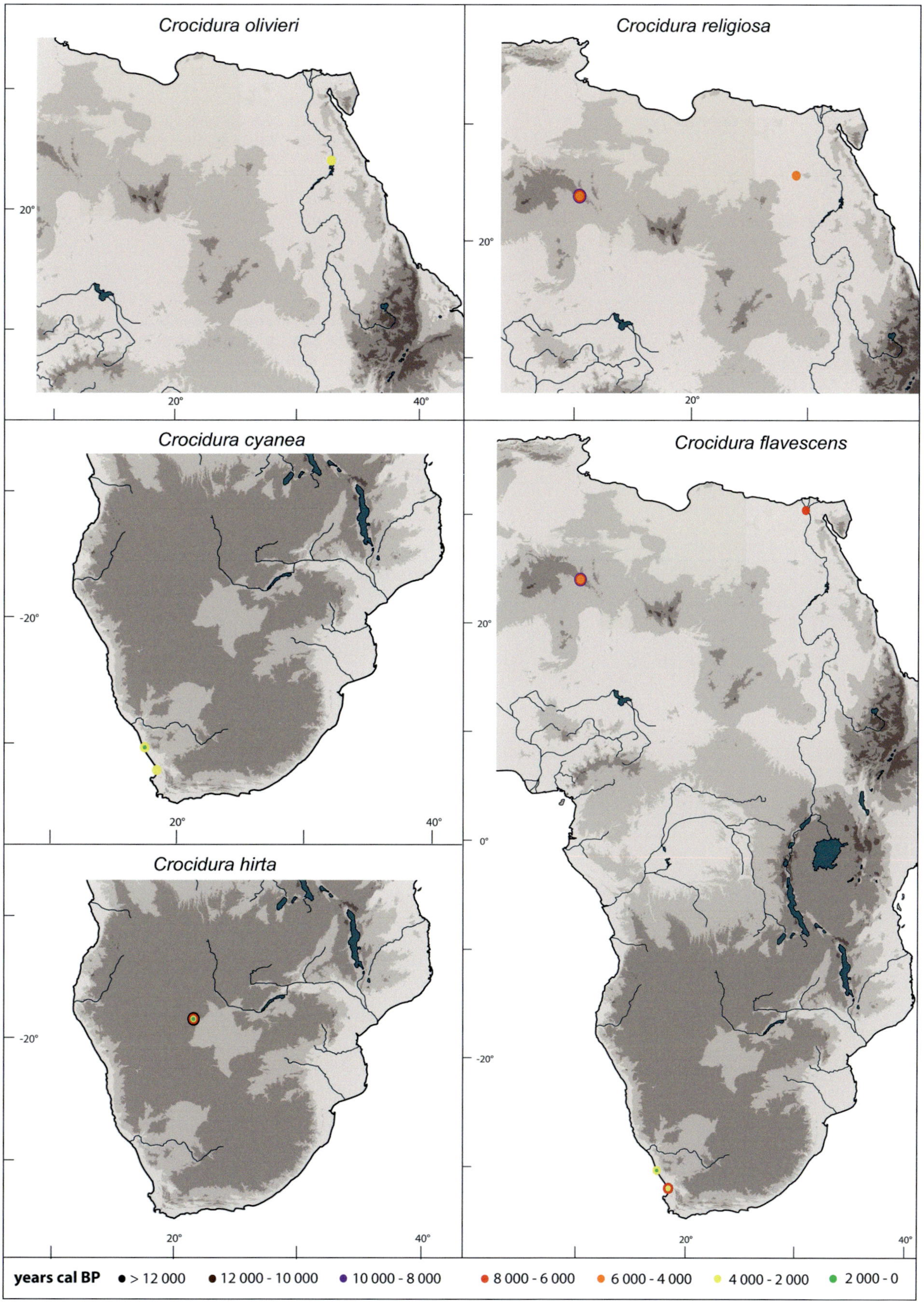

Crocidura olivieri
Crocidura religiosa
Crocidura cyanea
Crocidura flavescens
Crocidura hirta
20°
40°
-20°
0°
years cal BP
> 12 000
12 000 - 10 000
10 000 - 8 000
8 000 - 6 000
6 000 - 4 000
4 000 - 2 000
2 000 - 0

Crocidura olivieri

Country	Site N	Site
Egypt	135	Elephantine-necropolis

Crocidura religiosa

Country	Site N	Site
Algeria	21	Ti-n-Hanakaten
Egypt	128	Ein el Gazzareen

Crocidura cyanea

Country	Site N	Site
South Africa	721	Spoegrivier-a
	726	Steenbokfontein-2
	727	Steenbokfontein-3

Crocidura flavescens

Country	Site N	Site
Algeria	21	Ti-n-Hanakaten
Egypt	168	Merimde Benisâlame **C.f. deltae*
South Africa	721	Spoegrivier-a
	725	Steenbokfontein-1
	726	Steenbokfontein-2
	727	Steenbokfontein-3
	729	Steenbokfontein-4b

Crocidura hirta

Country	Site N	Site
Botswana	48	White Paintings RS-2ab *cf.
	49	White Paintings RS-2b-3ab *cf.

Suncus varilla

Country	Site N	Site
South Africa	721	Spoegrivier-a
	725	Steenbokfontein-1
	726	Steenbokfontein-2
	727	Steenbokfontein-3
	728	Steenbokfontein-4a
	729	Steenbokfontein-4b

Scutisorex somereni

Country	Site N	Site
Zaire	854	Matupi-I
	855	Matupi-II
	846	Kakontwe

Myosorex varius

Country	Site N	Site
South Africa	721	Spoegrivier-a
	725	Steenbokfontein-1
	726	Steenbokfontein-2
	727	Steenbokfontein-3
	728	Steenbokfontein-4a
	729	Steenbokfontein-4b

TALPIDAE

Talpa sp.

Country	Site N	Site
Morocco	420	Bou Saria B
	431	Kaf-taht-el-Ghar-C
	435	Kaf-taht-el-Ghar-G
	437	Kehf-el-Baroud
	438	Kehf-el-Baroud-aven

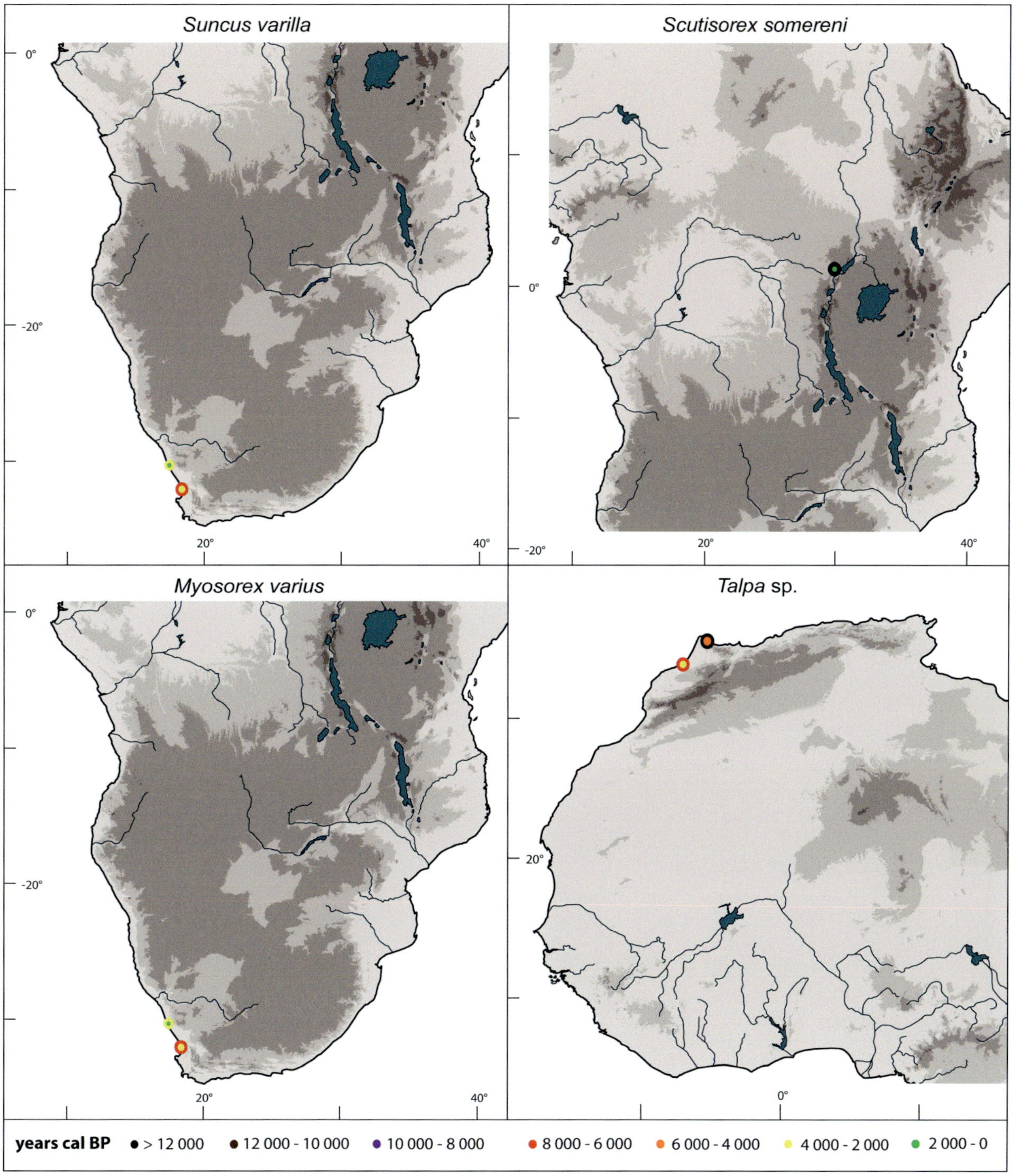
Suncus varilla
Scutisorex somereni
Myosorex varius
Talpa sp.
0°
-20°
20°
40°
years cal BP
> 12 000
12 000 - 10 000
10 000 - 8 000
8 000 - 6 000
6 000 - 4 000
4 000 - 2 000
2 000 - 0

Order Chiroptera

Country	Site N	SIte
Lesotho	304	Ha Makotoko
Libya	335	Ti-n-Thora Two Caves
Malawi	352	Malowa-2
Morocco	428	Ghar-Khal-Ib
	436	Kehf el Hammar
Mozambique	443	Manekweni-3
	445	Manekweni-5
South Africa	589	Doornfontein
	718	Sk4
Sudan	812	Shaqaduq-S1-B

Rousettus sp.

Country	Site N	SIte
Zaire	854	Matupi-I
	855	Matupi-II

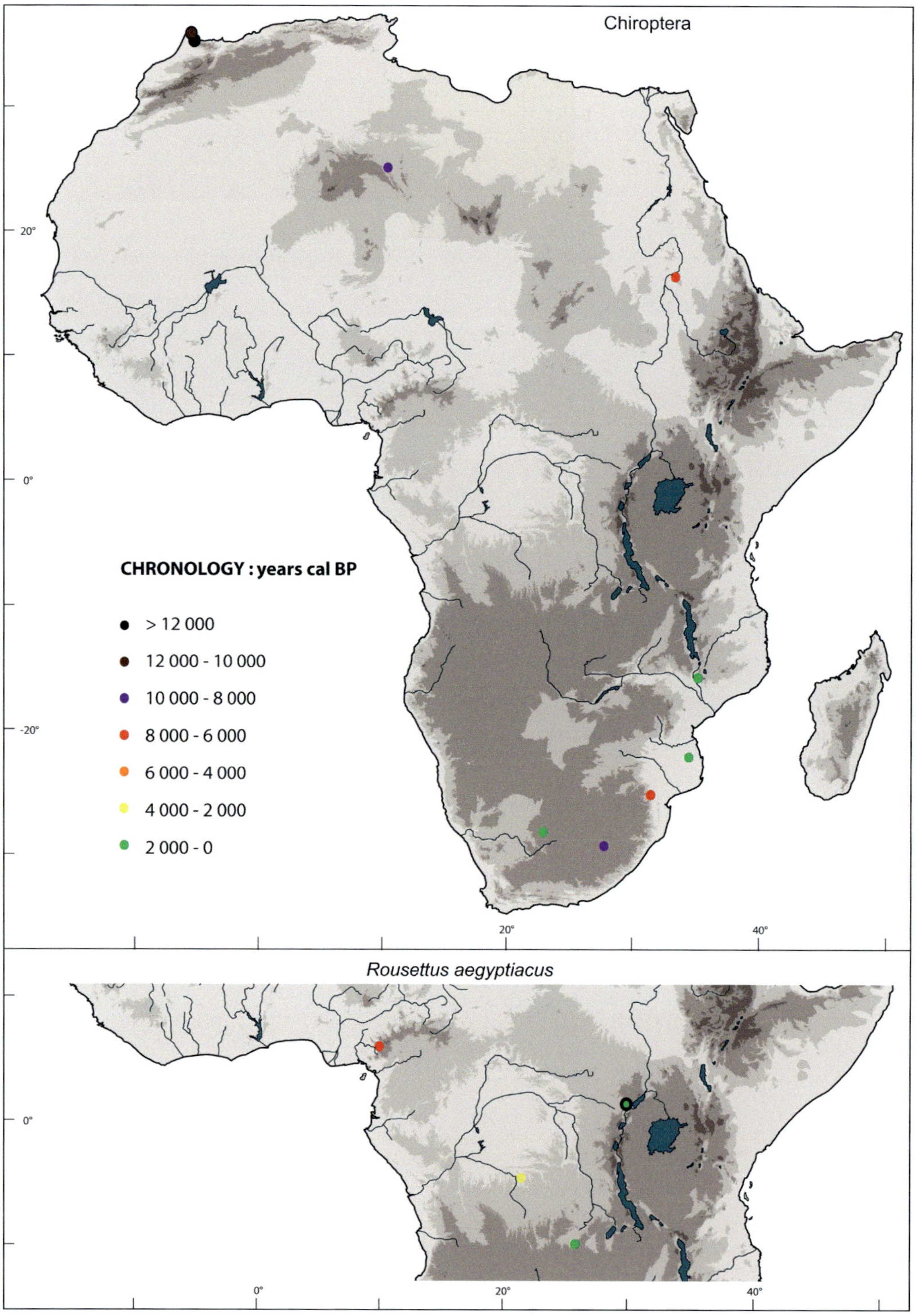

Rousettus aegyptiacus

Country	Site N	SIte
Cameroon	78	Shum Laka IIa+b
Zaire	842	Dimba
	848	Kiantapo
	854	Matupi-I
	855	Matupi-II

Rousettus angolensis

Country	Site N	SIte
Zaire	854	Matupi-I
	855	Matupi-II

Hypsignathus monstrosus

Country	Site N	SIte
Ghana	227	Asantamanso-ASV

Hipposideros commersoni

Country	Site N	SIte
Ghana	223	Anyinam-Area A-B-C

Taphozous perforatus

Country	Site N	SIte
Egypt	196	Tuna el-Gebel

Coleura afra

Country	Site N	SIte
Zaire	854	Matupi-I

Miniopterus schreibersii

Country	Site N	SIte
Zaire	854	Matupi-I

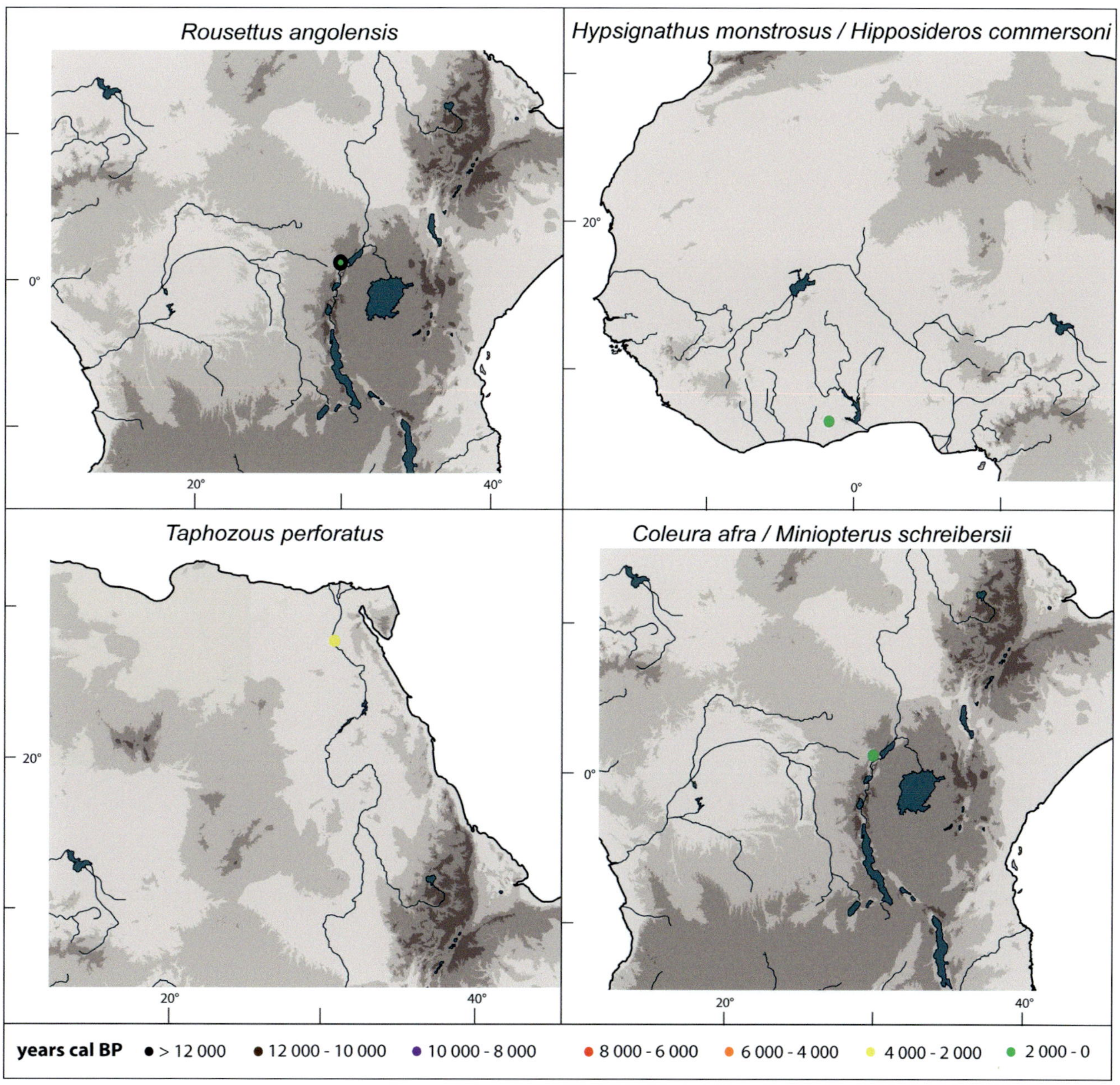

ORDER PHOLIDOTA

Manis temminckii

Country	Site N	Site
Botswana	26	Bosutwe-Mapungubwe
South Africa	603	Emambeni
	626	Jubilee Shelter-W
	663	Mhlwazini Cave
	669	Nelson Bay Cave- R
	699	Riversmead Shelter VI-IX
	722	Spoegrivier-b
	723	Spoegrivier-c

Manis gigantea

Country	Site N	Site
Ghana	225	Asantamanso-ASII/III

Manis tricuspis

Country	Site N	Site
Ghana	223	Anyinam-Area A-B-C
	227	Asantamanso-ASV

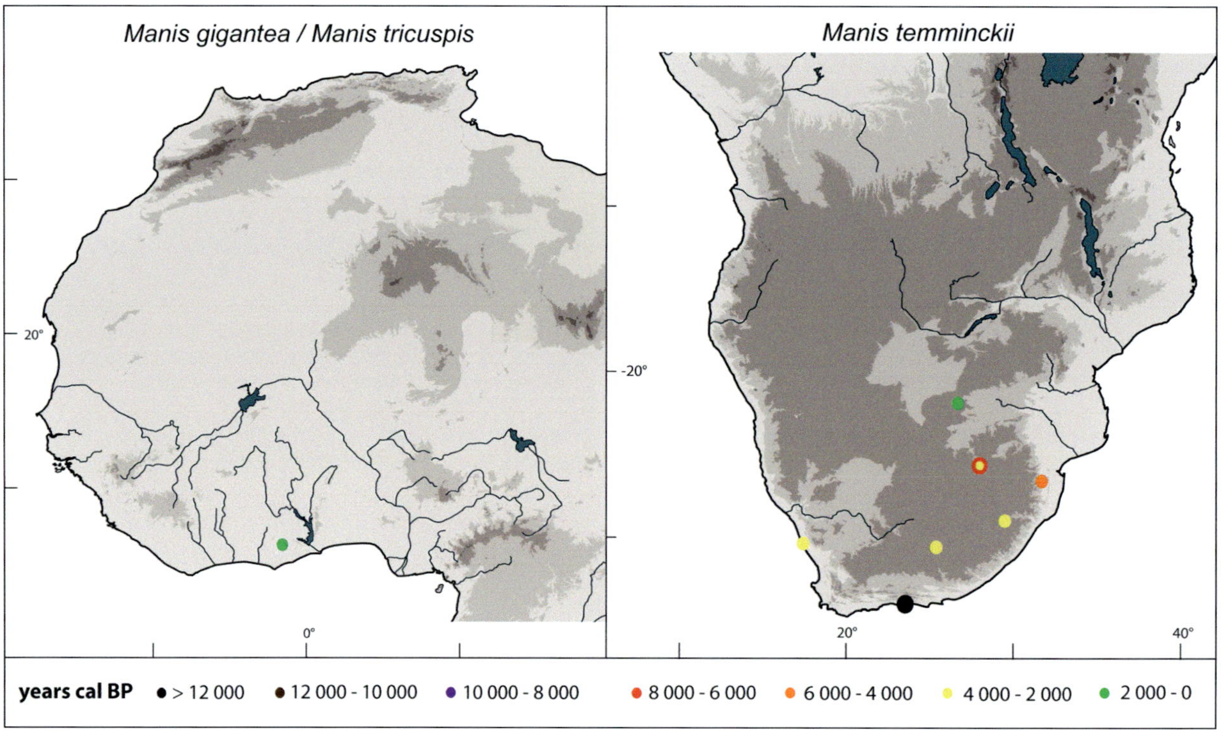

Order Carnivora

Felidae

Acinonyx jubatus

Country	Site N	Site	
Algeria	15	Medjez II	
	9	Dra-Mta-el-Ma-el-Abiod	
Kenya	259	Enkapune Ya Muto - Eb 4b	
	283	GvJm 62-LPNG	
South Africa	574	Bushman RS-LSA	* cf.

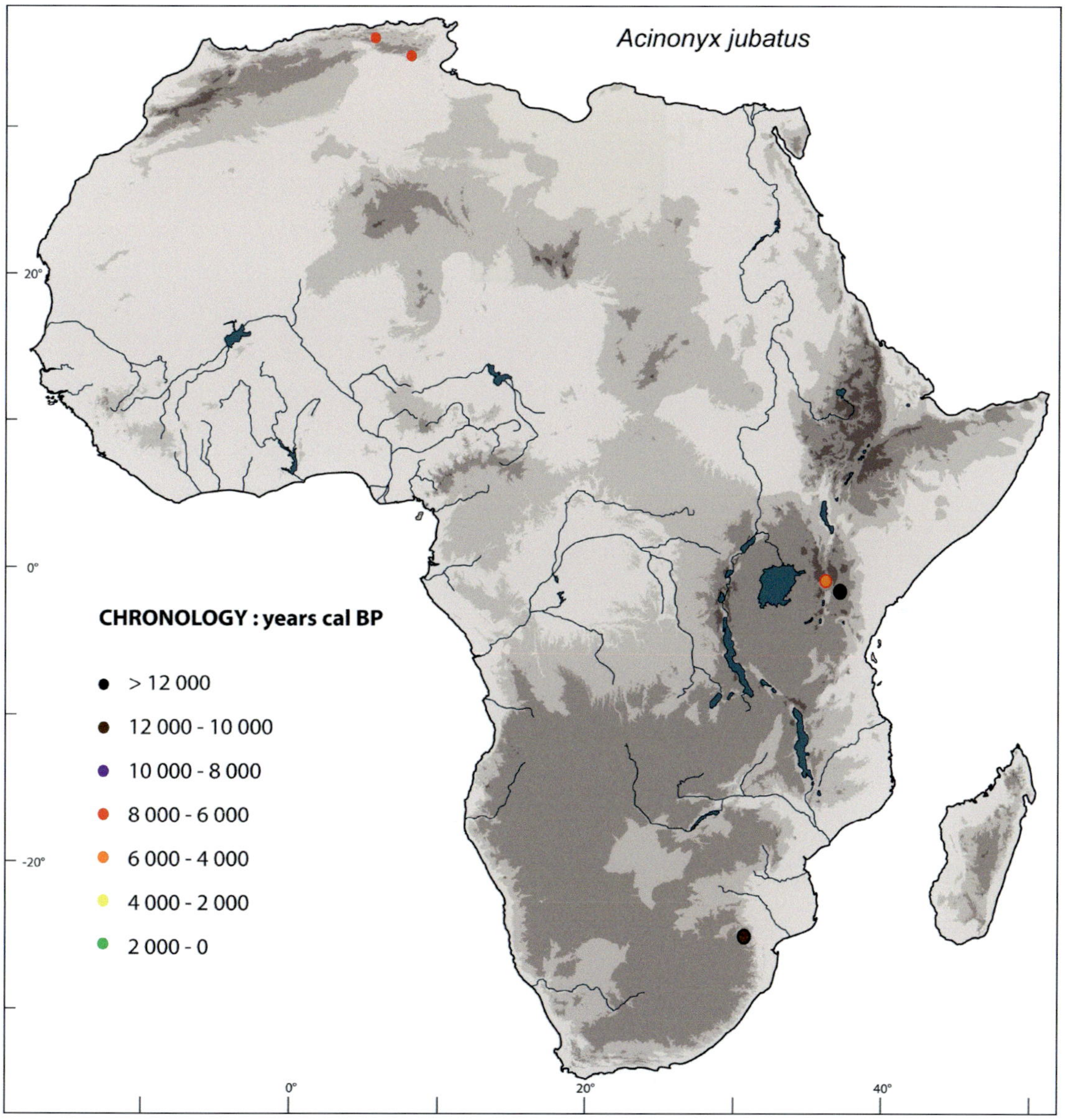

Caracal caracal

Country	Site N	Site	
Botswana	28	Bosutwe-Toutswe	
	32	Divuyu	
	48	White Paintings RS-2ab	* cf.
Egypt	93	Abu Ballas - 85/51-3	
Lesotho	308	Likoaeng-VII-IX	* cf.
	314	Muela	
	316	Sehonghong-a	
Morocco	422	El Harhoura 1	
Namibia	458	Namtib	* cf.
South Africa	563	Boomplass Cave-a	
	564	Boomplass Cave-b	
	565	Boomplass Cave-c	
	566	Boomplass Cave-d	

571	Buffelskloof RS-d	* cf.		676	Oudepost I
579	Byeneskranskop Cave 1d			707	Rose Cottage Cave-WI
580	Collingham Shelter			718	Sk4
595	Eland´s Bay Cave-AI			724	Spring Cave
596	Eland´s Bay Cave-RI			729	Steenbokfontein-4b
597	Eland´s Bay Cave-Wia			736	Tortoise cave-b
598	Eland´s Bay Cave-Wib			744	Voëlvlei
600	Eland´s Bay Cave-Wid			749	Witklip-3
601	Eland´s Bay Open-a			751	Wosi-early EIA
602	Eland´s Bay Open-b			753	Wosi-terminal EIA
604	Equus Cave 1A		Sudan	756	Abu Tabari 02/1
618	Hill X			765	El Kadada
620	HKLP Ib			766	El Kenger Middle/East
622	HKLP Vb	* cf.		793	Laqiya area - 82/29
627	Kasteelberg A			800	Mahal Teglinos K1-LG
628	Kasteelberg B			805	Saggai 1
663	Mhlwazini Cave			814	Sheikh el Amin
668	Nelson Bay Cave- PW			819	Um Direiwa
670	Nelson Bay Cave- W				

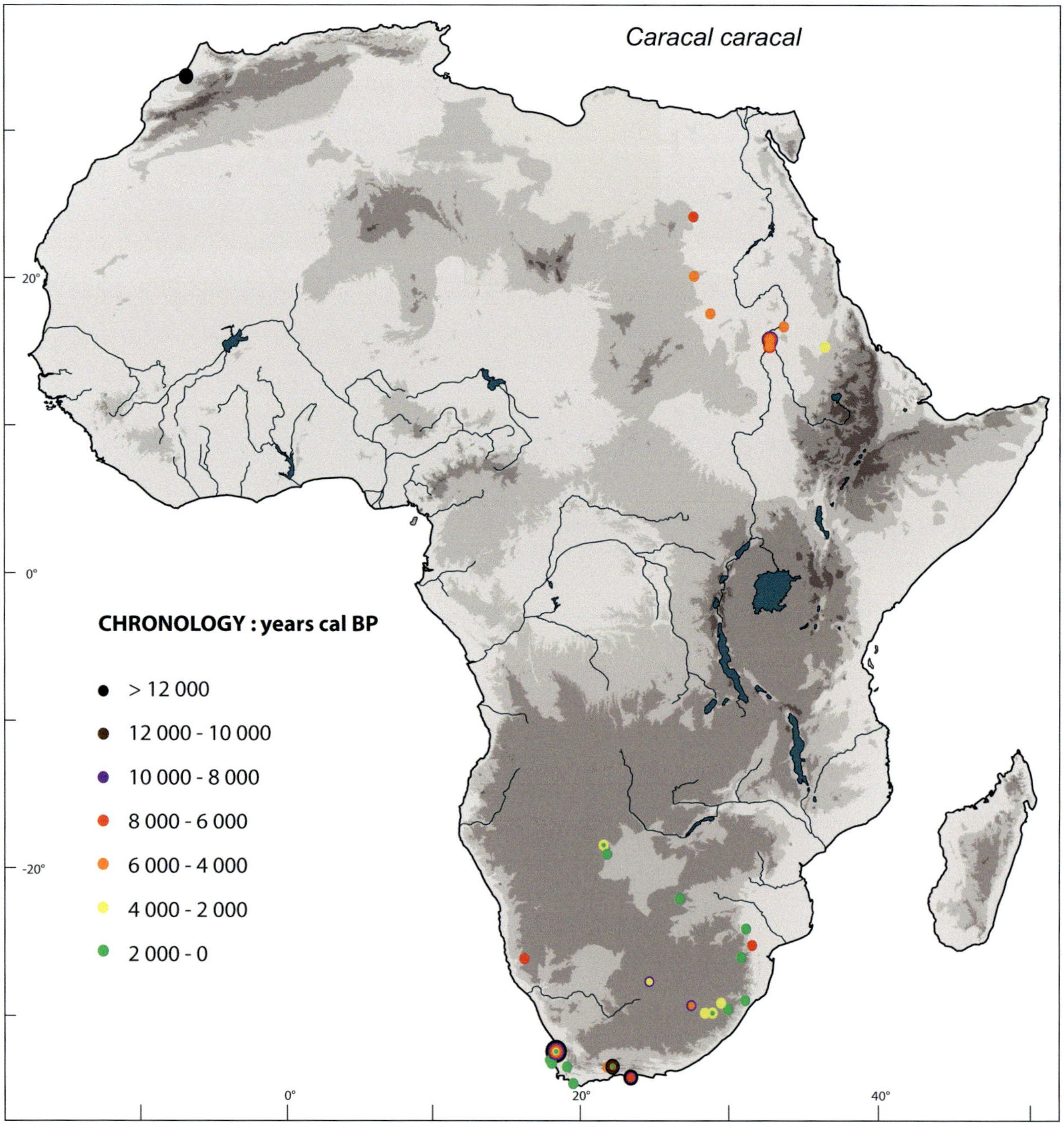

Felis serval

Country	Site N	Site
Congo	89	Tongo
Ethiopia	216	Moche Borago-1b
Kenya	258	Enkapune Ya Muto - Eb 4a
	259	Enkapune Ya Muto - Eb 4b
	260	Enkapune Ya Muto - Eb 5a
	261	Enkapune Ya Muto - Eb 5b
	275	GvJm 19-HA
	279	GvJm 22-LPG
	283	GvJm 62-LPNG
Mozambique	442	Manekweni-2-ABC
Nigeria	503	Gajiganna A
South Africa	674	Ondini
Sudan	783	Kadero
	815	Sheikh Mustafa
Zaire	848	Kiantapo
	855	Matupi-II

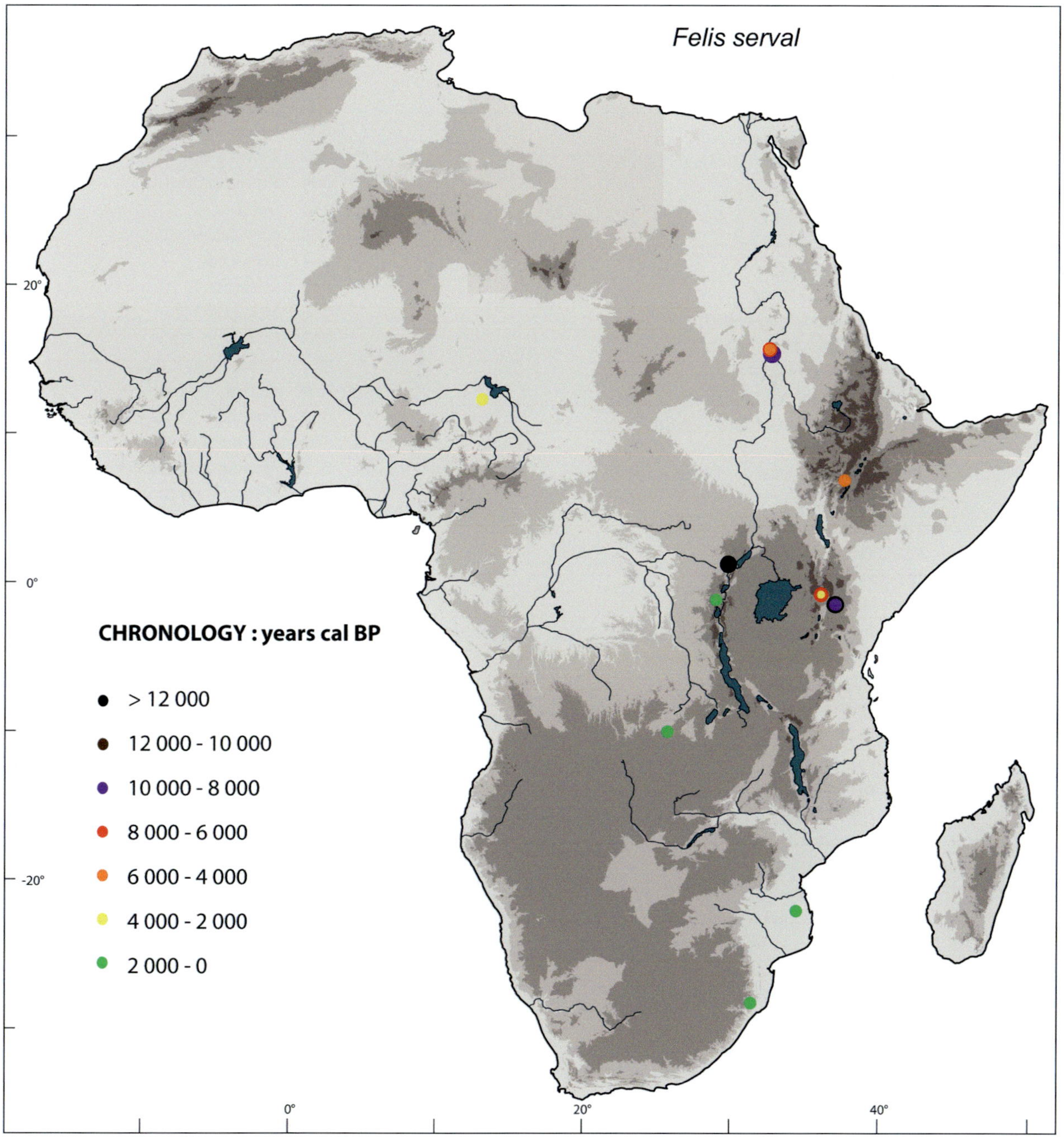

Felis sylvestris

Country	Site N	Site	
Botswana	32	Divuyu	
Djibouti	90	Asa Koma	
Egypt	109	Bir Kiseiba-EN	
	111	Bir Kiseiba-MN	
	165	Mahgar Dendera 2	
	94	Abu Ballas - 85/56	*cf.
Ghana	242	Mankuma	
Mauritania	393	Berouâga	
Morocco	430	Kaf-taht-el-Ghar-B	
	437	Kehf-el-Baroud	
	438	Kehf-el-Baroud-aven	
Nigeria	505	Gajiganna B2	
	530	Zilum	
South Africa	654	Mamba 1	
	752	Wosi-middle-late EIA	
	753	Wosi-terminal EIA	
Sudan	761	Debbat Bangdit	
	762	Debbat El Eheima	
	764	El Damer	
	765	El Kadada	
	766	El Kenger Middle/East	
	773	Esh Shaheinab	
	787	Khartoum Hopital	
	788	Khasm el Girba - Kassala	
	789	Khasm el Girba - KG14	
	791	Khasm el Girba - KG68	
	805	Saggai 1	
	808	Selima Sandsheet - 85/73-2	
	811	Shaqaduq-S1-A	
	813	Shaqaduq-S21	
	822	Upper Atbara Valley	

Felis sylvestris libyca

Country	Site N	Site	
Botswana	28	Bosutwe-Toutswe	
Chad	85	Koyom	
Egypt	100	Adaima-habitat	
	135	Elephantine-necropolis	
	154	Hierakonpolis HK6-Naq IC-IIA	*cf.
	169	Minshat Abu Omar	
	171	Napta Playa 77-1/4/5/6	*?
	172	Napta Playa 77-8	*?
	196	Tuna el-Gebel	
	99	Adaima-habitat	
Ghana	223	Anyinam-Area A-B-C	
	225	Asantamanso-ASII/III	
	227	Asantamanso-ASV	
Kenya	259	Enkapune Ya Muto - Eb 4b	
	260	Enkapune Ya Muto - Eb 5a	
	284	Kulchurdo	
Lesotho	312	Liphofung-CW	
	313	Liphofung-pCW	
	316	Sehonghong-a	
	323	Tloutle-pCW	
Libya	327	Hauah Fteah-Iberom	
	329	Hauah Fteah-Neolithic	
	330	In Aghelachem-RT,SF	*cf.
	333	Ti-n-Thora East	
	340	Uan Muhuggiag (MP)	
Mali	391	Windé Koroji SO	
Morocco	422	El Harhoura 1	
Namibia	448	Apollo 11-b	
	450	Brandberg-BBC	
	457	Haalenberg	
Niger	473	Gobero-K	
Nigeria	503	Gajiganna A	
South Africa	560	Blydefontein III-IV	
	562	Blydefontein VI-VII	
	565	Boomplass Cave-c	
	566	Boomplass Cave-d	
	568	Buffelskloof RS-a	*?
	572	Buffelskloof RS-e	*?
	578	Byeneskranskop Cave 1c	
	579	Byeneskranskop Cave 1d	
	580	Collingham Shelter	
	582	Colwinton Shelter-pCW	
	583	Connies Limpet Bar	
	586	Die Kelders	
	589	Doornfontein	
	595	Eland´s Bay Cave-Al	
	596	Eland´s Bay Cave-Rl	
	600	Eland´s Bay Cave-Wid	
	601	Eland´s Bay Open-a	
	602	Eland´s Bay Open-b	
	604	Equus Cave 1A	
	618	Hill X	
	628	Kasteelberg B	
	644	Leliehoek	
	646	Limerock 2	
	659	Marupale Hill	
	663	Mhlwazini Cave	
	670	Nelson Bay Cave- W	
	674	Ondini	
	680	Pancho's Kitchen Midden 6-7	
	690	Powerhouse Cave	*cf.
	698	Riversmead Shelter III-IV	
	699	Riversmead Shelter VI-IX	
	701	Rooikrans-CW	
	704	Rose Cottage Cave-A-Mn	
	707	Rose Cottage Cave-Wl	
	711	Serotwe Hill	
	718	Sk4	
	724	Spring Cave	
	727	Steenbokfontein-3	
	728	Steenbokfontein-4a	
	729	Steenbokfontein-4b	
	736	Tortoise cave-b	
	738	Tortoise cave-d	
	742	Ventershoek	
	744	Voëlvlei	
	747	Witklip-1	
	748	Witklip-2	
	749	Witklip-3	
	750	Witklip-4	
Sudan	767	El Mahalab	
	781	Jebel Umm Marrahi	
	783	Kadero	
	815	Sheikh Mustafa	
Zambia	860	Leopard's Hill-NI	
Zimbabwe	870	Redcliff Cave	

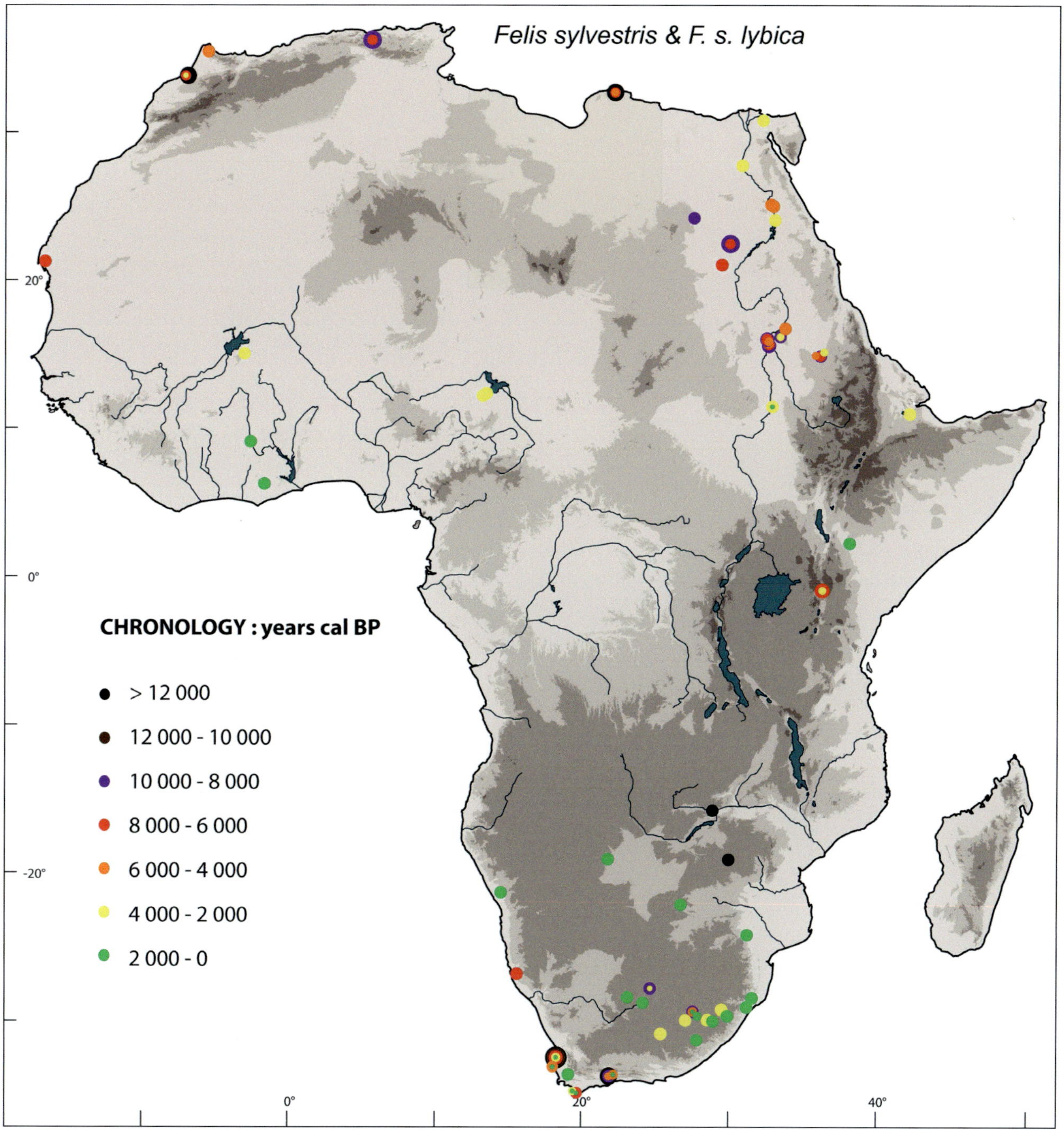
Felis sylvestris & F. s. lybica
CHRONOLOGY : years cal BP
> 12 000
12 000 - 10 000
10 000 - 8 000
8 000 - 6 000
6 000 - 4 000
4 000 - 2 000
2 000 - 0
20°
0°
-20°
0°
20°
40°

Felis sylvestris f. *catus*

Country	Site N	Site
Egypt	106	Berenike 95/9 Early Roman
	107	Berenike 95/97Late Roman
	108	Berenike 96/97 Ptolem.
	179	Quseir al-Qadim
Mali	371	Gao Ancien
Sudan	817	Soba

Felis aurata

Country	Site N	Site
Congo	88	Ntadi Yomba-level A
Ghana	232	Dawu

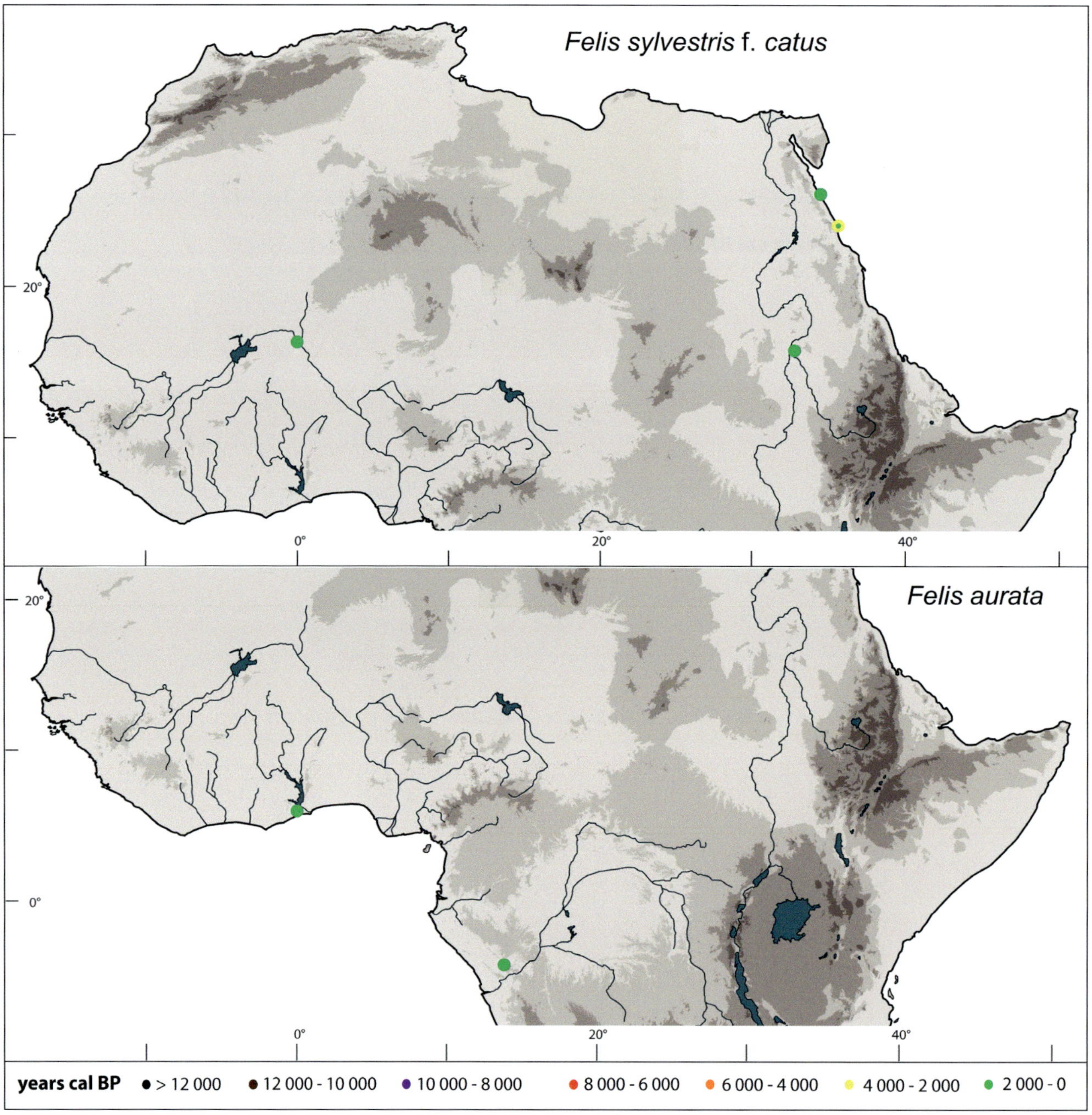

Felis nigripes

Country	Site N	Site
South Africa	604	Equus Cave 1A

Felis margarita

Country	Site N	Site
Morocco	422	El Harhoura 1

Felis chaus

Country	Site N	Site
Egypt	112	Buto/1-3
Algeria	15	Medjez II

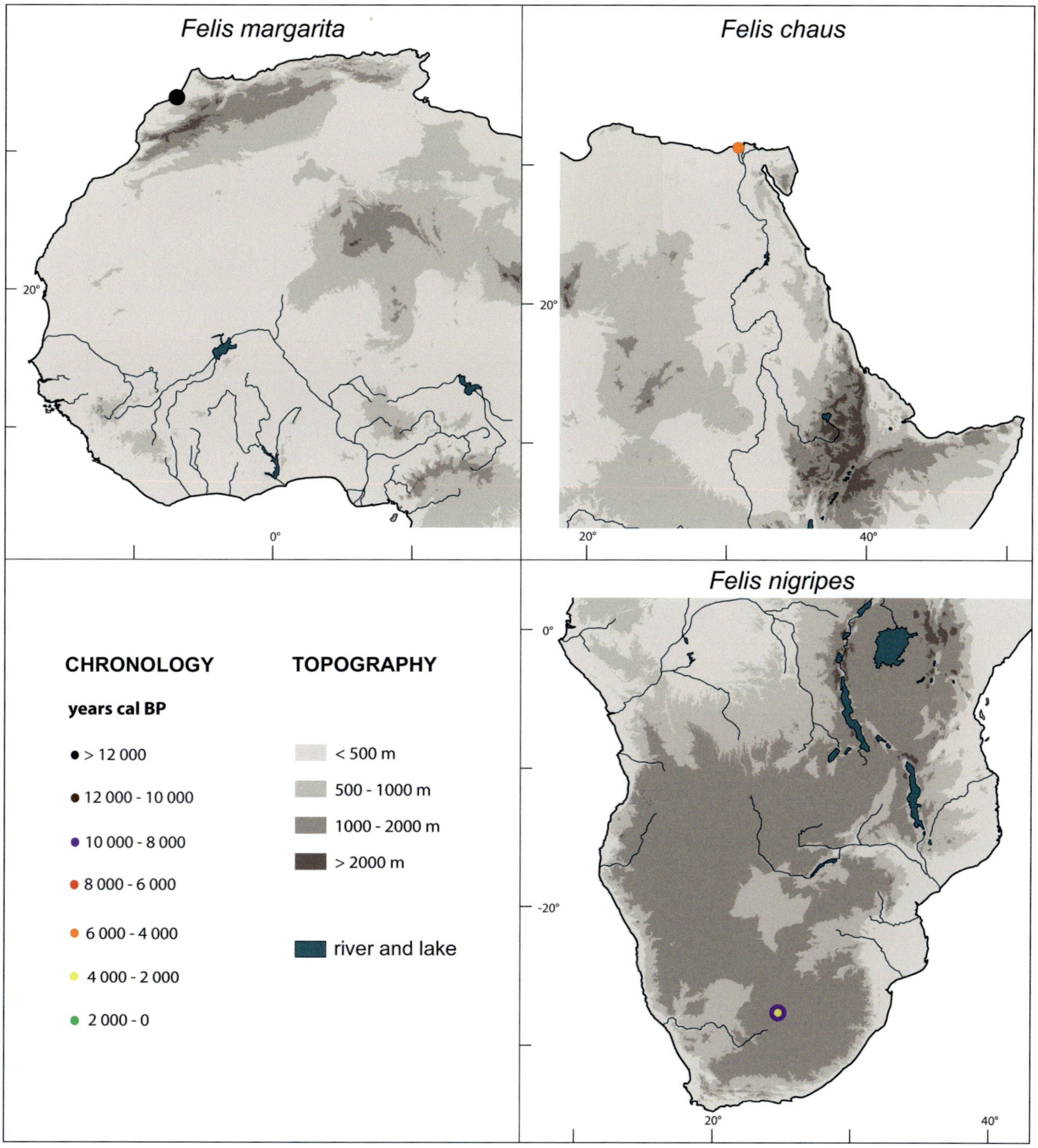

Panthera leo

Country	Site N	Site
Algeria	7	Damous el Ahmar
	9	Dra-Mta-el-Ma-el-Abiod
Botswana	42	Toteng1-SU2
Egypt	106	Berenike 95/9 Early Roman
	107	Berenike 95/97Late Roman
	168	Merimde Benisâlame
	174	Qantir
	197	Umm el-Qaab, Friedhof U
Kenya	275	GvJm 19-HA
	276	GvJm 19-HC
	278	GvJm 22-LHC
Libya	333	Ti-n-Thora East
Mali	363	Asselar
	374	Hassi el-Abiod
	385	Korounkorokalé-III
Mauritania	400	Dhar Tichitt Site 38
Morocco	422	El Harhoura 1
Mozambique	442	Manekweni-2-ABC
	444	Manekweni-4
South Africa	568	Buffelskloof RS-a
	589	Doornfontein
	615	Happy Rest
	617	Heuningneskrans-R
	618	Hill X
	638	Langsdraai
	641	Le6
	674	Ondini
	676	Oudepost I
	717	SK17
	718	Sk4
Sudan	762	Debbat El Eheima
	773	Esh Shaheinab
	785	Kerma * imported?
	787	Khartoum Hopital
	788	Khasm el Girba - Kassala
	805	Saggai 1
	817	Soba
Zaire	843	Ishango-AB
	845	Ishango-Z-Post-Em
	848	Kiantapo
Zimbabwe	870	Redcliff Cave

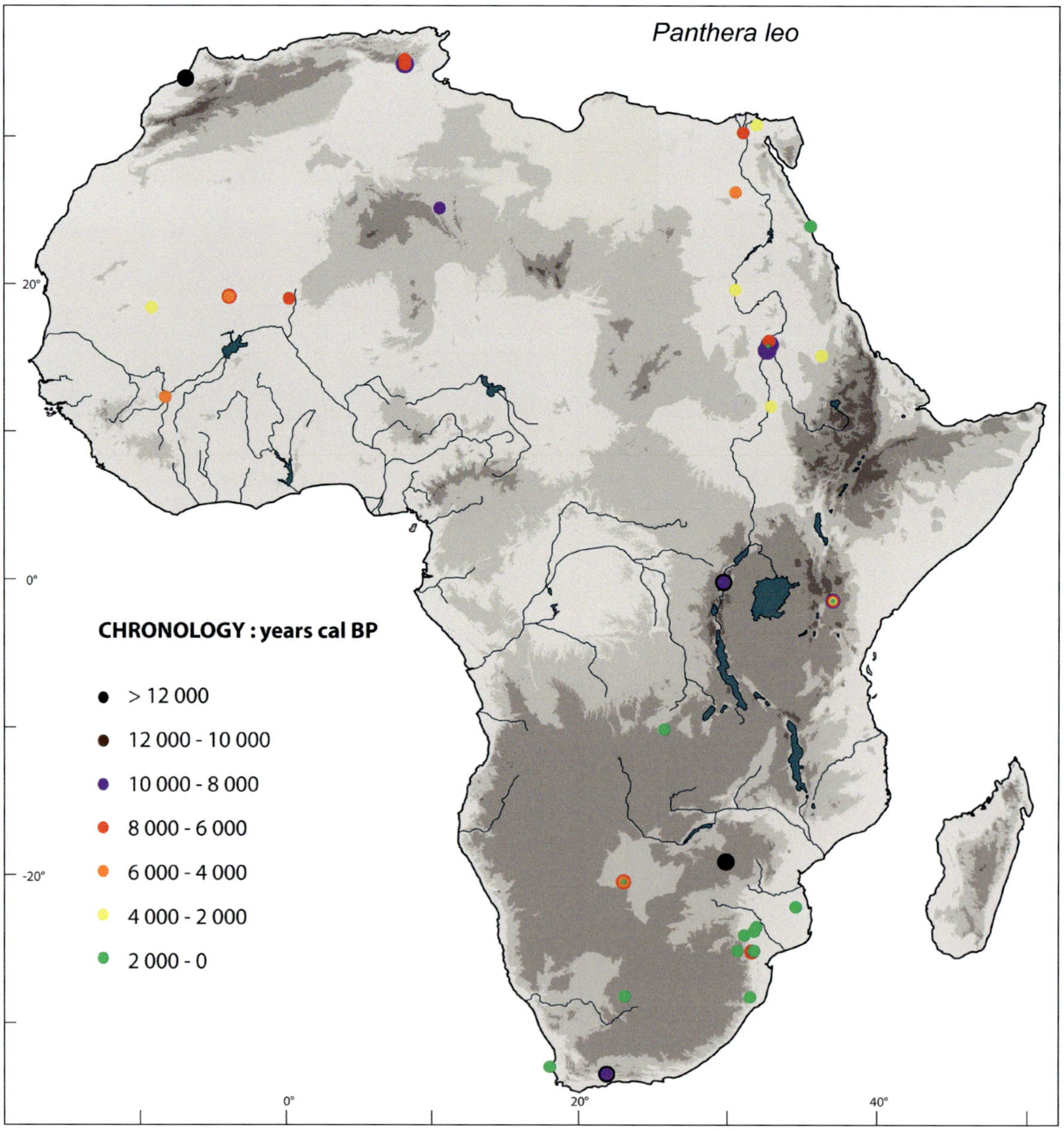

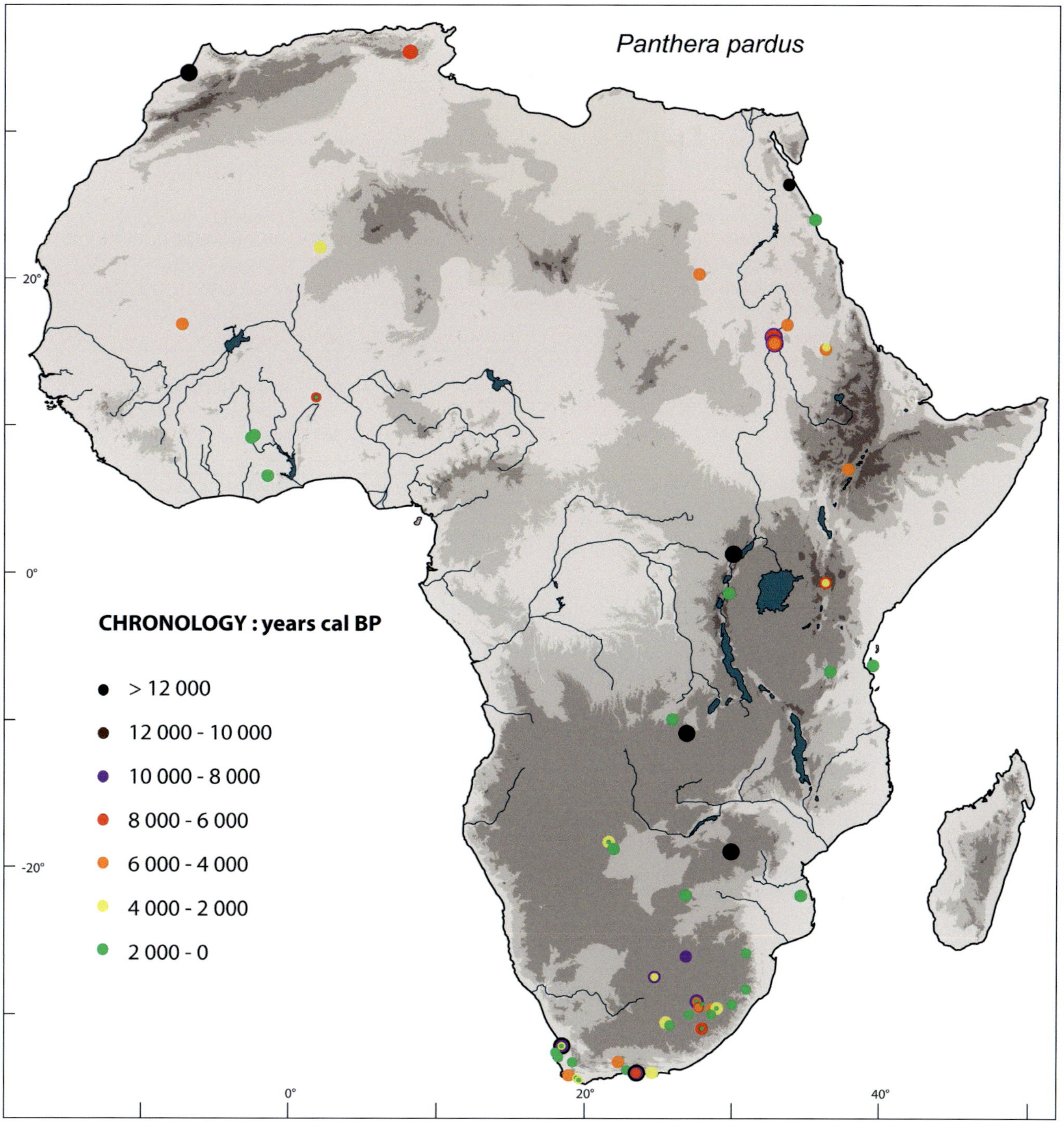

Panthera pardus

Country	Site N	Site
Algeria	2	A-88-28
	7	Damous el Ahmar
Botswana	26	Bosutwe-Mapungubwe
	29	Bosutwe-Zimbabwe
	32	Divuyu
	38	Nqoma
	48	White Paintings RS-1
	48	White Paintings RS-2ab
Burkina Faso	62	Péntènga 89/1-spits 4-10 *copro
Egypt	107	Berenike 95/97Late Roman
	185	Sodmein Cave older backfill
Ethiopia	216	Moche Borago-1b
Ghana	225	Asantamanso-ASII/III
	227	Asantamanso-ASV
	241	Mandara
	242	Mankuma
	246	Nyanga
Kenya	259	Enkapune Ya Muto - Eb 4b
	260	Enkapune Ya Muto - Eb 5a
	293	Naivasha Railway Shelter
Lesotho	303	Bolahla
	308	Likoaeng-VII-IX
	312	Liphofung-CW
	321	Tloutle-CWI
	323	Tloutle-pCW
Mauritania	402	DN4-BKH-S5
Morocco	422	El Harhoura 1
Mozambique	442	Manekweni-2-ABC
	445	Manekweni-5

Rwanda	535	Akameru-Cyinkomane	* intr.
South Africa	560	Blydefontein III-IV	
	565	Boomplass Cave-c	
	579	Byeneskranskop Cave 1d	
	580	Collingham Shelter	
	581	Colwinton Shelter-CW	
	582	Colwinton Shelter-pCW	
	586	Die Kelders	
	595	Eland´s Bay Cave-Al	
	596	Eland´s Bay Cave-Rl	
	597	Eland´s Bay Cave-Wia	
	598	Eland´s Bay Cave-Wib	
	604	Equus Cave 1A	
	609	Glentyre	
	622	HKLP Vb	
	628	Kasteelberg B	
	632	Klasies River Mouth-1II	
	652	Magogo-3	
	653	Malwula	
	668	Nelson Bay Cave- PW	
	670	Nelson Bay Cave- W	
	676	Oudepost I	
	678	Pancho's Kitchen Midden 1-2	
	699	Riversmead Shelter VI-IX	
	700	Rooiels	
	701	Rooikrans-CW	
	704	Rose Cottage Cave-A-Mn	
	707	Rose Cottage Cave-WI	
	733	Tienfontein 2	
	744	Voëlvlei	
Sudan	765	El Kadada	
	773	Esh Shaheinab	
	787	Khartoum Hopital	
	788	Khasm el Girba - Kassala	
	794	Laqiya area - 82/31	
	805	Saggai 1	
	814	Sheikh el Amin	
	815	Sheikh Mustafa	
	821	Upper Atbara hinterland	
Zaire	846	Kakontwe	
	848	Kiantapo	
	855	Matupi-II	
Zanzibar	867	Kizimkazi Dimbani	
Zimbabwe	870	Redcliff Cave	

VIVERRIDAE

Country	Site N	Site
Botswana	26	Bosutwe-Mapungubwe
	27	Bosutwe-Taukome
	28	Bosutwe-Toutswe
	34	Drotsky's Cave-LP
	46	Toteng3-SU4
	48	White Paintings RS-1
	48	White Paintings RS-2ab
	49	White Paintings RS-2b-3ab
Ethiopia	216	Moche Borago-1b
Ghana	245	Nyanawase/Akwamu Am.
Kenya	254	Ele Bor A-B
Lesotho	303	Bolahla
Mali	376	Jenné-Jeno - phase I/II
	377	Jenné-Jeno - phase III
	378	Jenné-Jeno - phase IV
Namibia	449	Apollo 11-c
	454	Fackelträger
	463	Tiras
South Africa	625	Jubilee Shelter-PW
	659	Marupale Hill
	722	Spoegrivier-b
	723	Spoegrivier-c
	725	Steenbokfontein-1
	744	Voëlvlei
Sudan	804	Rahib-80/87
	811	Shaqaduq-S1-A
	812	Shaqaduq-S1-B
	813	Shaqaduq-S21
Zaire	848	Kiantapo
	854	Matupi-I
Zimbabwe	870	Redcliff Cave

Civettictis civetta

Country	Site N	Site
Botswana	26	Bosutwe-Mapungubwe
	38	Nqoma
Cameroon	76	Shum Laka II c
	78	Shum Laka IIa+b
Chad	85	Koyom
Ghana	223	Anyinam-Area A-B-C
	225	Asantamanso-ASII/III
	227	Asantamanso-ASV
	233	Esiease - A-C-D
Kenya	254	Ele Bor A-B
	255	Ele Bor A-C
	259	Enkapune Ya Muto - Eb 4b
Mali	361	Akumbu-EIA
Senegal	541	Cubalel-III
South Africa	625	Jubilee Shelter-PW
	642	Le7a
	673	Ol20
	718	Sk4
Sudan	773	Esh Shaheinab
	805	Saggai 1
Zaire	848	Kiantapo
	854	Matupi-I

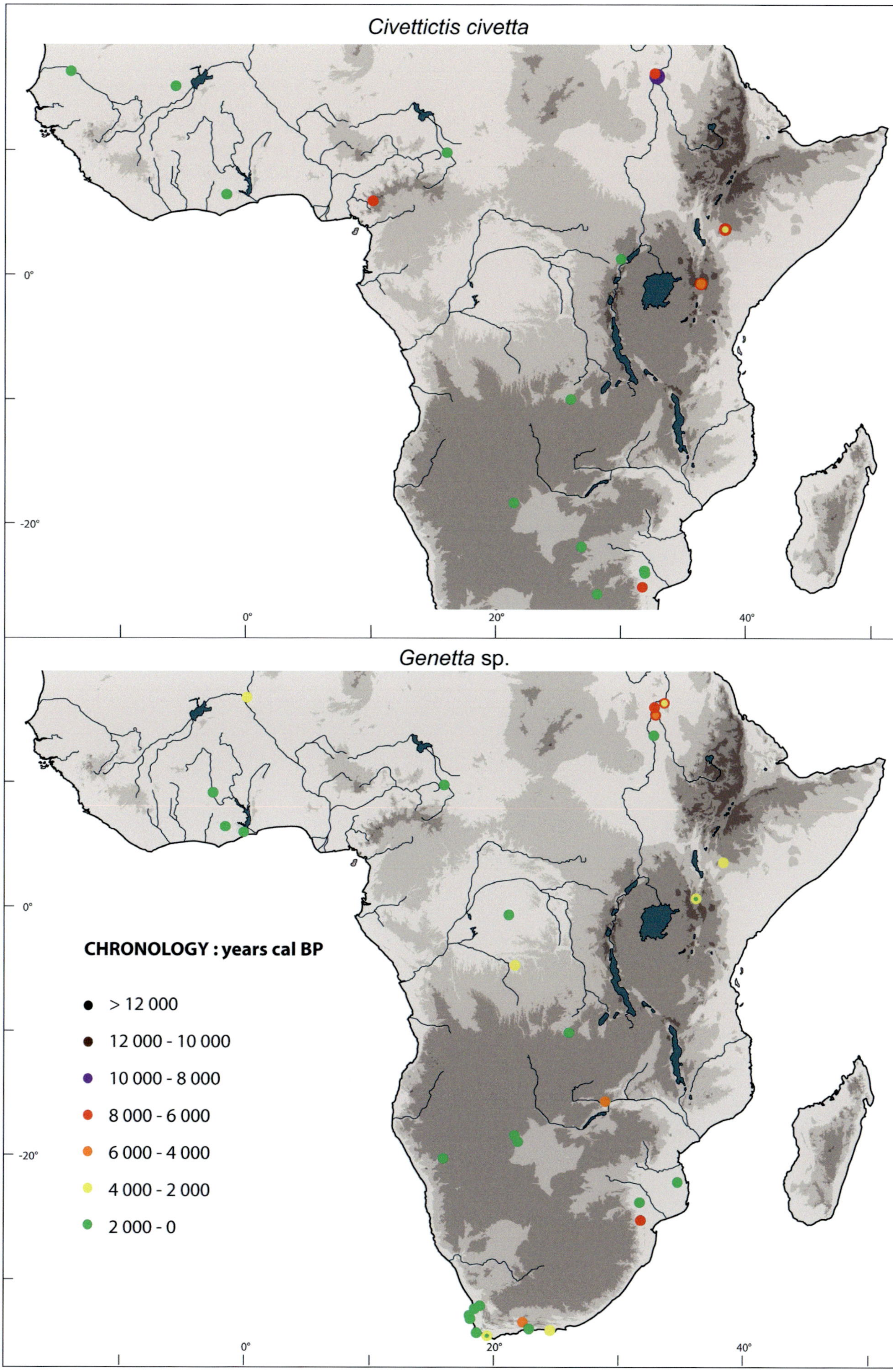
Civettictis civetta
0°
-20°
0°
20°
40°
Genetta sp.
0°
CHRONOLOGY : years cal BP
> 12 000
12 000 - 10 000
10 000 - 8 000
8 000 - 6 000
6 000 - 4 000
4 000 - 2 000
2 000 - 0
-20°
0°
20°
40°

Genetta sp.

Country	Site N	Site	
Botswana	32	Divuyu	
	48	White Paintings RS-1	
Chad	85	Koyom	
Ghana	223	Anyinam-Area A-B-C	
	225	Asantamanso-ASII/III	
	227	Asantamanso-ASV	
	232	Dawu	
	242	Mankuma	
Kenya	254	Ele Bor A-B	
	297	Ngenyn	
Mali	380	Karkarichinkat N-S	*?
Mozambique	442	Manekweni-2-ABC	
	444	Manekweni-4	
	445	Manekweni-5	
Namibia	455	Geduld-LIA	*cf.
South Africa	565	Boomplass Cave-c	
	583	Connies Limpet Bar	
	586	Die Kelders	
	600	Eland´s Bay Cave-Wid	
	609	Glentyre	
	628	Kasteelberg B	
	632	Klasies River Mouth-1II	*?
	634	Klein Kliphuis	
	640	Le2	
	671	Oakhurst	
	676	Oudepost I	
	717	SK17	
	718	Sk4	
	719	Smitswinkelbaai Cave	
	724	Spring Cave	
Sudan	773	Esh Shaheinab	
	780	Jebel et Tomat	
	811	Shaqaduq-S1-A	
	812	Shaqaduq-S1-B	
	814	Sheikh el Amin	
Zaire	842	Dimba	
	848	Kiantapo	
	857	Nkile	
Zambia	862	Leopard's Hill-NIII	

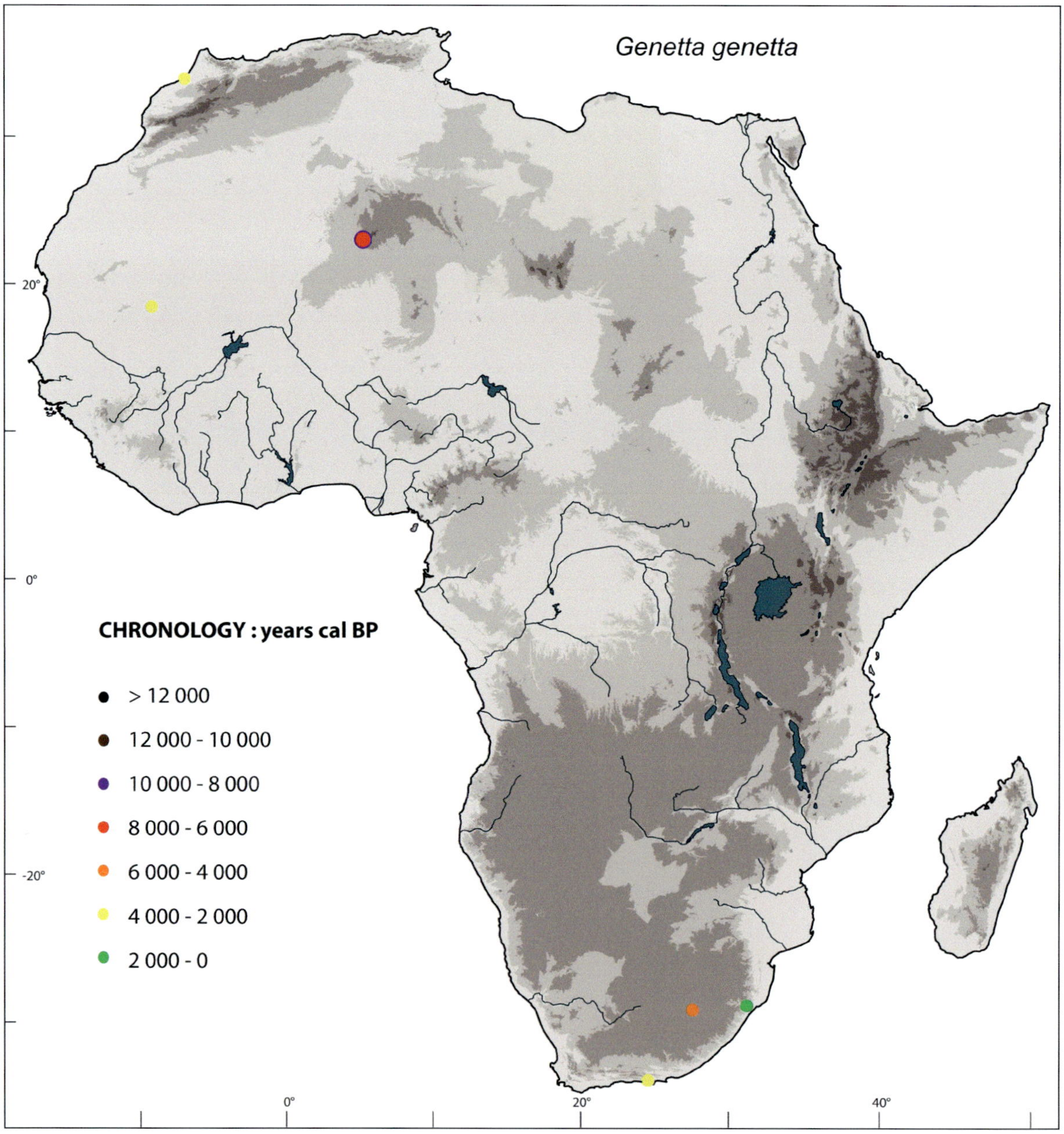

Genetta genetta

Country	Site N	Site
Algeria	5	Amekni
Mauritania	400	Dhar Tichitt Site 38
Morocco	438	Kehf-el-Baroud-aven
South Africa	633	Klasies River Mouth-5II
	644	Leliehoek
	751	Wosi-early EIA

Genetta tigrina

Country	Site N	Site
Kenya	259	Enkapune Ya Muto - Eb 4b

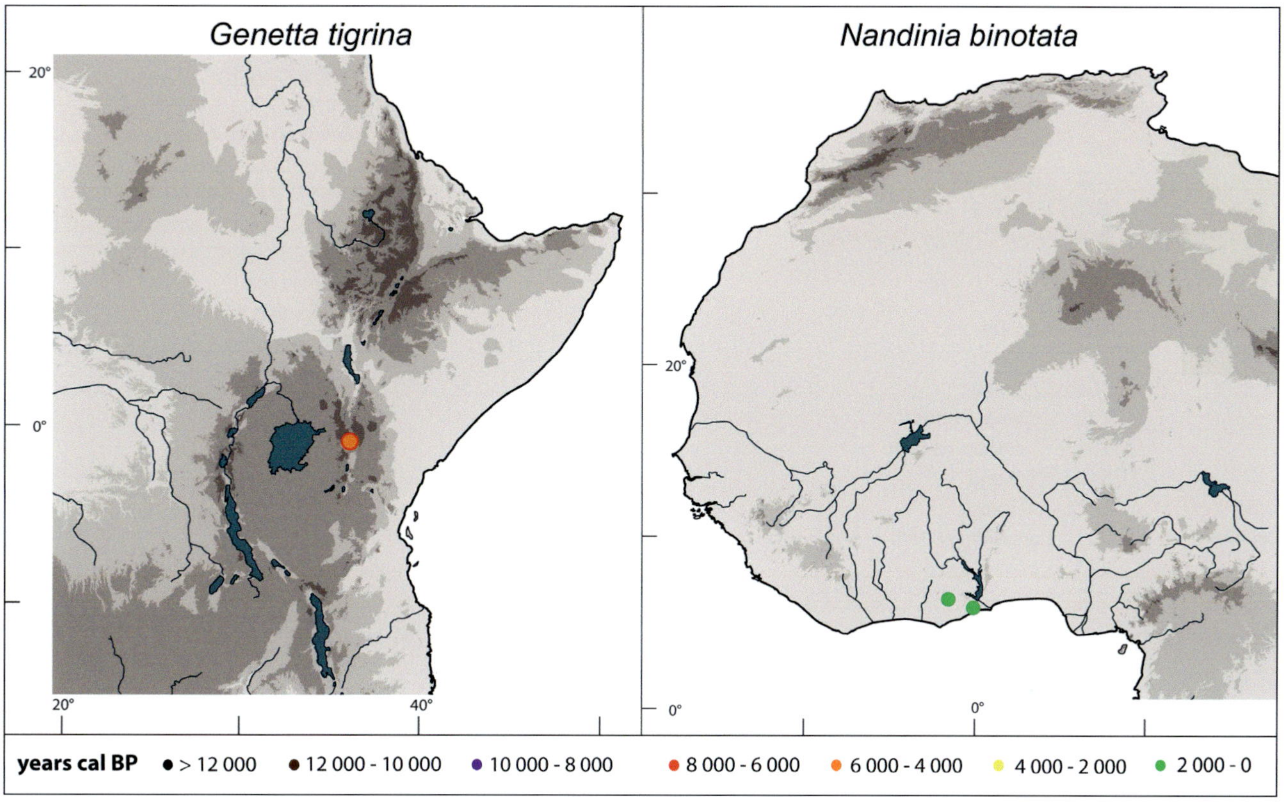

NANDINIIDAE

Nandinia binotata

Country	Site N	Site
Ghana	223	Anyinam-Area A-B-C
	227	Asantamanso-ASV
	232	Dawu

HERPESTIDAE

Country	Site N	Site	
Botswana	32	Divuyu	
	37	Matlapaneng	
	38	Nqoma	
Kenya	259	Enkapune Ya Muto - Eb 4b	
Lesotho	305	Lequetsoana	
	307	Likoaeng-III-V	
	308	Likoaeng-VII-IX	
	309	Likoaeng-XI	
	310	Likoaeng-XIII	
	316	Sehonghong-a	
Namibia	455	Geduld-LIA	*cf.
	456	Geduld-LSA	
Niger	473	Gobero-K	
South Africa	567	Border Cave	
	638	Langsdraai	
	691	Pr34	
	704	Rose Cottage Cave-A-Mn	
	718	Sk4	
	740	Twyfelpoort	
Sudan	754	Abu Darbein	
	757	Aneibis	
	764	El Damer	
	780	Jebel et Tomat	
	787	Khartoum Hopital	
	810	Shabona	
Zaire	848	Kiantapo	
Zambia	861	Leopard's Hill-NII	
	863	Leopard's Hill-PN	

Atilax paludinosus

Country	Site N	Site	
Chad	85	Koyom	
Ghana	225	Asantamanso-ASII/III	
	227	Asantamanso-ASV	
Lesotho	308	Likoaeng-VII-IX	
	316	Sehonghong-a	
	322	Tloutle-O	
South Africa	577	Byeneskranskop Cave 1b	
	580	Collingham Shelter	
	628	Kasteelberg B	
	654	Mamba 1	
	663	Mhlwazini Cave	
	670	Nelson Bay Cave- W	
	678	Pancho's Kitchen Midden 1-2	
	704	Rose Cottage Cave-A-Mn	
	706	Rose Cottage Cave-OIC	
	709	Sebatini Hill	
	734	Tienfontein 4A	*?
	753	Wosi-terminal EIA	
Zaire	855	Matupi-II	

Crossarchus sp.

Country	Site N	Site
Congo	87	Ntadi Yom ba-level B

Crossarchus alexandri

Country	Site N	Site
Zaire	854	Matupi-I

Bdeogale nigripes

Country	Site N	Site
Zaire	854	Matupi-I

Cynictis penicillata

Country	Site N	Site	
South Africa	559	Blydefontein I-II	
	560	Blydefontein III-IV	
	562	Blydefontein VI-VII	cf
	587	Dikbosch 1	
	588	Dikbosch 2	cf
	604	Equus Cave 1A	
	610	Green Elliott Shelter I-III	
	611	Green Elliott Shelter IV-VI	
	645	Limerock 1	cf
	646	Limerock 2	
	672	OFD1	?
	690	Powerhouse Cave	cf
	698	Riversmead Shelter III-IV	
	704	Rose Cottage Cave-A-Mn	
	733	Tienfontein 2	
	734	Tienfontein 4A	
	742	Ventershoek	

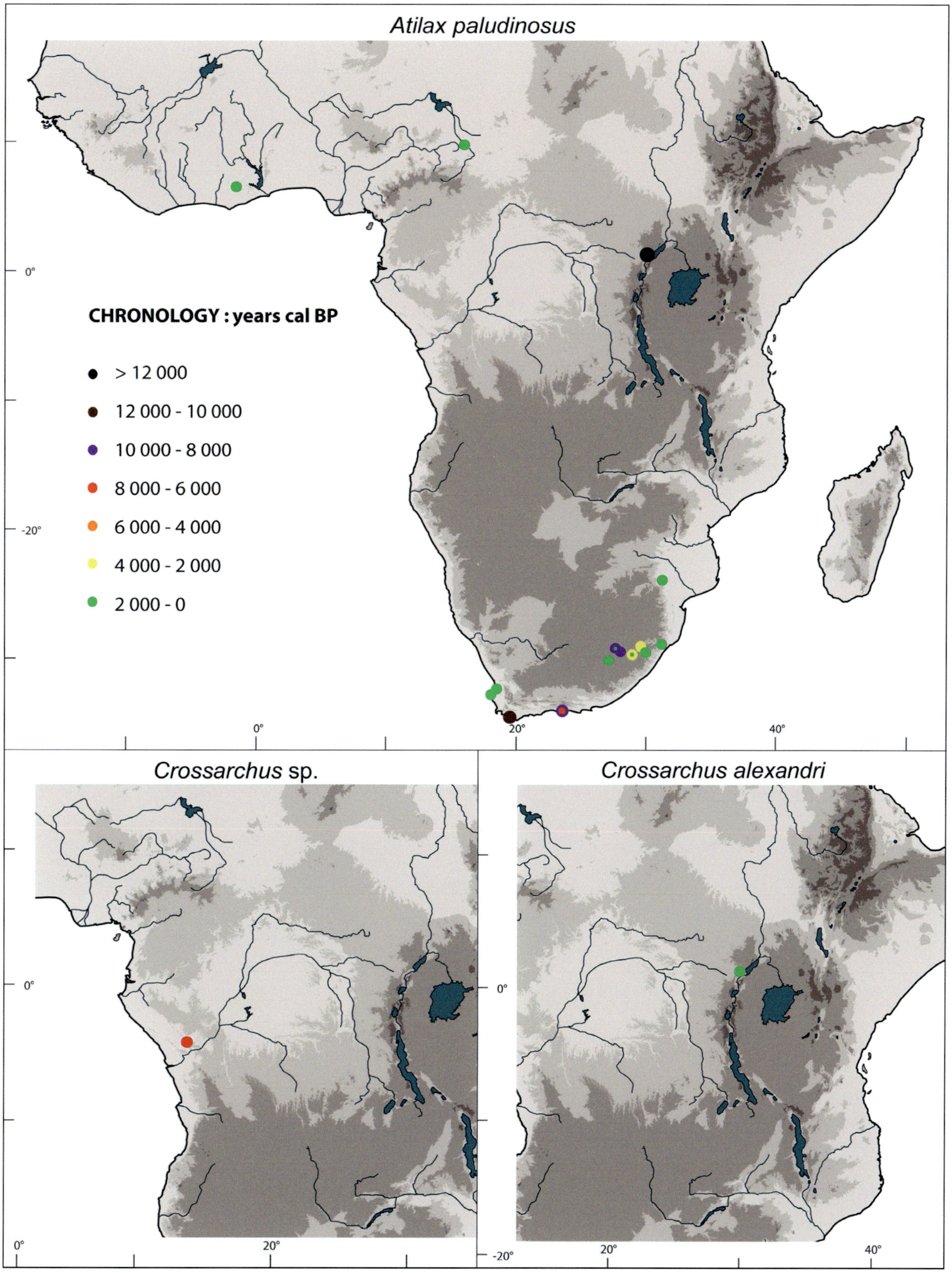
Atilax paludinosus
0°
-20°
0°
20°
40°
CHRONOLOGY : years cal BP
> 12 000
12 000 - 10 000
10 000 - 8 000
8 000 - 6 000
6 000 - 4 000
4 000 - 2 000
2 000 - 0
Crossarchus sp.
0°
0°
20°
Crossarchus alexandri
0°
-20°
20°
40°

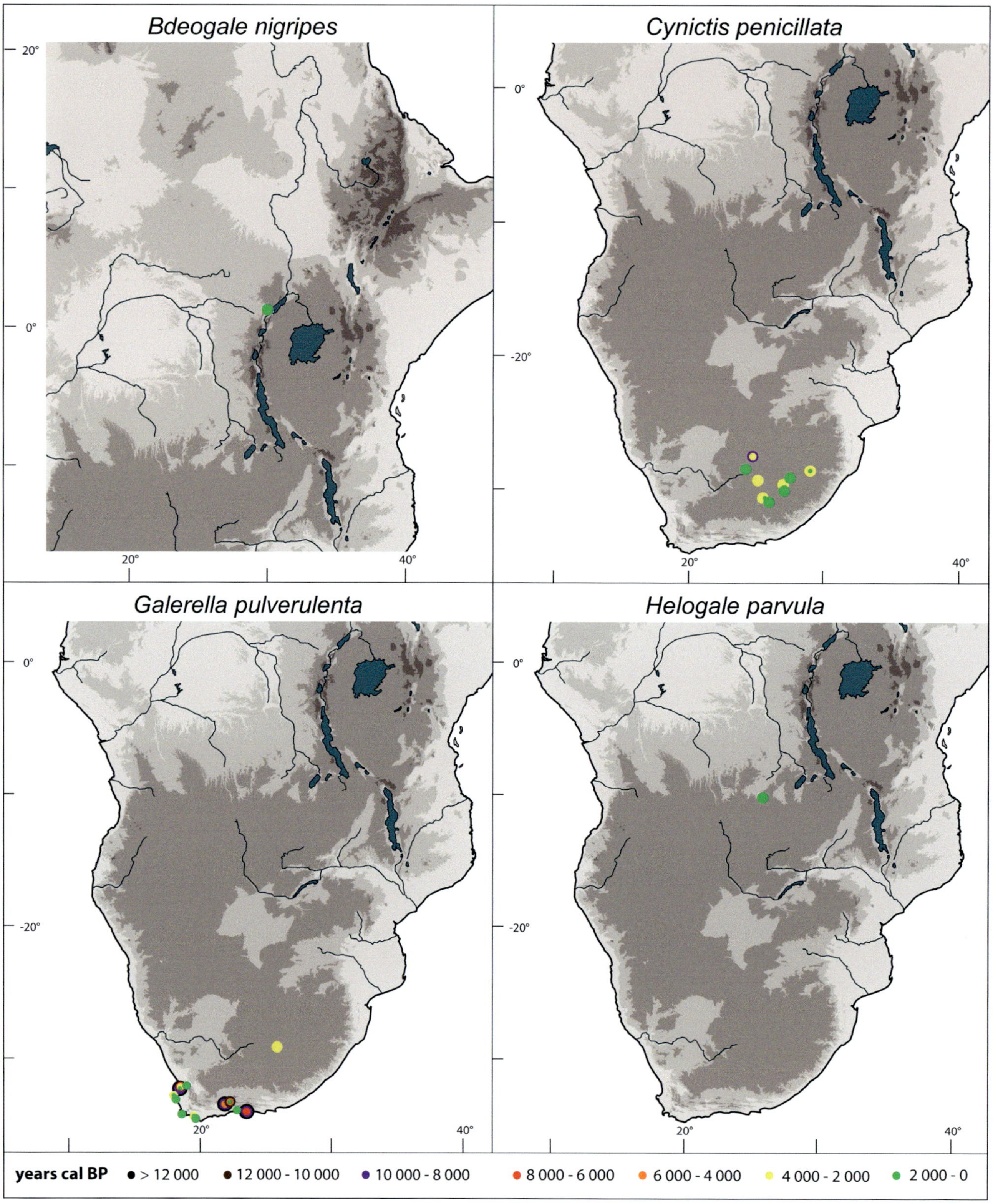

Helogale parvula

Country	Site N	Site
Zaire	848	Kiantapo

Galerella sanguinea

Country	Site N	Site	
Ghana	223	Anyinam-Area A-B-C	
Libya	333	Ti-n-Thora East	
South Africa	653	Malwula	?
Sudan	765	El Kadada	
	773	Esh Shaheinab	
	812	Shaqaduq-S1-B	
Zaire	848	Kiantapo	

Galerella pulverulenta

Country	Site N	Site	
South Africa	564	Boomplass Cave-b	
	565	Boomplass Cave-c	
	566	Boomplass Cave-d	
	568	Buffelskloof RS-a	?
	571	Buffelskloof RS-d	
	572	Buffelskloof RS-e	?
	579	Byeneskranskop Cave 1d	
	585	Deelpan	
	586	Die Kelders	
	595	Eland´s Bay Cave-Al	
	597	Eland´s Bay Cave-Wia	
	598	Eland´s Bay Cave-Wib	
	599	Eland´s Bay Cave-Wic	
	600	Eland´s Bay Cave-Wid	
	602	Eland´s Bay Open-b	
	609	Glentyre	
	613	Hail Stone Midden	
	627	Kasteelberg A	
	628	Kasteelberg B	
	634	Klein Kliphuis	
	669	Nelson Bay Cave- R	
	670	Nelson Bay Cave- W	
	676	Oudepost I	
	679	Pancho's Kitchen Midden 3-4	
	680	Pancho's Kitchen Midden 6-7	
	682	Paternoster3-4	cf
	719	Smitswinkelbaai Cave	
	724	Spring Cave	
	728	Steenbokfontein-4a	
	729	Steenbokfontein-4b	
	736	Tortoise cave-b	
	738	Tortoise cave-d	
	747	Witklip-1	
	748	Witklip-2	
	749	Witklip-3	

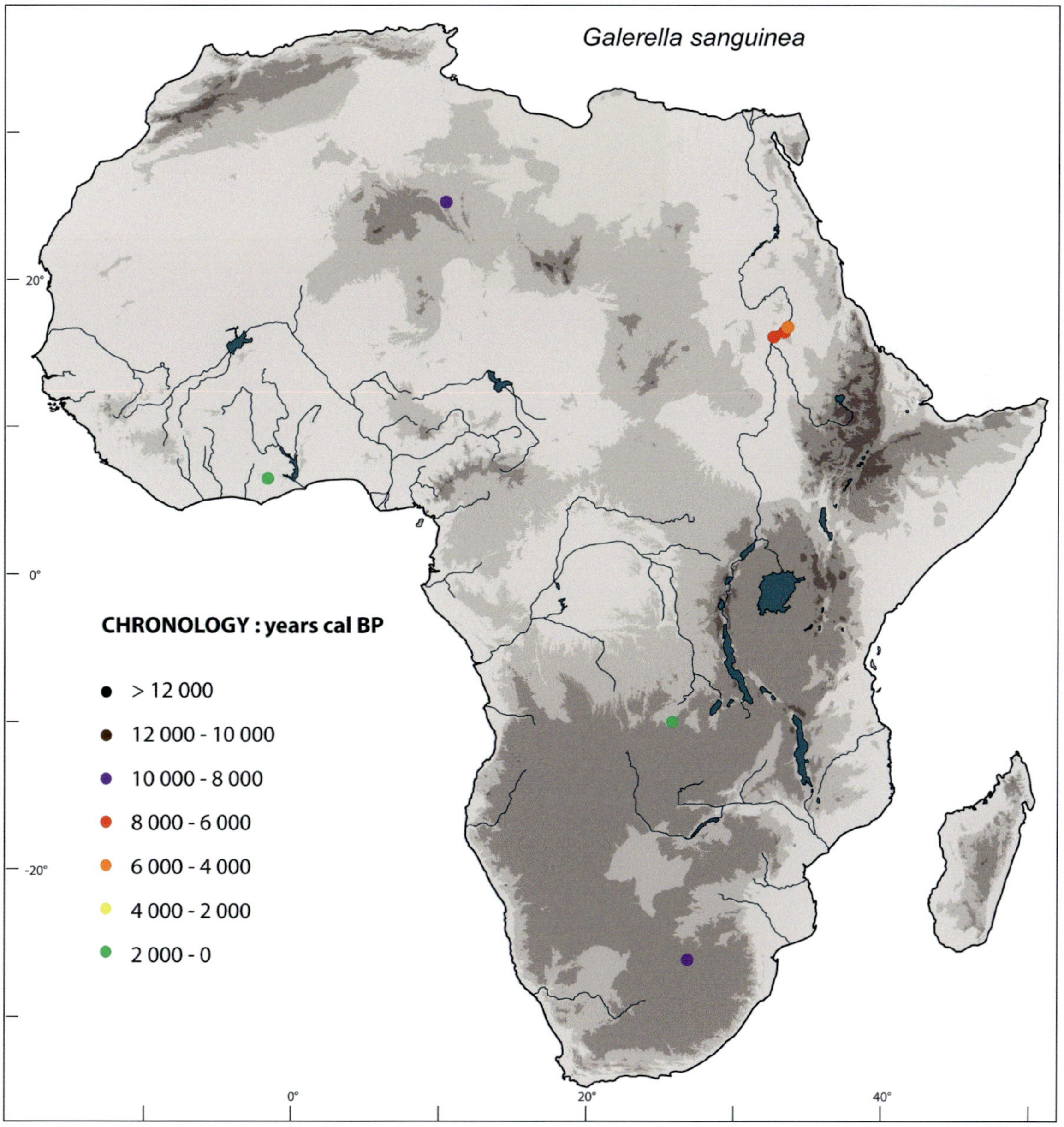

Herpestes ichneumon

Country	Site N	Site	
Algeria	14	Mankhor	
	15	Medjez II	
	21	Ti-n-Hanakaten	
	5	Amekni	
	9	Dra-Mta-el-Ma-el-Abiod	
Egypt	171	Napta Playa 77-1/4/5/6	
	196	Tuna el-Gebel	
Ghana	227	Asantamanso-ASV	
Libya	333	Ti-n-Thora East	
Mali	383	Kolima Sud-II	
Morocco	422	El Harhoura 1	
	426	Ghar-Khal IIIb-V	
Senegal	549	Tulel-Fobo	
South Africa	565	Boomplass Cave-c	
	566	Boomplass Cave-d	
	578	Byeneskranskop Cave 1c	cf
	579	Byeneskranskop Cave 1d	
	595	Eland´s Bay Cave-Al	
	599	Eland´s Bay Cave-Wic	
	601	Eland´s Bay Open-a	
	602	Eland´s Bay Open-b	
	604	Equus Cave 1A	
	627	Kasteelberg A	
	628	Kasteelberg B	
	653	Malwula	?
	668	Nelson Bay Cave- PW	
	669	Nelson Bay Cave- R	
	670	Nelson Bay Cave- W	
	672	OFD1	?
	676	Oudepost I	
	698	Riversmead Shelter III-IV	cf
	700	Rooiels	
	724	Spring Cave	
	729	Steenbokfontein-4b	
Sudan	761	Debbat Bangdit	?
	762	Debbat El Eheima	
	765	El Kadada	
Zambia	863	Leopard's Hill-PN	

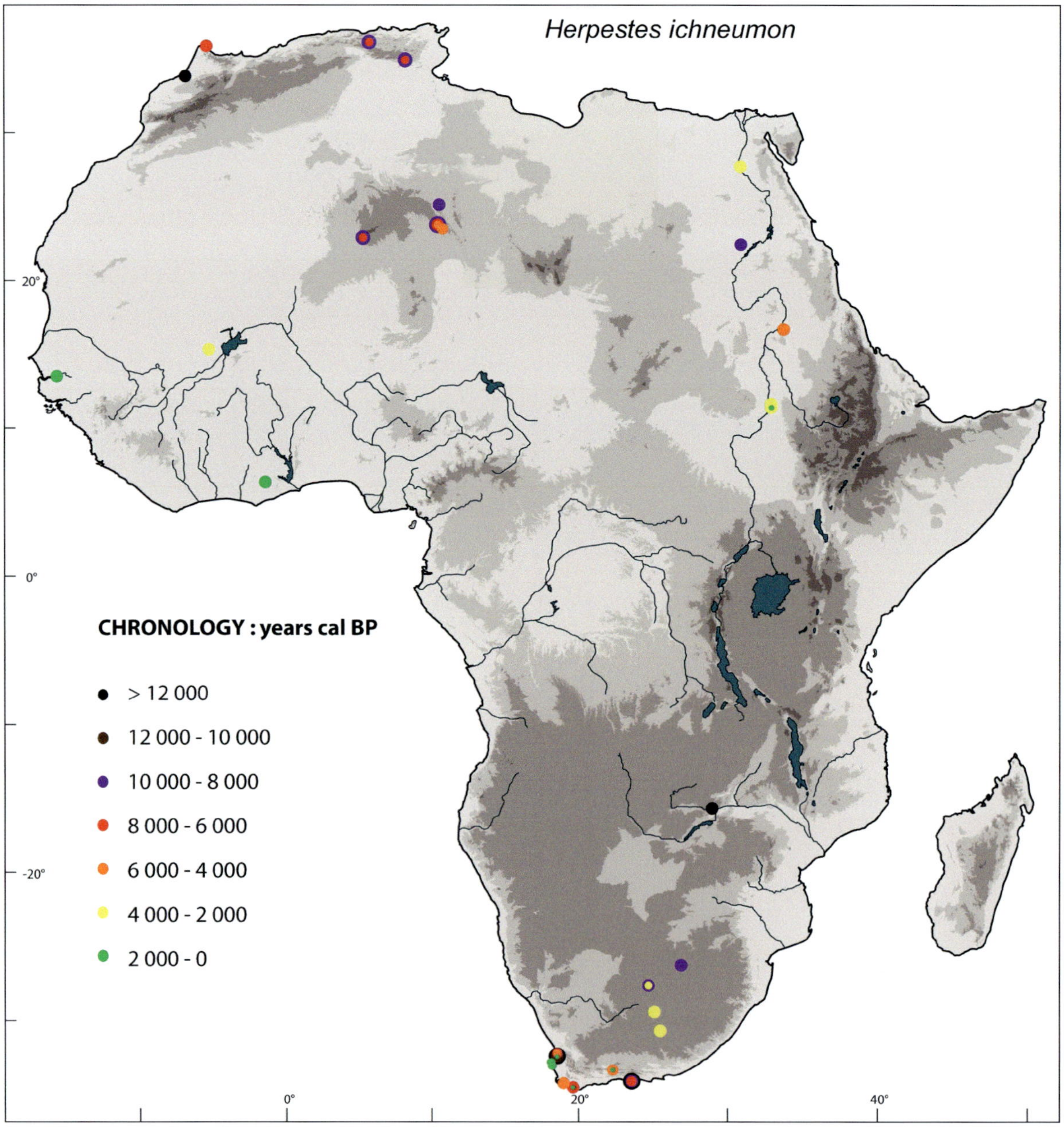

Herpestes naso

Country	Site N	Site
Zaire	848	Kiantapo

Ichneumia albicauda

Country	Site N	Site
Chad	85	Koyom
Ghana	246	Nyanga
Kenya	254	Ele Bor A-B

Mungos mungo

Country	Site N	Site
Chad	85	Koyom
Ghana	223	Anyinam-Area A-B-C
	225	Asantamanso-ASII/III
	227	Asantamanso-ASV
	233	Esiease - A-C-D
Sudan	765	El Kadada
	787	Khartoum Hopital
Zaire	848	Kiantapo
	853	Mashita Mbanza
Zambia	860	Leopard's Hill-NI
	863	Leopard's Hill-PN

Suricata suricatta

Country	Site N	Site	
Namibia	450	Brandberg-BBC	
	451	Brandberg-late LSA	
	452	Brandberg-LSA	
	453	Brandberg-MSA	
South Africa	622	HKLP Vb	cf
	707	Rose Cottage Cave-WI	

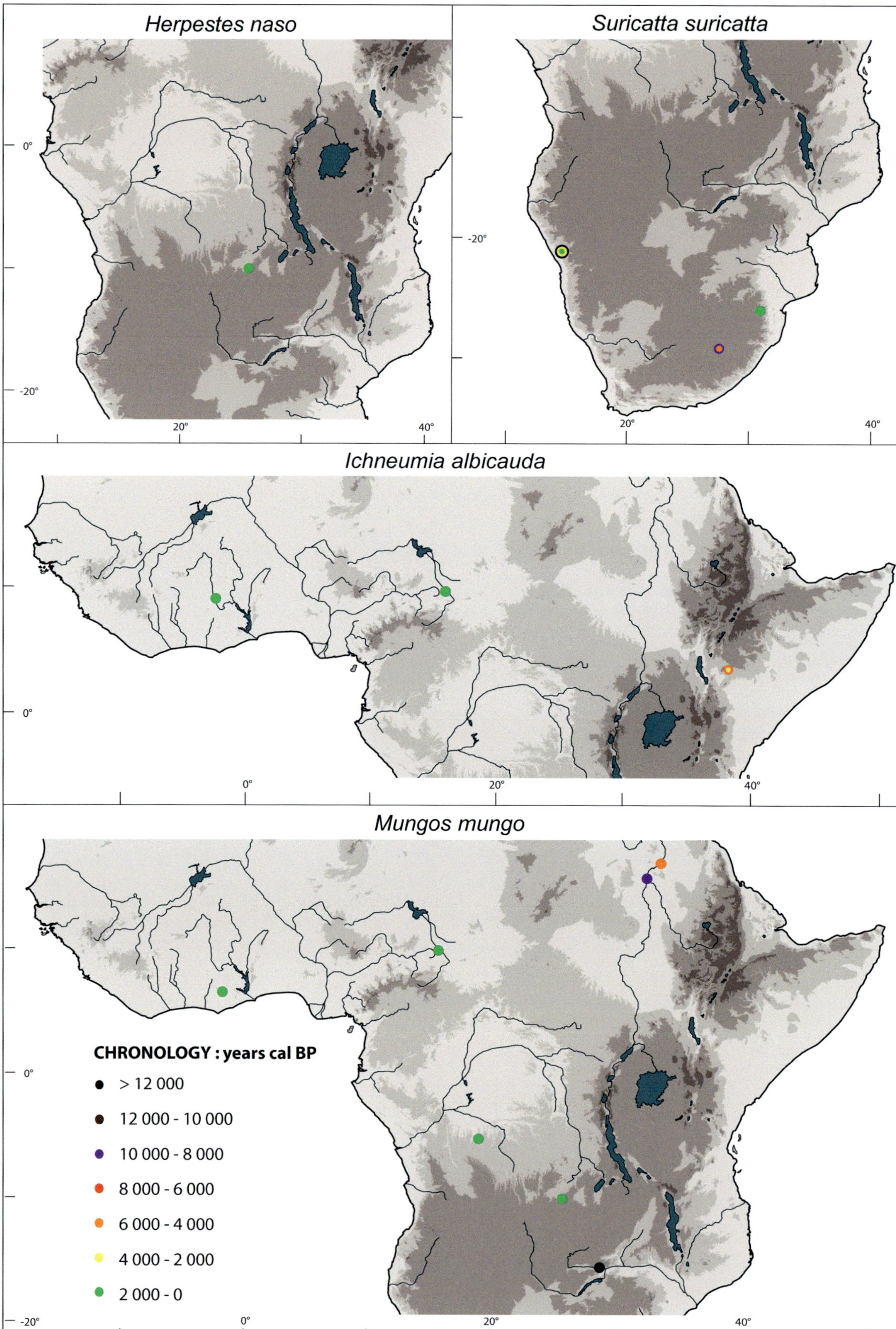
Herpestes naso
Suricatta suricatta
Ichneumia albicauda
Mungos mungo
0°
-20°
20°
40°
CHRONOLOGY : years cal BP
> 12 000
12 000 - 10 000
10 000 - 8 000
8 000 - 6 000
6 000 - 4 000
4 000 - 2 000
2 000 - 0

HYAENIDAE

Country	Site N	Site
Botswana	26	Bosutwe-Mapungubwe
	33	Drotsky's Cave-H
	35	Drotsky's Cave-P
	48	White Paintings RS-2ab
Burkina Faso	60	Péntènga 89/1-spit 13 *coprolith
	62	Péntènga 89/1-spits 4-10 *coprolith
	63	Saouga 94/120
Chad	81	Delebo
Libya	327	Hauah Fteah-Iberom
	328	Hauah Fteah-Lib-caps
Mali	363	Asselar
Morocco	426	Ghar-Khal IIIb-V
Mozambique	442	Manekweni-2-ABC
	444	Manekweni-4
Nigeria	510	Gilgila-lic
South Africa	593	Eiland Salt Works-a
	604	Equus Cave 1A

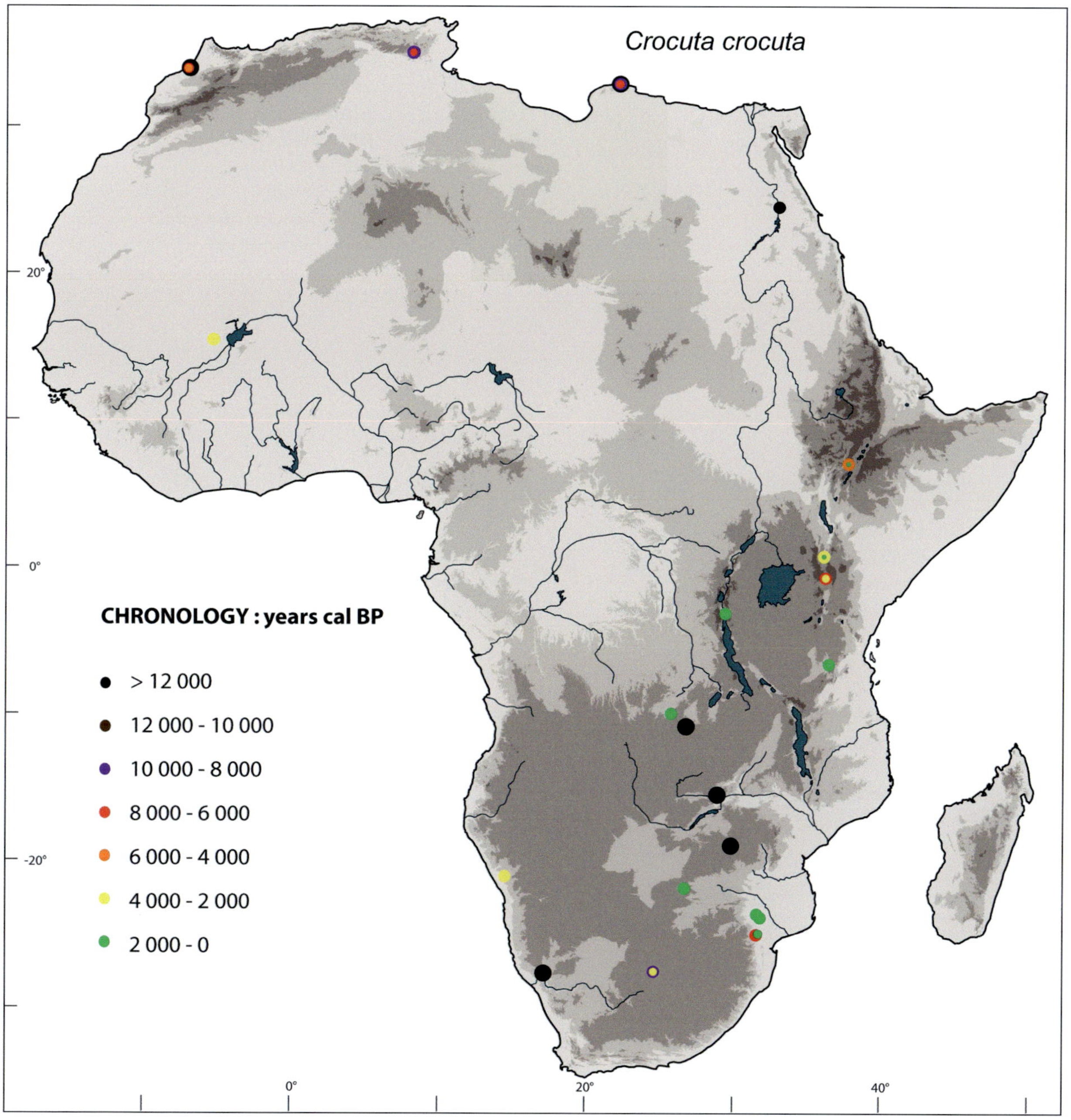

Crocuta crocuta

Country	Site N	Site	
Algeria	9	Dra-Mta-el-Ma-el-Abiod	
Botswana	26	Bosutwe-Mapungubwe	
Congo	86	Kawezi	
Egypt	162	Kom Ombo	
Ethiopia	216	Moche Borago-1b	
	219	Moche Borago-3	
Kenya	259	Enkapune Ya Muto - Eb 4b	
	260	Enkapune Ya Muto - Eb 5a	
	279	GvJm 22-LPG	
	283	GvJm 62-LPNG	
	293	Naivasha Railway Shelter	
	297	Ngenyn	
Libya	328	Hauah Fteah-Lib-caps	
Mali	382	Kolima Sud-I	
Morocco	422	El Harhoura 1	**C.c.spelea*
	437	Kehf-el-Baroud	
Namibia	449	Apollo 11-c	* *Crocuta* sp.
	452	Brandberg-LSA	
South Africa	604	Equus Cave 1A	
	664	Mo8	
	673	Ol20	
	717	SK17	
	718	Sk4	
Zaire	846	Kakontwe	
	848	Kiantapo	
Zambia	863	Leopard's Hill-PN	
Zimbabwe	870	Redcliff Cave	

Hyaena hyaena

Country	Site N	Site
Algeria	16	Meniet
	7	Damous el Ahmar
Egypt	109	Bir Kiseiba-EN
	111	Bir Kiseiba-MN
	161	Kharga dep.-Neolithic
	162	Kom Ombo
	168	Merimde Benisâlame
	199	Wadi el Bakht
	92	Abu Ballas - 85/50-1
	93	Abu Ballas - 85/51-3
Kenya	254	Ele Bor A-B
	259	Enkapune Ya Muto - Eb 4b
Niger	473	Gobero-K
Sudan	773	Esh Shaheinab
	787	Khartoum Hopital
	804	Rahib-80/87
	811	Shaqaduq-S1-A

Hyaena brunnea

Country	Site N	Site	
Botswana	28	Bosutwe-Toutswe	
	28	Bosutwe-Toutswe	cf
Lesotho	315	Ntloana Tsoana	cf
South Africa	563	Boomplass Cave-a	cf
	579	Byeneskranskop Cave 1d	cf
	580	Collingham Shelter	
	604	Equus Cave 1A	
	617	Heuningneskrans-R	
	625	Jubilee Shelter-PW	
	626	Jubilee Shelter-W	
	653	Malwula	?
	669	Nelson Bay Cave- R	cf
	676	Oudepost I	
	704	Rose Cottage Cave-A-Mn	
	721	Spoegrivier-a	
	722	Spoegrivier-b	

Proteles cristata

Country	Site N	Site	
Kenya	259	Enkapune Ya Muto - Eb 4b	
Lesotho	316	Sehonghong-a	cf
South Africa	556	2627 CD7	
	622	HKLP Vb	
	674	Ondini	
	692	QwaQwa Museum	
	710	Selongwe Hill	

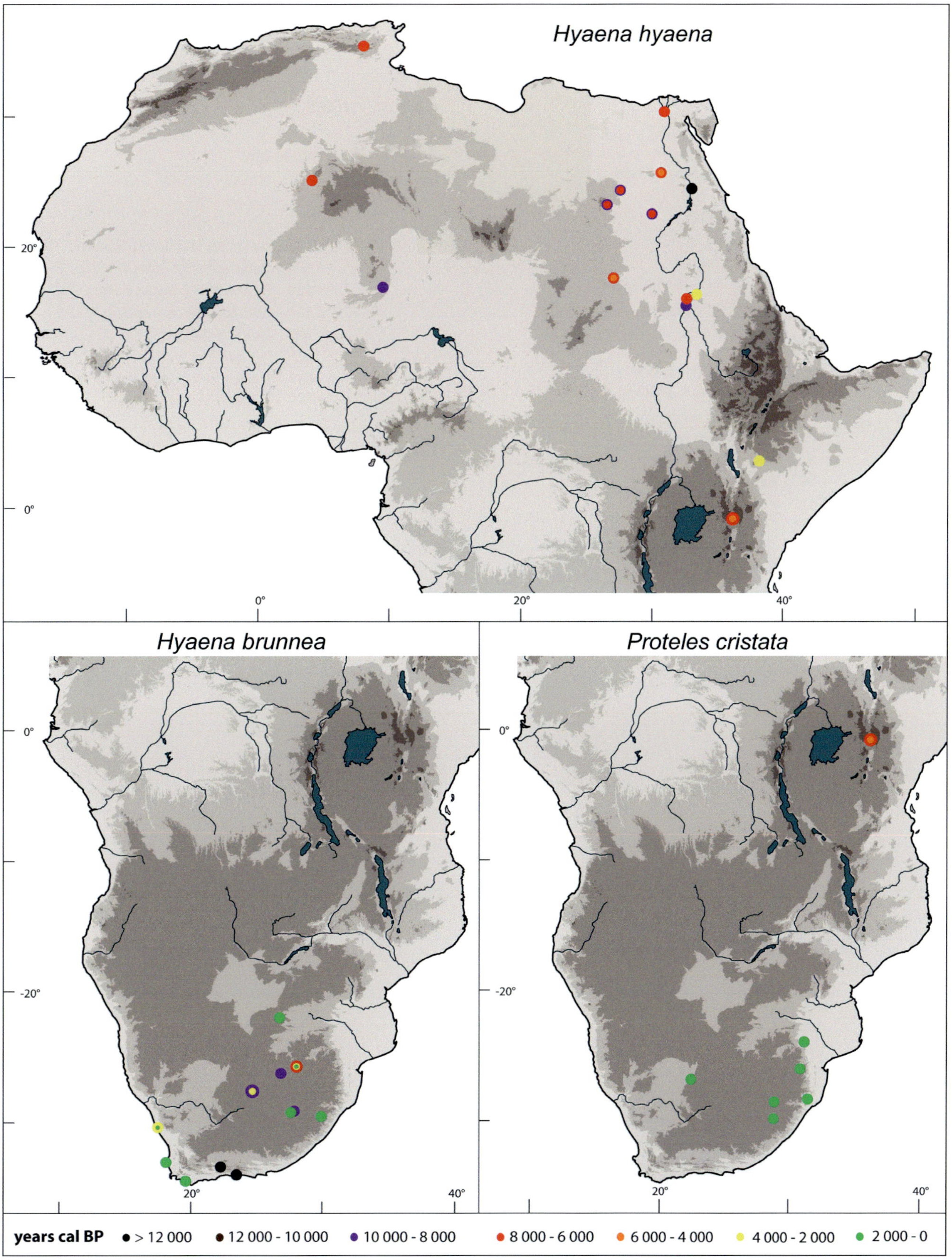
Hyaena hyaena
20°
0°
0°
20°
40°
Hyaena brunnea
0°
-20°
20°
40°
Proteles cristata
0°
-20°
20°
40°
years cal BP
> 12 000
12 000 - 10 000
10 000 - 8 000
8 000 - 6 000
6 000 - 4 000
4 000 - 2 000
2 000 - 0

Canidae

Vulpes sp.

Country	Site N	Site
Algeria	15	Medjez II
	9	Dra-Mta-el-Ma-el-Abiod
Botswana	32	Divuyu
	37	Matlapaneng
	38	Nqoma
Egypt	100	Adaima-habitat
	141	El-Tell el Iswid-Predynastic
	142	El-Tell el-Iswid A
	148	Fayoum Site 1
	153	Gilf Kebir
	168	Merimde Benisâlame
	201	Wadi Kubbaniya - E-78-3
	202	Wadi Kubbaniya - E-78-4
	203	Wadi Kubbaniya - E-81-1
	92	Abu Ballas - 85/50-1
	99	Adaima-habitat
Sudan	756	Abu Tabari 02/1
	763	Djarabona 84/13
	775	Gash Delta-K
	785	Kerma
	817	Soba
	831	Wadi Howar - 84/13
	832	Wadi Howar - 84/19

Vulpes vulpes

Country	Site N	Site
Algeria	10	Grotte Capeletti
	20	Tamar Hat
Egypt	100	Adaima-habitat
	106	Berenike 95/9 Early Roman
	107	Berenike 95/97Late Roman
	130	El Abadiya 2 South sector
	135	Elephantine-necropolis
	99	Adaima-habitat
Libya	327	Hauah Fteah-Iberom
	328	Hauah Fteah-Lib-caps
	329	Hauah Fteah-Neolithic

Vulpes rueppellii

Country	Site N	Site
Egypt	111	Bir Kiseiba-MN
	93	Abu Ballas - 85/51-3
	94	Abu Ballas - 85/56
Libya	333	Ti-n-Thora East
Sudan	800	Mahal Teglinos K1-LG

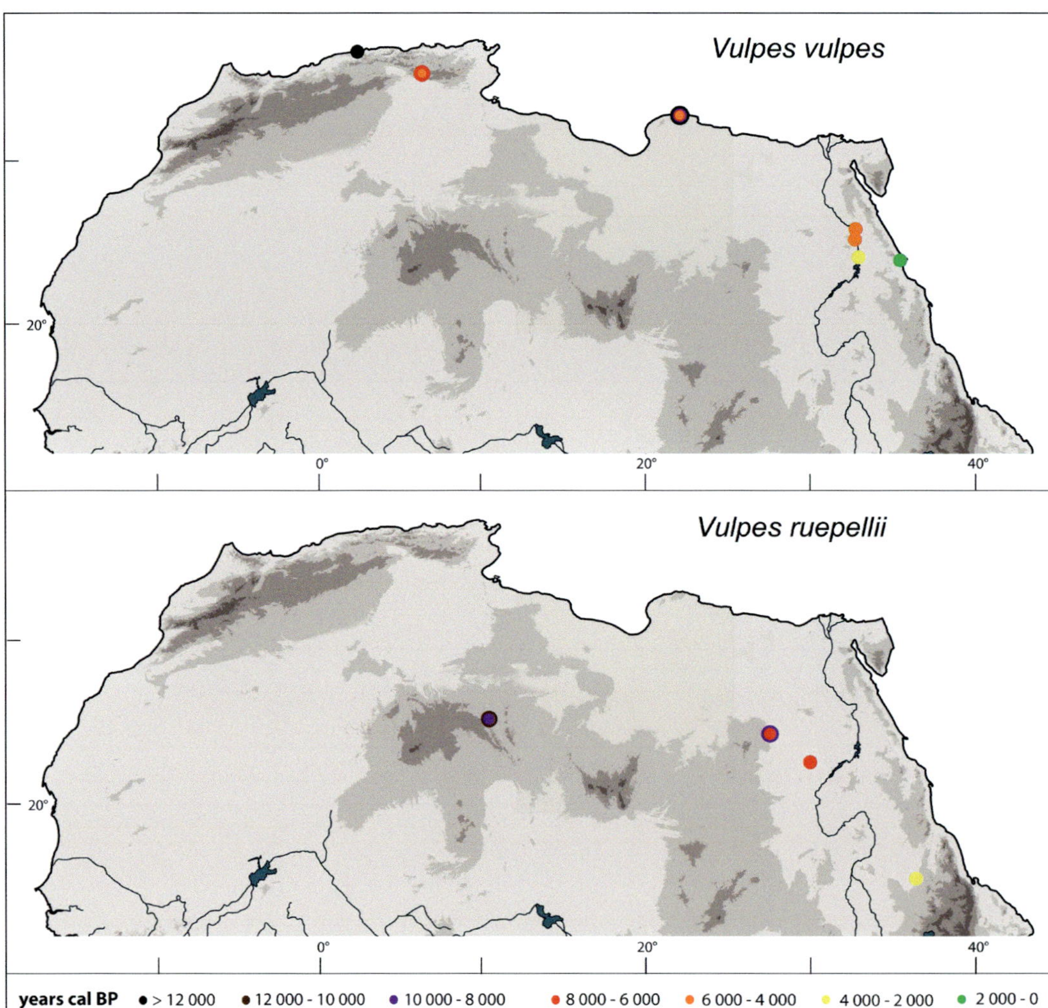

Vulpes chama

Country	Site N	Site	
South Africa	580	Collingham Shelter	
	587	Dikbosch 1	*?
	588	Dikbosch 2	
	589	Doornfontein	
	595	Eland´s Bay Cave-AI	
	604	Equus Cave 1A	
	625	Jubilee Shelter-PW	
	626	Jubilee Shelter-W	
	627	Kasteelberg A	
	628	Kasteelberg B	
	676	Oudepost I	
	690	Powerhouse Cave	
	699	Riversmead Shelter VI-IX	
	704	Rose Cottage Cave-A-Mn	
	742	Ventershoek	

Vulpes zerda

Country	Site N	Site
Egypt	157	Hierakonpolis HK6-Naq I-II *intr.
	168	Merimde Benisâlame
	94	Abu Ballas - 85/56

Vulpes pallida

Country	Site N	Site	
Burkina Faso	57	Oursi Village-MIA	
	63	Saouga 94/120	
	64	Saouga 95/7	
Mauritania	411	Tichitt - Akjinjeir phase	*cf.
	414	Tichitt - Goungou phase	*cf.

Vulpes atlantica

Country	Site N	Site
Morocco	421	Bou Saria C-D
	422	El Harhoura 1
	426	Ghar-Khal IIIb-V
	435	Kaf-taht-el-Ghar-G
	437	Kehf-el-Baroud
	438	Kehf-el-Baroud-aven

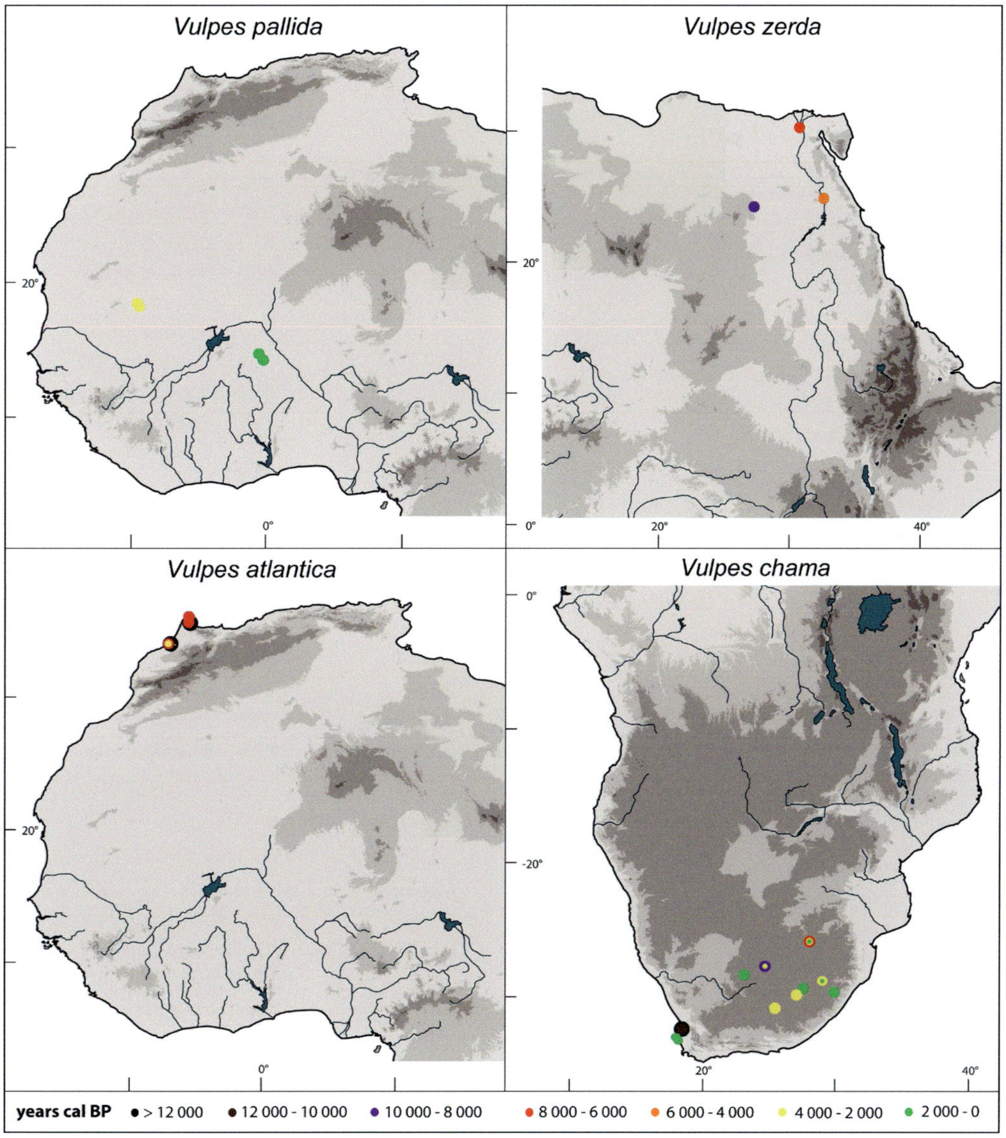

Canis aureus

Country	Site N	Site	
Algeria	15	Medjez II	*C.a.anthus
	20	Tamar Hat	
	4	Ain Mistehayia	
Chad	81	Delebo	
Egypt	109	Bir Kiseiba-EN	
	111	Bir Kiseiba-MN	
	121	Dakhleh Oasis, SMU	*C.a.lupaster
	138	Elkab-Epipaleo	*C.a.lupaster
	140	El-Tell el Iswid-Early Dynastic	
	143	El-Tell el-Iswid B	
	145	Fayoum G1-H1	*C.a.lupaster
	168	Merimde Benisâlame	
	171	Napta Playa 77-1/4/5/6	*?
	172	Napta Playa 77-8	
	93	Abu Ballas - 85/51-3	
	94	Abu Ballas - 85/56	
Libya	325	Et Tera	
	331	In Aghelachem-T1	*cf.
	333	Ti-n-Thora East	
	335	Ti-n-Thora Two Caves	
	336	Uan Afuda-US1-3	
	345	Uan Telocat (MP)	
Mali	374	Hassi el-Abiod	
Mauritania	405	Khatt Lemaiteg	
Morocco	422	El Harhoura 1	
	436	Kehf el Hammar	
	438	Kehf-el-Baroud-aven	
Niger	474	Gobero-T	
	488	Takene Bawat 2	
Sudan	764	El Damer	
	765	El Kadada	
	767	El Mahalab	
	777	Geili	*?
	781	Jebel Umm Marrahi	
	788	Khasm el Girba – Kassala	*?
	789	Khasm el Girba - KG14	*?
	800	Mahal Teglinos K1-LG	
	805	Saggai 1	
	814	Sheikh el Amin	
	815	Sheikh Mustafa	

Canis mesomelas

Country	Site N	Site	
Kenya	259	Enkapune Ya Muto - Eb 4b	
	260	Enkapune Ya Muto - Eb 5a	
	284	Kulchurdo	
Lesotho	304	Ha Makotoko	
	307	Likoaeng-III-V	
	308	Likoaeng-VII-IX	
	309	Likoaeng-XI	
	313	Liphofung-pCW	
	314	Muela	
	315	Ntloana Tsoana	
	316	Sehonghong-a	
	321	Tloutle-CWI	
	323	Tloutle-pCW	
Malawi	358	Matope Court-LIA	
South Africa	563	Boomplass Cave-a	*cf.
	564	Boomplass Cave-b	*cf.
	565	Boomplass Cave-c	*cf.
	566	Boomplass Cave-d	*cf.
	571	Buffelskloof RS-d	*cf.
	578	Byeneskranskop Cave 1c	*cf.?
	579	Byeneskranskop Cave 1d	*cf.
	580	Collingham Shelter	
	582	Colwinton Shelter-pCW	*cf.
	585	Deelpan	
	586	Die Kelders	
	595	Eland´s Bay Cave-AI	
	597	Eland´s Bay Cave-Wia	
	598	Eland´s Bay Cave-Wib	
	599	Eland´s Bay Cave-Wic	
	600	Eland´s Bay Cave-Wid	
	604	Equus Cave 1A	
	619	HKLP Ia	*cf.
	619	HKLP Ia	
	624	Jubilee Shelter-O	*cf.
	625	Jubilee Shelter-PW	*cf.
	626	Jubilee Shelter-W	*cf.
	631	Klasies River Mouth-1I	
	638	Langsdraai	
	648	Ma38	
	652	Magogo-3	
	654	Mamba 1	
	663	Mhlwazini Cave	
	667	Ndondondwane	
	668	Nelson Bay Cave- PW	
	669	Nelson Bay Cave- R	
	670	Nelson Bay Cave- W	
	673	Ol20	
	682	Paternoster3-4	*cf.
	683	Paternoster5-6	*cf.
	700	Rooiels	*cf.
	701	Rooikrans-CW	
	704	Rose Cottage Cave-A-Mn	
	706	Rose Cottage Cave-OIC	
	707	Rose Cottage Cave-WI	
	717	SK17	
	718	Sk4	
	725	Steenbokfontein-1	
	739	Tsh1	
	740	Twyfelpoort	
	752	Wosi-middle-late EIA	
	753	Wosi-terminal EIA	
Zaire	846	Kakontwe	
Zambia	863	Leopard's Hill-PN	
Zimbabwe	870	Redcliff Cave	

Canis adustus

Country	Site N	Site
Kenya	259	Enkapune Ya Muto - Eb 4b
South Africa	636	Klipspring
	650	Magogo-1
	713	SH16

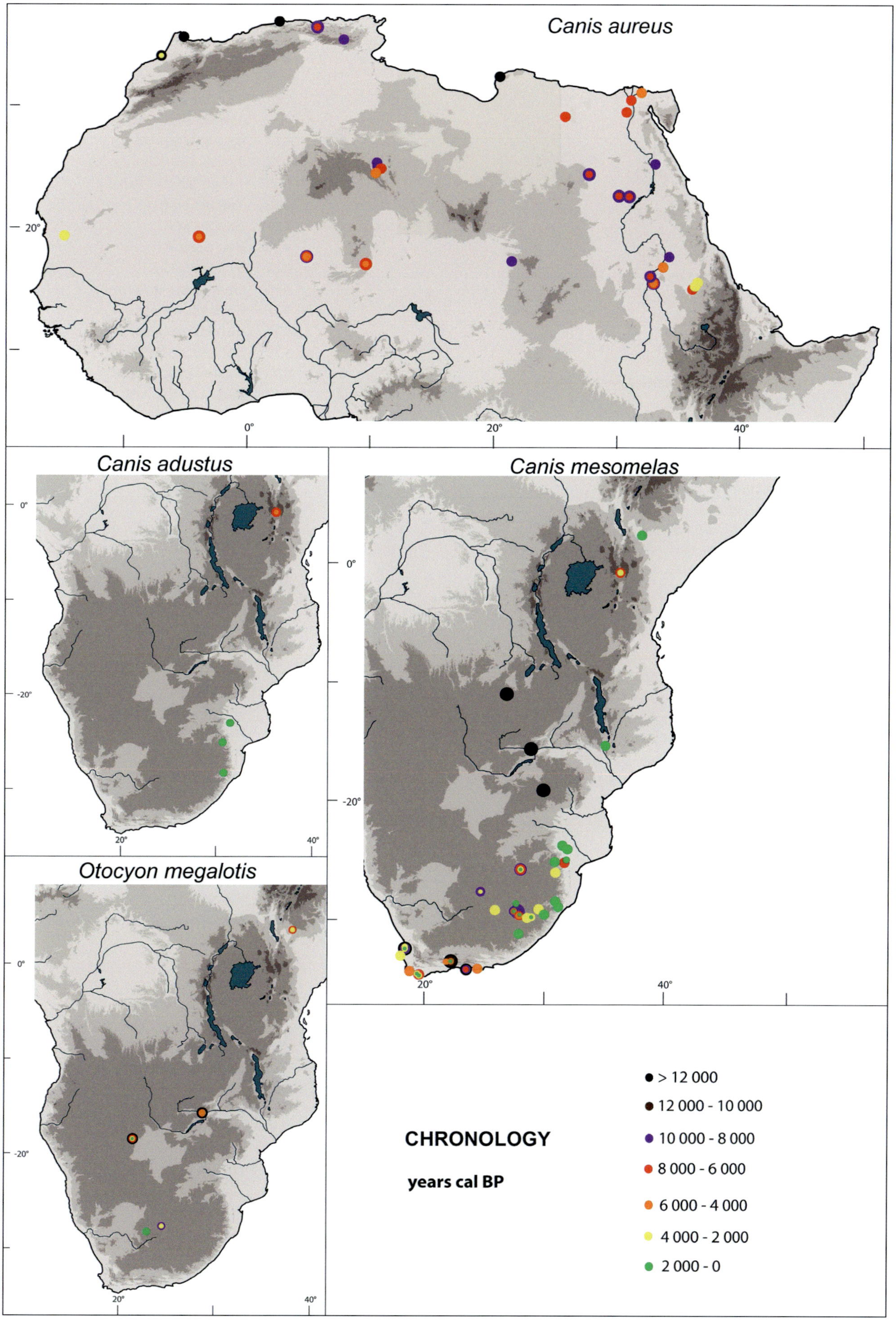
Canis aureus
20°
0°
20°
40°
Canis adustus
0°
-20°
20°
40°
Canis mesomelas
0°
-20°
20°
40°
Otocyon megalotis
0°
-20°
20°
40°
CHRONOLOGY
years cal BP
> 12 000
12 000 - 10 000
10 000 - 8 000
8 000 - 6 000
6 000 - 4 000
4 000 - 2 000
2 000 - 0

Canis lupus f. *familiaris*

Country	Site N	Site	
Botswana	40	Taukome	
	44	Toteng1-SU4	*cf.
Burkina Faso	52	Kissi 22	
	53	Kissi 40	
	54	Oursi hu-beero	
	55	Oursi Village-EIA	
	56	Oursi Village-LIA	
	57	Oursi Village-MIA	
	58	Oursi-IA	
	63	Saouga 94/120	
	64	Saouga 95/7	
Cameroon	68	Aissa Dugjé	
C.A.R.	79	Mondongué	
	80	Lingbangbo	
Egypt	100	Adaima-habitat	
	101	Ain Tirghi cemetery	
	102	Al-Ma´abda	
	103	Al-Ma´abda	
	105	Balat-Mastaba V	
	107	Berenike 95/97Late Roman	
	112	Buto/1-3	
	113	Buto-6	
	114	Buto-9-12	
	120	Dahschûr	
	123	Dakhleh Ouasis, Roman	
	129	El Abadiya 2 North sector	
	130	El Abadiya 2 South sector	
	135	Elephantine-necropolis	
	137	Elephantine-Satet temple	
	141	El-Tell el Iswid-Predynastic	
	142	El-Tell el-Iswid A	
	154	Hierakonpolis HK6-Naq IC-IIA	
	155	Hierakonpolis HK6-Naq IIA-B	
	156	Hierakonpolis HK6-Naq III	
	157	Hierakonpolis HK6-Naq I-II	
	164	Maadi	
	168	Merimde Benisâlame	
	169	Minshat Abu Omar	
	172	Napta Plata 77-8	
	175	Qasr el-Sagha-1	
	188	Tell el-Dab´a - b	
	191	Toukh	
	196	Tuna el-Gebel	
	197	Umm el-Qaab, Friedhof U	
	98	Adaima-burials	
	99	Adaima-habitat	
Ghana	225	Asantamanso-ASII/III	
	227	Asantamanso-ASV	
	230	Bono Manso	
	231	Daboya	
	232	Dawu	
	236	Frederiksgave plantation	
	241	Mandara	
	242	Mankuma	
	245	Nyanawase/Akwamu Am.	
	246	Nyanga	
Kenya	272	Gogo Falls-I	
	273	Gogo Falls-II-III	
Libya	326	Hauah Fteah- Historic	
	328	Hauah Fteah-Lib-caps	
	329	Hauah Fteah-Neolithic	
Mali	360	Akumbu-AK3	
	361	Akumbu-EIA	
	362	Akumbu-MIA	
	364	Dia-Shoma Horizon I	
	371	Gao Ancien	
	372	Gao Saney	
	376	Jenné-Jeno - phase I/II	
	377	Jenné-Jeno - phase III	
	378	Jenné-Jeno - phase IV	
	389	Tongo Maaré Diabel	
Mauritania	408	Tegdaoust-b	
Morocco	419	Bou Saria A	
	420	Bou Saria B	
	421	Bou Saria C-D	
	426	Ghar-Khal IIIb-V	
	427	Ghar-Khal II-IIIa	
	431	Kaf-taht-el-Ghar-C	
	437	Kehf-el-Baroud	
	438	Kehf-el-Baroud-aven	
Mozambique	442	Manekweni-2-ABC	
Niger	471	Chin Tafidet	
Nigeria	501	Dorota	
	502	Elkido North	
	503	Gajiganna A	
	504	Gajiganna B1-lower	
	507	Galaga	
	510	Gilgila-Iic	
	516	Labe Kanuri	
	518	Mege I	
	525	Ngala I	
	530	Zilum	
Rhodesia	532	Harleigh Farm-1	
	533	Harleigh Farm-2	
Senegal	538	Cubalel-Ia	*intr.
	539	Cubalel-Ib	
	540	Cubalel-II	
	541	Cubalel-III	
	542	Ndiamon-Badat	
	544	Sintiou-Bara	
	546	Siouré-Ib	
	547	Siouré-II	
	548	Siouré-IV, V	
	549	Tulel-Fobo	
South Africa	636	Klipspring	
	650	Magogo-1	
	651	Magogo-2	
	654	Mamba 1	
	659	Marupale Hill	
	667	Ndondondwane	
	702	Rooikrans-pCW	
	708	Schroda	
	720	Sonkoanini Hill	
	751	Wosi-early EIA	
	752	Wosi-middle-late EIA	
	753	Wosi-terminal EIA	
Sudan	759	Berber	
	761	Debbat Bangdit	
	762	Debbat El Eheima	
	765	El Kadada	
	773	Esh Shaheinab	
	778	Gism el-Arba	
	780	Jebel et Tomat	
	783	Kadero	*?
	785	Kerma	

Country	Site N	Site
	793	Laqiya area - 82/29
	797	Mahal Teglinos K1-CG
	800	Mahal Teglinos K1-LG
	817	Soba
	829	Wadi Howar - 84/1
Zaire	857	Nkile
Zanzibar	867	Kizimkazi Dimbani
Zimbabwe	869	Muozi

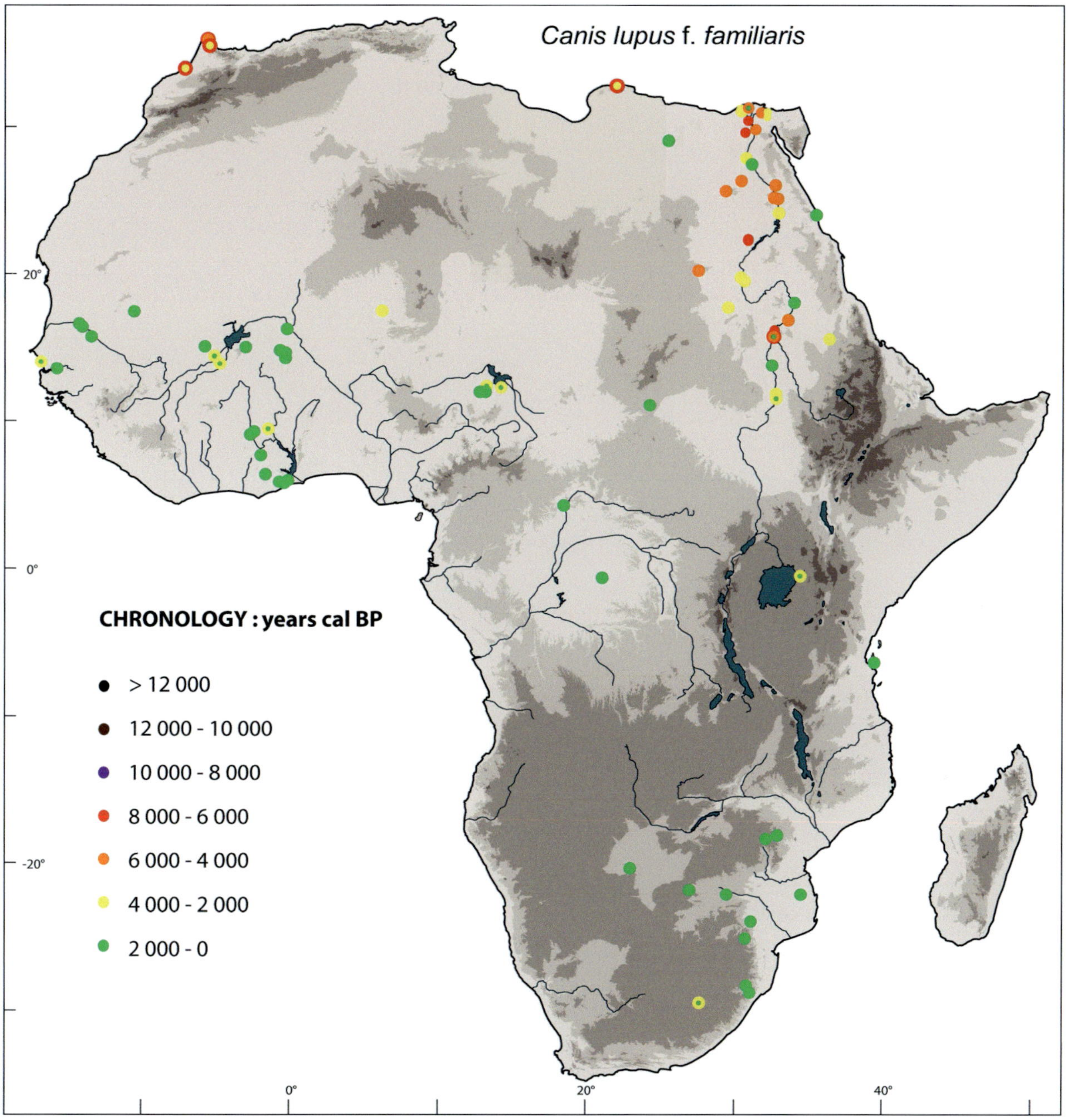

Otocyon megalotis

Country	Site N	Site
Botswana	48	White Paintings RS-2ab
	48	White Paintings RS-1
	49	White Paintings RS-2b-3ab
Kenya	254	Ele Bor A-B
	255	Ele Bor A-C
South Africa	589	Doornfontein
	604	Equus Cave 1A
Zambia	862	Leopard's Hill-NIII
	863	Leopard's Hill-PN

Lycaon pictus

Country	Site N	Site
Egypt	185	Sodmein Cave older backfill
Kenya	259	Enkapune Ya Muto - Eb 4b
Lesotho	316	Sehonghong-a
Niger	473	Gobero-K
South Africa	585	Deelpan
	718	Sk4
Sudan	793	Laqiya area - 82/29
	805	Saggai 1
	821	Upper Atbara hinterland
West. Sahara	836	Izriten
Zaire	845	Ishango-Z-Post-Em
Zimbabwe	870	Redcliff Cave

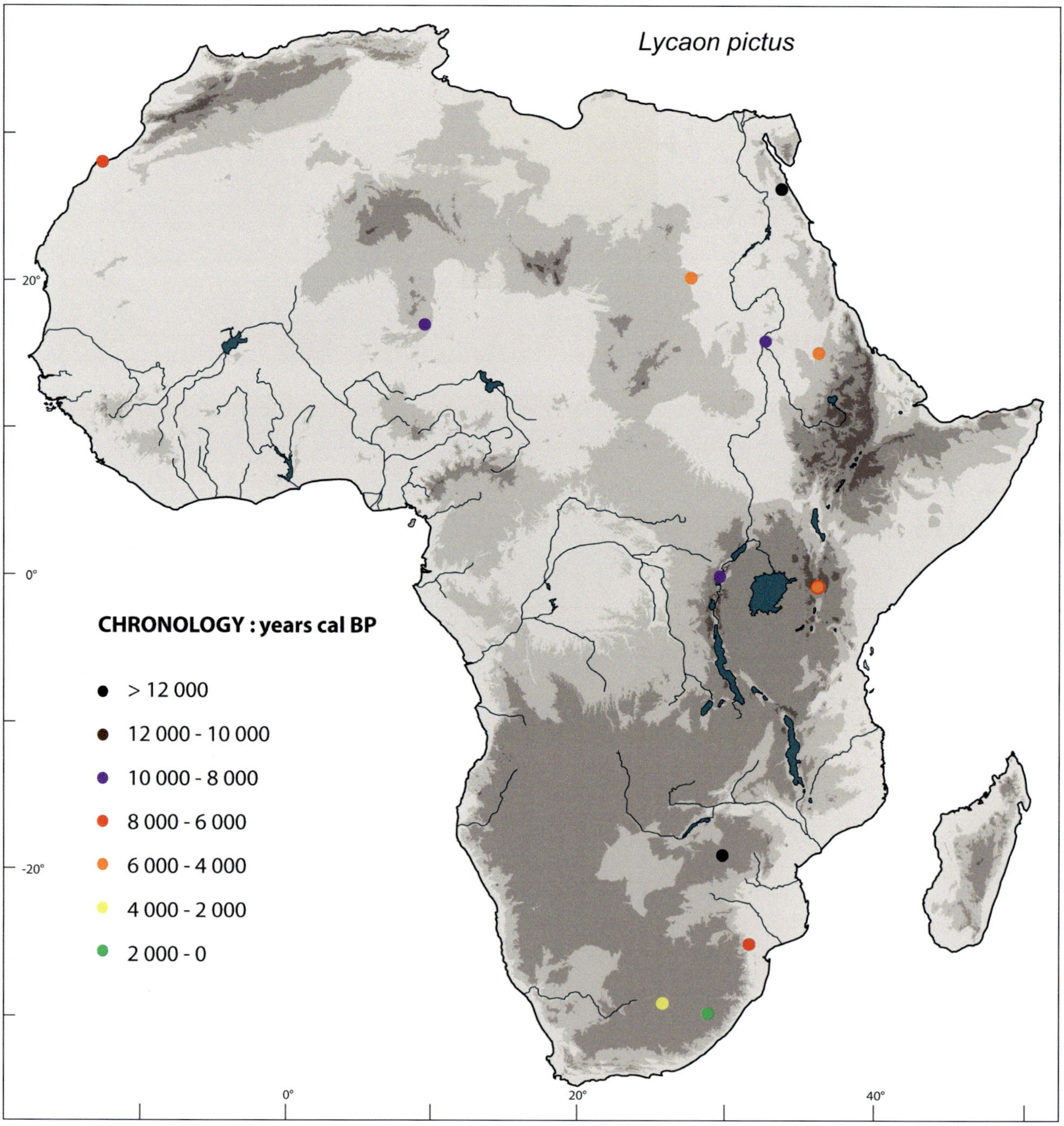

MUSTELIDAE

Mustelidae

Country	Site N	Site
Botswana	38	Nqoma
Libya	327	Hauah Fteah-Iberom
Sudan	756	Abu Tabari 02/1
	767	El Mahalab

Mustela putorius

Country	Site N	Site
Morocco	422	El Harhoura 1

Ictonyx libyca

Country	Site N	Site
Egypt	168	Merimde Benisâlame
Mali	362	Akumbu-MIA

Ictonyx striatus

Country	Site N	Site	
Burundi	66	Ruzizi Plain	
South Africa	580	Collingham Shelter	*cf.
	586	Die Kelders	
	595	Eland´s Bay Cave-AI	
	620	HKLP Ib	*cf.
	620	HKLP Ib	
	628	Kasteelberg B	
	639	Large Rock Shelter	
	676	Oudepost I	
	679	Pancho's Kitchen Midden 3-4	
	718	Sk4	
	724	Spring Cave	
	736	Tortoise cave-b	
	738	Tortoise cave-d	

Poecilogale albinucha

Country	Site N	Site
Zaire	848	Kiantapo

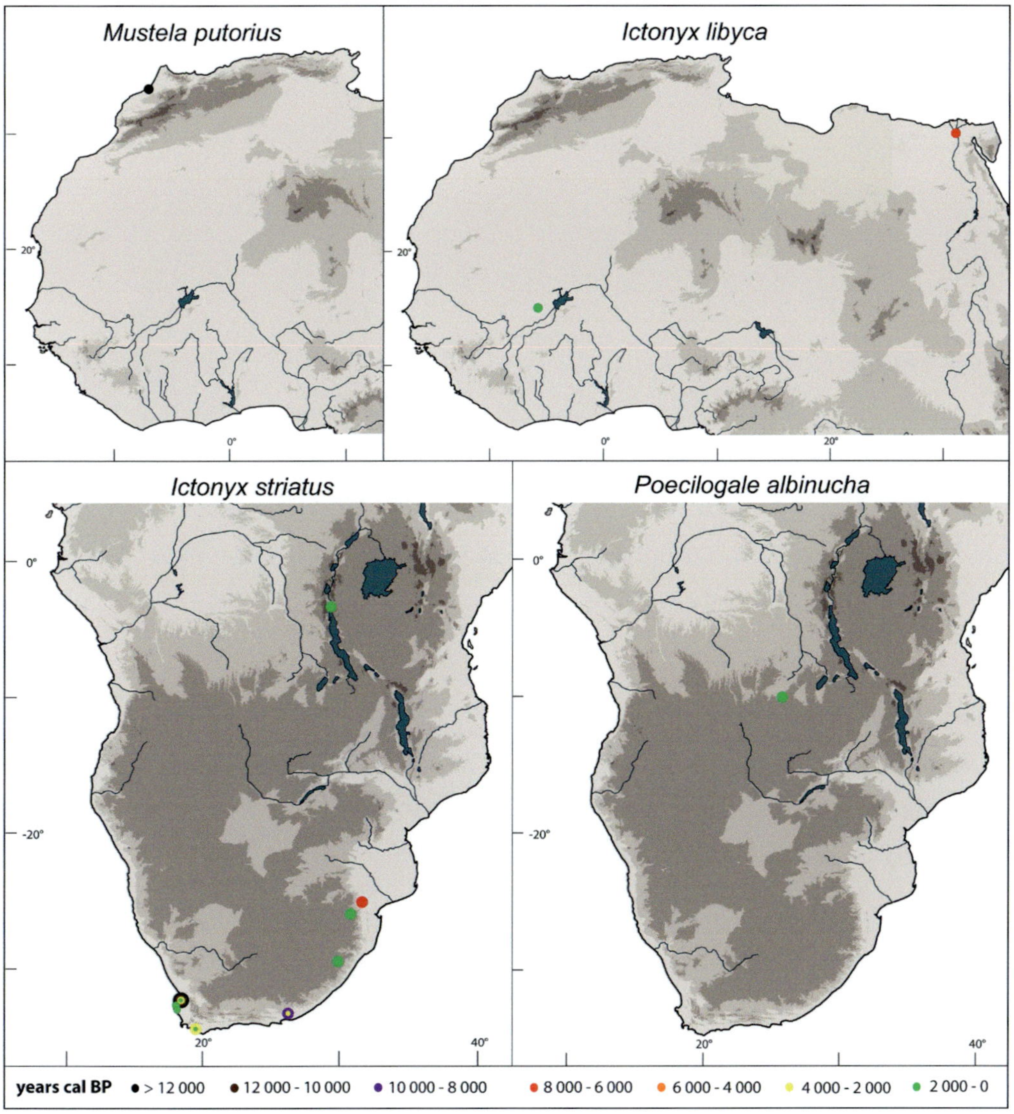

Mellivora capensis

Country	Site N	Site	
Botswana	49	White Paintings RS-2b-3ab	
Egypt	184	Sodmein Cave - lev 5 (MP5)	
Kenya	258	Enkapune Ya Muto - Eb 4a	
	259	Enkapune Ya Muto - Eb 4b	
	260	Enkapune Ya Muto - Eb 5a	
	276	GvJm 19-HC	
	277	GvJm 19-TLPG	
Lesotho	310	Likoaeng-XIII	*cf.
Mauritania	400	Dhar Tichitt Site 38	
South Africa	565	Boomplass Cave-c	*cf.
	578	Byeneskranskop Cave 1c	
	586	Die Kelders	
	595	Eland´s Bay Cave-AI	
	596	Eland´s Bay Cave-RI	
	598	Eland´s Bay Cave-Wib	
	600	Eland´s Bay Cave-Wid	
	602	Eland´s Bay Open-b	
	604	Equus Cave 1A	
	626	Jubilee Shelter-W	
	627	Kasteelberg A	
	628	Kasteelberg B	
	653	Malwula	
	668	Nelson Bay Cave- PW	
	670	Nelson Bay Cave- W	*?
	676	Oudepost I	
	679	Pancho's Kitchen Midden 3-4	
	690	Powerhouse Cave	
	736	Tortoise cave-b	
	749	Witklip-3	
Sudan	765	El Kadada	
	773	Esh Shaheinab	
	777	Geili	
	783	Kadero	
	788	Khasm el Girba - Kassala	
	812	Shaqaduq-S1-B	
	821	Upper Atbara hinterland	
Zaire	848	Kiantapo	
Zambia	862	Leopard's Hill-NIII	
	863	Leopard's Hill-PN	

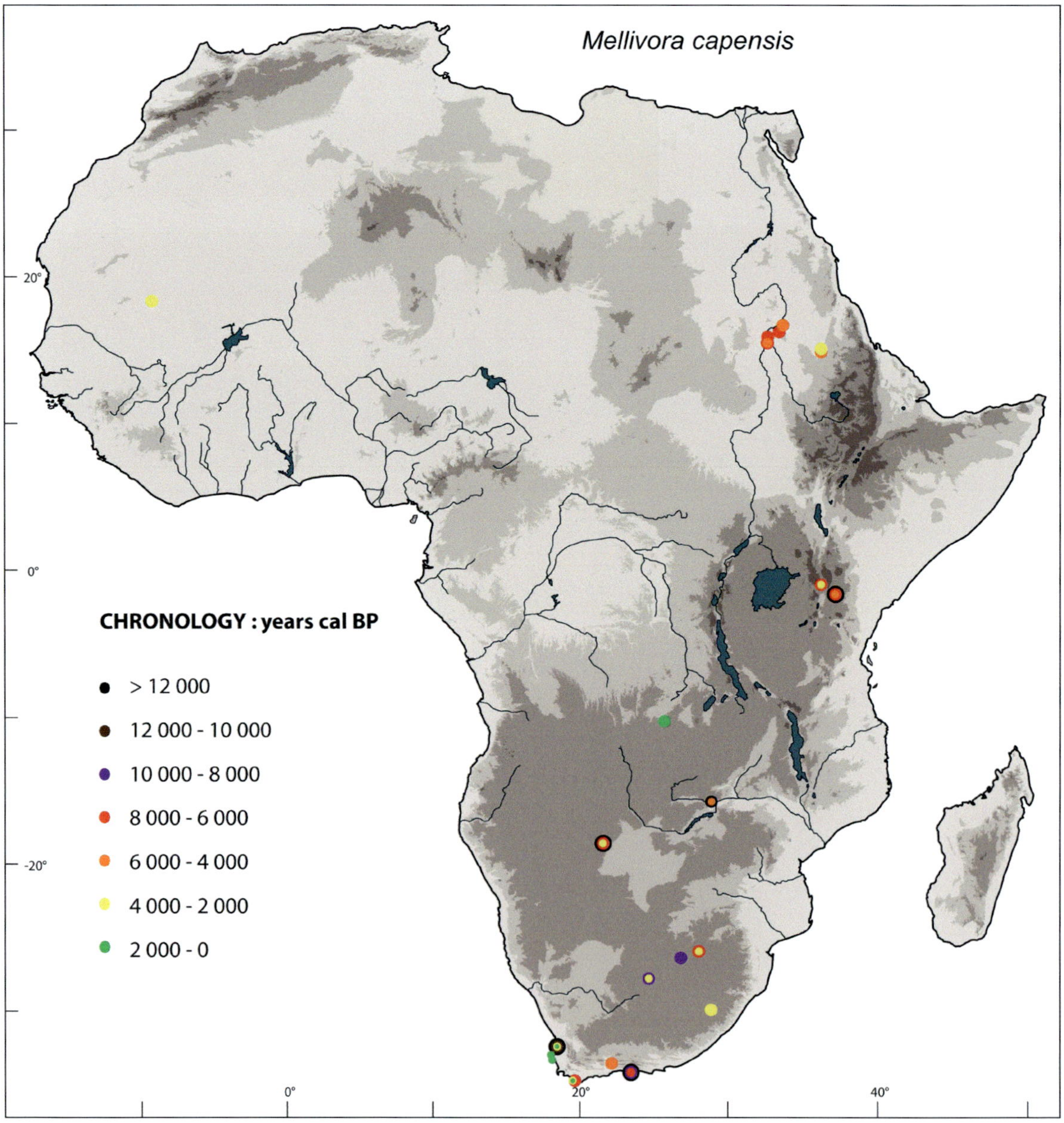

Aonyx capensis

Country	Site N	Site	
Lesotho	309	Likoaeng-XI	
	310	Likoaeng-XIII	
Mali	374	Hassi el-Abiod	
South Africa	579	Byeneskranskop Cave 1d	
	580	Collingham Shelter	
	585	Deelpan	
	632	Klasies River Mouth-1II	
	642	Le7a	
	670	Nelson Bay Cave- W	
	700	Rooiels	
	707	Rose Cottage Cave-WI	
	733	Tienfontein 2	
	734	Tienfontein 4A	*?
	742	Ventershoek	
	752	Wosi-middle-late EIA	
	753	Wosi-terminal EIA	
Zaire	843	Ishango-AB	
Zambia	860	Leopard's Hill-NI	
	863	Leopard's Hill-PN	
Zimbabwe	870	Redcliff Cave	*cf.

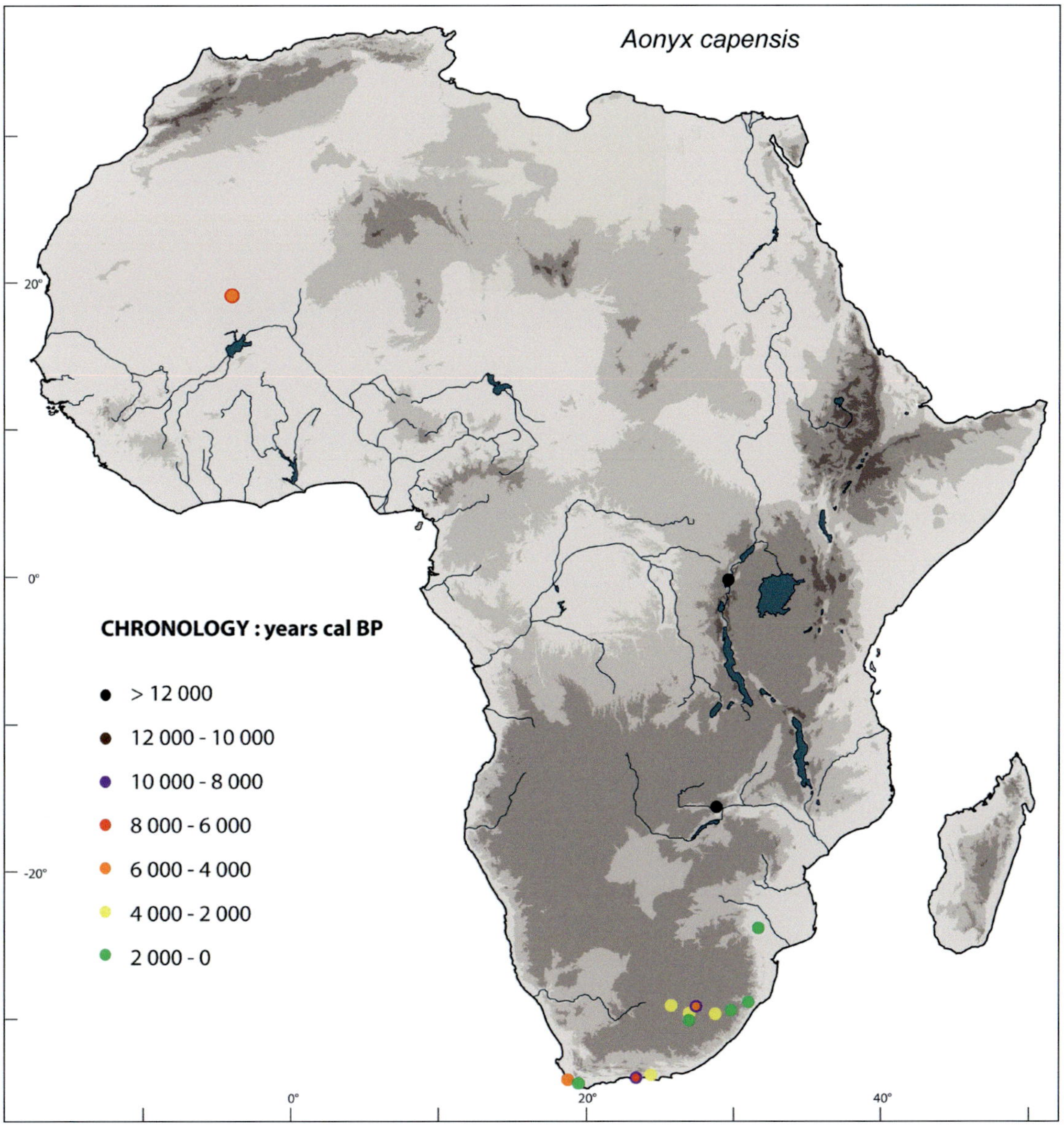

Lutra maculicollis

Country	Site N	Site
Egypt	118	Catfish Cave-L4
	166	Makhadma 2
	167	Makhadma 4
Mali	374	Hassi el-Abiod
Zaire	843	Ishango-AB
	858	Sanga

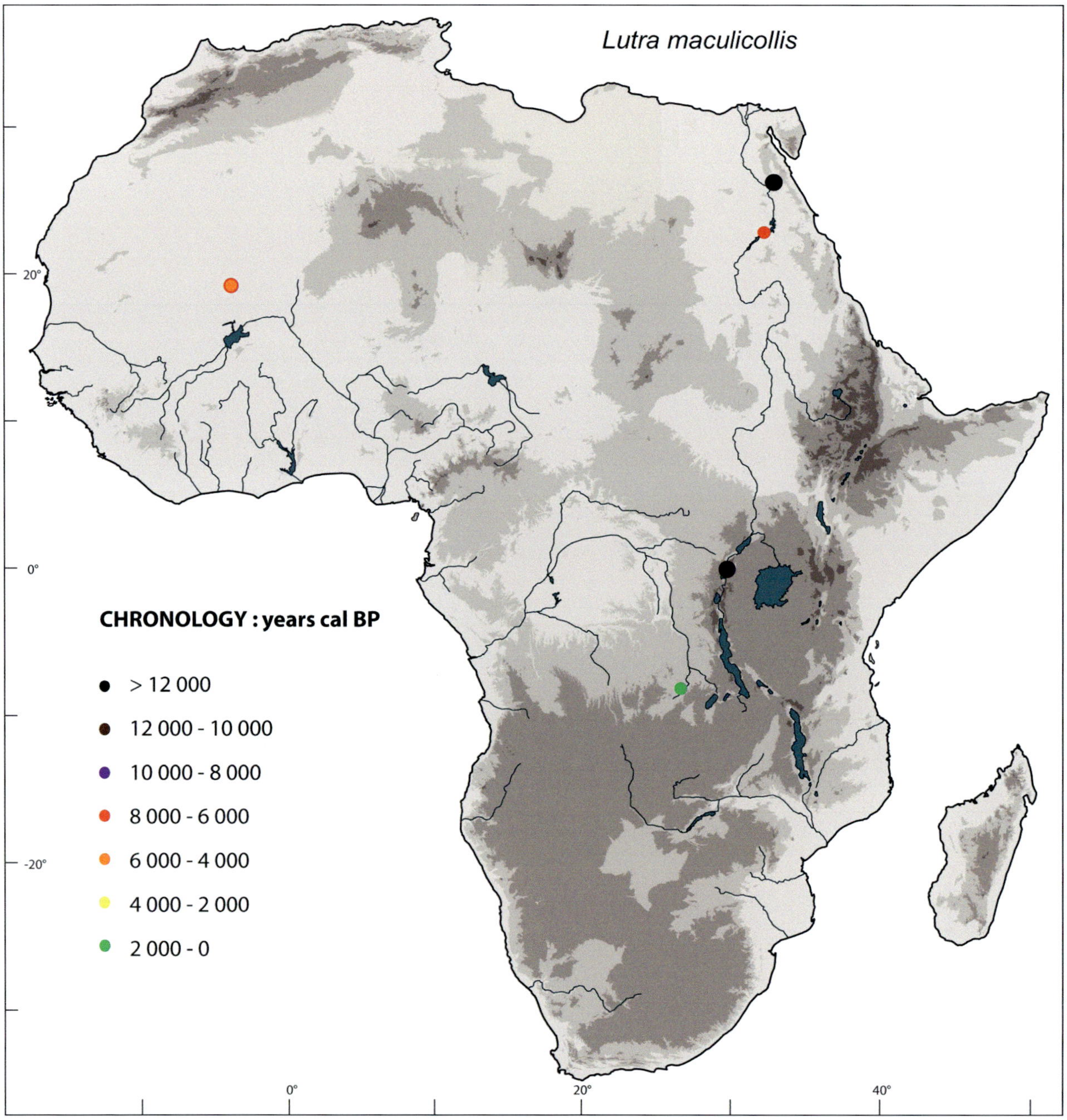

PHOCIDAE

Mirounga leonina

Country	Site N	Site
Namibia	462	Steenbras Bay
South Africa	630	Klasies River Mouth-1DIII
	668	Nelson Bay Cave- PW
	670	Nelson Bay Cave- W

Monachus monachus

Country	Site N	Site
South Africa	586	Die Kelders
West. Sahara	840	Site F

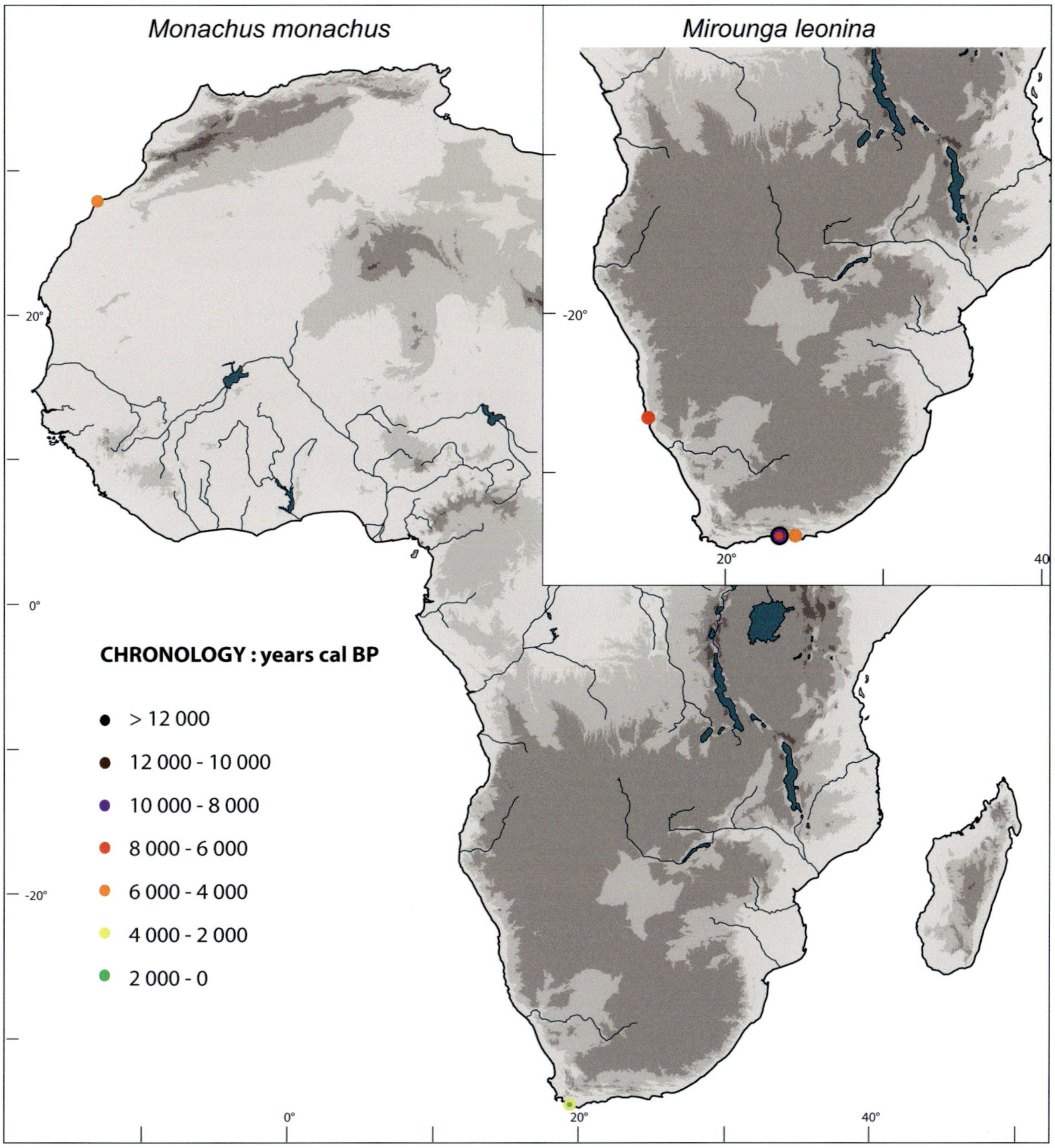

Otariidae

Arctocephalus pusillus

Country	Site N	Site	
Namibia	456	Geduld-LSA	
	462	Steenbras Bay	
South Africa	578	Byeneskranskop Cave 1c	
	579	Byeneskranskop Cave 1d	
	590	Drie Susters	
	591	Drie Susters Main	
	595	Eland´s Bay Cave-AI	
	596	Eland´s Bay Cave-RI	
	597	Eland´s Bay Cave-Wia	
	598	Eland´s Bay Cave-Wib	
	599	Eland´s Bay Cave-Wic	
	600	Eland´s Bay Cave-Wid	
	601	Eland´s Bay Open-a	
	602	Eland´s Bay Open-b	
	608	Forest Hall	
	613	Hail Stone Midden	
	627	Kasteelberg A	
	628	Kasteelberg B	
	630	Klasies River Mouth-1DIII	
	631	Klasies River Mouth-1I	
	632	Klasies River Mouth-1II	
	633	Klasies River Mouth-5II	
	637	Kreeftebaai	
	668	Nelson Bay Cave- PW	
	669	Nelson Bay Cave- R	*?
	670	Nelson Bay Cave- W	
	676	Oudepost I	
	678	Pancho's Kitchen Midden 1-2	
	680	Pancho's Kitchen Midden 6-7	
	681	Paternoster1-2	
	682	Paternoster3-4	
	683	Paternoster5-6	
	700	Rooiels	
	719	Smitswinkelbaai Cave	
	721	Spoegrivier-a	
	722	Spoegrivier-b	
	723	Spoegrivier-c	
	724	Spring Cave	
	725	Steenbokfontein-1	
	726	Steenbokfontein-2	
	727	Steenbokfontein-3	
	728	Steenbokfontein-4a	
	729	Steenbokfontein-4b	
	735	Tortoise cave-a	
	736	Tortoise cave-b	
	737	Tortoise cave-c	
	738	Tortoise cave-d	
	743	Vlaeberg	
	749	Witklip-3	
	750	Witklip-4	

Arctocephalus gazella

Country	Site N	Site
South Africa	668	Nelson Bay Cave- PW

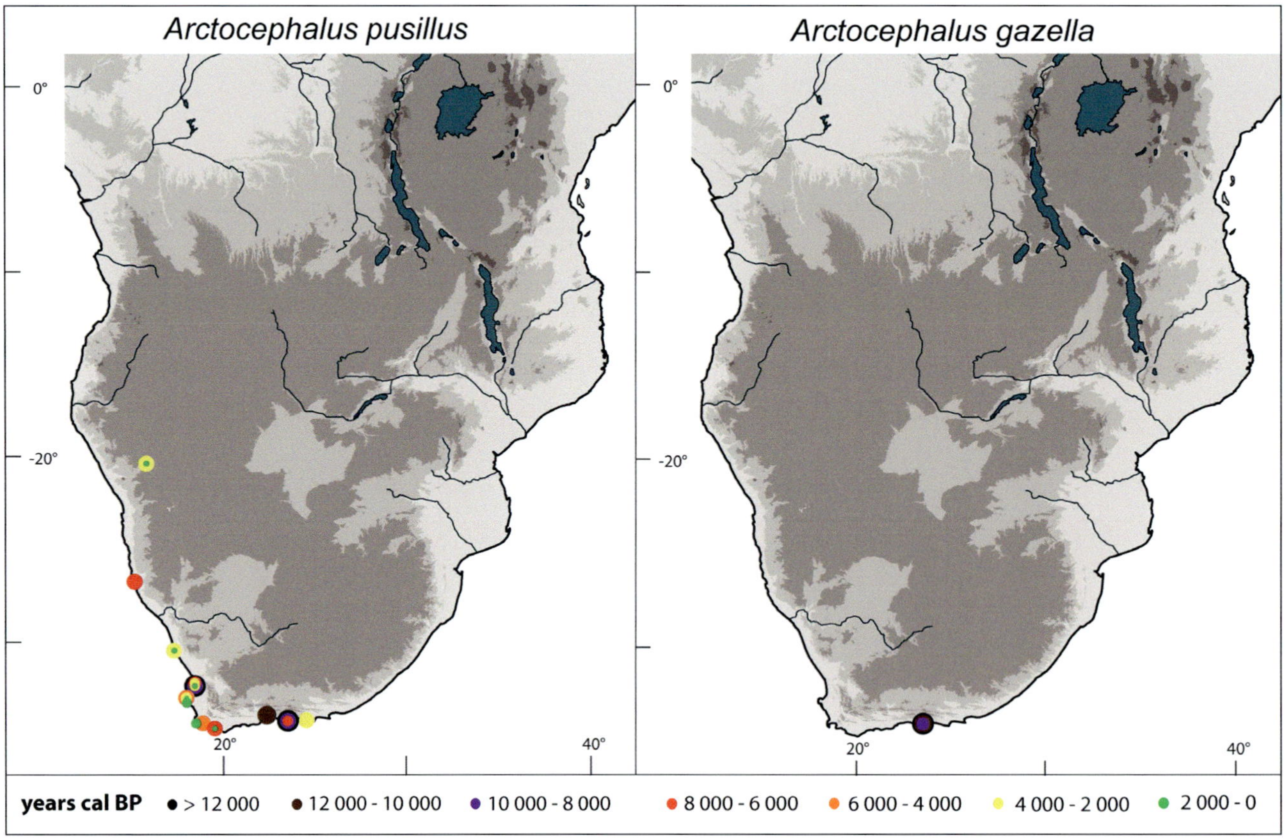

URSIDAE

Ursus sp.

Country	Site N	Site
Algeria	19	Takouatz Guerrissène
	7	Damous el Ahmar
Morocco	419	Bou Saria A
	421	Bou Saria C-D
	426	Ghar-Khal IIIb-V
	427	Ghar-Khal II-IIIa
	431	Kaf-taht-el-Ghar-C
	432	Kaf-taht-el-Ghar-D
	433	Kaf-taht-el-Ghar-E
	434	Kaf-taht-el-Ghar-F
	435	Kaf-taht-el-Ghar-G
	437	Kehf-el-Baroud

Ursus arctos

Country	Site N	Site
Algeria	20	Tamar Hat
	8	Djurdjura
Libya	329	Hauah Fteah-Neolithic
Morocco	436	Kehf el Hammar

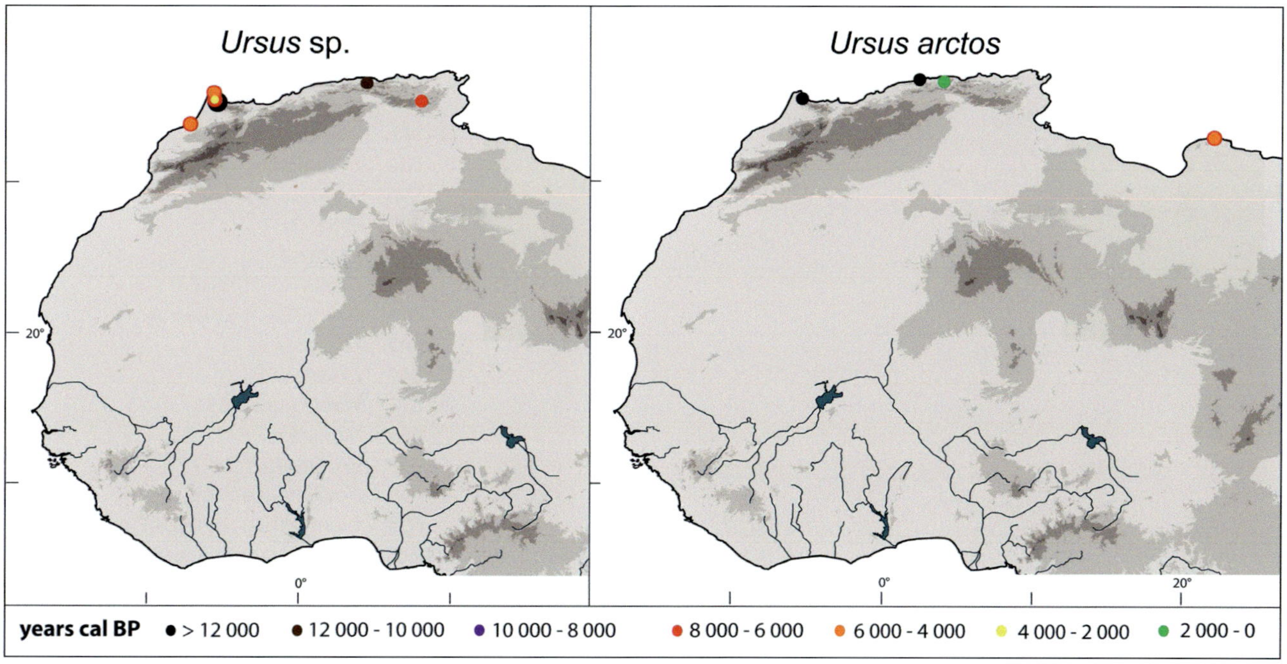

ORDER PERISSODACTYLA

EQUIDAE

Equus sp.

Country	Site N	Site
Algeria	21	Ti-n-Hanakaten
	6	Columnata
	9	Dra-Mta-el-Ma-el-Abiod
Botswana	28	Bosutwe-Toutswe
	37	Matlapaneng
	39	Qogana
	48	White Paintings RS-2ab
	48	White Paintings RS-1
	49	White Paintings RS-2b-3ab
Burkina Faso	54	Oursi hu-beero
	56	Oursi Village-LIA
	57	Oursi Village-MIA
	64	Saouga 95/7
Cameroon	69	Aissa Dugjé-b
Djibouti	90	Asa Koma
Egypt	106	Berenike 95/9 Early Roman
	107	Berenike 95/97Late Roman
	122	Dakhleh Ouasis, Ba
	144	El-Zarkeh
	156	Hierakonpolis HK6-Naq III
	179	Quseir al-Qadim
Ethiopia	207	Axum-D
	208	Axum-K
Ghana	228	Begho
	242	Mankuma
Kenya	247	Akira-N
	249	Akira-SPN
	253	Ele Bor A-A
	254	Ele Bor A-B
	269	GaJi12
	270	GaJi2
	272	Gogo Falls-I
	274	Gogo Falls-sq26-29
	276	GvJm 19-HC
	277	GvJm 19-TLPG
	278	GvJm 22-LHC
	279	GvJm 22-LPG
	280	GvJm 46-LPG
	282	GvJm 62-LPG
	283	GvJm 62-LPNG
	284	Kulchurdo
	287	Lowasera-a
	288	Lowasera-b
	289	Lowasera-c
	294	Narosura
	302	Sambo Ngige
Libya	326	Hauah Fteah- Historic
	327	Hauah Fteah-Iberom
	329	Hauah Fteah-Neolithic
	336	Uan Afuda-US1-3
	346	Wadi Tannezzuft-00/195
Mali	361	Akumbu-EIA
	362	Akumbu-MIA
	373	Hamdallahi
Mauritania	400	Dhar Tichitt Site 38
	401	Dhar Tichitt Site 46
	405	Khatt Lemaiteg
Morocco	422	El Harhoura 1
	423	El Harhoura 2
	424	El Mnasra
	433	Kaf-taht-el-Ghar-E
	435	Kaf-taht-el-Ghar-G
	437	Kehf-el-Baroud
	438	Kehf-el-Baroud-aven
Namibia	450	Brandberg-BBC
	452	Brandberg-LSA
	455	Geduld-LIA
	456	Geduld-LSA
	457	Haalenberg
Niger	467	Adrar Bous-T
	488	Takene Bawat 2
South Africa	559	Blydefontein I-II
	560	Blydefontein III-IV
	561	Blydefontein V
	562	Blydefontein VI-VII
	563	Boomplass Cave-a
	564	Boomplass Cave-b
	565	Boomplass Cave-c
	566	Boomplass Cave-d
	573	Burchell´s Shelter
	581	Colwinton Shelter-CW
	584	De Hoop
	587	Dikbosch 1
	588	Dikbosch 2
	589	Doornfontein
	590	Drie Susters
	606	Ficus B
	609	Glentyre
	610	Green Elliott Shelter I-III
	611	Green Elliott Shelter IV-VI
	645	Limerock 1
	646	Limerock 2
	656	Mapungubwe-Hill
	659	Marupale Hill
	689	Pont-Drift-Zhizo
	690	Powerhouse Cave
	694	Ravenscraig Shelter-pCW
	695	Ravenscraig Shelter-R
	697	Riversmead Shelter I-II
	698	Riversmead Shelter III-IV
	705	Rose Cottage Cave-DB2 *intr.
	708	Schroda
	742	Ventershoek
Sudan	755	Abu Hugar
	771	El Zakiab
	774	Gash Delta-AA
	775	Gash Delta-K
	787	Khartoum Hopital
	819	Um Direiwa
West. Sahara	841	Site H

Equus burchellii

Country	Site N	Site	
Angola	23	Benfica	
	24	Kamabanga	
	25	Kitala	
Botswana	26	Bosutwe-Mapungubwe	
	27	Bosutwe-Taukome	
	28	Bosutwe-Toutswe	
	29	Bosutwe-Zimbabwe	
	31	Commando Kop	
	35	Drotsky's Cave-P	
	38	Nqoma	
	40	Taukome	
	48	White Paintings RS-1	
	48	White Paintings RS-2ab	
	49	White Paintings RS-2b-3ab	
Kenya	247	Akira-N	
	249	Akira-SPN	
	250	Causeway Site	
	251	Crescent Island main	
	257	Engaruani	*cf.
	272	Gogo Falls-I	
	273	Gogo Falls-II-III	
	275	GvJm 19-HA	
	276	GvJm 19-HC	
	277	GvJm 19-TLPG	
	278	GvJm 22-LHC	
	279	GvJm 22-LPG	
	280	GvJm 46-LPG	
	282	GvJm 62-LPG	
	283	GvJm 62-LPNG	
	290	Maasai Gorge Shelter	
	292	Marula Rock Shelter	
	293	Naivasha Railway Shelter	
	294	Narosura	
	296	Ngamuriak	
	297	Ngenyn	*cf.
	298	Prolonged Drift	
Lesotho	304	Ha Makotoko	
	315	Ntloana Tsoana	
	321	Tloutle-CWI	
	323	Tloutle-pCW	
	324	Tloutle-R	
Libya	325	Et Tera	
Malawi	351	Malowa-1	
	352	Malowa-2	
	353	Malowa-3	
	354	Malowa-4	
	355	Malowa-5	
	356	Matope Court-EIA	
	358	Matope Court-LIA	
Mozambique	442	Manekweni-2-ABC	
	444	Manekweni-4	
Rhodesia	533	Harleigh Farm-2	
	534	Mabveni	
South Africa	556	2627 CD7	
	567	Border Cave	*cf.
	574	Bushman RS-LSA	
	575	Bushman RS-MSA	
	585	Deelpan	
	589	Doornfontein	
	593	Eiland Salt Works-a	
	594	Eiland Salt Works-b	
	604	Equus Cave 1A	
	605	Ficus A	
	607	Ficus C	
	614	Hapi Pan	
	615	Happy Rest	
	616	Heuningneskrans-I/O	
	617	Heuningneskrans-R	
	618	Hill X	
	625	Jubilee Shelter-PW	
	626	Jubilee Shelter-W	
	640	Le2	
	641	Le6	
	642	Le7a	
	643	Le7b	
	648	Ma38	
	649	Ma4	
	653	Malwula	
	655	Mapotini Hill	
	659	Marupale Hill	
	664	Mo8	
	673	Ol20	
	677	Pa8.1	
	684	Ph1	
	685	Ph6	
	686	Ph9	
	688	Pont-Drift-LKA	
	691	Pr34	
	696	Rhenosterkloof	
	704	Rose Cottage Cave-A-Mn	
	705	Rose Cottage Cave-DB2	
	710	Selongwe Hill	
	711	Serotwe Hill	
	713	SH16	
	714	Sh27	
	715	Shankare Hill-a	
	716	Shankare Hill-b	
	717	SK17	
	718	Sk4	
	720	Sonkoanini Hill	
	721	Spoegrivier-a	
	732	Ti6	
	739	Tsh1	
	753	Wosi-terminal EIA	
Tanzania	835	Mumba	
Zaire	843	Ishango-AB	*?
	845	Ishango-Z-Post-Em	*?
	846	Kakontwe	
	848	Kiantapo	
Zambia	860	Leopard's Hill-NI	
	861	Leopard's Hill-NII	
	862	Leopard's Hill-NIII	
	863	Leopard's Hill-PN	
Zimbabwe	870	Redcliff Cave	
	872	Simamwe	

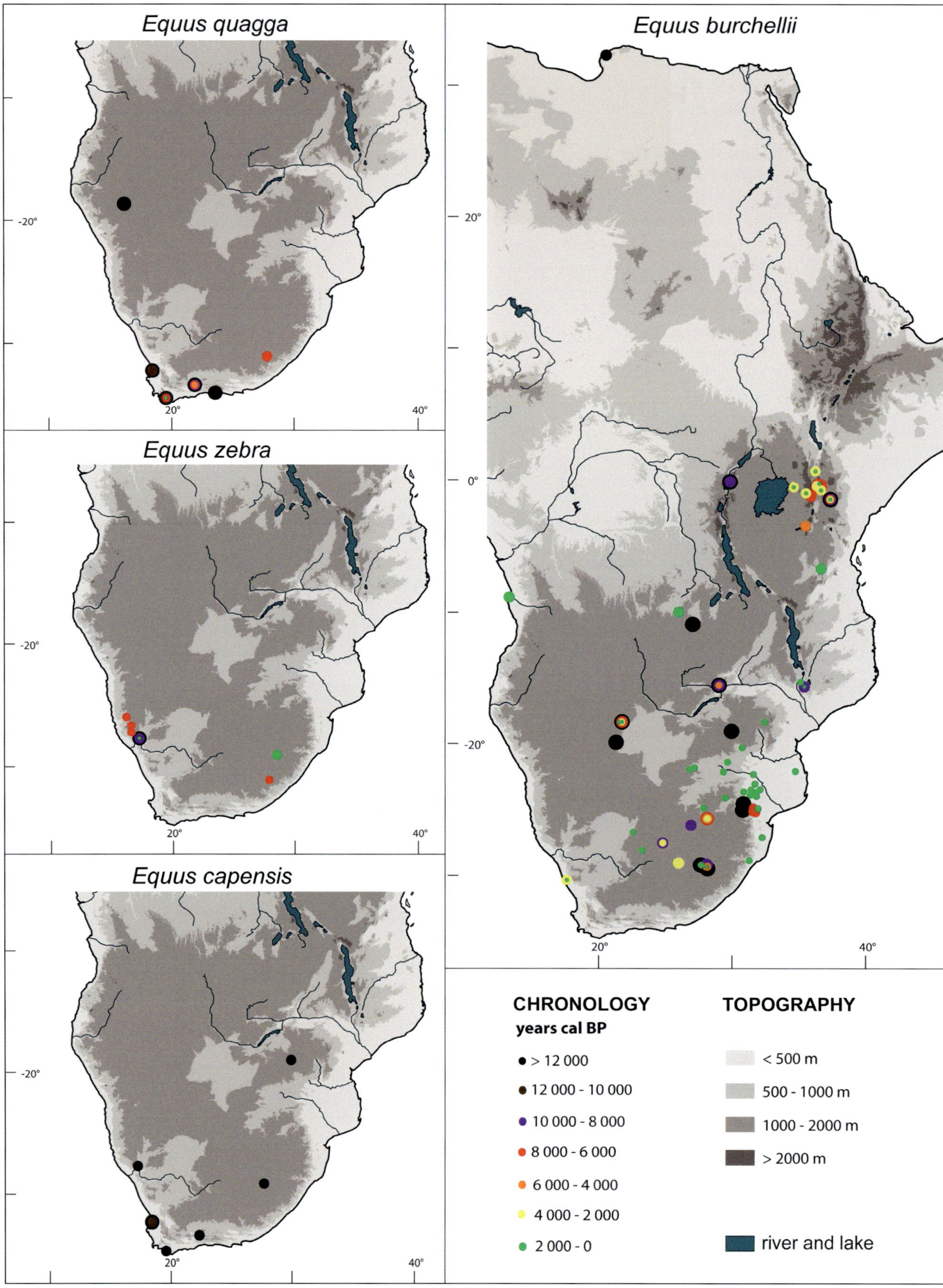

Equus quagga
-20°
20°
40°
Equus zebra
-20°
20°
40°
Equus capensis
-20°
20°
40°
Equus burchellii
20°
0°
-20°
20°
40°
CHRONOLOGY
years cal BP
> 12 000
12 000 - 10 000
10 000 - 8 000
8 000 - 6 000
6 000 - 4 000
4 000 - 2 000
2 000 - 0
TOPOGRAPHY
< 500 m
500 - 1000 m
1000 - 2000 m
> 2000 m
river and lake

Equus quagga

Country	Site N	Site
Namibia	460	Oshigambo Peninsula
South Africa	568	Buffelskloof RS-a *or *E.zebra*
	569	Buffelskloof RS-b *or E.zebra
	570	Buffelskloof RS-c *or E.zebra
	571	Buffelskloof RS-d *or E.zebra
	576	Byeneskranskop Cave 1a *cf.
	577	Byeneskranskop Cave 1b *cf.
	578	Byeneskranskop Cave 1c *cf.
	579	Byeneskranskop Cave 1d *cf.
	581	Colwinton Shelter-CW
	595	Eland´s Bay Cave-Al
	596	Eland´s Bay Cave-RI
	669	Nelson Bay Cave- R *cf.

Equus zebra

Country	Site N	Site
Namibia	446	Aar I and II *cf.
	447	Apollo 11-a **E.z.* cf.*hartmannae*
	448	Apollo 11-b **E.z.* cf.*hartmannae*
	449	Apollo 11-c **E.z.* cf.*hartmannae*
	458	Namtib *cf.
	461	Pokkenbank *cf.
South Africa	581	Colwinton Shelter-CW *cf.
	730	Tandjesberg

Equus capensis

Country	Site N	Site
Namibia	449	Apollo 11-c
South Africa	563	Boomplass Cave-a
	576	Byeneskranskop Cave 1a *cf.
	595	Eland´s Bay Cave-Al
	596	Eland´s Bay Cave-RI
	705	Rose Cottage Cave-DB2
Zimbabwe	870	Redcliff Cave

Equus grevyi

Country	Site N	Site
Egypt	121	Dakhleh Oasis, SMU
	122	Dakhleh Ouasis, Ba
Ethiopia	209	Gademotta
Kenya	255	Ele Bor A-C
	273	Gogo Falls-II-III
	274	Gogo Falls-sq26-29
	277	GvJm 19-TLPG
	279	GvJm 22-LPG
	280	GvJm 46-LPG
	283	GvJm 62-LPNG

Equus africanus

Country	Site N	Site
Egypt	137	Elephantine-Satet temple *or asinus
	162	Kom Ombo
	164	Maadi
	185	Sodmein Cave older backfill
Libya	333	Ti-n-Thora East
	339	Uan Muhuggiag (LP)
Sudan	790	Khasm el Girba - KG15-16
	794	Laqiya area - 82/31
	800	Mahal Teglinos K1-LG

Equus mauritanicus

Country	Site N	Site
Algeria	4	Ain Mistehayia
	7	Damous el Ahmar *?
	13	Kef Zoura D
	20	Tamar Hat
Morocco	422	El Harhoura 1
West. Sahara	837	Sebkha Amtal

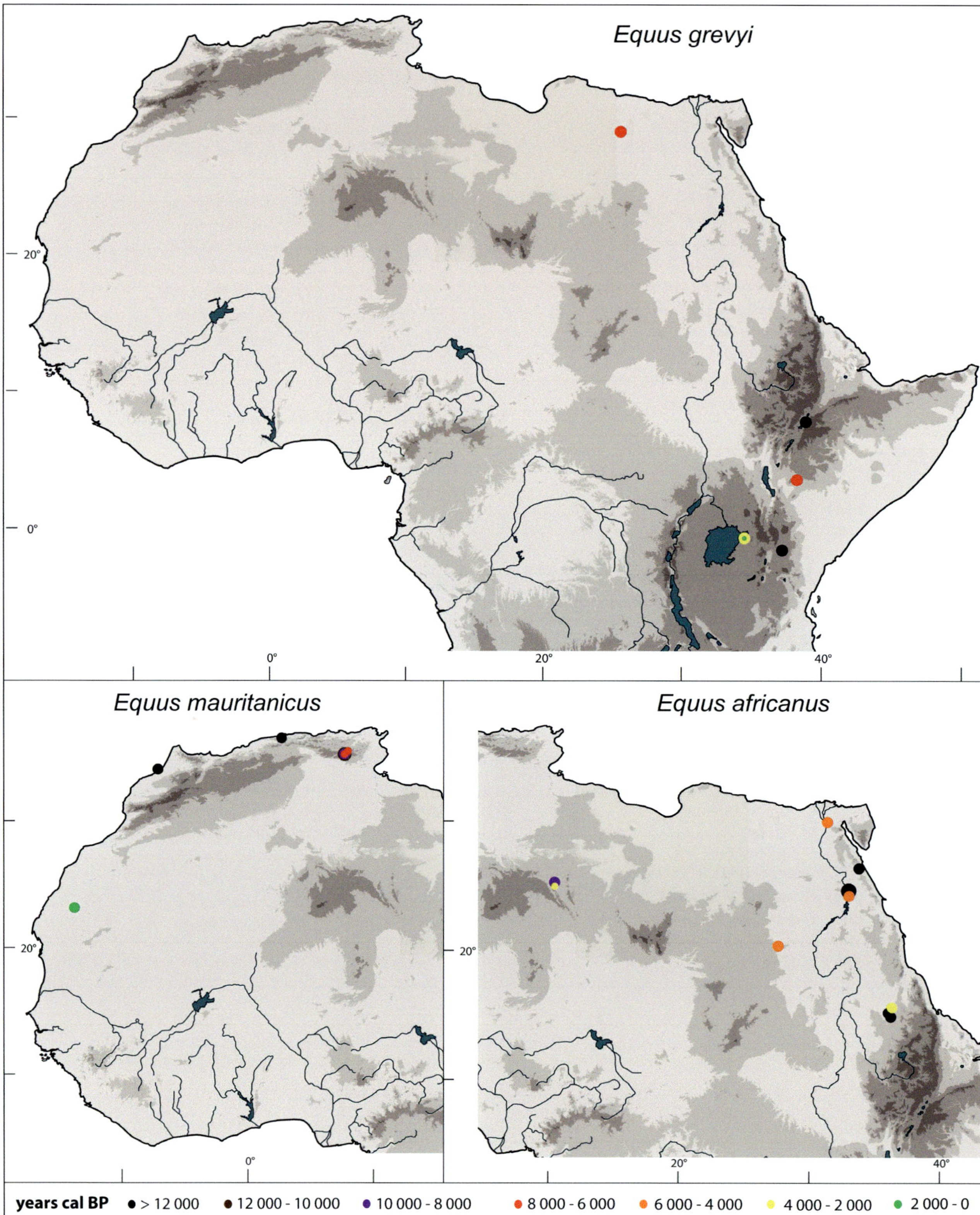
Equus grevyi
20°
0°
0°
20°
40°
Equus mauritanicus
20°
0°
Equus africanus
20°
20°
40°
years cal BP
> 12 000
12 000 - 10 000
10 000 - 8 000
8 000 - 6 000
6 000 - 4 000
4 000 - 2 000
2 000 - 0

Equus przewalskii f.caballus

Country	Site N	Site	
Burkina Faso	63	Saouga 94/120	
	64	Saouga 95/7	
Cameroon	68	Aissa Dugjé	
Egypt	106	Berenike 95/9 Early Roman	
	114	Buto-9-12	
	114	Buto-9-12	
	123	Dakhleh Ouasis, Roman	
	174	Qantir	
	187	Tell el-Dab´a	
	188	Tell el-Dab´a - b	
	189	Tell Heboua	
Ghana	231	Daboya	
Mali	361	Akumbu-EIA	
	373	Hamdallahi	
	377	Jenné-Jeno - phase III	*cf.
	378	Jenné-Jeno - phase IV	*cf.
South Africa	659	Marupale Hill	
Sudan	760	Buhen	
	806	Sai	

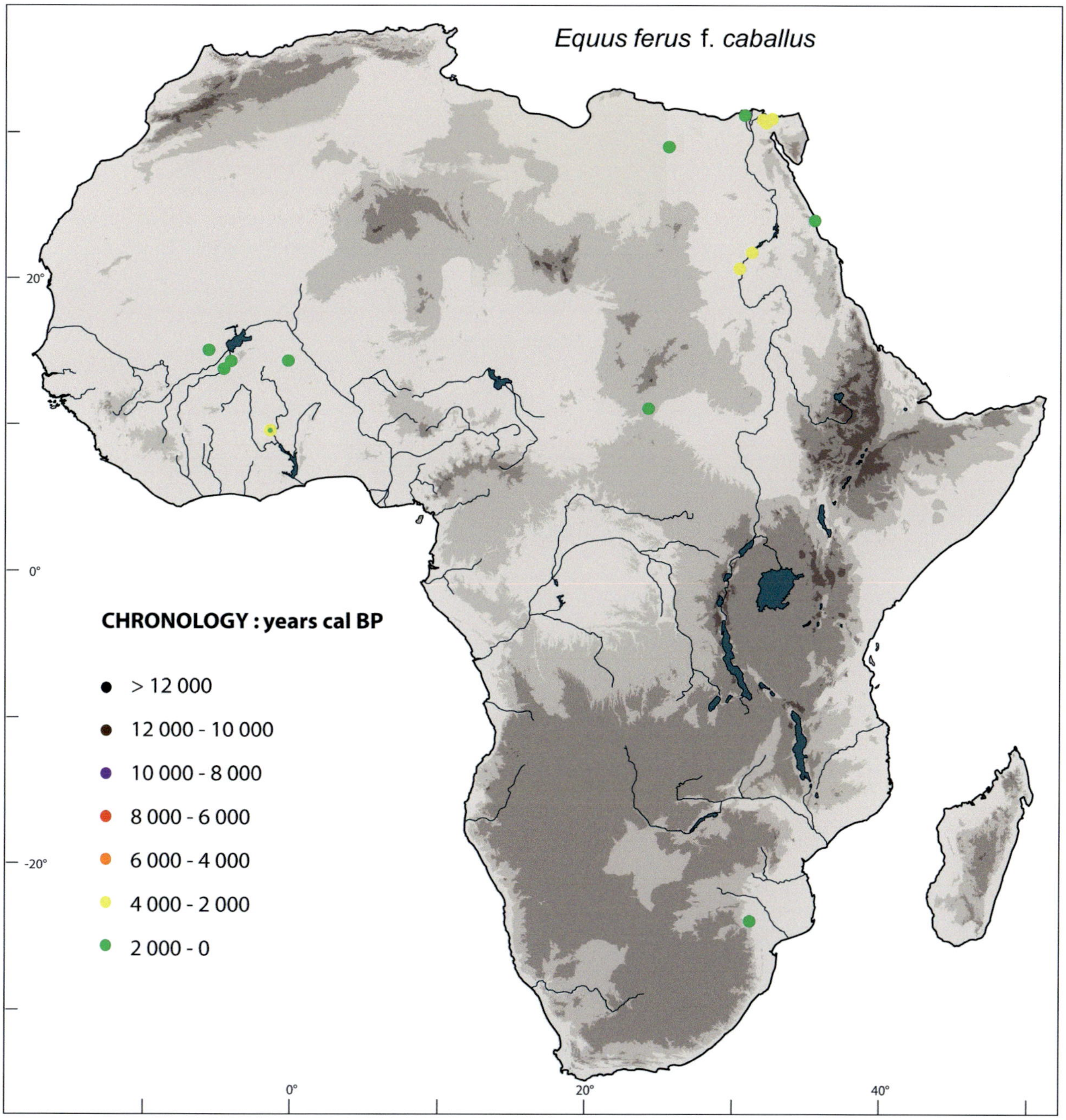

Equus africanus f. asinus

Country	Site N	Site	
Egypt	100	Adaima-habitat	
	107	Berenike 95/97Late Roman	
	108	Berenike 96/97 Ptolem.	
	113	Buto-6	
	114	Buto-9-12	
	123	Dakhleh Ouasis, Roman	
	135	Elephantine-necropolis	
	136	Elephantine-NW Chnum T	
	140	El-Tell el Iswid-Early Dynastic	
	141	El-Tell el Iswid-Predynastic	
	142	El-Tell el-Iswid A	
	143	El-Tell el-Iswid B	
	144	El-Zarkeh	
	154	Hierakonpolis HK6-Naq IC-IIA	
	156	Hierakonpolis HK6-Naq III	
	160	Kharga dep.-graeco-roman	
	164	Maadi	
	169	Minshat Abu Omar	
	170	Mons Claudianus	
	174	Qantir	
	187	Tell el-Dab´a	*mule
	188	Tell el-Dab´a - b	
	95	Abu Sha´ar-4C	
	96	Abu Sha´ar-5C	
	97	Abusir	
	99	Adaima-habitat	
Ethiopia	220	Quiha	*?
Ghana	231	Daboya	
Kenya	294	Narosura	
Lesotho	316	Sehonghong-a	
Senegal	538	Cubalel-Ia	*cf.
	539	Cubalel-Ib	*cf.
	546	Siouré-Ib	*cf.
	548	Siouré-IV, V	*cf.
	552	Walaldé-W2	
South Africa	674	Ondini	
Sudan	778	Gism el-Arba	
	785	Kerma	
	817	Soba	
	825	Wadi Hariq 01/4	
	826	Wadi Hariq 97/5	
	827	Wadi Hariq 97/7	
	829	Wadi Howar - 84/1	
West. Sahara	837	Sebkha Amtal	

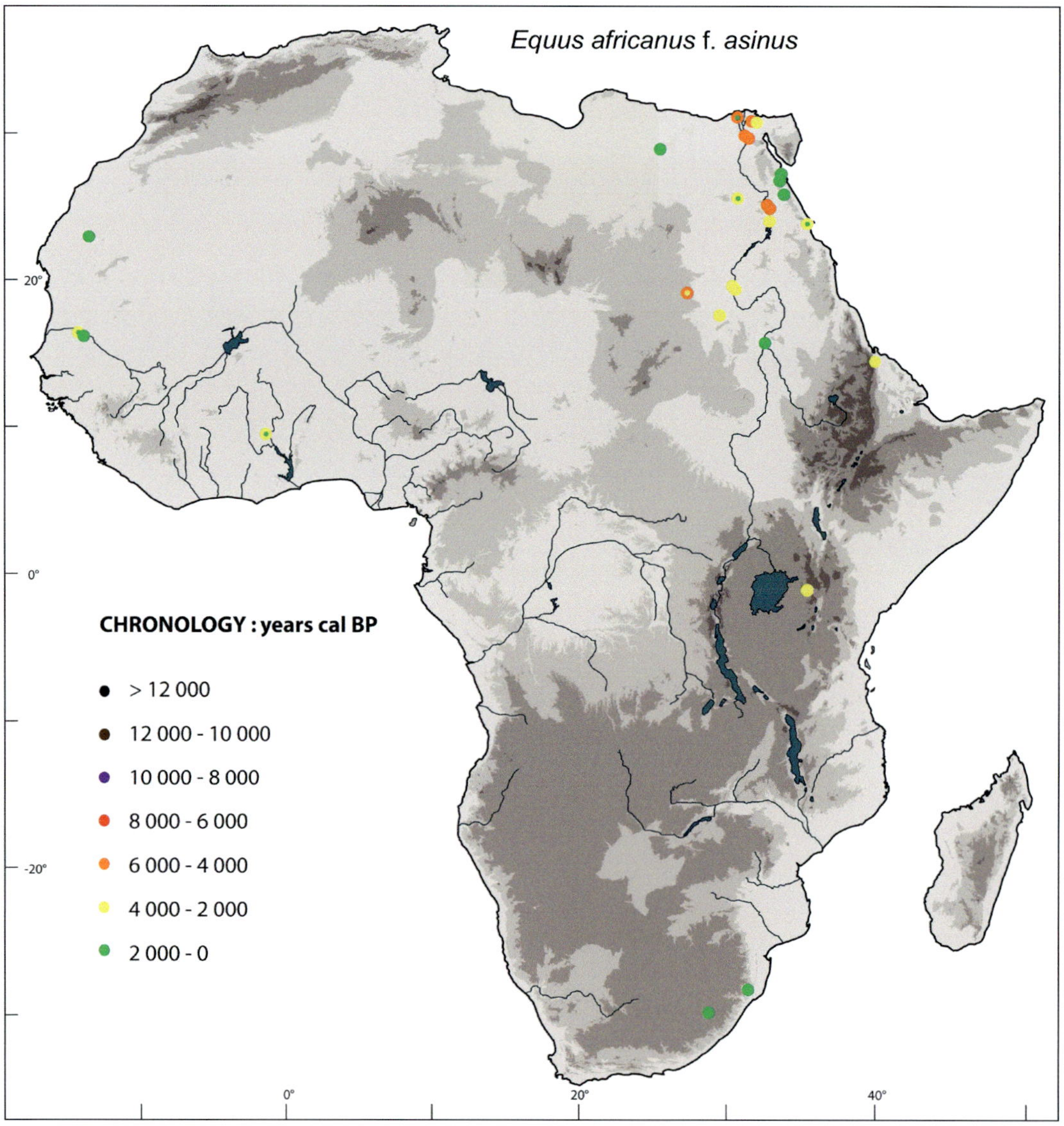

RHINOCEROTIDAE

Rhinocerotidae (indet.)

Country	Site N	Site
Algeria	11	Gueldaman Akbou
	21	Ti-n-Hanakaten
Botswana	26	Bosutwe-Mapungubwe
	34	Drotsky's Cave-LP
	49	White Paintings RS-2b-3ab
Kenya	272	Gogo Falls-I
	273	Gogo Falls-II-III
	274	Gogo Falls-sq26-29
	279	GvJm 22-LPG
Mauritania	395	Dhar Tichitt- highest beach
	396	Dhar Tichitt- latest phase
	399	Dhar Tichitt- post beach
Morocco	422	El Harhoura 1
Mozambique	444	Manekweni-4
	445	Manekweni-5
Namibia	455	Geduld-LIA
Niger	467	Adrar Bous-T
South Africa	566	Boomplass Cave-d
	587	Dikbosch 1
	594	Eiland Salt Works-b
	595	Eland´s Bay Cave-AI
	596	Eland´s Bay Cave-RI
	641	Le6
	642	Le7a
	643	Le7b
	664	Mo8
	673	Ol20
	690	Powerhouse Cave
	698	Riversmead Shelter III-IV
	713	SH16
	717	SK17
	720	Sonkoanini Hill
	721	Spoegrivier-a
Sudan	773	Esh Shaheinab
	781	Jebel Umm Marrahi
	787	Khartoum Hopital
	809	Selima Sandsheet - 85/80
	816	Singa
Zaire	843	Ishango-AB
	845	Ishango-Z-Post-Em
Zambia	861	Leopard's Hill-NII

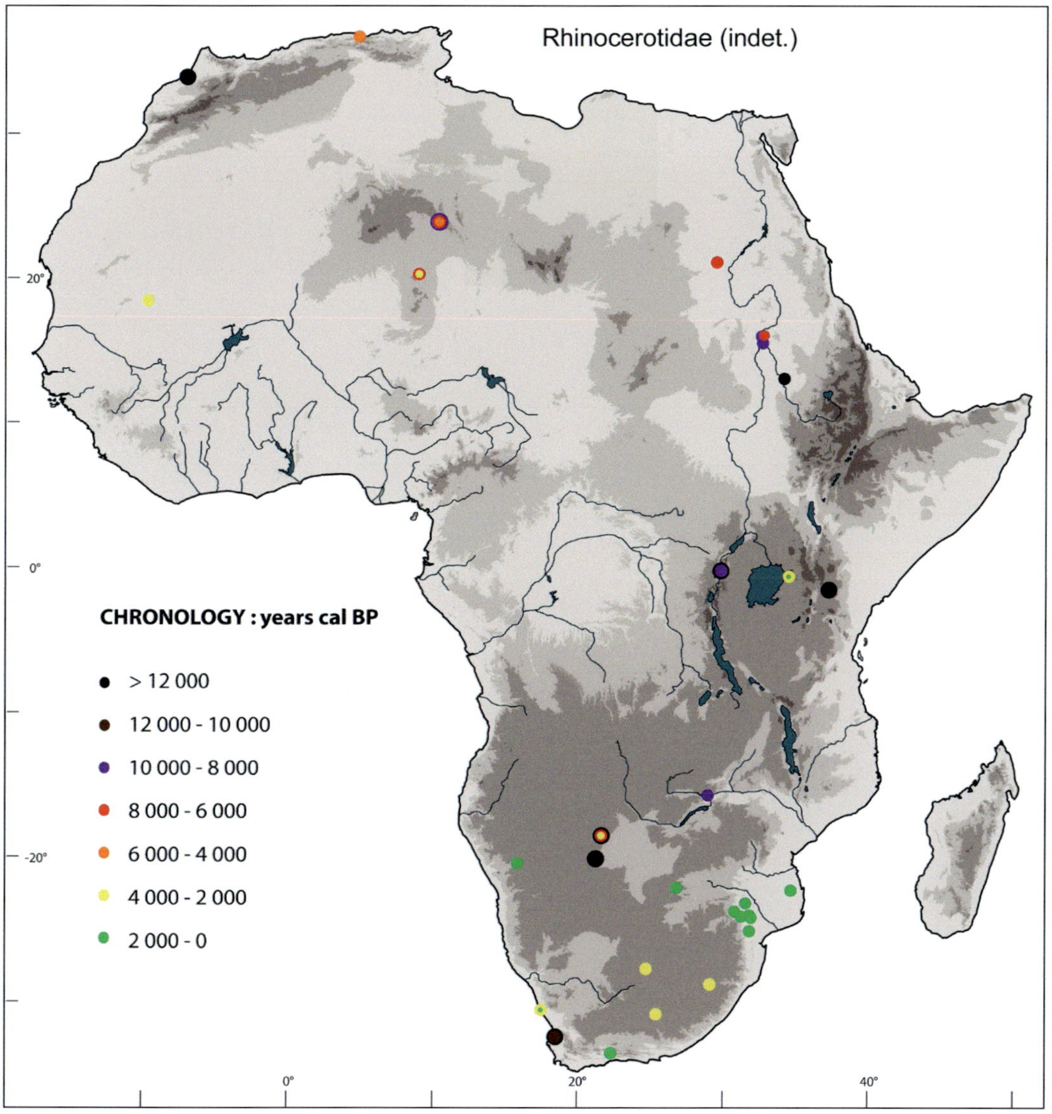

Ceratotherium simum

Country	Site N	Site	
Algeria	7	Damous el Ahmar	
Botswana	26	Bosutwe-Mapungubwe	
Kenya	298	Prolonged Drift	
Mali	374	Hassi el-Abiod	
Mauritania	393	Berouâga	
	394	Chami-A2	
	397	Dhar Tichitt- lowest beach	
	398	Dhar Tichitt- middle beach	
	414	Tichitt - Goungou phase	
Morocco	438	Kehf-el-Baroud-aven	
	439	Rouazi necropolis	
Sudan	755	Abu Hugar	*cf.
	773	Esh Shaheinab	
Tanzania	835	Mumba	

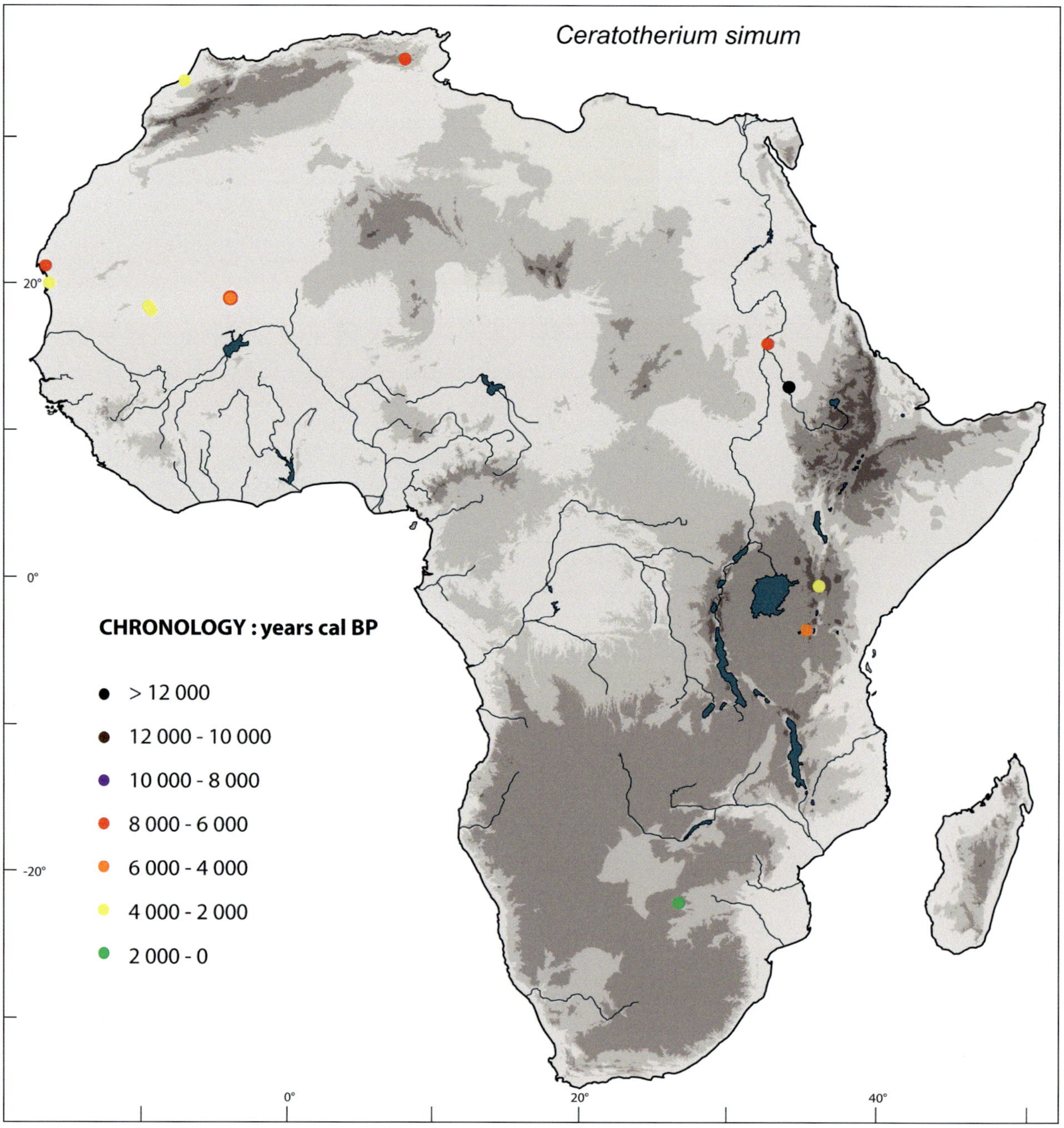

Diceros bicornis

Country	Site N	Site	
Botswana	26	Bosutwe-Mapungubwe	
Congo	87	Ntadi Yom ba-level B	
Kenya	267	GaJi1	*?
	270	GaJi2	
	294	Narosura	
	296	Ngamuriak	
Mali	374	Hassi el-Abiod	
	375	In Begaouaen	
Namibia	454	Fackelträger	
South Africa	568	Buffelskloof RS-a	*cf.
	569	Buffelskloof RS-b	*?
	576	Byeneskranskop Cave 1a	
	577	Byeneskranskop Cave 1b	
	578	Byeneskranskop Cave 1c	
	579	Byeneskranskop Cave 1d	
	586	Die Kelders	
	591	Drie Susters Main	
	596	Eland´s Bay Cave-RI	
	627	Kasteelberg A	* intr?
	628	Kasteelberg B	
	676	Oudepost I	
	688	Pont-Drift-LKA	
	738	Tortoise cave-d	
Sudan	757	Aneibis	
	765	El Kadada	*cf.
	773	Esh Shaheinab	
Tanzania	835	Mumba	* ?

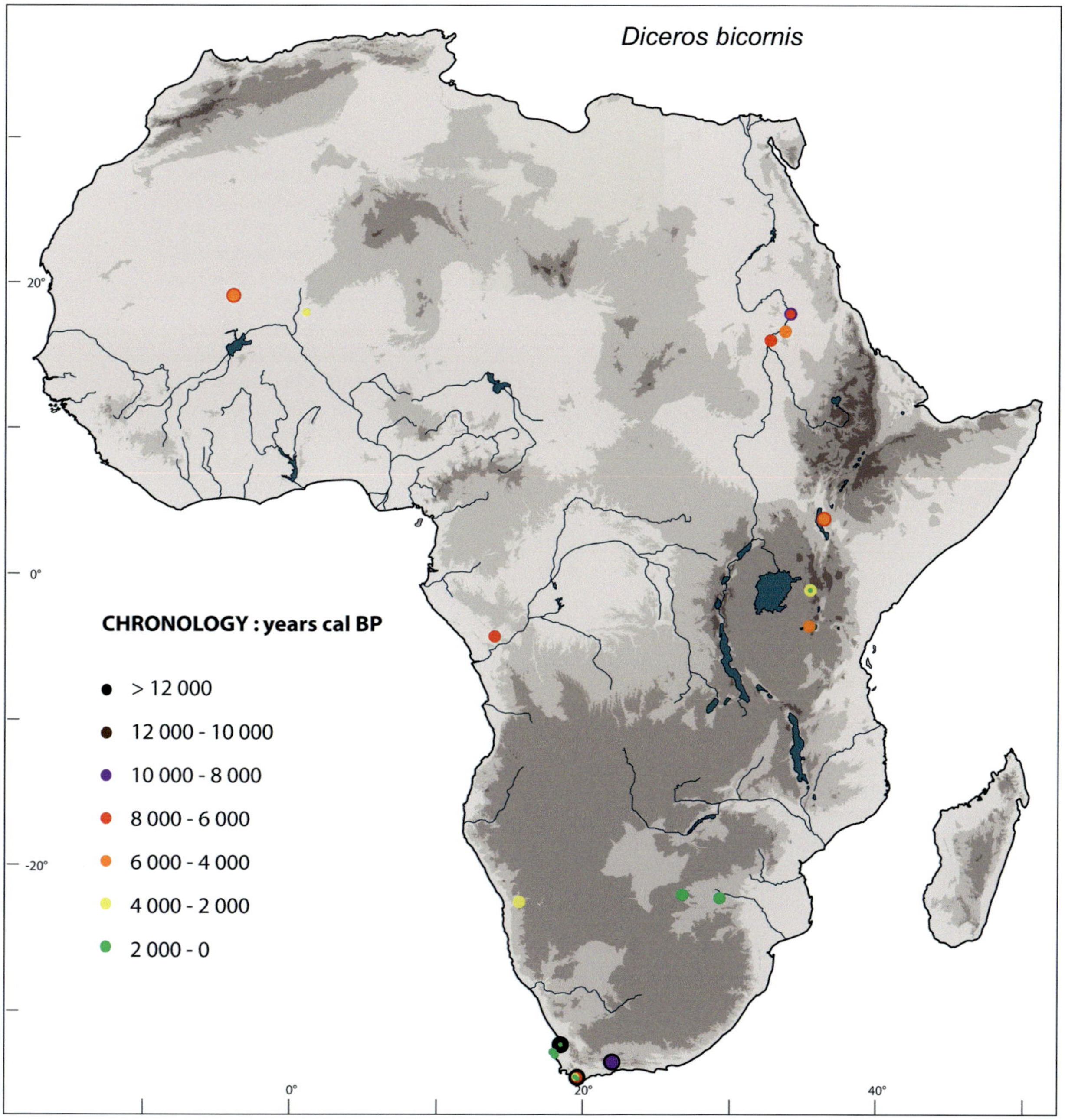

ORDER ARTIODACTYLA

SUIDAE

Sus scrofa

Country	Site N	Site
Algeria	10	Grotte Capeletti
	15	Medjez II
	20	Tamar Hat
	9	Dra-Mta-el-Ma-el-Abiod
Egypt	103	Al-Ma´abda
	112	Buto/1-3
	123	Dakhleh Ouasis, Roman
	168	Merimde Benisâlame
	191	Toukh * S.s. aff.palustris
Ethiopia	208	Axum-K
Libya	326	Hauah Fteah- Historic
	329	Hauah Fteah-Neolithic
Morocco	421	Bou Saria C-D
	422	El Harhoura 1
	423	El Harhoura 2
	426	Ghar-Khal IIIb-V
	429	Kaf-taht-el-Ghar-A
	430	Kaf-taht-el-Ghar-B
	431	Kaf-taht-el-Ghar-C
	433	Kaf-taht-el-Ghar-E
	434	Kaf-taht-el-Ghar-F
	435	Kaf-taht-el-Ghar-G
	437	Kehf-el-Baroud
	438	Kehf-el-Baroud-aven
Niger	466	Adrar Bous-K

Sus scrofa f. domestica

Country	Site N	Site
Egypt	100	Adaima-habitat
	102	Al-Ma´abda
	106	Berenike 95/9 Early Roman
	107	Berenike 95/97Late Roman
	112	Buto/1-3
	113	Buto-6
	114	Buto-9-12
	129	El Abadiya 2 North sector
	130	El Abadiya 2 South sector
	133	El Omari
	135	Elephantine-necropolis
	137	Elephantine-Satet temple
	140	El-Tell el Iswid-Early Dynastic
	141	El-Tell el Iswid-Predynastic
	142	El-Tell el-Iswid A
	143	El-Tell el-Iswid B
	144	El-Zarkeh
	154	Hierakonpolis HK6-Naq IC-IIA
	155	Hierakonpolis HK6-Naq IIA-B
	156	Hierakonpolis HK6-Naq III
	164	Maadi
	168	Merimde Benisâlame
	169	Minshat Abu Omar
	170	Mons Claudianus
	179	Quseir al-Qadim
	188	Tell el-Dab´a - b
	196	Tuna el-Gebel
	206	Workmen Village
	95	Abu Sha´ar-4C
	96	Abu Sha´ar-5C
	98	Adaima-burials
	99	Adaima-habitat
Ghana	223	Anyinam-Area A-B-C
	227	Asantamanso-ASV
	229	Bibease Plantation
	231	Daboya
	233	Esiease - A-C-D
	245	Nyanawase/Akwamu Am.
Lesotho	313	Liphofung-pCW
	316	Sehonghong-a
Morocco	419	Bou Saria A
	420	Bou Saria B
	421	Bou Saria C-D
	425	Ghar-Khal I
	426	Ghar-Khal IIIb-V
	427	Ghar-Khal II-IIIa
	429	Kaf-taht-el-Ghar-A
	430	Kaf-taht-el-Ghar-B
	431	Kaf-taht-el-Ghar-C
	433	Kaf-taht-el-Ghar-E
	437	Kehf-el-Baroud
	438	Kehf-el-Baroud-aven
South Africa	629	Kekane
	676	Oudepost I
	738	Tortoise cave-d
Zaire	857	Nkile

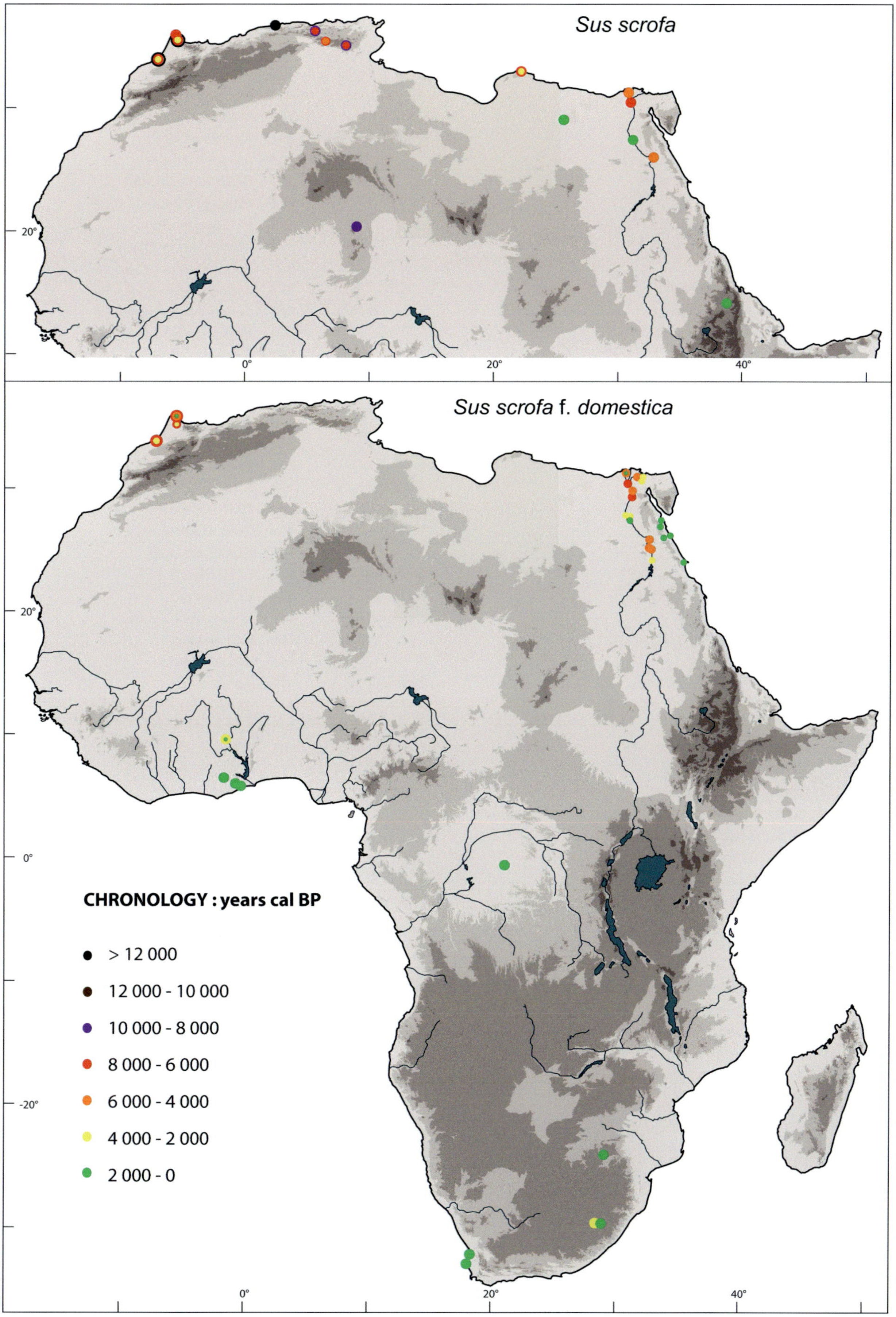
Sus scrofa
20°
0°
20°
40°
Sus scrofa f. domestica
20°
0°
-20°
CHRONOLOGY : years cal BP
> 12 000
12 000 - 10 000
10 000 - 8 000
8 000 - 6 000
6 000 - 4 000
4 000 - 2 000
2 000 - 0
0°
20°
40°

Phacochoerus aethiopicus

Country	Site N	Site	
Algeria	11	Gueldaman Akbou	
	5	Amekni	
Angola	24	Kamabanga	
Botswana	26	Bosutwe-Mapungubwe	
	28	Bosutwe-Toutswe	
	38	Nqoma	
	48	White Paintings RS-2ab	
	49	White Paintings RS-2b-3ab	
Burkina Faso	52	Kissi 22	
	54	Oursi hu-beero	
	56	Oursi Village-LIA	
	57	Oursi Village-MIA	
	58	Oursi-IA	
	61	Péntènga 89/1-spits 15-21	
	63	Saouga 94/120	
	64	Saouga 95/7	
Cameroon	70	Blé A	
	73	Blé E	
Chad	81	Delebo	
Egypt	121	Dakhleh Oasis, SMU	
Ethiopia	216	Moche Borago-1b	
Ghana	241	Mandara	
	242	Mankuma	
Kenya	251	Crescent Island main	
	254	Ele Bor A-B	
	256	Ele Bor A-D	
	257	Engaruani	
	260	Enkapune Ya Muto - Eb 5a	
	266	FxJj12-N	
	267	GaJi1	
	269	GaJi12	
	270	GaJi2	
	272	Gogo Falls-I	
	273	Gogo Falls-II-III	
	274	Gogo Falls-sq26-29	
	275	GvJm 19-HA	
	276	GvJm 19-HC	
	277	GvJm 19-TLPG	
	278	GvJm 22-LHC	
	279	GvJm 22-LPG	
	292	Marula Rock Shelter	
	293	Naivasha Railway Shelter	
	294	Narosura	
	296	Ngamuriak	
	297	Ngenyn	
	298	Prolonged Drift	
	299	Salasun-LSA	
Lesotho	304	Ha Makotoko	
	305	Lequetsoana	
	310	Likoaeng-XIII	
	314	Muela	
	315	Ntloana Tsoana	
	316	Sehonghong-a	
	321	Tloutle-CWI	
	322	Tloutle-O	
Libya	333	Ti-n-Thora East	
	340	Uan Muhuggiag (MP)	
	345	Uan Telocat (MP)	
Malawi	350	Chikumba	
	351	Malowa-1	
	352	Malowa-2	
	353	Malowa-3	
	354	Malowa-4	
	355	Malowa-5	
Mali	361	Akumbu-EIA	
	368	erg Ine Sakane	
	369	Est-Hassi el Abiod	
	374	Hassi el-Abiod	
	375	In Begaouaen	
	376	Jenné-Jeno - phase I/II	
	377	Jenné-Jeno - phase III	
	377	Jenné-Jeno - phase III	
	378	Jenné-Jeno - phase IV	
	380	Karkarichinkat N-S	
	381	Kobadi	
	385	Korounkorokalé-III	
	387	Tessalit	
	388	Tiabel Goudiodié	
	390	Windé Koroji I,III	
	391	Windé Koroji SO	
Mauritania	394	Chami-A2	
	402	DN4-BKH-S5	
	403	Draa Malichigdane	
	405	Khatt Lemaiteg	
Morocco	422	El Harhoura 1	
	424	El Mnasra	
Mozambique	442	Manekweni-2-ABC	
	444	Manekweni-4	
	445	Manekweni-5	
Namibia	447	Apollo 11-a	
	449	Apollo 11-c	
Niger	465	Adrar Bous 10	
	467	Adrar Bous-T	
	471	Chin Tafidet	
	474	Gobero-T	
	475	In Aridal 1	
	478	Mentes 13	
	481	Mentes 4	
	482	Mentes 5	
	483	Mentes 6	
	484	Mentes 7	
	487	Takene Bawat 1	
	488	Takene Bawat 2	
	489	Takene Bawat 5	
	497	TMY3	
Nigeria	498	Bukarkurari-I	
	503	Gajiganna A	
	504	Gajiganna B1-lower	
	505	Gajiganna B2	
	507	Galaga	
	511	Kariari	
	512	Kelumeri	
	514	Kursakata-IA	
	518	Mege I	
	524	NA 93/42	
	526	Ngala II	
	527	Ngala IIIa	
Rwanda	536	Muganza	
Senegal	546	Siouré-Ib	
	547	Siouré-II	
South Africa	563	Boomplass Cave-a	
	568	Buffelskloof RS-a	
	570	Buffelskloof RS-c	*?

Country	No.	Site	Note
	574	Bushman RS-LSA	
	575	Bushman RS-MSA	
	578	Byeneskranskop Cave 1c	
	581	Colwinton Shelter-CW	
	594	Eiland Salt Works-b	
	604	Equus Cave 1A	
	610	Green Elliott Shelter I-III	
	611	Green Elliott Shelter IV-VI	
	615	Happy Rest	
	616	Heuningneskrans-I/O	
	617	Heuningneskrans-R	
	618	Hill X	
	622	HKLP Vb	
	622	HKLP Vb	*cf.
	623	Icon	
	624	Jubilee Shelter-O	
	625	Jubilee Shelter-PW	
	626	Jubilee Shelter-W	
	636	Klipspring	
	638	Langsdraai	
	640	Le2	
	641	Le6	
	642	Le7a	
	644	Leliehoek	
	648	Ma38	
	650	Magogo-1	
	655	Mapotini Hill	
	657	Mapungubwe-K2/TS3	*P.a./suid
	659	Marupale Hill	
	663	Mhlwazini Cave	
	668	Nelson Bay Cave- PW	
	669	Nelson Bay Cave- R	
	673	Ol20	
	688	Pont-Drift-LKA	
	691	Pr34	
	697	Riversmead Shelter I-II	
	698	Riversmead Shelter III-IV	*?
	699	Riversmead Shelter VI-IX	
	701	Rooikrans-CW	
	704	Rose Cottage Cave-A-Mn	
	705	Rose Cottage Cave-DB2	*intr.
	706	Rose Cottage Cave-OIC	
	707	Rose Cottage Cave-WI	
	709	Sebatini Hill	
	710	Selongwe Hill	
	711	Serotwe Hill	
	714	Sh27	
	716	Shankare Hill-b	
	717	SK17	
	718	Sk4	
	730	Tandjesberg	
	739	Tsh1	
	740	Twyfelpoort	
	742	Ventershoek	
	752	Wosi-middle-late EIA	
	753	Wosi-terminal EIA	
Sudan	754	Abu Darbein	
	756	Abu Tabari 02/1	
	757	Aneibis	
	761	Debbat Bangdit	
	762	Debbat El Eheima	
	764	El Damer	
	765	El Kadada	
	767	El Mahalab	
	771	El Zakiab	
	773	Esh Shaheinab	
	775	Gash Delta-K	
	781	Jebel Umm Marrahi	
	782	Kabbashi	
	783	Kadero	
	787	Khartoum Hopital	
	788	Khasm el Girba - Kassala	
	789	Khasm el Girba - KG14	
	800	Mahal Teglinos K1-LG	
	801	Mahal Teglinos K1-MG	
	804	Rahib-80/87	
	805	Saggai 1	
	810	Shabona	
	811	Shaqaduq-S1-A	
	812	Shaqaduq-S1-B	
	813	Shaqaduq-S21	
	814	Sheikh el Amin	
	815	Sheikh Mustafa	
	817	Soba	
	819	Um Direiwa	
	821	Upper Atbara hinterland	
	822	Upper Atbara Valley	
	828	Wadi Howar - 80/73	
Tanzania	835	Mumba	
Zaire	843	Ishango-AB	
	844	Ishango-C	
	845	Ishango-Z-Post-Em	
	846	Kakontwe	
	848	Kiantapo	
	853	Mashita Mbanza	
	855	Matupi-II	
Zambia	860	Leopard's Hill-NI	
	861	Leopard's Hill-NII	
	862	Leopard's Hill-NIII	
	863	Leopard's Hill-PN	
Zimbabwe	870	Redcliff Cave	
	871	Ruanga Ruin	

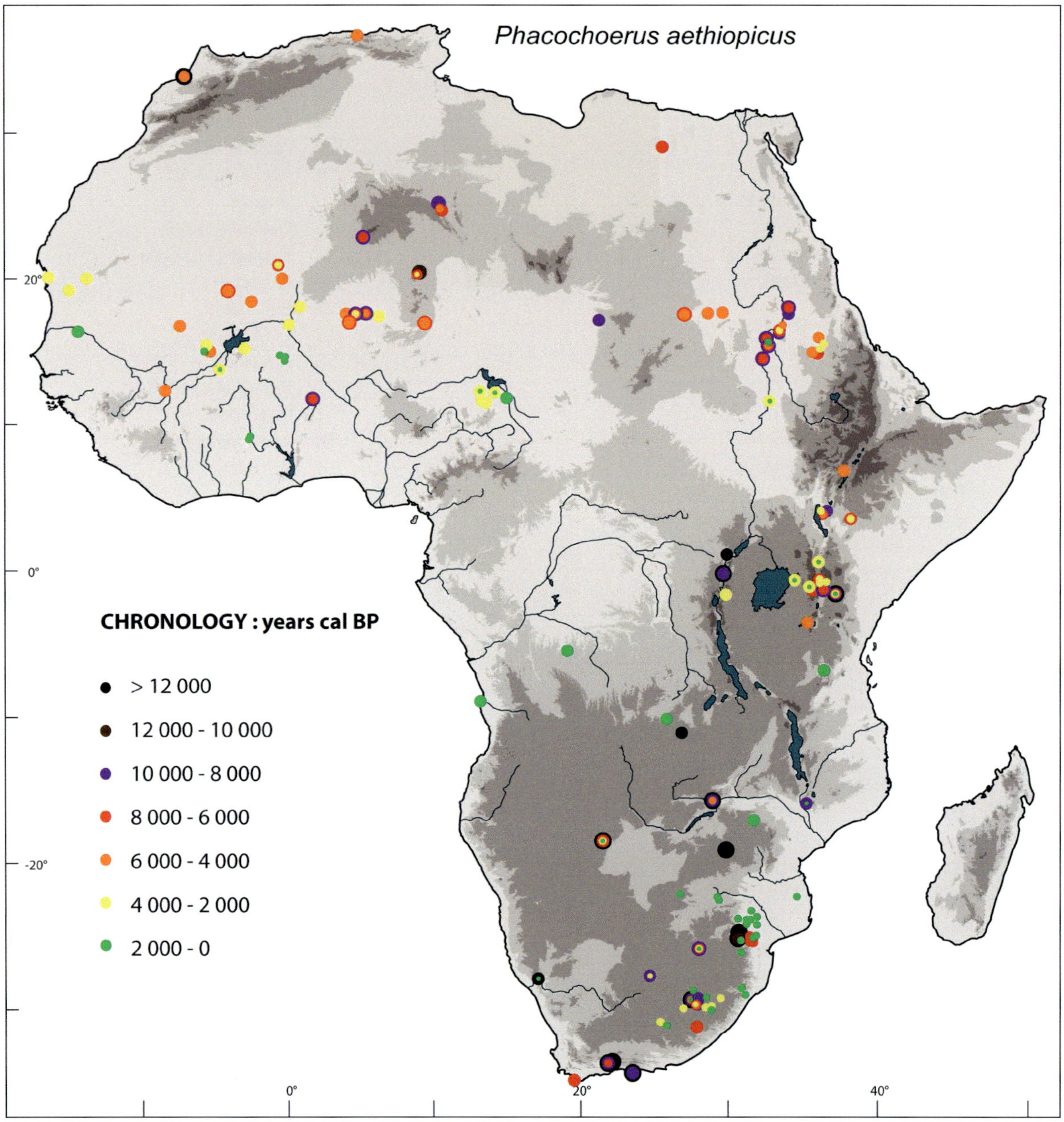

Hylochoerus meinertzhageni

Country	Site N	Site
Cameroon	67	Abeke
	76	Shum Laka II c
	77	Shum Laka II d
	78	Shum Laka IIa+b
Congo	89	Tongo
Kenya	258	Enkapune Ya Muto - Eb 4a
	260	Enkapune Ya Muto - Eb 5a
	292	Marula Rock Shelter
Zaire	843	Ishango-AB
	854	Matupi-I
	855	Matupi-II

Potamochoerus sp.

Country	Site N	Site
Rhodesia	533	Harleigh Farm-2

Potamochoerus larvatus

Country	Site N	Site
Ethiopia	216	Moche Borago-1b
Mozambique	441	Manekweni-1
	442	Manekweni-2-ABC
	443	Manekweni-3
	444	Manekweni-4
	445	Manekweni-5
South Africa	609	Glentyre
	671	Oakhurst
Sudan	780	Jebel et Tomat
Tanzania	835	Mumba

Metriodiochoerus

Country	Site N	Site
Zimbabwe	870	Redcliff Cave

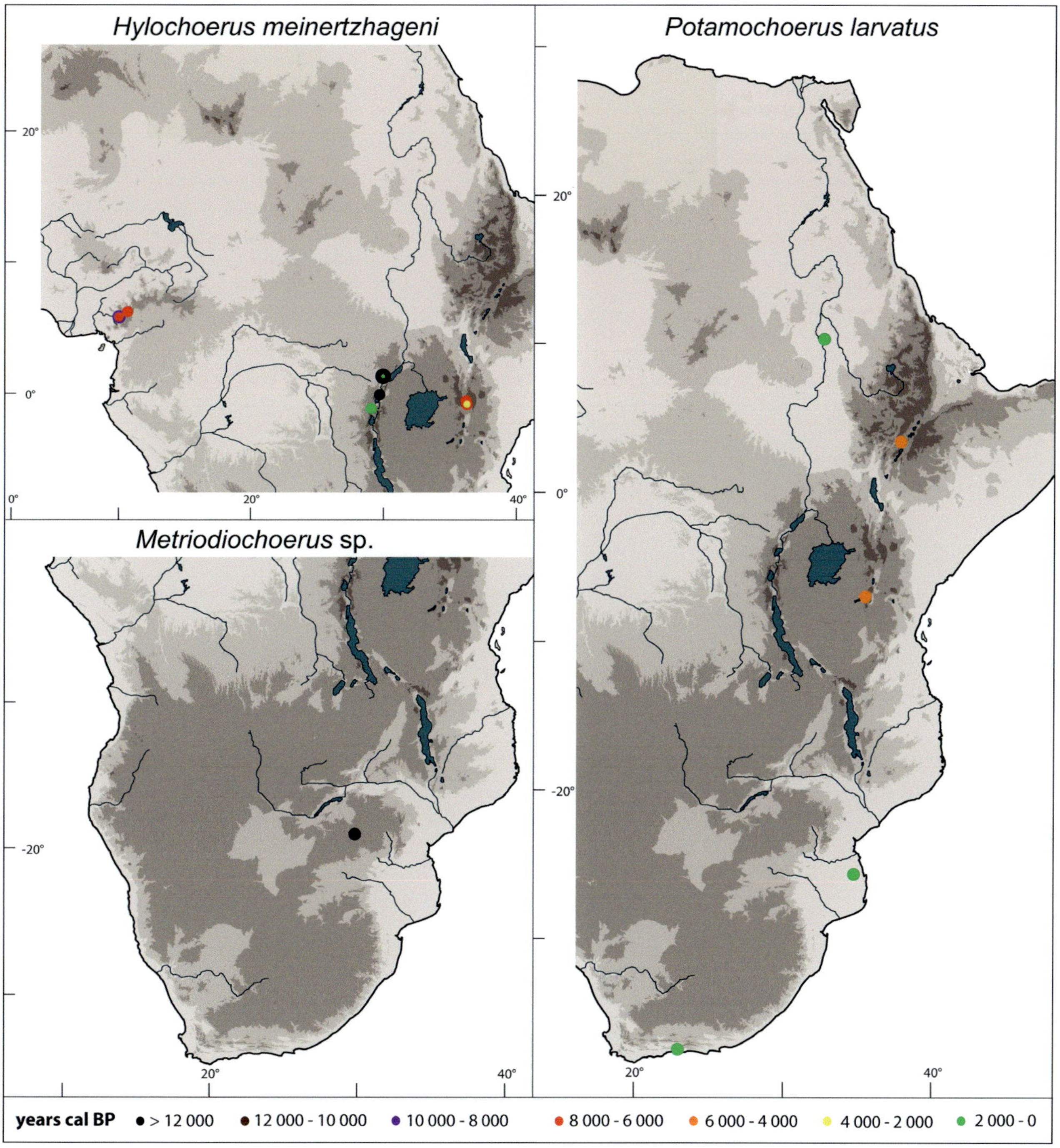

Potamochoerus porcus

Country	Site N	Site	
Botswana	49	White Paintings RS-2b-3ab	
Cameroon	67	Abeke	
C.A.R.	79	Mondongué	
	80	Lingbangbo	
Chad	81	Delebo	*?
Congo	87	Ntadi Yom ba-level B	
Ghana	223	Anyinam-Area A-B-C	
	225	Asantamanso-ASII/III	
	227	Asantamanso-ASV	
	230	Bono Manso	
	233	Esiease - A-C-D	
	241	Mandara	
Kenya	251	Crescent Island main	
	258	Enkapune Ya Muto - Eb 4a	
	259	Enkapune Ya Muto - Eb 4b	
	260	Enkapune Ya Muto - Eb 5a	
	273	Gogo Falls-II-III	
	275	GvJm 19-HA	
	278	GvJm 22-LHC	
	297	Ngenyn	
Lesotho	305	Lequetsoana	
Malawi	358	Matope Court-LIA	
South Africa	564	Boomplass Cave-b	

	567	Border Cave	
	577	Byeneskranskop Cave 1b	
	578	Byeneskranskop Cave 1c	
	579	Byeneskranskop Cave 1d	
	580	Collingham Shelter	
	603	Emambeni	
	616	Heuningneskrans-I/O	
	632	Klasies River Mouth-1II	*?
	633	Klasies River Mouth-5II	
	639	Large Rock Shelter	
	641	Le6	
	642	Le7a	
	648	Ma38	
	652	Magogo-3	
	668	Nelson Bay Cave- PW	
	669	Nelson Bay Cave- R	
	670	Nelson Bay Cave- W	
	675	Onink	
	751	Wosi-early EIA	
	752	Wosi-middle-late EIA	
	753	Wosi-terminal EIA	
Sudan	789	Khasm el Girba - KG14	
Zaire	842	Dimba	
	843	Ishango-AB	
	844	Ishango-C	
	845	Ishango-Z-Post-Em	
	846	Kakontwe	
	848	Kiantapo	
	854	Matupi-I	
	855	Matupi-II	
	856	Ngovo	

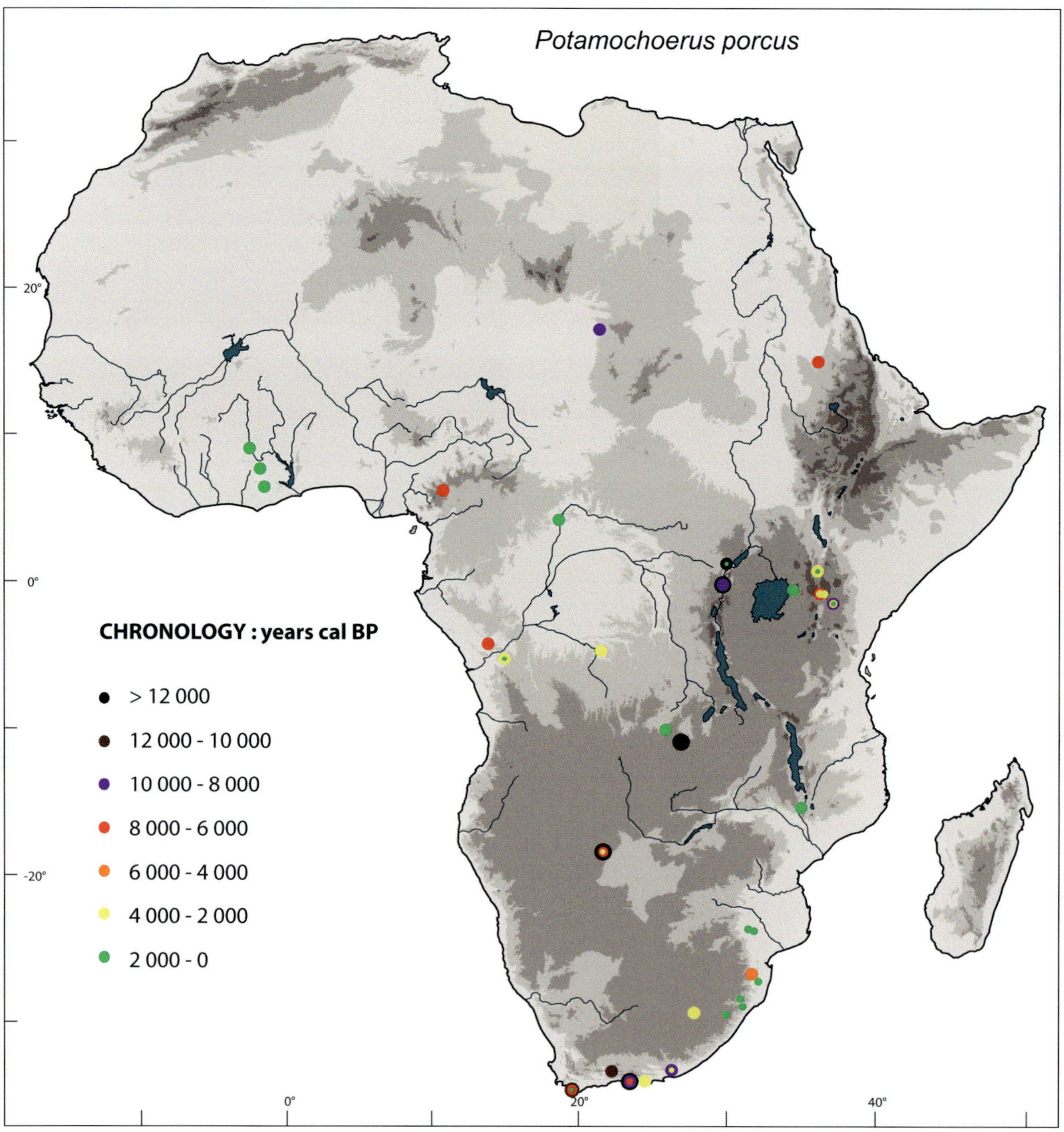

HIPPOPOTAMIDAE

Hippopotamus amphibius

Country	Site N	Site
Algeria	12	In Guezzam South
Botswana	26	Bosutwe-Mapungubwe
	27	Bosutwe-Taukome
	28	Bosutwe-Toutswe
	32	Divuyu
	39	Qogana
	42	Toteng1-SU2
	46	Toteng3-SU4
Cameroon	75	Nkang
Chad	81	Delebo
	85	Koyom
Djibouti	90	Asa Koma
Egypt	100	Adaima-habitat
	107	Berenike 95/97Late Roman
	112	Buto/1-3
	113	Buto-6
	114	Buto-9-12
	122	Dakhleh Ouasis, Ba
	133	El Omari
	135	Elephantine-necropolis
	137	Elephantine-Satet temple
	138	Elkab-Epipaleo
	139	Elkab-necropolis
	140	El-Tell el Iswid-Early Dynastic
	141	El-Tell el Iswid-Predynastic
	142	El-Tell el-Iswid A
	143	El-Tell el-Iswid B
	145	Fayoum G1-H1
	146	Fayoum G3-H2
	147	Fayoum G5-G4
	154	Hierakonpolis HK6-Naq IC-IIA
	156	Hierakonpolis HK6-Naq III
	157	Hierakonpolis HK6-Naq I-II
	162	Kom Ombo
	164	Maadi
	166	Makhadma 2
	167	Makhadma 4
	168	Merimde Benisâlame
	169	Minshat Abu Omar
	175	Qasr el-Sagha-1
	177	Qasr el-Sagha-3
	188	Tell el-Dab´a - b
	197	Umm el-Qaab, Friedhof U
	99	Adaima-habitat
Ethiopia	207	Axum-D
	209	Gademotta
Ghana	231	Daboya
	244	Ntereso
Kenya	250	Causeway Site
	251	Crescent Island main
	257	Engaruani
	267	GaJi1
	269	GaJi12
	270	GaJi2
	272	Gogo Falls-I
	273	Gogo Falls-II-III
	273	Gogo Falls-II-III
	274	Gogo Falls-sq26-29
	289	Lowasera-c
Malawi	356	Matope Court-EIA
	357	Matope Court-IA
	359	Namichimba
Mali	369	Est-Hassi el Abiod
	374	Hassi el-Abiod
	377	Jenné-Jeno - phase III
	381	Kobadi
	382	Kolima Sud-I
	383	Kolima Sud-II
	386	Mansa Musa * traded tusks
	388	Tiabel Goudiodié
Mauritania	395	Dhar Tichitt- highest beach
	398	Dhar Tichitt- middle beach
	401	Dhar Tichitt Site 46
	414	Tichitt - Goungou phase
	415	Tichitt - Khimiya phase
Niger	466	Adrar Bous-K
	467	Adrar Bous-T
	470	Awzerkat-1
	471	Chin Tafidet
	473	Gobero-K
	474	Gobero-T
	477	Mentes 1
	478	Mentes 13
	481	Mentes 4
	482	Mentes 5
	483	Mentes 6
	484	Mentes 7
	486	Mentes nord
	487	Takene Bawat 1
	488	Takene Bawat 2
	489	Takene Bawat 5
	490	Takene Bawat 6
	492	Tarada-T
	497	TMY3
Nigeria	514	Kursakata-IA
	526	Ngala II
	527	Ngala IIIa
Senegal	549	Tulel-Fobo
South Africa	568	Buffelskloof RS-a
	574	Bushman RS-LSA
	579	Byeneskranskop Cave 1d
	586	Die Kelders
	595	Eland´s Bay Cave-AI
	597	Eland´s Bay Cave-Wia
	598	Eland´s Bay Cave-Wib
	600	Eland´s Bay Cave-Wid
	608	Forest Hall
	609	Glentyre
	614	Hapi Pan
	616	Heuningneskrans-I/O
	631	Klasies River Mouth-1I
	632	Klasies River Mouth-1II
	633	Klasies River Mouth-5II
	643	Le7b
	650	Magogo-1
	654	Mamba 1
	661	Mgoduyanuka
	667	Ndondondwane
	668	Nelson Bay Cave- PW
	670	Nelson Bay Cave- W

	673	Ol20
	674	Ondini
	676	Oudepost I
	688	Pont-Drift-LKA
	700	Rooiels
	713	SH16
	718	Sk4
	751	Wosi-early EIA
	752	Wosi-middle-late EIA
	753	Wosi-terminal EIA
Sudan	754	Abu Darbein
	755	Abu Hugar
	756	Abu Tabari 02/1
	757	Aneibis
	758	Awlad el Imam
	761	Debbat Bangdit
	762	Debbat El Eheima
	764	El Damer
	765	El Kadada
	768	El Nofalab
	771	El Zakiab
	772	Erg of Ennedi
	773	Esh Shaheinab
	774	Gash Delta-AA
	775	Gash Delta-K
	776	Gebel Tageru
	777	Geili
	778	Gism el-Arba
	781	Jebel Umm Marrahi
	782	Kabbashi
	783	Kadero
	785	Kerma
	787	Khartoum Hopital
	790	Khasm el Girba - KG15-16
	791	Khasm el Girba - KG68
	804	Rahib-80/87
	805	Saggai 1
	810	Shabona
	814	Sheikh el Amin
	815	Sheikh Mustafa
	816	Singa
	822	Upper Atbara Valley
	828	Wadi Howar - 80/73
	834	W-Gebel Rahib
Tanzania	835	Mumba
Zaire	843	Ishango-AB
	844	Ishango-C
	845	Ishango-Z-Post-Em
Zambia	863	Leopard's Hill-PN

Hippopotamus lemerlei

Country	Site N	Site
Madagascar	349	Madagascar

Hippopotamus madagascariensis

Country	Site N	Site
Madagascar	349	Madagascar

Hippopotamus laloumena

Country	Site N	Site
Madagascar	348	Fompona

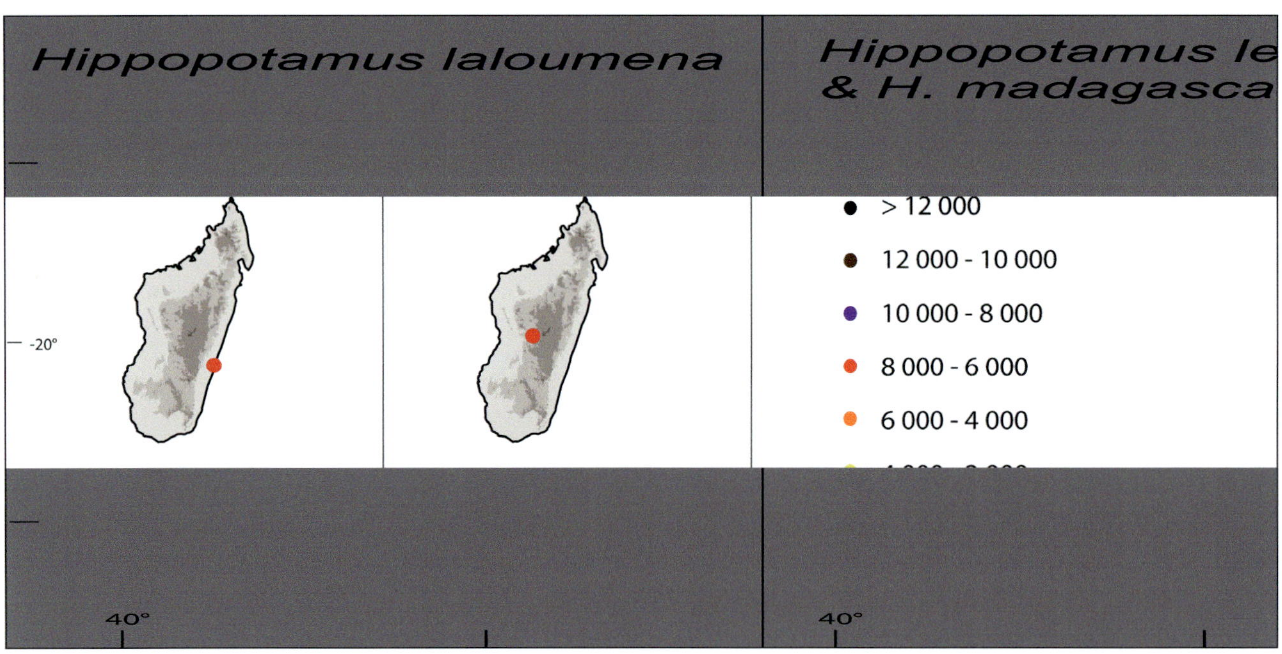

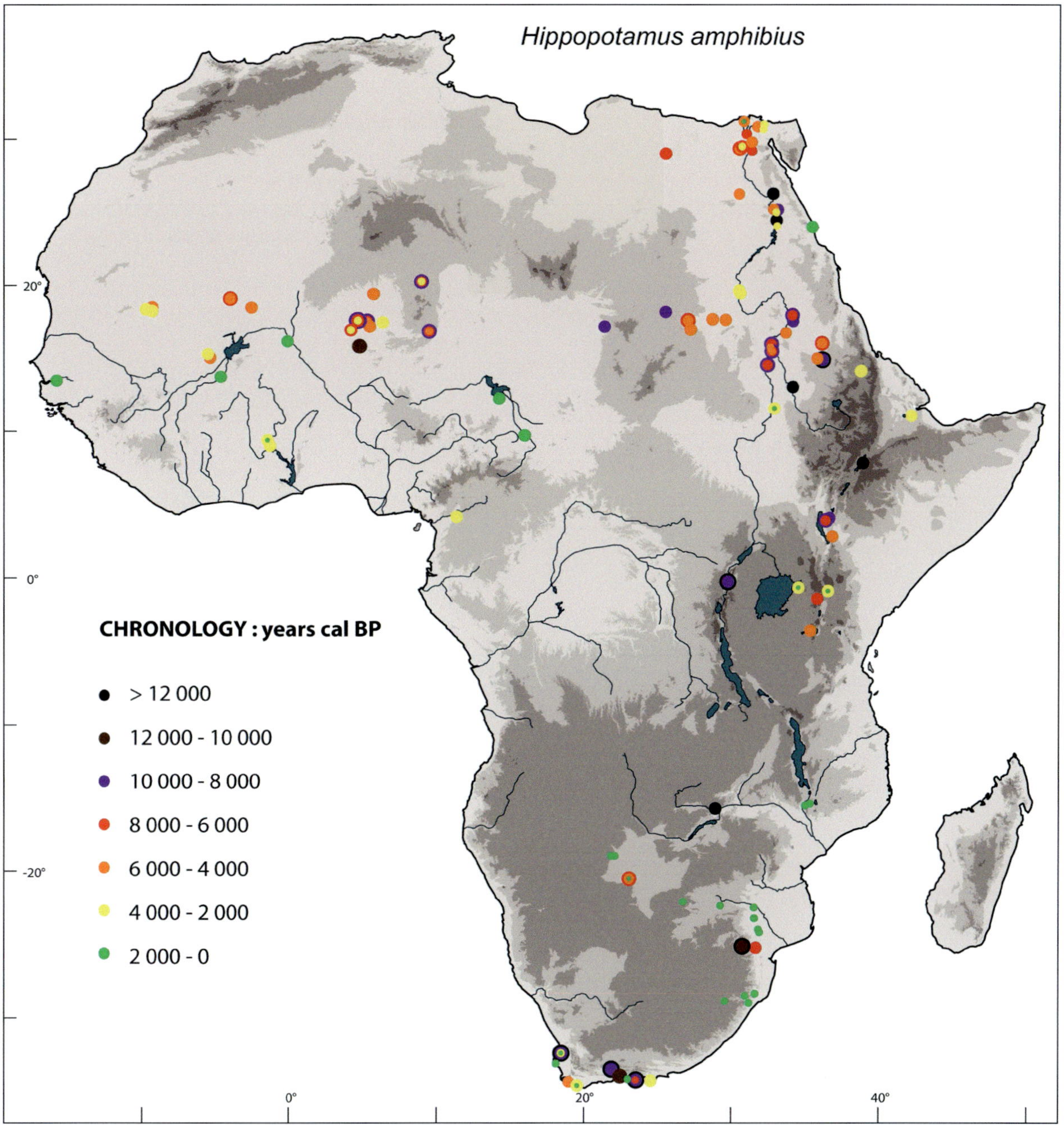
Hippopotamus amphibius
CHRONOLOGY : years cal BP
> 12 000
12 000 - 10 000
10 000 - 8 000
8 000 - 6 000
6 000 - 4 000
4 000 - 2 000
2 000 - 0
20°
0°
-20°
0°
20°
40°

Camelidae

Camelus sp.

Country	Site N	Site	
Algeria	9	Dra-Mta-el-Ma-el-Abiod	
	15	Medjez II	*intr.?

Camelus bactrianus

Country	Site N	Site
Egypt	106	Berenike 95/9 Early Roman
	107	Berenike 95/97Late Roman

Camelus dromedarius

Country	Site N	Site	
Egypt	114	Buto-9-12	
	123	Dakhleh Ouasis, Roman	
	144	El-Zarkeh	
	160	Kharga dep.-graeco-roman	
	170	Mons Claudianus	
	179	Quseir al-Qadim	
	95	Abu Sha´ar-4C	
	96	Abu Sha´ar-5C	
Ethiopia	212	Laga Oda-b	
Kenya	254	Ele Bor A-B	
Mauritania	407	Tegdaoust-a	
	408	Tegdaoust-b	
	409	Tegdaoust-c	
Senegal	546	Siouré-Ib	
	552	Walaldé-W2	*?
Sudan	817	Soba	
West. Sahara	837	Sebkha Amtal	*intr.?
	838	Sebkha Laasailia	
	841	Site H	*intr.

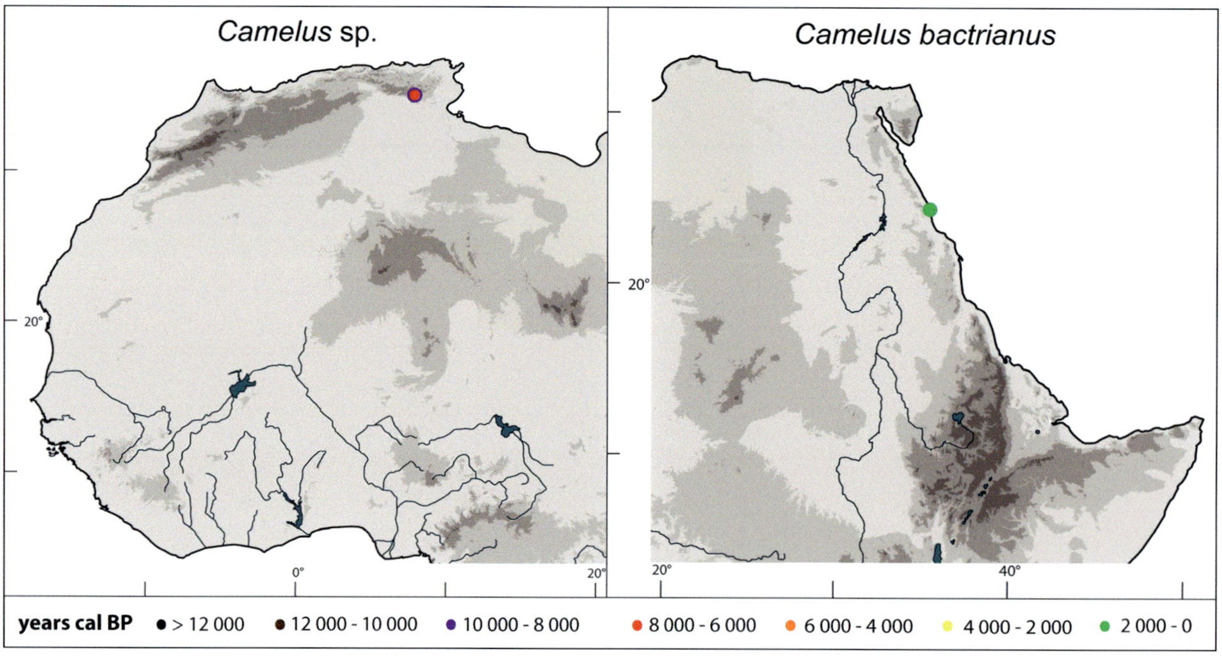

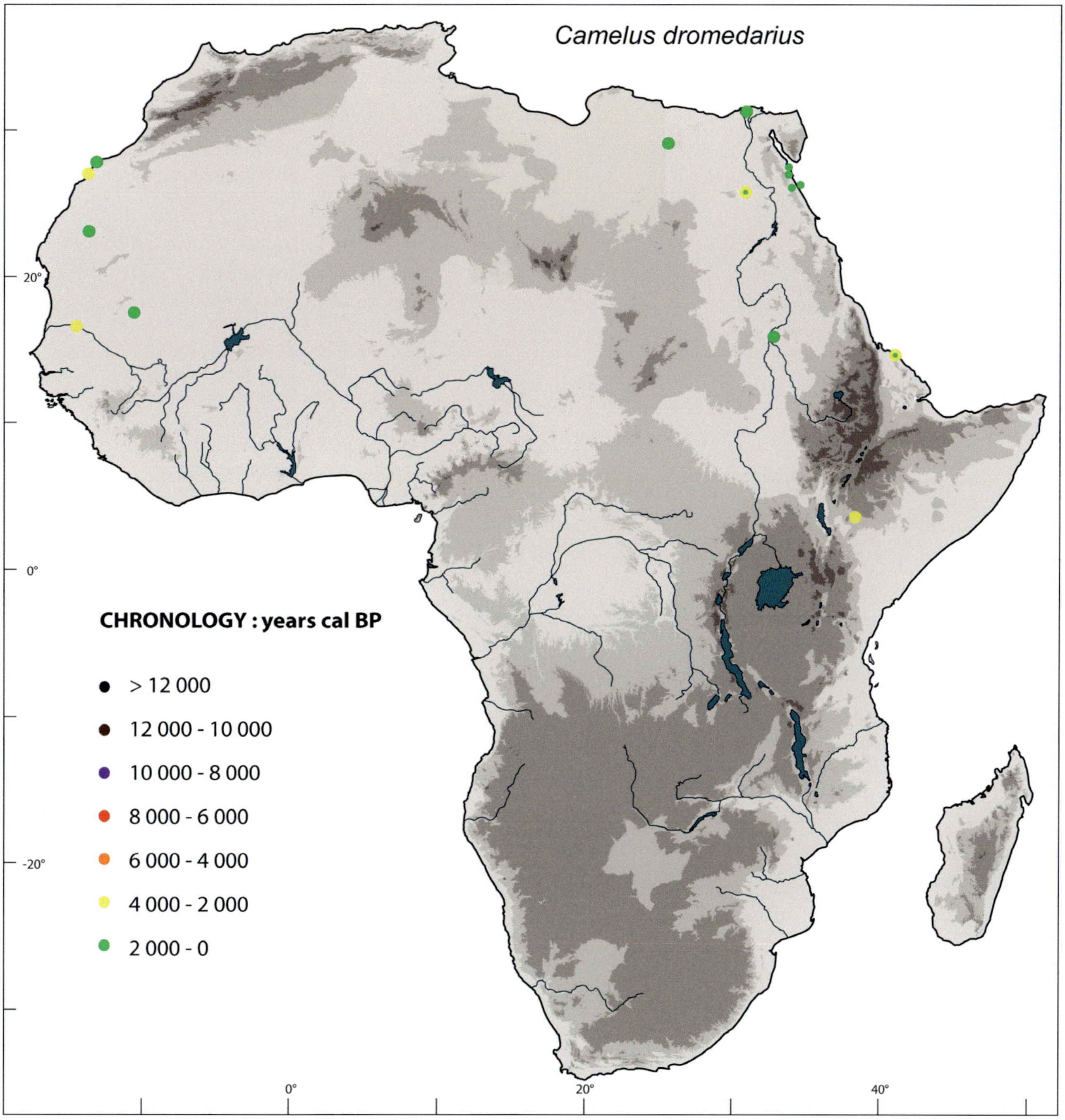
Camelus dromedarius
CHRONOLOGY : years cal BP
> 12 000
12 000 - 10 000
10 000 - 8 000
8 000 - 6 000
6 000 - 4 000
4 000 - 2 000
2 000 - 0
20°
0°
-20°
0°
20°
40°

BOVIDAE

Antilopinae

Neotragus moschatus

Country	Site N	Site
Kenya	297	Ngenyn
Malawi	352	Malowa-2
	353	Malowa-3
Mozambique	442	Manekweni-2-ABC
	443	Manekweni-3
	444	Manekweni-4
	445	Manekweni-5
Zanzibar	867	Kizimkazi Dimbani

Neotragus pygmaeus

Country	Site N	Site
Ghana	223	Anyinam-Area A-B-C
	224	Anyinam-Area D
	225	Asantamanso-ASII/III
	227	Asantamanso-ASV
	229	Bibease Plantation
	232	Dawu
	233	Esiease - A-C-D
	238	Kintampo RS K6-K
	245	Nyanawase/Akwamu Am.

Neotragus batesi

Country	Site N	Site
Zaire	854	Matupi-I
	855	Matupi-II

Oreotragus oreotragus

Country	Site N	Site	
Botswana	26	Bosutwe-Mapungubwe	
	28	Bosutwe-Toutswe	
	31	Commando Kop	
Kenya	272	Gogo Falls-I	
	273	Gogo Falls-II-III	
	275	GvJm 19-HA	
	276	GvJm 19-HC	
	277	GvJm 19-TLPG	
	278	GvJm 22-LHC	
	279	GvJm 22-LPG	
Lesotho	304	Ha Makotoko	
	305	Lequetsoana	
	308	Likoaeng-VII-IX	
	309	Likoaeng-XI	
	310	Likoaeng-XIII	
	312	Liphofung-CW	
	313	Liphofung-pCW	
	314	Muela	
	315	Ntloana Tsoana	
	316	Sehonghong-a	
	321	Tloutle-CWI	
	322	Tloutle-O	
	323	Tloutle-pCW	
Malawi	358	Matope Court-LIA	
Namibia	446	Aar I and II	
	447	Apollo 11-a	
	448	Apollo 11-b	
	449	Apollo 11-c	
	452	Brandberg-LSA	
	453	Brandberg-MSA	
	457	Haalenberg	
	458	Namtib	
	461	Pokkenbank	
South Africa	559	Blydefontein I-II	
	563	Boomplass Cave-a	
	564	Boomplass Cave-b	
	565	Boomplass Cave-c	
	566	Boomplass Cave-d	
	568	Buffelskloof RS-a	
	570	Buffelskloof RS-c	
	571	Buffelskloof RS-d	
	572	Buffelskloof RS-e	
	574	Bushman RS-LSA	
	576	Byeneskranskop Cave 1a	
	578	Byeneskranskop Cave 1c	* ?
	579	Byeneskranskop Cave 1d	
	580	Collingham Shelter	
	581	Colwinton Shelter-CW	
	582	Colwinton Shelter-pCW	
	586	Die Kelders	
	589	Doornfontein	
	595	Eland´s Bay Cave-AI	
	616	Heuningneskrans-I/O	
	617	Heuningneskrans-R	
	618	Hill X	
	619	HKLP Ia	
	620	HKLP Ib	
	624	Jubilee Shelter-O	
	625	Jubilee Shelter-PW	
	626	Jubilee Shelter-W	
	653	Malwula	
	658	Mapungubwe-ST	
	660	Mauermanshoek	
	663	Mhlwazini Cave	
	675	Onink	
	689	Pont-Drift-Zhizo	
	694	Ravenscraig Shelter-pCW	
	695	Ravenscraig Shelter-R	
	700	Rooiels	
	704	Rose Cottage Cave-A-Mn	
	706	Rose Cottage Cave-OIC	
	707	Rose Cottage Cave-WI	
	711	Serotwe Hill	
	714	Sh27	
	715	Shankare Hill-a	
	716	Shankare Hill-b	
	719	Smitswinkelbaai Cave	
	740	Twyfelpoort	
	741	Umgungundlovu	
	742	Ventershoek	
	744	Voëlvlei	
Sudan	780	Jebel et Tomat	
Tanzania	835	Mumba	
Zaire	855	Matupi-II	
Zimbabwe	870	Redcliff Cave	

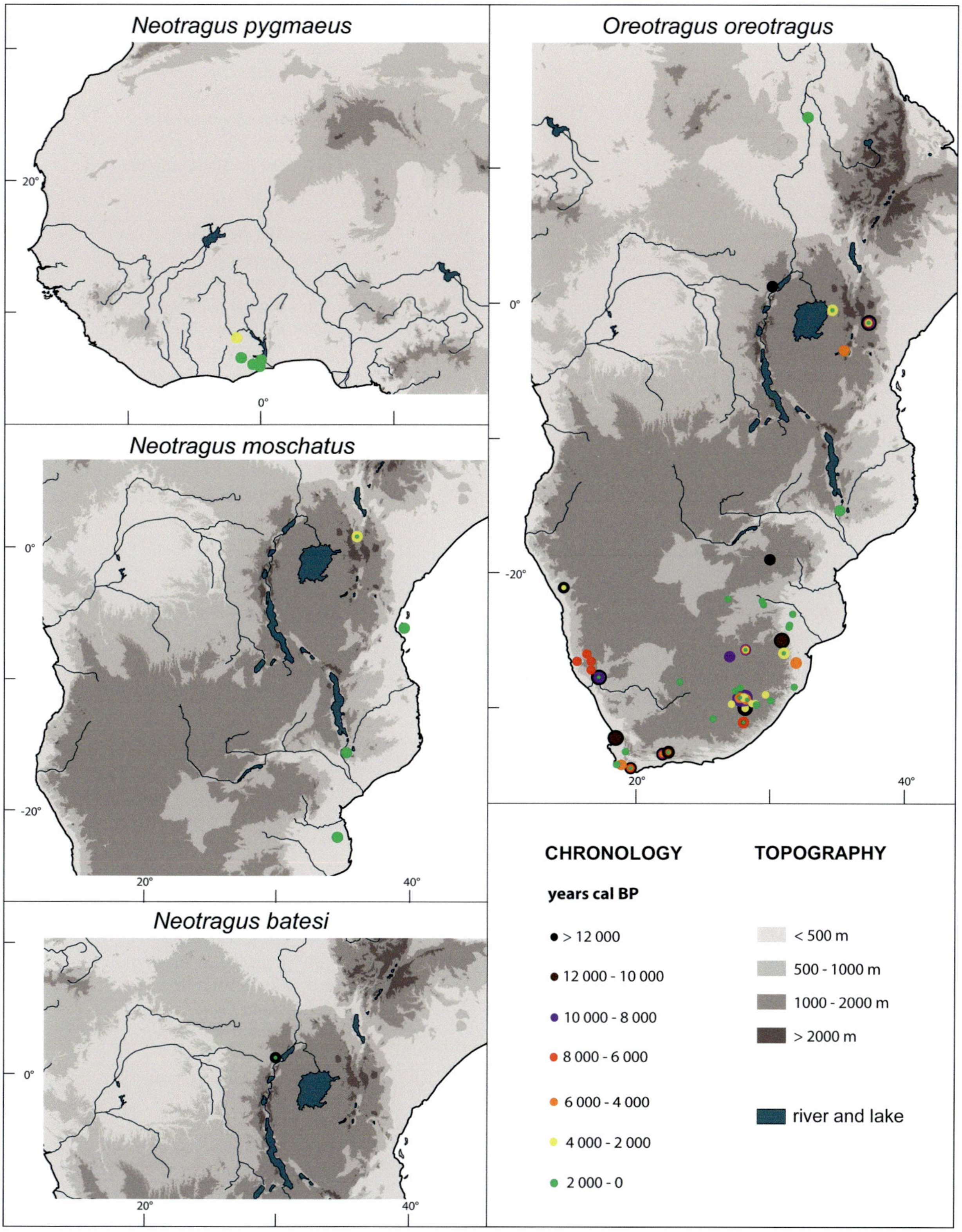
Neotragus pygmaeus
20°
0°
Neotragus moschatus
0°
-20°
20°
40°
Neotragus batesi
0°
20°
40°
Oreotragus oreotragus
0°
-20°
20°
40°
CHRONOLOGY
years cal BP
> 12 000
12 000 - 10 000
10 000 - 8 000
8 000 - 6 000
6 000 - 4 000
4 000 - 2 000
2 000 - 0
TOPOGRAPHY
< 500 m
500 - 1000 m
1000 - 2000 m
> 2000 m
river and lake

Ourebia ourebi

Country	Site N	Site	
Burkina Faso	58	Oursi-IA	
	61	Péntènga 89/1-spits 15-21	
	63	Saouga 94/120	
	64	Saouga 95/7	
Cameroon	74	Djaba Hosséré	
Chad	85	Koyom	
Kenya	259	Enkapune Ya Muto - Eb 4b	
	260	Enkapune Ya Muto - Eb 5a	
	261	Enkapune Ya Muto - Eb 5b	
	272	Gogo Falls-I	
	273	Gogo Falls-II-III	
	274	Gogo Falls-sq26-29	
	275	GvJm 19-HA	
	276	GvJm 19-HC	
	277	GvJm 19-TLPG	
	278	GvJm 22-LHC	
	279	GvJm 22-LPG	
	280	GvJm 46-LPG	
	292	Marula Rock Shelter	
Lesotho	304	Ha Makotoko	
	313	Liphofung-pCW	
	314	Muela	
	316	Sehonghong-a	
Malawi	357	Matope Court-IA	
Mali	380	Karkarichinkat N-S	
	383	Kolima Sud-II	
Nigeria	504	Gajiganna B1-lower	
	514	Kursakata-IA	
Senegal	540	Cubalel-II	
	549	Tulel-Fobo	
South Africa	580	Collingham Shelter	
	603	Emambeni	
	625	Jubilee Shelter-PW	
	626	Jubilee Shelter-W	
	663	Mhlwazini Cave	
	667	Ndondondwane	
	668	Nelson Bay Cave- PW	
	670	Nelson Bay Cave- W	
	675	Onink	
	718	Sk4	
Sudan	754	Abu Darbein	
	757	Aneibis	
	761	Debbat Bangdit	
	762	Debbat El Eheima	
	764	El Damer	
	765	El Kadada	
	766	El Kenger Middle/East	
	773	Esh Shaheinab	
	777	Geili	
	781	Jebel Umm Marrahi	
	787	Khartoum Hopital	
	788	Khasm el Girba - Kassala	
	789	Khasm el Girba - KG14	
	805	Saggai 1	
	812	Shaqaduq-S1-B	
	813	Shaqaduq-S21	
	822	Upper Atbara Valley	
Zaire	843	Ishango-AB	
	845	Ishango-Z-Post-Em	
	848	Kiantapo	
	854	Matupi-I	
	855	Matupi-II	
	855	Matupi-II	*?
Zimbabwe	870	Redcliff Cave	

Raphicerus sp.

Country	Site N	Site	
Botswana	26	Bosutwe-Mapungubwe	
Namibia	454	Fackelträger	
South Africa	563	Boomplass Cave-a	
	564	Boomplass Cave-b	
	565	Boomplass Cave-c	
	566	Boomplass Cave-d	
	568	Buffelskloof RS-a	
	569	Buffelskloof RS-b	
	570	Buffelskloof RS-c	
	571	Buffelskloof RS-d	
	572	Buffelskloof RS-e	
	576	Byeneskranskop Cave 1a	
	577	Byeneskranskop Cave 1b	
	578	Byeneskranskop Cave 1c	
	579	Byeneskranskop Cave 1d	
	583	Connies Limpet Bar	
	590	Drie Susters	
	591	Drie Susters Main	
	592	Driebos	
	595	Eland´s Bay Cave-Al	
	596	Eland´s Bay Cave-RI	
	597	Eland´s Bay Cave-Wia	
	598	Eland´s Bay Cave-Wib	
	599	Eland´s Bay Cave-Wic	
	600	Eland´s Bay Cave-Wid	
	601	Eland´s Bay Open-a	
	602	Eland´s Bay Open-b	
	603	Emambeni	
	609	Glentyre	
	613	Hail Stone Midden	
	621	HKLP Va	*cf.
	622	HKLP Vb	*cf.
	627	Kasteelberg A	
	628	Kasteelberg B	
	628	Kasteelberg B	
	653	Malwula	
	668	Nelson Bay Cave- PW	
	669	Nelson Bay Cave- R	
	670	Nelson Bay Cave- W	
	671	Oakhurst	
	678	Pancho's Kitchen Midden 1-2	
	679	Pancho's Kitchen Midden 3-4	
	680	Pancho's Kitchen Midden 6-7	
	683	Paternoster5-6	
	700	Rooiels	
	718	Sk4	
	719	Smitswinkelbaai Cave	

724 Spring Cave
725 Steenbokfontein-1
726 Steenbokfontein-2
727 Steenbokfontein-3
728 Steenbokfontein-4a
729 Steenbokfontein-4b
735 Tortoise cave-a
737 Tortoise cave-c
738 Tortoise cave-d
744 Voëlvlei
747 Witklip-1
748 Witklip-2
749 Witklip-3
750 Witklip-4
870 Redcliff Cave

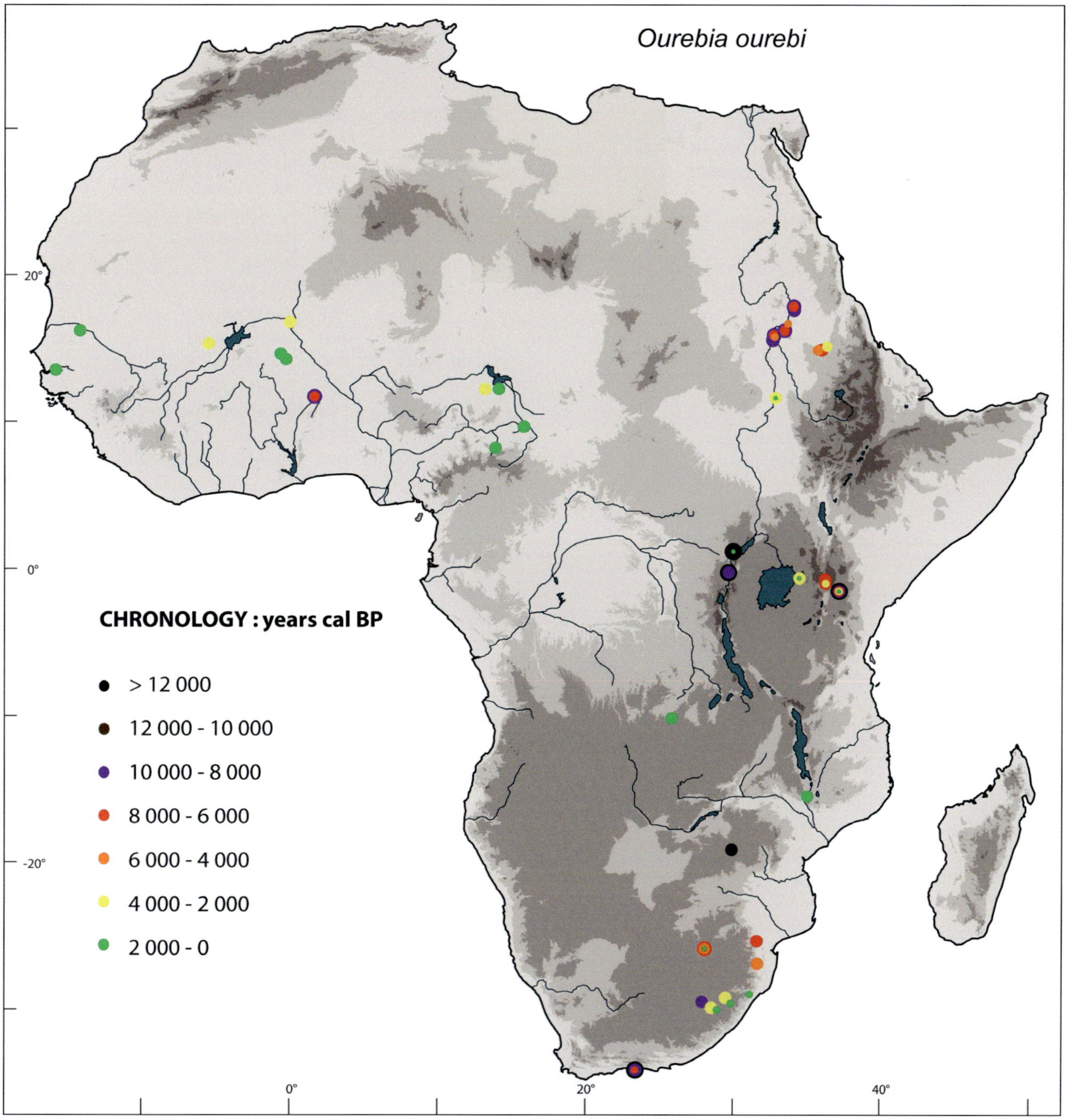

Raphicerus melanotis

Country	Site N	Site	
South Africa	572	Buffelskloof RS-e	
	577	Byeneskranskop Cave 1b	
	578	Byeneskranskop Cave 1c	
	579	Byeneskranskop Cave 1d	
	586	Die Kelders	
	595	Eland´s Bay Cave-AI	
	597	Eland´s Bay Cave-Wia	
	598	Eland´s Bay Cave-Wib	
	599	Eland´s Bay Cave-Wic	
	600	Eland´s Bay Cave-Wid	
	628	Kasteelberg B	
	631	Klasies River Mouth-1I	
	632	Klasies River Mouth-1II	
	633	Klasies River Mouth-5II	
	634	Klein Kliphuis	*cf.
	660	Mauermanshoek	
	676	Oudepost I	
	736	Tortoise cave-b	
	738	Tortoise cave-d	

Raphicerus sharpei

Country	Site N	Site
Botswana	26	Bosutwe-Mapungubwe
South Africa	710	Selongwe Hill
	718	Sk4
	720	Sonkoanini Hill
Zambia	863	Leopard's Hill-PN

Raphicerus campestris

Country	Site N	Site	
Botswana	26	Bosutwe-Mapungubwe	
	27	Bosutwe-Taukome	
	28	Bosutwe-Toutswe	*cf.
	28	Bosutwe-Toutswe	
	29	Bosutwe-Zimbabwe	
	31	Commando Kop	
	34	Drotsky's Cave-LP	
	41	Toteng1-SU1	
	42	Toteng1-SU2	
	43	Toteng1-SU3	*cf.
	43	Toteng1-SU3	
	44	Toteng1-SU4	
	48	White Paintings RS-1	
	48	White Paintings RS-2ab	
	49	White Paintings RS-2b-3ab	
Kenya	250	Causeway Site	
	251	Crescent Island main	
	259	Enkapune Ya Muto - Eb 4b	
	260	Enkapune Ya Muto - Eb 5a	
	261	Enkapune Ya Muto - Eb 5b	
	275	GvJm 19-HA	
	276	GvJm 19-HC	
	277	GvJm 19-TLPG	
	278	GvJm 22-LHC	
	279	GvJm 22-LPG	
	280	GvJm 46-LPG	
	293	Naivasha Railway Shelter	
Lesotho	304	Ha Makotoko	
	307	Likoaeng-III-V	
	308	Likoaeng-VII-IX	
	309	Likoaeng-XI	
	310	Likoaeng-XIII	
	313	Liphofung-pCW	
	314	Muela	
	315	Ntloana Tsoana	
	316	Sehonghong-a	
	321	Tloutle-CWI	
	322	Tloutle-O	
	323	Tloutle-pCW	
Malawi	357	Matope Court-IA	
	359	Namichimba	*cf.
Mozambique	441	Manekweni-1	
	442	Manekweni-2-ABC	
	444	Manekweni-4	
	445	Manekweni-5	
Namibia	454	Fackelträger	
	457	Haalenberg	
	458	Namtib	*cf.
	461	Pokkenbank	*cf.
	462	Steenbras Bay	*cf.
Rhodesia	531	Chamabvefva	*cf.
South Africa	560	Blydefontein III-IV	
	562	Blydefontein VI-VII	
	571	Buffelskloof RS-d	
	574	Bushman RS-LSA	
	579	Byeneskranskop Cave 1d	
	580	Collingham Shelter	
	581	Colwinton Shelter-CW	
	582	Colwinton Shelter-pCW	
	585	Deelpan	
	586	Die Kelders	
	589	Doornfontein	
	594	Eiland Salt Works-b	
	595	Eland´s Bay Cave-AI	
	597	Eland´s Bay Cave-Wia	
	598	Eland´s Bay Cave-Wib	
	599	Eland´s Bay Cave-Wic	
	600	Eland´s Bay Cave-Wid	
	604	Equus Cave 1A	
	610	Green Elliott Shelter I-III	
	611	Green Elliott Shelter IV-VI	
	618	Hill X	
	619	HKLP Ia	
	619	HKLP Ia	*cf.
	620	HKLP Ib	*cf.
	620	HKLP Ib	
	622	HKLP Vb	
	622	HKLP Vb	*cf.
	625	Jubilee Shelter-PW	
	626	Jubilee Shelter-W	
	628	Kasteelberg B	
	638	Langsdraai	
	640	Le2	
	641	Le6	
	642	Le7a	
	643	Le7b	
	659	Marupale Hill	
	660	Mauermanshoek	
	663	Mhlwazini Cave	
	673	Ol20	
	674	Ondini	

	676	Oudepost I
	680	Pancho's Kitchen Midden 6-7
	684	Ph1
	688	Pont-Drift-LKA
	689	Pont-Drift-Zhizo
	691	Pr34
	698	Riversmead Shelter III-IV
	699	Riversmead Shelter VI-IX
	701	Rooikrans-CW
	702	Rooikrans-pCW
	704	Rose Cottage Cave-A-Mn
	706	Rose Cottage Cave-OIC
	707	Rose Cottage Cave-WI
	710	Selongwe Hill
	711	Serotwe Hill
	713	SH16
	715	Shankare Hill-a
	716	Shankare Hill-b
	718	Sk4
	720	Sonkoanini Hill
	725	Steenbokfontein-1
	726	Steenbokfontein-2
	727	Steenbokfontein-3
	728	Steenbokfontein-4a
	729	Steenbokfontein-4b
	736	Tortoise cave-b
	736	Tortoise cave-b
	738	Tortoise cave-d
	739	Tsh1
	740	Twyfelpoort
	741	Umgungundlovu
	742	Ventershoek
	749	Witklip-3
Tanzania	835	Mumba
Zambia	864	Namakala

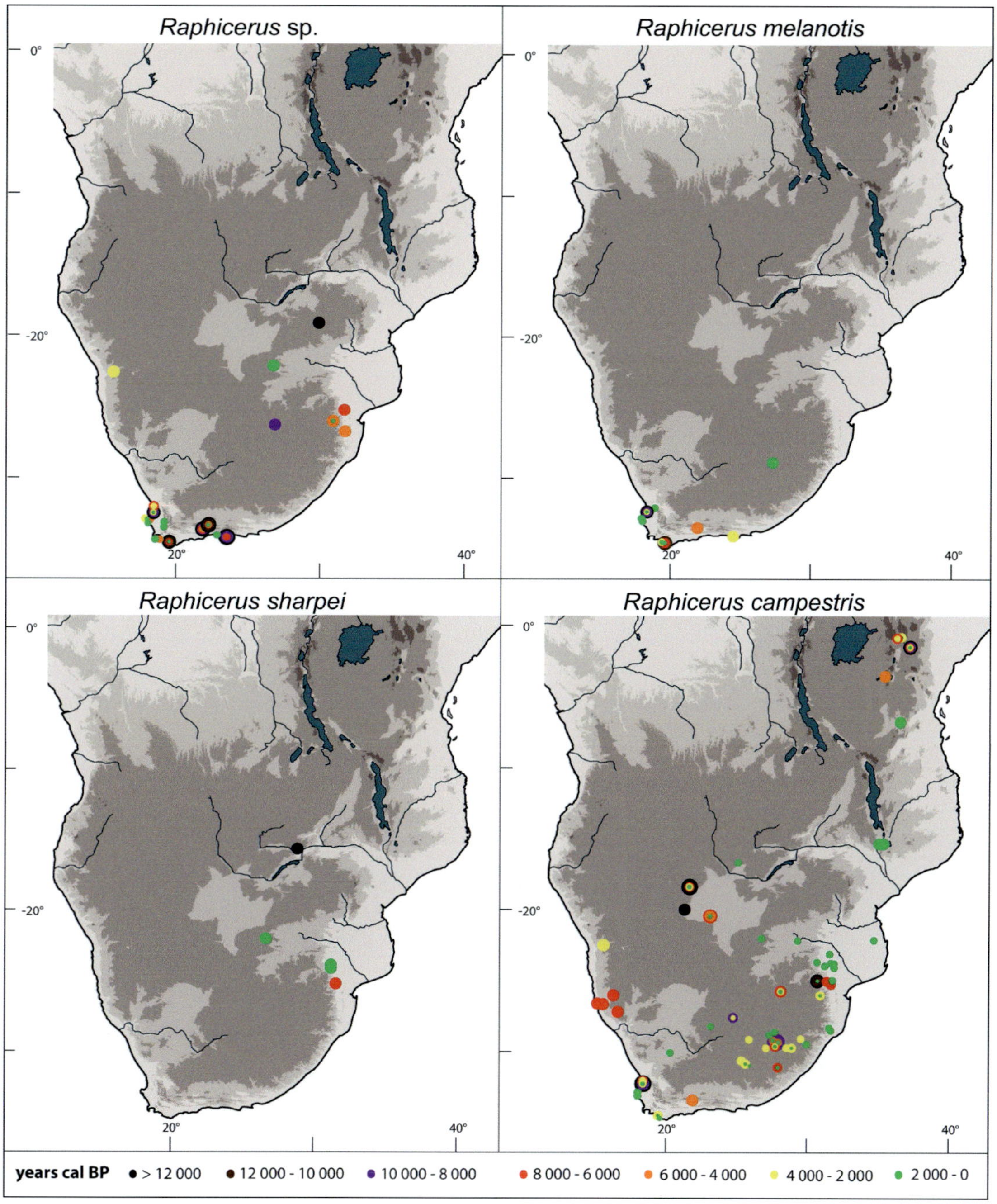

Antidorcas sp.

Country	Site N	Site
South Africa	563	Boomplass Cave-a

Antidorcas bondi

Country	Site N	Site
South Africa	617	Heuningneskrans-R
	705	Rose Cottage Cave-DB2
Zimbabwe	870	Redcliff Cave

Antidorcas marsupialis

Country	Site N	Site	
Botswana	26	Bosutwe-Mapungubwe	
	28	Bosutwe-Toutswe	
Lesotho	304	Ha Makotoko	
	305	Lequetsoana	
	314	Muela	
	315	Ntloana Tsoana	
	321	Tloutle-CWI	
	322	Tloutle-O	
	323	Tloutle-pCW	
Namibia	446	Aar I and II	
	447	Apollo 11-a	
	449	Apollo 11-c	
	450	Brandberg-BBC	
	451	Brandberg-late LSA	
	452	Brandberg-LSA	
	453	Brandberg-MSA	
	454	Fackelträger	*cf.
	454	Fackelträger	
	458	Namtib	
	459	Nos	
	460	Oshigambo Peninsula	
	461	Pokkenbank	
	462	Steenbras Bay	
	463	Tiras	
South Africa	585	Deelpan	
	589	Doornfontein	
	604	Equus Cave 1A	
	610	Green Elliott Shelter I-III	
	611	Green Elliott Shelter IV-VI	
	617	Heuningneskrans-R	
	634	Klein Kliphuis	
	644	Leliehoek	
	660	Mauermanshoek	
	663	Mhlwazini Cave	
	669	Nelson Bay Cave- R	*cf.
	689	Pont-Drift-Zhizo	*cf.
	701	Rooikrans-CW	
	704	Rose Cottage Cave-A-Mn	
	705	Rose Cottage Cave-DB2	
	706	Rose Cottage Cave-OIC	
	707	Rose Cottage Cave-WI	
	730	Tandjesberg	
	740	Twyfelpoort	
	742	Ventershoek	
Zimbabwe	870	Redcliff Cave	

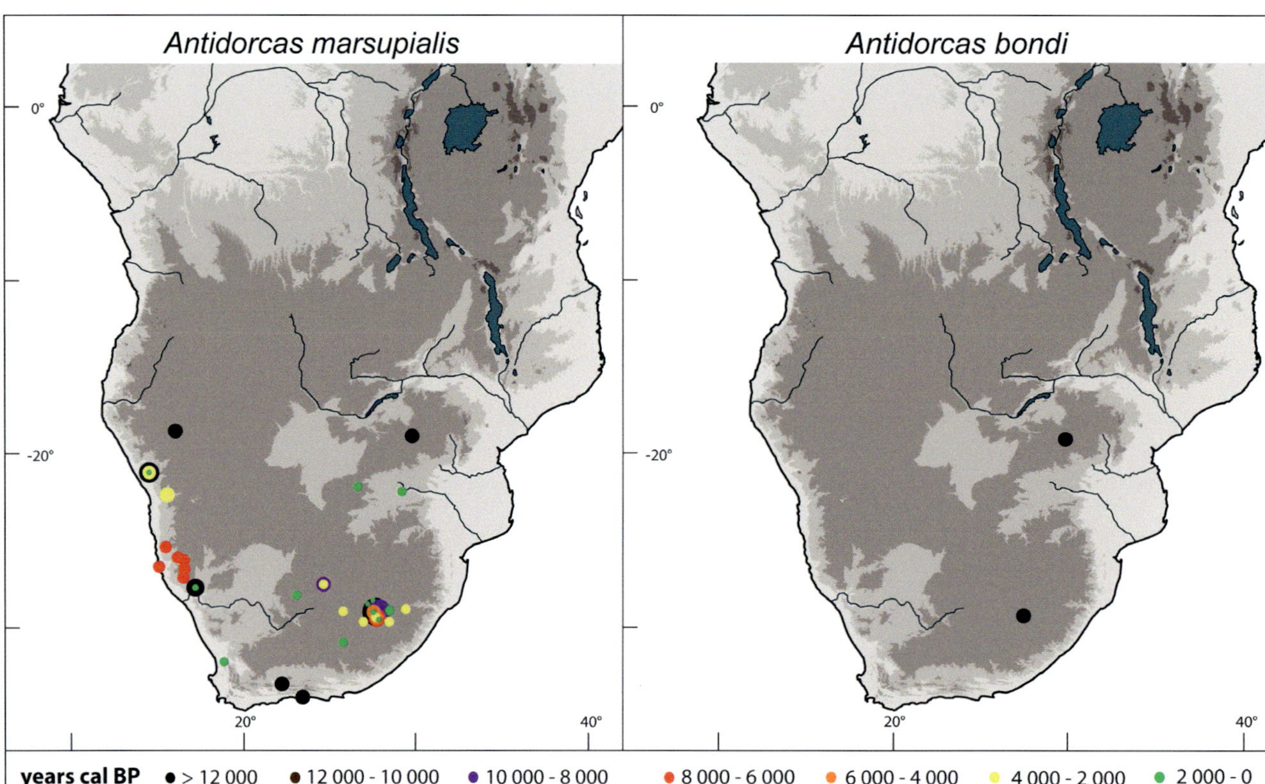

Gazella sp.

Country	Site N	Site	
Algeria	10	Grotte Capeletti	
	12	In Guezzam South	
	13	Kef Zoura D	
	15	Medjez II	
	18	Oued Telidjene A	
	21	Ti-n-Hanakaten	
	22	Zmeilet el-Barka	
	4	Ain Mistehayia	
	7	Damous el Ahmar	
Chad	81	Delebo	
	83	Gabrong-b	
	84	Gabrong-d	
Egypt	115	Catfish Cave-L1	
	145	Fayoum G1-H1	
	146	Fayoum G3-H2	
	147	Fayoum G5-G4	
	148	Fayoum Site 1	
	149	Fayoum Site 2	
	150	Fayoum Site 3	
	151	Fayoum Site 4	
	152	Fayoum Site 5	
	153	Gilf Kebir	
	168	Merimde Benisâlame	
	171	Napta Playa 77-1/4/5/6	
	172	Napta Playa 77-8	
	173	Napta Playa E77-3/7	*intr.
	174	Qantir	
	188	Tell el-Dab´a - b	
	196	Tuna el-Gebel	
	198	Wadi el Akhdar	
	199	Wadi el Bakht	
Ethiopia	215	Moche Borago-1a	
	216	Moche Borago-1b	
Kenya	253	Ele Bor A-A	
	254	Ele Bor A-B	
	254	Ele Bor A-B	
	254	Ele Bor A-B	
	255	Ele Bor A-C	
	255	Ele Bor A-C	
	255	Ele Bor A-C	
	256	Ele Bor A-D	
	257	Engaruani	
	284	Kulchurdo	
	297	Ngenyn	
Libya	325	Et Tera	*intr.
	326	Hauah Fteah- Historic	
	327	Hauah Fteah-Iberom	
	328	Hauah Fteah-Lib-caps	
	329	Hauah Fteah-Neolithic	
	332	Murzuq	
Mali	368	erg Ine Sakane	
	369	Est-Hassi el Abiod	
	374	Hassi el-Abiod	
	387	Tessalit	
Mauritania	393	Berouâga	
	396	Dhar Tichitt- latest phase	
	399	Dhar Tichitt- post beach	
	400	Dhar Tichitt Site 38	
	401	Dhar Tichitt Site 46	
	402	DN4-BKH-S5	
	409	Tegdaoust-c	
Morocco	419	Bou Saria A	
	420	Bou Saria B	
	421	Bou Saria C-D	
	423	El Harhoura 2	
	425	Ghar-Khal I	
	426	Ghar-Khal IIIb-V	
	427	Ghar-Khal II-IIIa	
	429	Kaf-taht-el-Ghar-A	
	430	Kaf-taht-el-Ghar-B	
	431	Kaf-taht-el-Ghar-C	
	432	Kaf-taht-el-Ghar-D	
	433	Kaf-taht-el-Ghar-E	
	434	Kaf-taht-el-Ghar-F	
	435	Kaf-taht-el-Ghar-G	
	436	Kehf el Hammar	
	437	Kehf-el-Baroud	
	438	Kehf-el-Baroud-aven	
Niger	465	Adrar Bous 10	
	467	Adrar Bous-T	
	468	Arlit-a	
	469	Arlit-b	
	473	Gobero-K	
	484	Mentes 7	
	490	Takene Bawat 6	
	497	TMY3	
Senegal	543	Ogo region	
Sudan	755	Abu Hugar	
	756	Abu Tabari 02/1	
	763	Djarabona 84/13	
	773	Esh Shaheinab	
	775	Gash Delta-K	
	788	Khasm el Girba - Kassala	
	797	Mahal Teglinos K1-CG	
	798	Mahal Teglinos K1-EG	
	799	Mahal Teglinos K1-JM	
	800	Mahal Teglinos K1-LG	
	801	Mahal Teglinos K1-MG	
	802	Mahal Teglinos K1-PG	
	804	Rahib-80/87	
	817	Soba	
	824	Wadi Hariq 01/1	
	824	Wadi Hariq 01/1	
	825	Wadi Hariq 01/4	
	826	Wadi Hariq 97/5	
	827	Wadi Hariq 97/7	
	829	Wadi Howar - 84/1	
	831	Wadi Howar - 84/13	
	832	Wadi Howar - 84/19	
West. Sahara	837	Sebkha Amtal	

Gazella atlantica

Country	Site N	Site
Morocco	422	El Harhoura 1

Gazella cuvieri

Country	Site N	Site	
Algeria	6	Columnata	*aff.
	9	Dra-Mta-el-Ma-el-Abiod	
Morocco	422	El Harhoura 1	
	426	Ghar-Khal IIIb-V	
	431	Kaf-taht-el-Ghar-C	
	433	Kaf-taht-el-Ghar-E	
	435	Kaf-taht-el-Ghar-G	
	437	Kehf-el-Baroud	

Gazella leptoceros

Country	Site N	Site	
Egypt	121	Dakhleh Oasis, SMU	
	124	Dakhleh Ouasis,OK	
	126	Eastpan 95-1/1	
	162	Kom Ombo	* ?

Gazella (Nanger) soemmerringii

Country	Site N	Site
Egypt	137	Elephantine-Satet temple
	185	Sodmein Cave older backfill
Sudan	789	Khasm el Girba - KG14
	790	Khasm el Girba - KG15-16
	822	Upper Atbara Valley

Gazella dorcas

Country	Site N	Site	
Algeria	16	Meniet	
	2	A-88-28	
	3	A-88-35	
	5	Amekni	
	9	Dra-Mta-el-Ma-el-Abiod	
Djibouti	90	Asa Koma	
Egypt	100	Adaima-habitat	
	106	Berenike 95/9 Early Roman	
	107	Berenike 95/97Late Roman	
	109	Bir Kiseiba-EN	
	111	Bir Kiseiba-MN	
	121	Dakhleh Oasis, SMU	
	122	Dakhleh Ouasis, Ba	
	123	Dakhleh Ouasis, Roman	
	124	Dakhleh Ouasis,OK	
	126	Eastpan 95-1/1	
	127	Eastpan 95-2	
	129	El Abadiya 2 North sector	
	130	El Abadiya 2 South sector	
	135	Elephantine-necropolis	
	137	Elephantine-Satet temple	
	138	Elkab-Epipaleo	
	139	Elkab-necropolis	
	156	Hierakonpolis HK6-Naq III	
	161	Kharga dep.-Neolithic	
	162	Kom Ombo	
	163	Lobo - 81/55	
	165	Mahgar Dendera 2	
	170	Mons Claudianus	
	175	Qasr el-Sagha-1	
	176	Qasr el-Sagha-2	
	181	Selima Sandsheet - 85/79	
	182	Shenshef	
	183	Shuwikhat 1	
	184	Sodmein Cave - lev 5 (MP5)	
	186	Sodmein Cave y. backfill	
	191	Toukh	* G. isabella
	193	Tree Shelter-3	
	199	Wadi el Bakht	
	201	Wadi Kubbaniya - E-78-3	
	202	Wadi Kubbaniya - E-78-4	
	203	Wadi Kubbaniya - E-81-1	
	205	Willmann´s camp - 81/61	
	91	Abu Ballas - 83/39	
	92	Abu Ballas - 85/50-1	
	93	Abu Ballas - 85/51-3	
	94	Abu Ballas - 85/56	
	95	Abu Sha´ar-4C	
	99	Adaima-habitat	
Libya	330	In Aghelachem-RT,SF	
	331	In Aghelachem-T1	
	333	Ti-n-Thora East	
	335	Ti-n-Thora Two Caves	
	338	Uan Muhuggiag (LA)	
	339	Uan Muhuggiag (LP)	
	345	Uan Telocat (MP)	
	347	Wadi Tannezzuft-96/129	
Mali	363	Asselar	
	368	erg Ine Sakane	
	369	Est-Hassi el Abiod	
	374	Hassi el-Abiod	
	380	Karkarichinkat N-S	
	387	Tessalit	
Mauritania	397	Dhar Tichitt- lowest beach	*cf.
	398	Dhar Tichitt- middle beach	*cf.
	402	DN4-BKH-S5	
	405	Khatt Lemaiteg	
	406	Khneivissa	*cf.
	411	Tichitt - Akjinjeir phase	*cf.
	412	Tichitt - Arriane phase	
	413	Tichitt - Chebka phase	*cf.
	414	Tichitt - Goungou phase	
	415	Tichitt - Khimiya phase	*cf.
	416	Tichitt - Naguez phase	
Morocco	427	Ghar-Khal II-IIIa	
	437	Kehf-el-Baroud	
Niger	495	Tin Ouaffadene	*cf.
Sudan	767	El Mahalab	
	778	Gism el-Arba	
	780	Jebel et Tomat	*cf.
	785	Kerma	
	790	Khasm el Girba - KG15-16	
	793	Laqiya area - 82/29	
	794	Laqiya area - 82/31	
	808	Selima Sandsheet - 85/73-2	
	814	Sheikh el Amin	
	815	Sheikh Mustafa	
	817	Soba	
	824	Wadi Hariq 01/1	
	826	Wadi Hariq 97/5	
	827	Wadi Hariq 97/7	
West. Sahara	837	Sebkha Amtal	
	838	Sebkha Laasailia	
	839	Sebkha Mahariat	

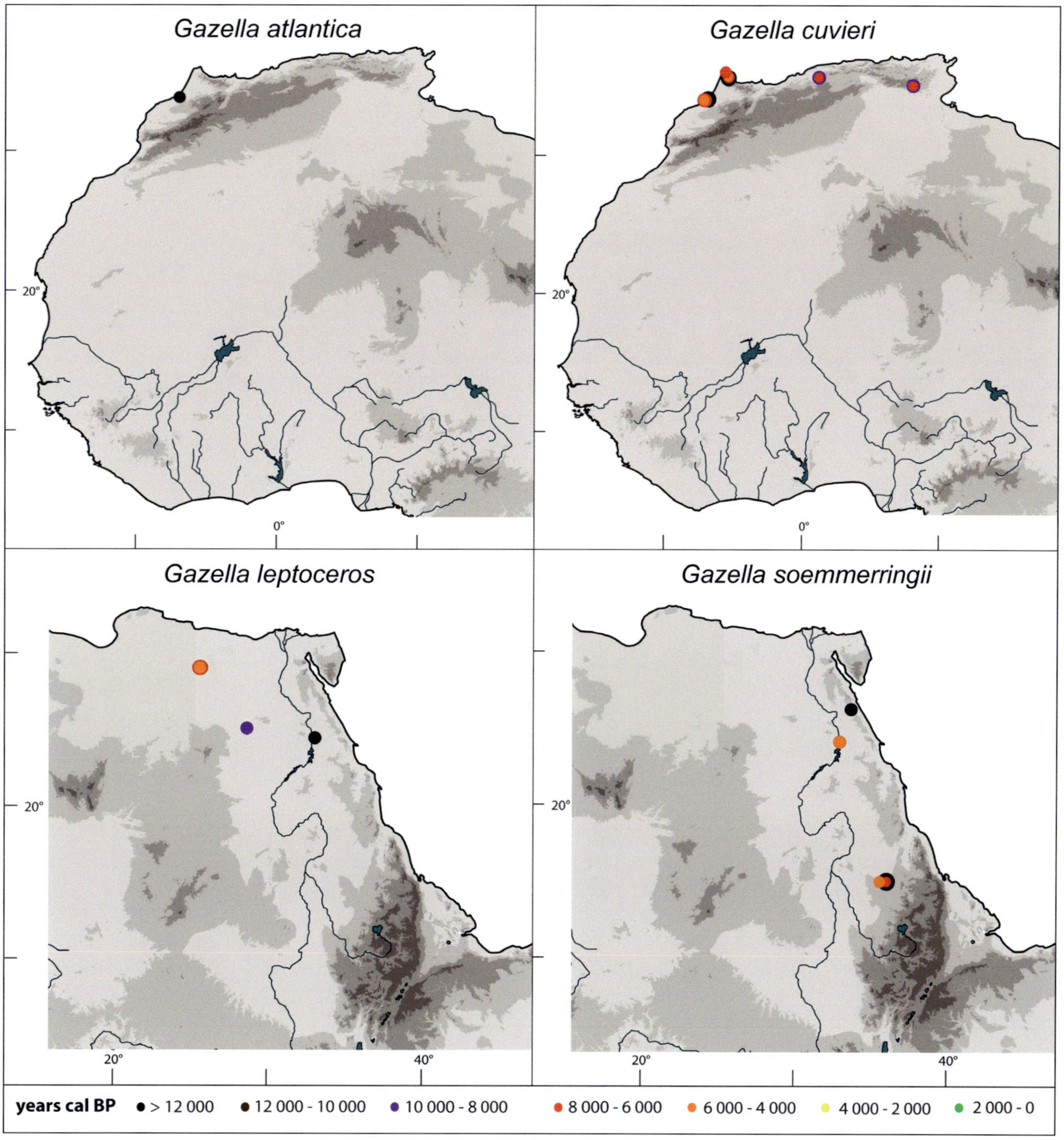

Gazella (Nanger) dama

Country	Site N	Site
Egypt	109	Bir Kiseiba-EN
	111	Bir Kiseiba-MN
	180	Selima Sandsheet - 85/78
	181	Selima Sandsheet - 85/79
	199	Wadi el Bakht
	205	Willmann´s camp - 81/61
	92	Abu Ballas - 85/50-1
	93	Abu Ballas - 85/51-3
	94	Abu Ballas - 85/56
Libya	333	Ti-n-Thora East
	335	Ti-n-Thora Two Caves
	345	Uan Telocat (MP)
Mali	361	Akumbu-EIA
	362	Akumbu-MIA
	368	erg Ine Sakane
	369	Est-Hassi el Abiod
	374	Hassi el-Abiod
	387	Tessalit
Mauritania	393	Berouâga
	400	Dhar Tichitt Site 38
	401	Dhar Tichitt Site 46
	405	Khatt Lemaiteg
Niger	473	Gobero-K
Sudan	756	Abu Tabari 02/1
	778	Gism el-Arba
	785	Kerma
	793	Laqiya area - 82/29
	794	Laqiya area - 82/31
	804	Rahib-80/87
	808	Selima Sandsheet - 85/73-2
	815	Sheikh Mustafa
	824	Wadi Hariq 01/1

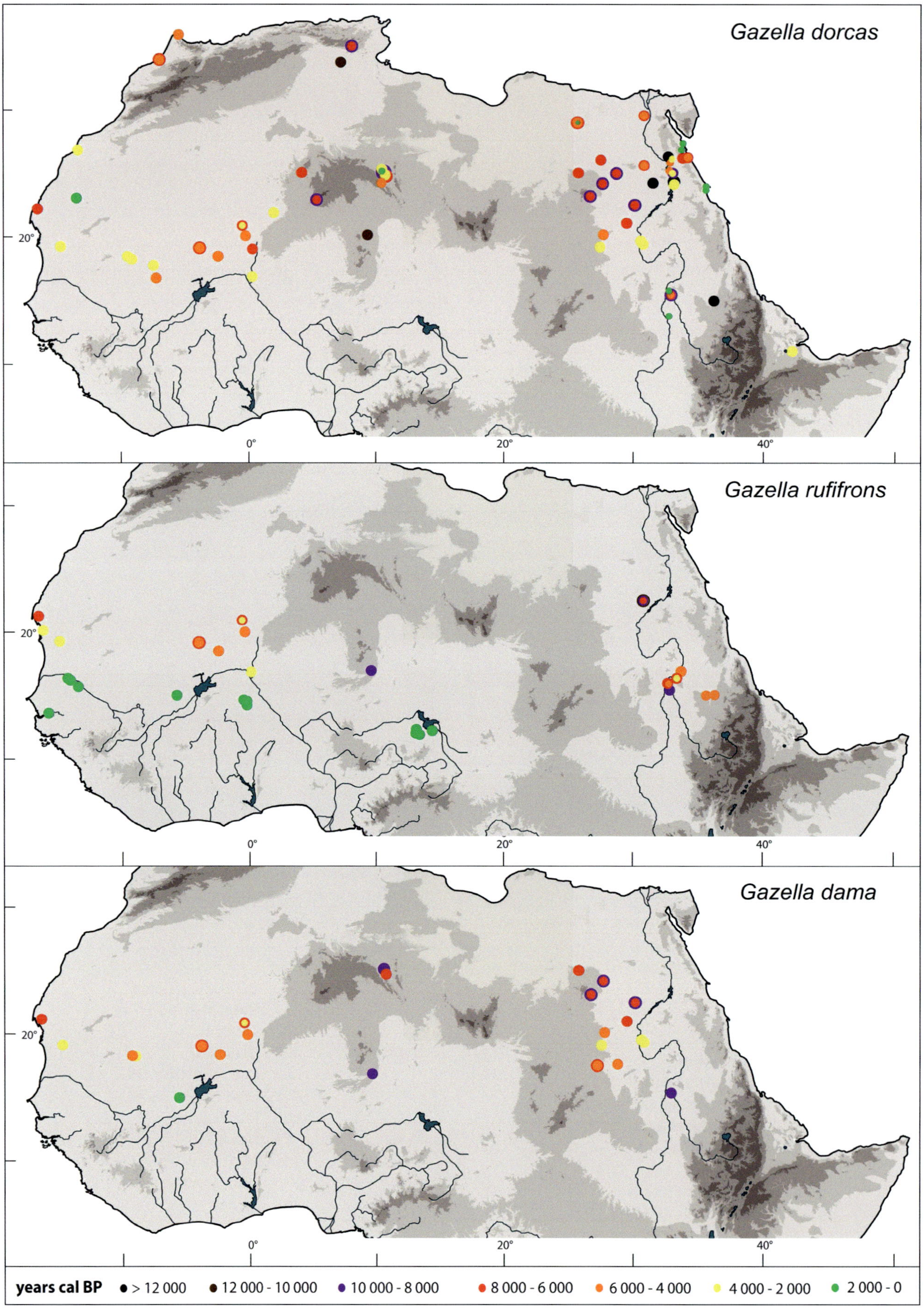
Gazella dorcas
Gazella rufifrons
Gazella dama
20°
0°
20°
40°
years cal BP
> 12 000
12 000 - 10 000
10 000 - 8 000
8 000 - 6 000
6 000 - 4 000
4 000 - 2 000
2 000 - 0

Gazella rufifrons

Country	Site N	Site	
Burkina Faso	52	Kissi 22	
	53	Kissi 40	
	56	Oursi Village-LIA	
	57	Oursi Village-MIA	
	58	Oursi-IA	
	63	Saouga 94/120	
	64	Saouga 95/7	
Egypt	171	Napta Playa 77-1/4/5/6	
	172	Napta Playa 77-8	
	173	Napta Playa E77-3/7	
Mali	360	Akumbu-AK3	
	361	Akumbu-EIA	
	362	Akumbu-MIA	
	368	erg Ine Sakane	
	369	Est-Hassi el Abiod	
	374	Hassi el-Abiod	
	380	Karkarichinkat N-S	*cf.
	387	Tessalit	
Mauritania	393	Berouâga	
	394	Chami-A2	
	405	Khatt Lemaiteg	
Niger	473	Gobero-K	
Nigeria	502	Elkido North	
	507	Galaga	
	516	Labe Kanuri	
	520	Mege III	
Senegal	538	Cubalel-Ia	
	540	Cubalel-II	
	541	Cubalel-III	
	544	Sintiou-Bara	
	548	Siouré-IV, V	
	549	Tulel-Fobo	
Sudan	765	El Kadada	
	766	El Kenger Middle/East	
	773	Esh Shaheinab	
	777	Geili	
	788	Khasm el Girba - Kassala	
	811	Shaqaduq-S1-A	
	812	Shaqaduq-S1-B	
	815	Sheikh Mustafa	
	821	Upper Atbara hinterland	
	822	Upper Atbara Valley	

Gazella (Nanger) granti

Country	Site N	Site
Kenya	247	Akira-N
	267	GaJi1
	270	GaJi2
	273	Gogo Falls-II-III
	278	GvJm 22-LHC
	279	GvJm 22-LPG
	292	Marula Rock Shelter
	293	Naivasha Railway Shelter
	294	Narosura
	295	Ndabibi
	298	Prolonged Drift
	299	Salasun-LSA
Sudan	815	Sheikh Mustafa
Tanzania	835	Mumba

Gazella thomsonii

Country	Site N	Site
Kenya	251	Crescent Island main
	259	Enkapune Ya Muto - Eb 4b
	273	Gogo Falls-II-III
	275	GvJm 19-HA
	276	GvJm 19-HC
	277	GvJm 19-TLPG
	278	GvJm 22-LHC
	279	GvJm 22-LPG
	280	GvJm 46-LPG
	283	GvJm 62-LPNG
	292	Marula Rock Shelter
	293	Naivasha Railway Shelter
	294	Narosura
	295	Ndabibi
	297	Ngenyn
	298	Prolonged Drift
	299	Salasun-LSA
	301	Salasun-PN
Mozambique	442	Manekweni-2-ABC
	443	Manekweni-3
	444	Manekweni-4
	445	Manekweni-5
Sudan	814	Sheikh el Amin
	815	Sheikh Mustafa
Tanzania	835	Mumba
Zambia	863	Leopard's Hill-PN

Litocranius walleri

Country	Site N	Site
Kenya	253	Ele Bor A-A
	254	Ele Bor A-B
	255	Ele Bor A-C

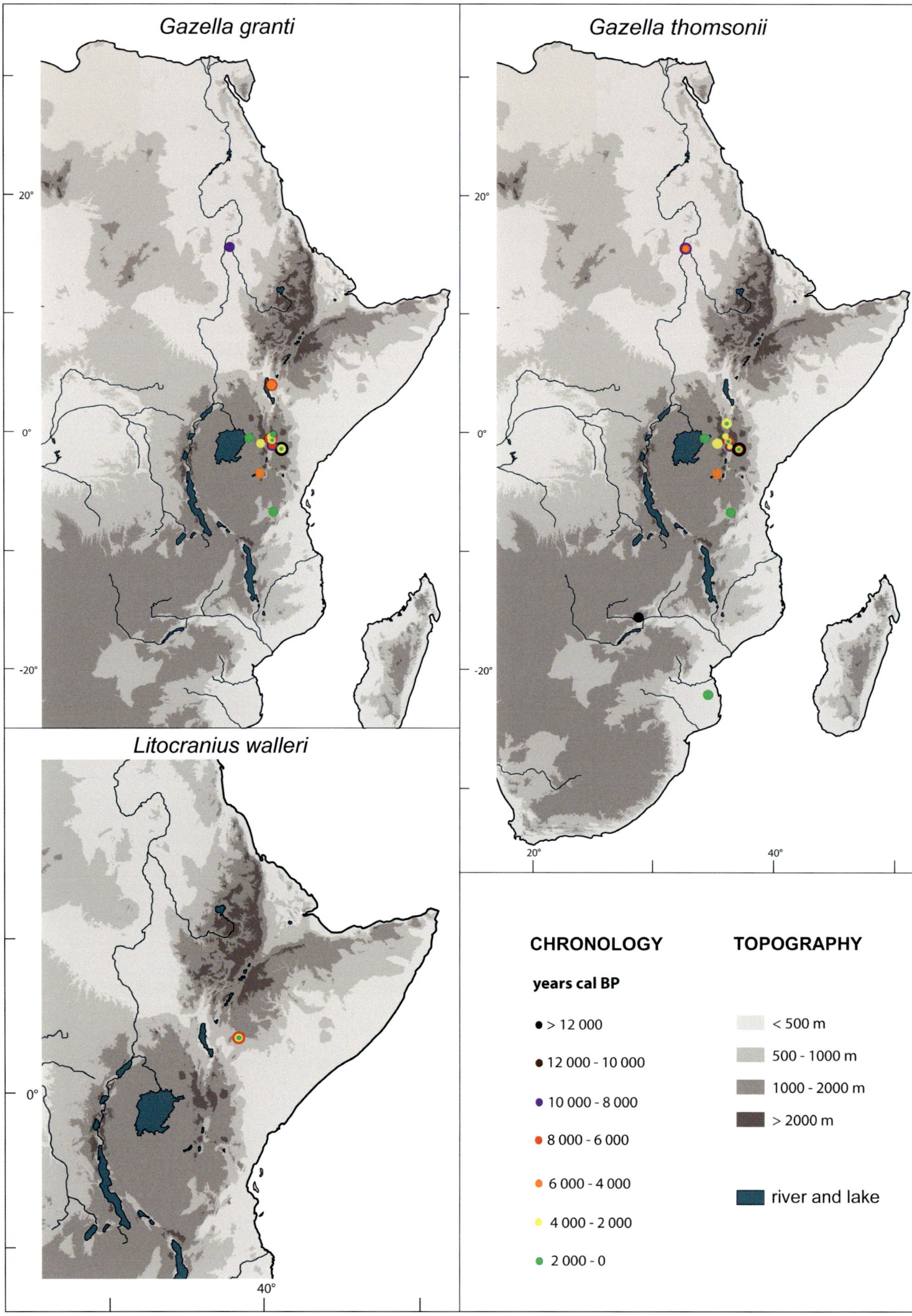
Gazella granti
Gazella thomsonii
Litocranius walleri
20°
0°
-20°
40°
CHRONOLOGY
years cal BP
> 12 000
12 000 - 10 000
10 000 - 8 000
8 000 - 6 000
6 000 - 4 000
4 000 - 2 000
2 000 - 0
TOPOGRAPHY
< 500 m
500 - 1000 m
1000 - 2000 m
> 2000 m
river and lake

Madoqua sp.

Country	Site N	Site
Kenya	251	Crescent Island main
	252	Dongodien
	270	GaJi2
	294	Narosura
	296	Ngamuriak
	297	Ngenyn
Sudan	775	Gash Delta-K
Tanzania	835	Mumba

Madoqua kirkii

Country	Site N	Site
Kenya	259	Enkapune Ya Muto - Eb 4b
	275	GvJm 19-HA
	276	GvJm 19-HC
	278	GvJm 22-LHC
	282	GvJm 62-LPG
	283	GvJm 62-LPNG
	297	Ngenyn
Namibia	455	Geduld-LIA

Madoqua saltiana

Country	Site N	Site
Djibouti	90	Asa Koma
Sudan	754	Abu Darbein
	757	Aneibis
	764	El Damer
	797	Mahal Teglinos K1-CG
	798	Mahal Teglinos K1-EG
	799	Mahal Teglinos K1-JM
	800	Mahal Teglinos K1-LG
	802	Mahal Teglinos K1-PG

Madoqua guentheri

Country	Site N	Site	
Kenya	253	Ele Bor A-A	*cf.
	254	Ele Bor A-B	*cf.
	255	Ele Bor A-C	*cf.
	256	Ele Bor A-D	*cf.
	297	Ngenyn	

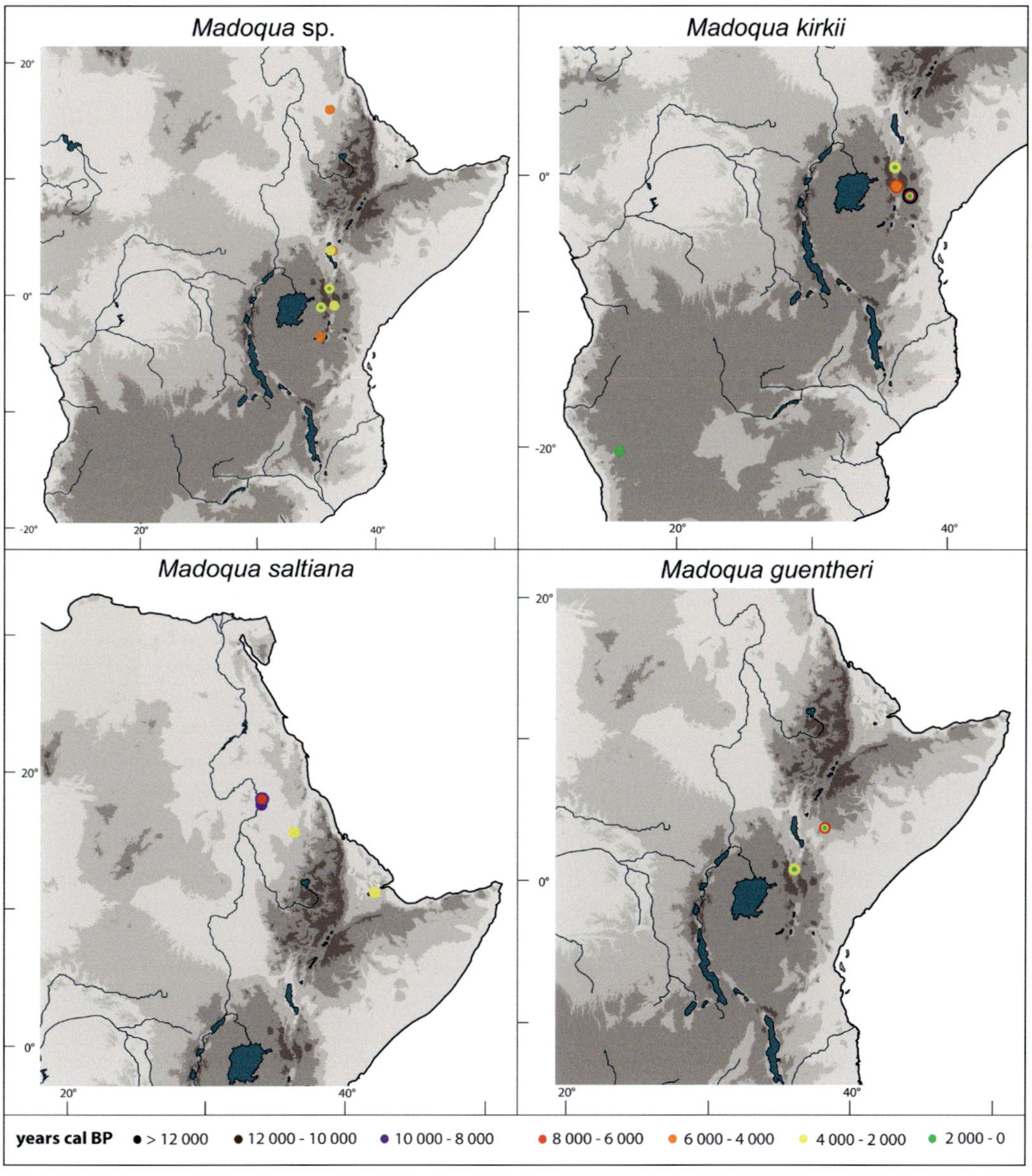

Bovinae

Taurotragus sp.

Country	Site N	Site	
Libya	327	Hauah Fteah-Iberom	
Mali	380	Karkarichinkat N-S	*?
Zaire	843	Ishango-AB	
	845	Ishango-Z-Post-Em	

Taurotragus oryx

Country	Site N	Site	
Botswana	26	Bosutwe-Mapungubwe	
	28	Bosutwe-Toutswe	
	32	Divuyu	
	37	Matlapaneng	
	38	Nqoma	
	39	Qogana	
	49	White Paintings RS-2b-3ab	
Kenya	251	Crescent Island main	
	254	Ele Bor A-B	
	255	Ele Bor A-C	
	256	Ele Bor A-D	
	258	Enkapune Ya Muto - Eb 4a	
	273	Gogo Falls-II-III	
	274	Gogo Falls-sq26-29	
	284	Kulchurdo	
	293	Naivasha Railway Shelter	
	297	Ngenyn	
	298	Prolonged Drift	
Lesotho	303	Bolahla	
	304	Ha Makotoko	
	305	Lequetsoana	
	307	Likoaeng-III-V	
	309	Likoaeng-XI	
	310	Likoaeng-XIII	
	311	Likoaeng-XVII	
	312	Liphofung-CW	
	313	Liphofung-pCW	
	314	Muela	
	315	Ntloana Tsoana	
	316	Sehonghong-a	
	321	Tloutle-CWI	
	322	Tloutle-O	
	323	Tloutle-pCW	
Mozambique	441	Manekweni-1	
	442	Manekweni-2-ABC	
	444	Manekweni-4	
	445	Manekweni-5	
Namibia	460	Oshigambo Peninsula	
South Africa	563	Boomplass Cave-a	
	564	Boomplass Cave-b	
	568	Buffelskloof RS-a	
	569	Buffelskloof RS-b	
	570	Buffelskloof RS-c	*?
	574	Bushman RS-LSA	
	576	Byeneskranskop Cave 1a	
	577	Byeneskranskop Cave 1b	
	579	Byeneskranskop Cave 1d	
	580	Collingham Shelter	
	581	Colwinton Shelter-CW	
	582	Colwinton Shelter-pCW	
	583	Connies Limpet Bar	
	594	Eiland Salt Works-b	
	595	Eland´s Bay Cave-Al	
	596	Eland´s Bay Cave-Rl	
	597	Eland´s Bay Cave-Wia	
	602	Eland´s Bay Open-b	
	604	Equus Cave 1A	
	609	Glentyre	*?
	610	Green Elliott Shelter I-III	
	611	Green Elliott Shelter IV-VI	
	620	HKLP Ib	*cf.
	620	HKLP Ib	
	621	HKLP Va	*cf.
	622	HKLP Vb	*cf.
	622	HKLP Vb	
	623	Icon	
	627	Kasteelberg A	
	628	Kasteelberg B	
	631	Klasies River Mouth-1I	
	641	Le6	
	642	Le7a	
	659	Marupale Hill	
	663	Mhlwazini Cave	
	668	Nelson Bay Cave- PW	
	669	Nelson Bay Cave- R	
	676	Oudepost I	
	679	Pancho's Kitchen Midden 3-4	
	680	Pancho's Kitchen Midden 6-7	
	681	Paternoster1-2	*cf.
	694	Ravenscraig Shelter-pCW	
	701	Rooikrans-CW	
	703	Roosfontein	
	704	Rose Cottage Cave-A-Mn	
	705	Rose Cottage Cave-DB2	
	706	Rose Cottage Cave-OIC	
	707	Rose Cottage Cave-WI	
	719	Smitswinkelbaai Cave	
	730	Tandjesberg	
	738	Tortoise cave-d	
	742	Ventershoek	
Tanzania	835	Mumba	
Zaire	846	Kakontwe	
Zambia	860	Leopard's Hill-NI	
	862	Leopard's Hill-NIII	
	864	Namakala	
Zimbabwe	870	Redcliff Cave	

Taurotragus derbianus

Country	Site N	Site
Cameroon	74	Djaba Hosséré
Mali	368	erg Ine Sakane
	374	Hassi el-Abiod
Mauritania	400	Dhar Tichitt Site 38
	401	Dhar Tichitt Site 46
	405	Khatt Lemaiteg
Niger	473	Gobero-K
Senegal	543	Ogo region

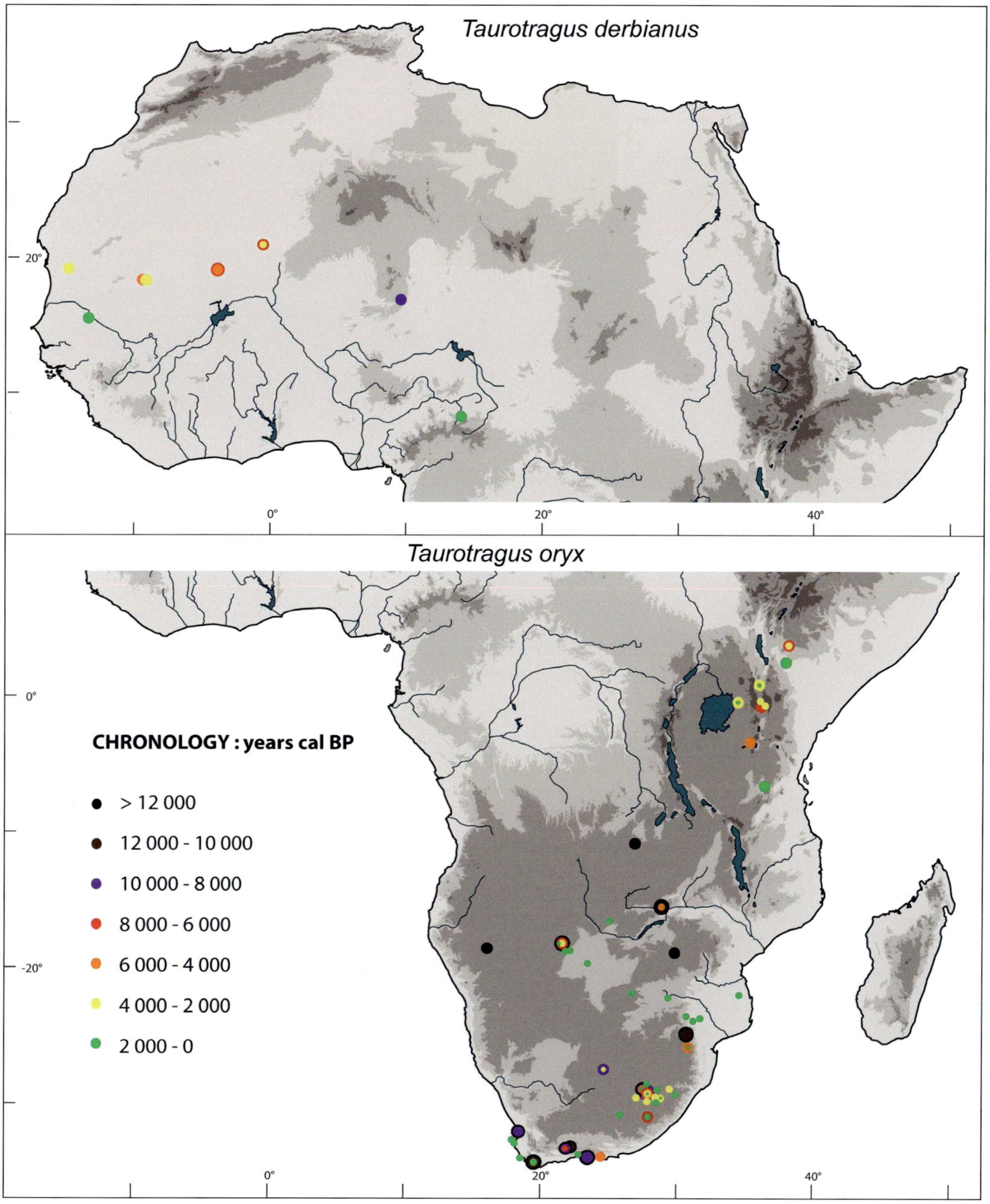

Tragelaphus eurycerus

Country	Site N	Site
Ghana	225	Asantamanso-ASII/III
Zaire	855	Matupi-II

Tragelaphus angasii

Country	Site N	Site
Mozambique	441	Manekweni-1
	442	Manekweni-2-ABC
	443	Manekweni-3
	444	Manekweni-4
	445	Manekweni-5
South Africa	650	Magogo-1
	667	Ndondondwane

Tragelaphus imberbis

Country	Site N	Site
Kenya	258	Enkapune Ya Muto - Eb 4a
	272	Gogo Falls-I
	273	Gogo Falls-II-III

Tragelaphus buxtoni

Country	Site N	Site	
Ethiopia	216	Moche Borago-1b	*cf.
	219	Moche Borago-3	*cf.

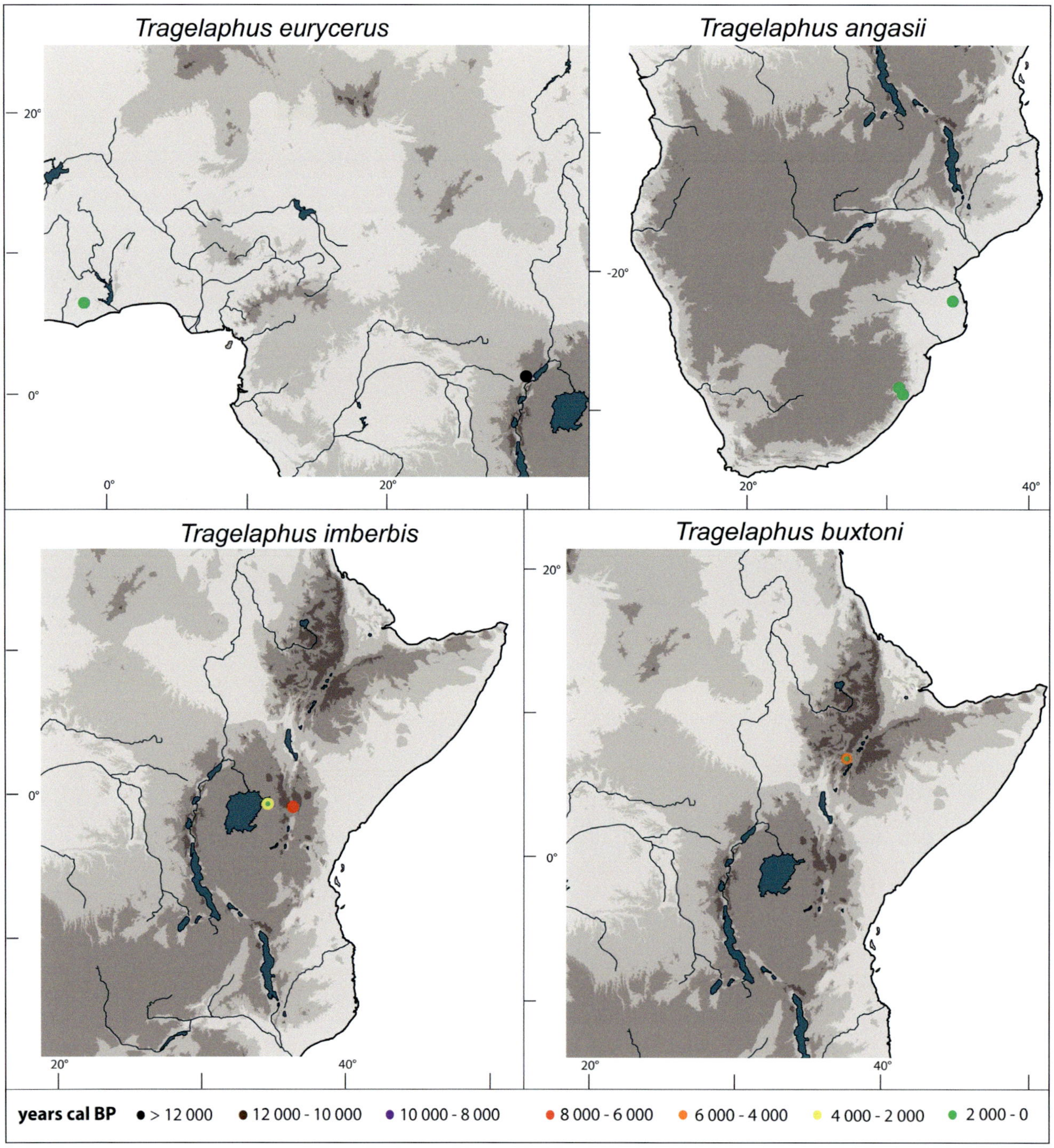

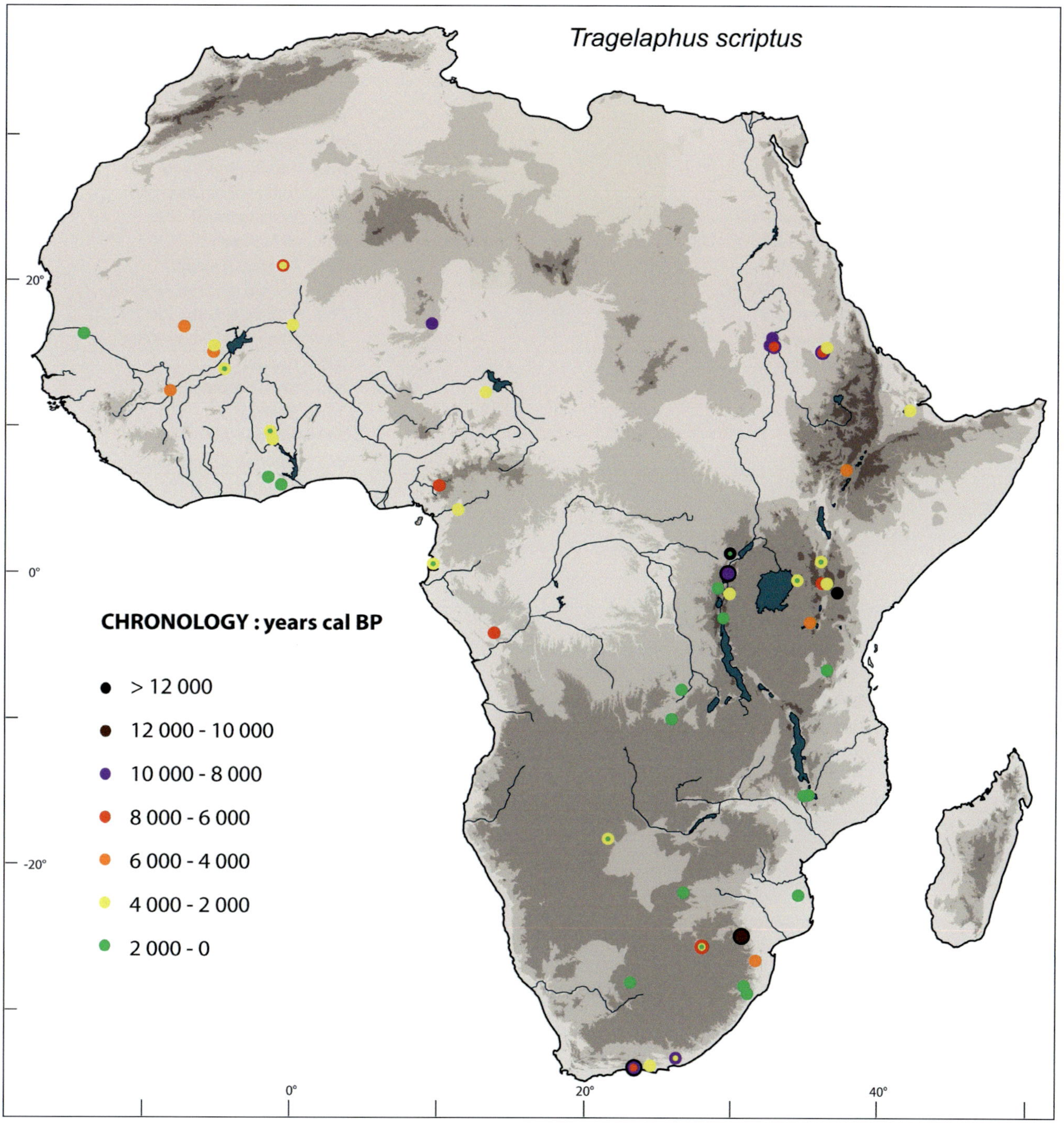

Tragelaphus scriptus

Country	Site N	Site	
Botswana	27	Bosutwe-Taukome	
	48	White Paintings RS-2ab	
Burundi	66	Ruzizi Plain	
Cameroon	75	Nkang	
	76	Shum Laka II c	*?
Congo	87	Ntadi Yom ba-level B	
	89	Tongo	
Djibouti	90	Asa Koma	
Ethiopia	216	Moche Borago-1b	*cf.
	217	Moche Borago-1c	*cf.
Gabon	221	Okala	* or C. sylvicultor
Ghana	223	Anyinam-Area A-B-C	
	225	Asantamanso-ASII/III	
	227	Asantamanso-ASV	
	231	Daboya	
	244	Ntereso	
	245	Nyanawase/Akwamu Am.	
Kenya	251	Crescent Island main	
	258	Enkapune Ya Muto - Eb 4a	
	259	Enkapune Ya Muto - Eb 4b	
	260	Enkapune Ya Muto - Eb 5a	
	261	Enkapune Ya Muto - Eb 5b	
	273	Gogo Falls-II-III	
	274	Gogo Falls-sq26-29	
	277	GvJm 19-TLPG	
	293	Naivasha Railway Shelter	
	297	Ngenyn	

Country	Site N	Site	
Malawi	358	Matope Court-LIA	
	359	Namichimba	
Mali	368	erg Ine Sakane	
	376	Jenné-Jeno - phase I/II	
	377	Jenné-Jeno - phase III	
	380	Karkarichinkat N-S	
	383	Kolima Sud-II	
	385	Korounkorokalé-III	
	388	Tiabel Goudiodié	
Mauritania	402	DN4-BKH-S5	
Mozambique	441	Manekweni-1	
	442	Manekweni-2-ABC	
	443	Manekweni-3	
	444	Manekweni-4	
	445	Manekweni-5	
Niger	473	Gobero-K	
Nigeria	505	Gajiganna B2	
Rwanda	536	Muganza	*cf.
Senegal	540	Cubalel-II	
	541	Cubalel-III	
South Africa	574	Bushman RS-LSA	
	589	Doornfontein	*T.s.sylvaticus
	603	Emambeni	
	625	Jubilee Shelter-PW	
	626	Jubilee Shelter-W	
	633	Klasies River Mouth-5II	
	639	Large Rock Shelter	
	650	Magogo-1	
	654	Mamba 1	
	668	Nelson Bay Cave- PW	
	670	Nelson Bay Cave- W	
	753	Wosi-terminal EIA	
Sudan	767	El Mahalab	
	787	Khartoum Hopital	
	788	Khasm el Girba - Kassala	
	789	Khasm el Girba - KG14	
	791	Khasm el Girba - KG68	
	805	Saggai 1	
	815	Sheikh Mustafa	
Tanzania	835	Mumba	
Zaire	843	Ishango-AB	
	845	Ishango-Z-Post-Em	
	848	Kiantapo	
	854	Matupi-I	
	855	Matupi-II	
	858	Sanga	

Tragelaphus strepsiceros

Country	Site N	Site	
Botswana	26	Bosutwe-Mapungubwe	
	28	Bosutwe-Toutswe	*cf.
	28	Bosutwe-Toutswe	
	31	Commando Kop	
	33	Drotsky's Cave-H	
	49	White Paintings RS-2b-3ab	
Egypt	184	Sodmein Cave - lev 5 (MP5)	
	185	Sodmein Cave older backfill	
Ghana	223	Anyinam-Area A-B-C	
Mozambique	442	Manekweni-2-ABC	
	443	Manekweni-3	
	444	Manekweni-4	
	445	Manekweni-5	
Niger	473	Gobero-K	
South Africa	563	Boomplass Cave-a	
	564	Boomplass Cave-b	
	568	Buffelskloof RS-a	
	574	Bushman RS-LSA	
	589	Doornfontein	
	604	Equus Cave 1A	
	616	Heuningneskrans-I/O	
	617	Heuningneskrans-R	
	618	Hill X	
	622	HKLP Vb	*cf.
	642	Le7a	
	643	Le7b	
	648	Ma38	
	653	Malwula	
	659	Marupale Hill	*cf.
	664	Mo8	
	684	Ph1	
	691	Pr34	
	706	Rose Cottage Cave-OIC	
	710	Selongwe Hill	
	711	Serotwe Hill	
	713	SH16	
	716	Shankare Hill-b	
	720	Sonkoanini Hill	
	739	Tsh1	
Sudan	754	Abu Darbein	
	757	Aneibis	
	764	El Damer	
	765	El Kadada	
	773	Esh Shaheinab	
	777	Geili	
	788	Khasm el Girba - Kassala	
	791	Khasm el Girba - KG68	
	805	Saggai 1	
	812	Shaqaduq-S1-B	
	821	Upper Atbara hinterland	
Tanzania	835	Mumba	
Zaire	843	Ishango-AB	
	846	Kakontwe	
Zambia	860	Leopard's Hill-NI	
Zimbabwe	870	Redcliff Cave	

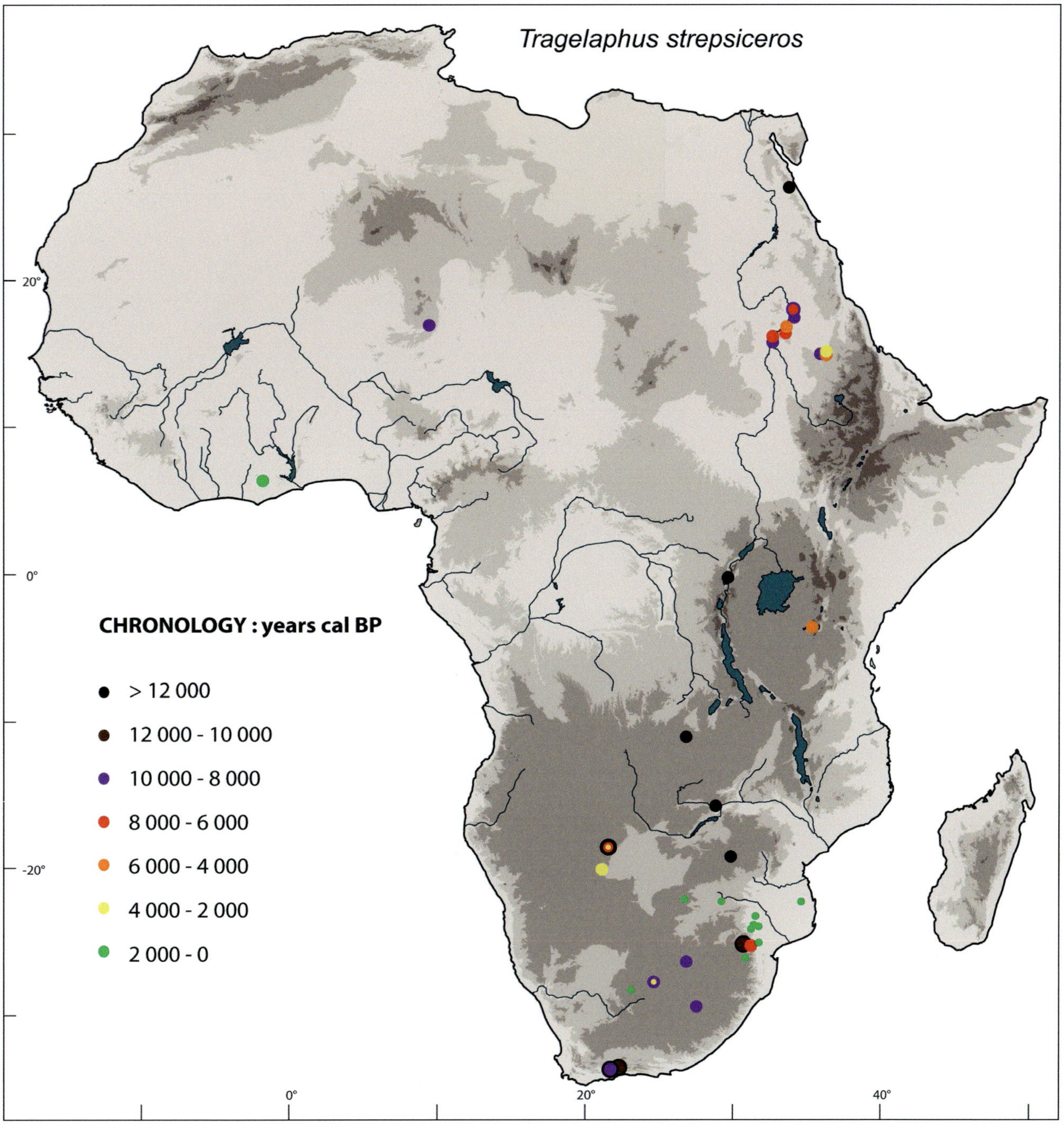

Tragelaphus spekii

Country	Site N	Site
Botswana	28	Bosutwe-Toutswe
	42	Toteng1-SU2
C.A.R.	79	Mondongué
Mali	369	Est-Hassi el Abiod
	374	Hassi el-Abiod
	381	Kobadi
	382	Kolima Sud-I
	383	Kolima Sud-II
	388	Tiabel Goudiodié
Namibia	460	Oshigambo Peninsula
Niger	473	Gobero-K
Rwanda	535	Akameru-Cyinkomane
Sudan	787	Khartoum Hopital
Zaire	843	Ishango-AB
	855	Matupi-II

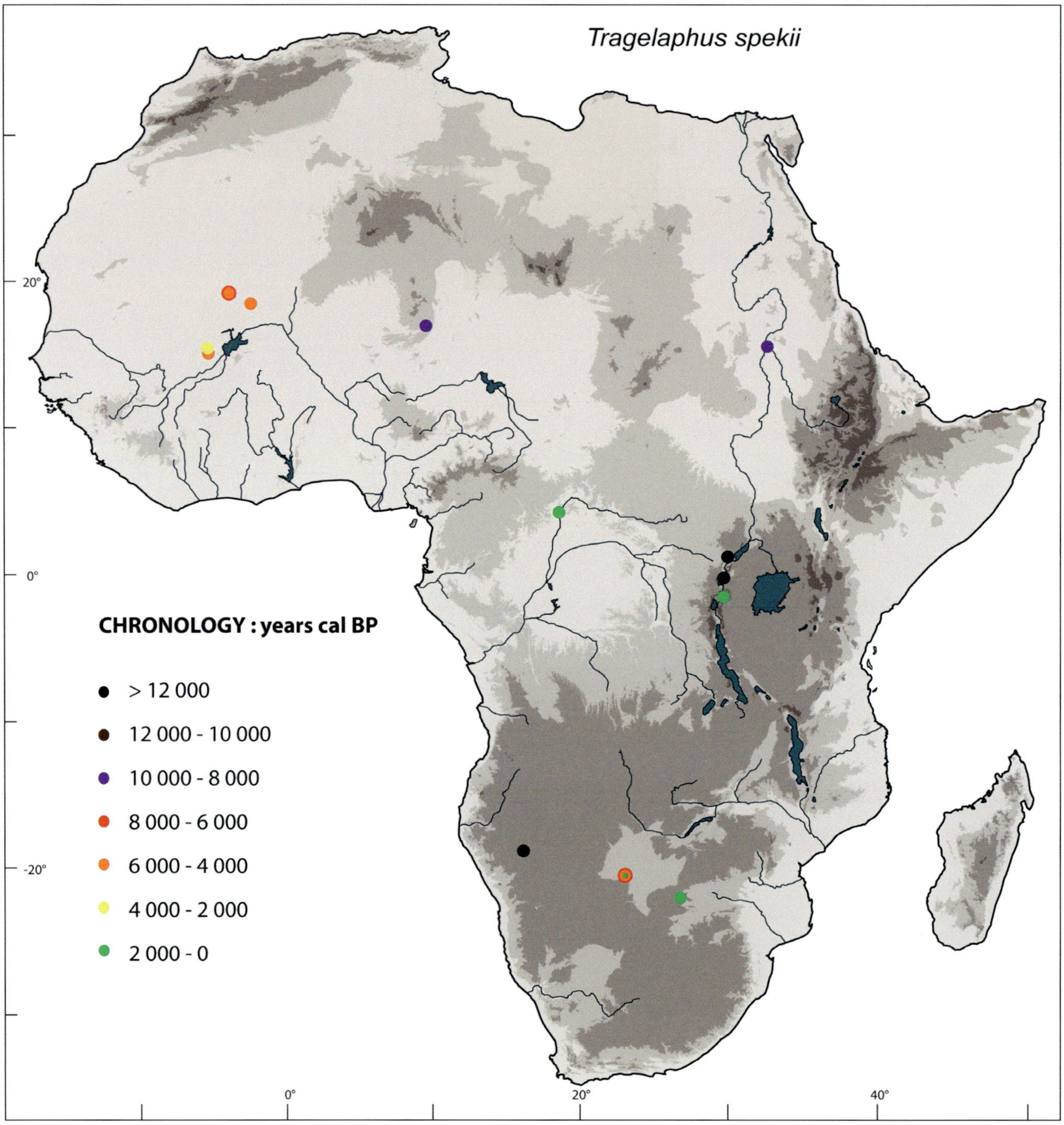
Tragelaphus spekii
CHRONOLOGY : years cal BP
> 12 000
12 000 - 10 000
10 000 - 8 000
8 000 - 6 000
6 000 - 4 000
4 000 - 2 000
2 000 - 0
20°
0°
-20°
0°
20°
40°

Bubalus (Homoioceras) antiquus

Country	Site N	Site
Algeria	16	Meniet
	5	Amekni
	9	Dra-Mta-el-Ma-el-Abiod

Pelorovis antiquus

Country	Site N	Site	
Algeria	11	Gueldaman Akbou	
	15	Medjez II	
	6	Columnata	
Egypt	122	Dakhleh Ouasis, Ba	
Kenya	257	Engaruani	*cf.
	279	GvJm 22-LPG	
Mauritania	394	Chami-A2	
Morocco	422	El Harhoura 1	
South Africa	563	Boomplass Cave-a	
	668	Nelson Bay Cave- PW	
	669	Nelson Bay Cave- R	
Sudan	755	Abu Hugar	
	809	Selima Sandsheet - 85/80	
	828	Wadi Howar - 80/73	
Zaire	843	Ishango-AB	
	844	Ishango-C	

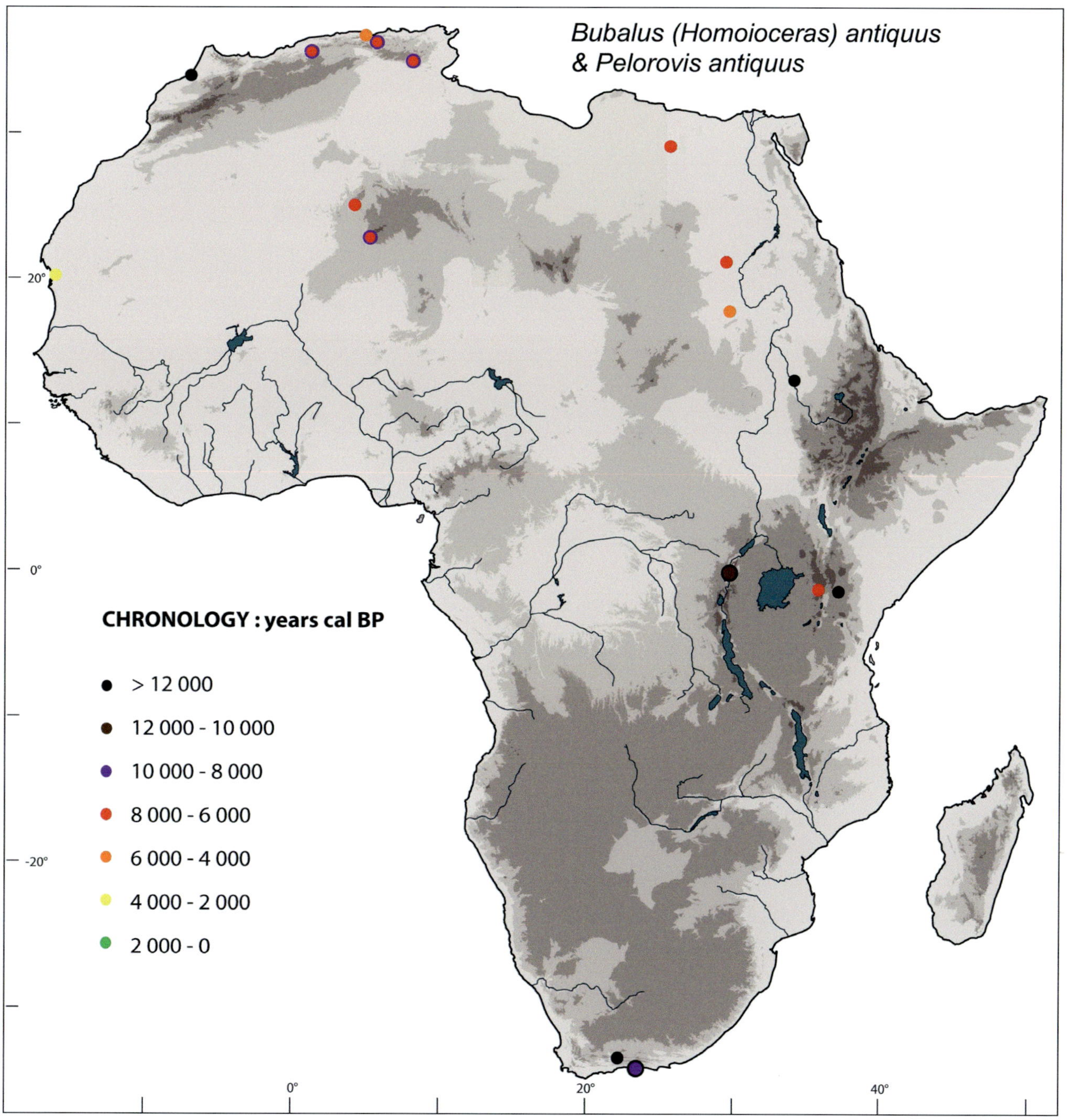

Bos primigenius

Country	Site N	Site	
Algeria	11	Gueldaman Akbou	
	13	Kef Zoura D	
	14	Mankhor	*or B. cf. bubaloides
	15	Medjez II	
	4	Ain Mistehayia	*or H. antiquus
	6	Columnata	
	9	Dra-Mta-el-Ma-el-Abiod	
Egypt	112	Buto/1-3	
	113	Buto-6	
	114	Buto-9-12	
	122	Dakhleh Ouasis, Ba	
	131	El Abadyia-A-B	
	132	El Abadyia-C1-C2	
	138	Elkab-Epipaleo	
	145	Fayoum G1-H1	
	147	Fayoum G5-G4	
	154	Hierakonpolis HK6-Naq IC-IIA	
	162	Kom Ombo	
	166	Makhadma 2	
	167	Makhadma 4	
	168	Merimde Benisâlame	
	183	Shuwikhat 1	
	197	Umm el-Qaab, Friedhof U	
	201	Wadi Kubbaniya - E-78-3	
	202	Wadi Kubbaniya - E-78-4	
	203	Wadi Kubbaniya - E-81-1	
	99	Adaima-habitat	
Mauritania	394	Chami-A2	
Morocco	421	Bou Saria C-D	
	422	El Harhoura 1	
	424	El Mnasra	
	426	Ghar-Khal IIIb-V	
	433	Kaf-taht-el-Ghar-E	*?
	434	Kaf-taht-el-Ghar-F	
	435	Kaf-taht-el-Ghar-G	
	437	Kehf-el-Baroud	
Sudan	823	Wadi Halfa	

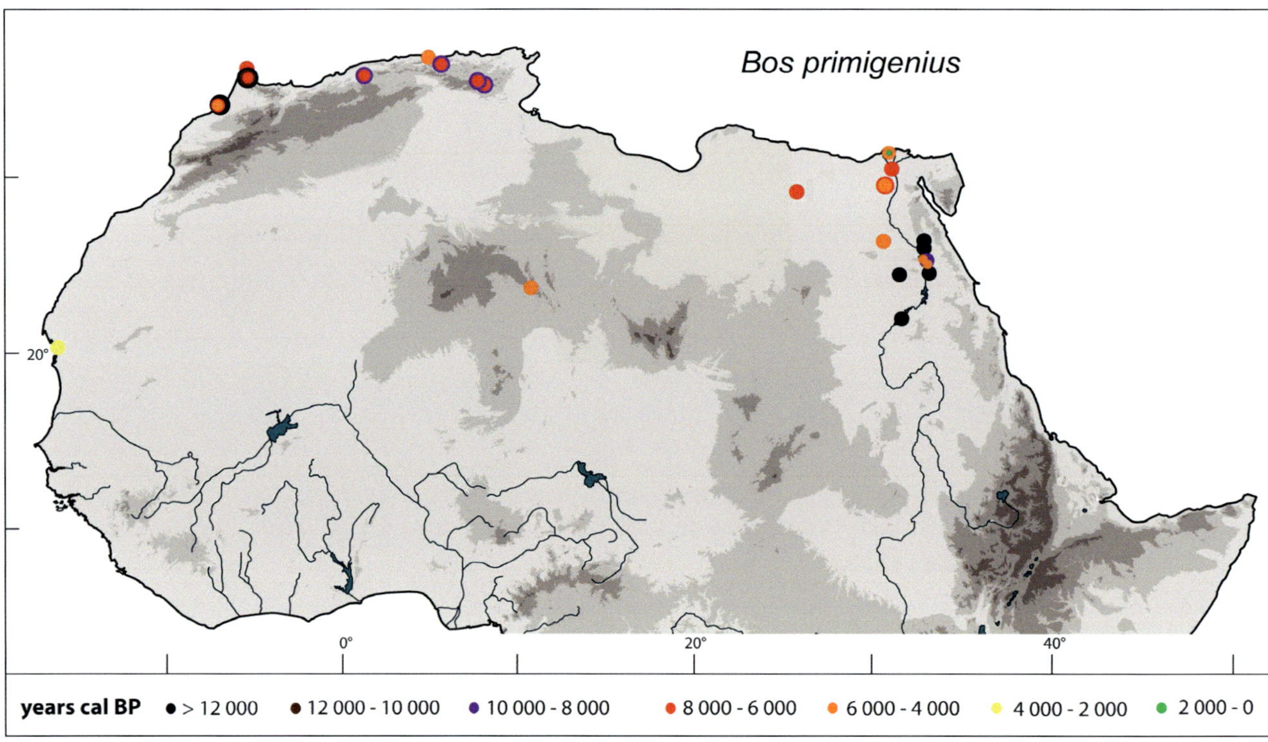

Bos primigenius f. taurus

Country	Site N	Site	
Algeria	1	A-88-15	
	11	Gueldaman Akbou *B.brachyceros	
	14	Mankhor	
	2	A-88-28	
	21	Ti-n-Hanakaten	
	7	Damous el Ahmar	
Angola	24	Kamabanga	
	25	Kitala	
Botswana	26	Bosutwe-Mapungubwe	
	27	Bosutwe-Taukome	
	28	Bosutwe-Toutswe	
	29	Bosutwe-Zimbabwe	
	30	Broadhurst	
	31	Commando Kop	
	32	Divuyu	
	36	Matanga	
	38	Nqoma	
	40	Taukome	
	42	Toteng1-SU2	*cf.
	42	Toteng1-SU2	
	43	Toteng1-SU3	*cf.
	43	Toteng1-SU3	
Burkina Faso	52	Kissi 22	
	53	Kissi 40	
	54	Oursi hu-beero	
	55	Oursi Village-EIA	
	56	Oursi Village-LIA	
	57	Oursi Village-MIA	
	58	Oursi-IA	
	63	Saouga 94/120	
	64	Saouga 95/7	
Burundi	66	Ruzizi Plain	
Cameroon	70	Blé A	
	71	Blé B	
	72	Blé C	
	73	Blé E	
Chad	82	Enneri Bardagué	
	83	Gabrong-b	
Congo	89	Tongo	
Djibouti	90	Asa Koma	
Egypt	100	Adaima-habitat	
	102	Al-Ma´abda	
	103	Al-Ma´abda	
	105	Balat-Mastaba V	
	106	Berenike 95/9 Early Roman	
	107	Berenike 95/97Late Roman	
	109	Bir Kiseiba	* Putative cattle
	112	Buto/1-3	
	113	Buto-6	
	114	Buto-9-12	
	120	Dahschûr	
	121	Dakhleh Oasis, SMU	
	123	Dakhleh Ouasis, Roman	
	124	Dakhleh Ouasis,OK	
	127	Eastpan 95-2	
	129	El Abadiya 2 North sector	
	130	El Abadiya 2 South sector	
	133	El Omari	
	134	El Salamuni	
	135	Elephantine-necropolis	
	136	Elephantine-NW Chnum T	
	137	Elephantine-Satet temple	
	139	Elkab-necropolis	
	140	El-Tell el Iswid-Early Dynastic	
	141	El-Tell el Iswid-Predynastic	
	142	El-Tell el-Iswid A	
	143	El-Tell el-Iswid B	
	146	Fayoum G3-H2	
	147	Fayoum G5-G4	
	153	Gilf Kebir	
	154	Hierakonpolis HK6-Naq IC-IIA	
	155	Hierakonpolis HK6-Naq IIA-B	
	156	Hierakonpolis HK6-Naq III	
	157	Hierakonpolis HK6-Naq I-II	
	160	Kharga dep.-graeco-roman	
	164	Maadi	
	165	Mahgar Dendera 2	
	168	Merimde Benisâlame	
	169	Minshat Abu Omar	
	170	Mons Claudianus	
	171	Napta Playa 77-1/4/5/6	* put. cattle
	175	Qasr el-Sagha-1	
	177	Qasr el-Sagha-3	
	179	Quseir al-Qadim	
	182	Shenshef	
	188	Tell el-Dab´a - b	
	191	Toukh	* B. brachyceros
	196	Tuna el-Gebel	
	197	Umm el-Qaab, Friedhof U	
	199	Wadi el Bakht	
	206	Workmen Village	
	95	Abu Sha´ar-4C	
	98	Adaima-burials	
Ethiopia	207	Axum-D	
	208	Axum-K	
	210	Gobedra	
	211	Laga Oda-a	
	213	Lake Besala	
	214	Lalibela	
	220	Quiha	
Ghana	223	Anyinam-Area A-B-C	
	224	Anyinam-Area D	
	225	Asantamanso-ASII/III	
	227	Asantamanso-ASV	
	228	Begho	
	229	Bibease Plantation	
	231	Daboya	
	231	Daboya	
	232	Dawu	
	233	Esiease - A-C-D	
	238	Kintampo RS K6-K	
	241	Mandara	
	242	Mankuma	
	243	Mumute	
	244	Ntereso	
	245	Nyanawase/Akwamu Am.	
	246	Nyanga	
Kenya	247	Akira-N	
	248	Akira-PIA	
	249	Akira-SPN	
	250	Causeway Site	
	251	Crescent Island main	
	252	Dongodien	*cf.

Country	No.	Site	Note
	261	Enkapune Ya Muto - Eb 5b	
	263	Enkapune Ya Muto BSS-HTL	
	272	Gogo Falls-I	
	273	Gogo Falls-II-III	
	274	Gogo Falls-sq26-29	
	276	GvJm 19-HC	
	278	GvJm 22-LHC	
	281	GvJm 62-H	
	284	Kulchurdo	
	285	Lemek North-East	*Bos sp.
	290	Maasai Gorge Shelter	
	291	Maringishu	
	293	Naivasha Railway Shelter	
	294	Narosura	
	295	Ndabibi	
	296	Ngamuriak	
	297	Ngenyn	
	298	Prolonged Drift	
	299	Salasun-LSA	
	300	Salasun-PIA	
	301	Salasun-PN	
	302	Sambo Ngige	*Bos sp.
Lesotho	303	Bolahla	
	305	Lequetsoana	
	306	Likoaeng-I	
	313	Liphofung-pCW	
	314	Muela	
	316	Sehonghong-a	
	323	Tloutle-pCW	
Libya	330	In Aghelachem-RT,SF	
	334	Ti-n-Thora North	
	339	Uan Muhuggiag (LP)	
	340	Uan Muhuggiag (MP)	
	343	Uan Telocat (EP)	
	345	Uan Telocat (MP)	
Malawi	357	Matope Court-IA	
Mali	360	Akumbu-AK3	*Bos sp.
	361	Akumbu-EIA	*Bos sp.
	362	Akumbu-MIA	*Bos sp.
	364	Dia-Shoma Horizon I	
	365	Dia-Shoma Horizon II	
	366	Dia-Shoma Horizon III+IV	
	367	Dia-Shoma Horizon V	
	368	erg Ine Sakane	
	369	Est-Hassi el Abiod	
	372	Gao Saney	*Bos sp.
	373	Hamdallahi	
	374	Hassi el-Abiod	
	379	Karkarichinkat KN05	
	381	Kobadi	
	382	Kolima Sud-I	
	383	Kolima Sud-II	
	387	Tessalit	
	389	Tongo Maaré Diabel	
	390	Windé Koroji I,III	
	392	Zampia-Tumulus 7	
Mauritania	400	Dhar Tichitt Site 38	
	401	Dhar Tichitt Site 46	
	402	DN4-BKH-S5	
	404	Ile de Tidra	
	405	Khatt Lemaiteg	
	406	Khneivissa	
	407	Tegdaoust-a	
	408	Tegdaoust-b	
	409	Tegdaoust-c	
	410	Tegdaoust-d	
	411	Tichitt - Akjinjeir phase	
	412	Tichitt - Arriane phase	
	413	Tichitt - Chebka phase	
	414	Tichitt - Goungou phase	
	415	Tichitt - Khimiya phase	
	416	Tichitt - Naguez phase	
	417	Tichitt - Nkhal phase	
Morocco	419	Bou Saria A	
	420	Bou Saria B	
	421	Bou Saria C-D	
	424	El Mnasra	*?
	425	Ghar-Khal I	
	426	Ghar-Khal IIIb-V	
	427	Ghar-Khal II-IIIa	
	429	Kaf-taht-el-Ghar-A	
	430	Kaf-taht-el-Ghar-B	
	431	Kaf-taht-el-Ghar-C	
	432	Kaf-taht-el-Ghar-D	
	433	Kaf-taht-el-Ghar-E	*?
	437	Kehf-el-Baroud	
	438	Kehf-el-Baroud-aven	
Mozambique	441	Manekweni-1	
	442	Manekweni-2-ABC	
	443	Manekweni-3	
	444	Manekweni-4	
	445	Manekweni-5	
Niger	464	Adrar Bous	
	467	Adrar Bous-T	
	468	Arlit-a	
	469	Arlit-b	
	471	Chin Tafidet	
	474	Gobero-T	
	489	Takene Bawat 5	
	496	TMY1	
	497	TMY3	
Nigeria	498	Bukarkurari-I	
	499	Bukarkurari-II	
	500	Daima	
	501	Dorota	
	502	Elkido North	
	503	Gajiganna A	
	504	Gajiganna B1-lower	
	505	Gajiganna B2	
	506	Gajiganna C	
	507	Galaga	
	508	Gilgila-Iia	
	509	Gilgila-Iib	
	510	Gilgila-Iic	
	511	Kariari	
	512	Kelumeri	
	513	Kursakata	
	514	Kursakata-IA	
	515	Kursakata-LSA	
	516	Labe Kanuri	
	517	Labe NA 97/24	
	518	Mege I	
	519	Mege II	
	520	Mege III	
	521	Mege IV	
	522	NA 93/10	
	523	NA 93/36	
	524	NA 93/42	

Country	No.	Site	Note
	525	Ngala I	
	526	Ngala II	
	527	Ngala IIIa	
	528	Ngala IIIb	
	529	Ngala IV	
	530	Zilum	
Rhodesia	531	Chamabvefva	
	532	Harleigh Farm-1	
	533	Harleigh Farm-2	
	534	Mabveni	
Rwanda	535	Akameru-Cyinkomane	
	537	Ryamurari	
Senegal	538	Cubalel-Ia	*Bos sp.
	539	Cubalel-Ib	*Bos sp.
	540	Cubalel-II	*Bos sp.
	541	Cubalel-III	*Bos sp.
	542	Ndiamon-Badat	
	544	Sintiou-Bara	*Bos t. /indicus ?
	545	Siouré-Ia	*Bos sp.
	546	Siouré-Ib	*Bos sp.
	547	Siouré-II	*Bos sp.
	548	Siouré-IV, V	*Bos sp.
	549	Tulel-Fobo	
	550	Walaldé-W1/1-19	
	551	Walaldé-W1/20-25	
	552	Walaldé-W2	
South Africa	556	2627 CD7	
	557	Badfontein Valley	
	586	Die Kelders	
	591	Drie Susters Main	
	593	Eiland Salt Works-a	
	594	Eiland Salt Works-b	
	605	Ficus A	
	606	Ficus B	
	607	Ficus C	
	614	Hapi Pan	
	615	Happy Rest	
	618	Hill X	
	619	HKLP Ia	
	623	Icon	
	627	Kasteelberg A	*cf.
	628	Kasteelberg B	*cf.
	629	Kekane	
	635	Klingbeil 2530 AB 1	
	636	Klipspring	
	637	Kreeftebaai	
	638	Langsdraai	
	639	Large Rock Shelter	
	640	Le2	
	641	Le6	
	642	Le7a	
	643	Le7b	
	647	Lydenburg Heads site	
	649	Ma4	
	650	Magogo-1	
	651	Magogo-2	
	652	Magogo-3	
	654	Mamba 1	
	656	Mapungubwe-Hill	
	657	Mapungubwe-K2/TS3	
	658	Mapungubwe-ST	
	659	Marupale Hill	
	661	Mgoduyanuka	
	665	Msuluzi Confluence	
	667	Ndondondwane	
	673	Ol20	
	674	Ondini	
	676	Oudepost I	
	688	Pont-Drift-LKA	
	689	Pont-Drift-Zhizo	
	692	QwaQwa Museum	
	693	Randjies	
	696	Rhenosterkloof	
	702	Rooikrans-pCW	
	704	Rose Cottage Cave-A-Mn	
	708	Schroda	
	709	Sebatini Hill	
	710	Selongwe Hill	
	715	Shankare Hill-a	
	716	Shankare Hill-b	
	717	SK17	
	719	Smitswinkelbaai Cave	
	720	Sonkoanini Hill	
	741	Umgungundlovu	
	743	Vlaeberg	
	745	Westbury Rock Shelter	
	746	Wildebeestfontein	
	751	Wosi-early EIA	
	752	Wosi-middle-late EIA	
	753	Wosi-terminal EIA	
Sudan	761	Debbat Bangdit	
	762	Debbat El Eheima	
	765	El Kadada	
	766	El Kenger Middle/East	
	768	El Nofalab	
	771	El Zakiab	
	772	Erg of Ennedi	
	773	Esh Shaheinab	
	775	Gash Delta-K	
	776	Gebel Tageru	
	777	Geili	
	778	Gism el-Arba	
	780	Jebel et Tomat	
	783	Kadero	
	784	Kadruka	
	785	Kerma	
	793	Laqiya area - 82/29	
	797	Mahal Teglinos K1-CG	
	798	Mahal Teglinos K1-EG	
	799	Mahal Teglinos K1-JM	
	800	Mahal Teglinos K1-LG	
	801	Mahal Teglinos K1-MG	
	802	Mahal Teglinos K1-PG	
	811	Shaqaduq-S1-A	
	814	Sheikh el Amin	
	817	Soba	
	819	Um Direiwa	
	821	Upper Atbara hinterland	
	822	Upper Atbara Valley	
	824	Wadi Hariq 01/1	
	825	Wadi Hariq 01/4	
	826	Wadi Hariq 97/5	
	827	Wadi Hariq 97/7	
	829	Wadi Howar - 84/1	
	830	Wadi Howar - 84/12-1	
	831	Wadi Howar - 84/13	
	832	Wadi Howar - 84/19	
West. Sahara	837	Sebkha Amtal	*Bos sp.

Zambia	859	Bulila	869	Muozi	*cf.
	865	Nanga	869	Muozi	
	866	Salumano	871	Ruanga Ruin	
Zanzibar	867	Kizimkazi Dimbani	872	Simamwe	
Zimbabwe	868	Mtanye			

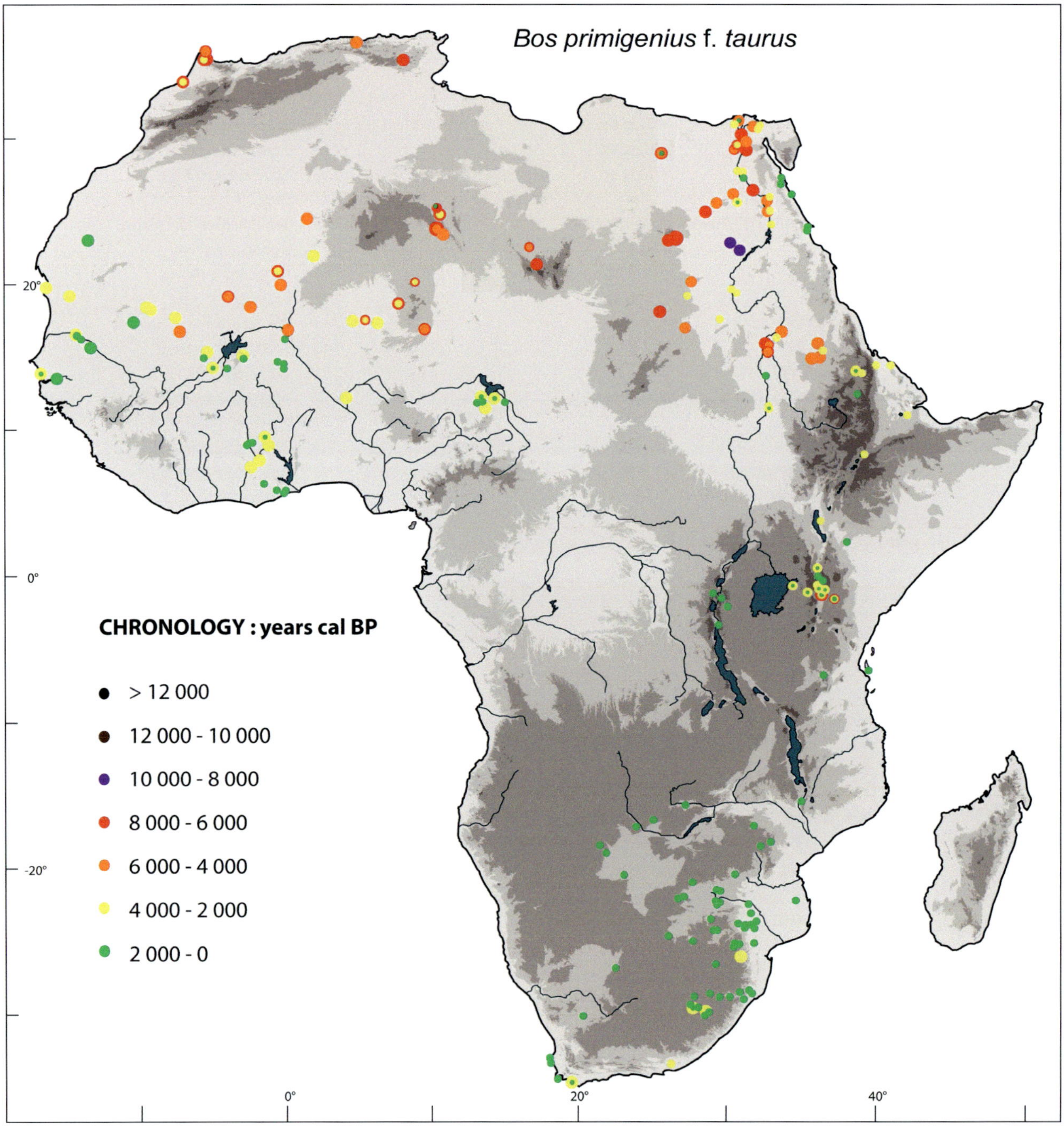

Syncerus caffer

Country	Site N	Site	
Botswana	26	Bosutwe-Mapungubwe	
	27	Bosutwe-Taukome	
	28	Bosutwe-Toutswe	
	29	Bosutwe-Zimbabwe	
	31	Commando Kop	
	32	Divuyu	
	37	Matlapaneng	
	38	Nqoma	
	39	Qogana	
	41	Toteng1-SU1	
	42	Toteng1-SU2	
	42	Toteng1-SU2	*cf.
	43	Toteng1-SU3	
	44	Toteng1-SU4	
Burkina Faso	65	Tin Akof	
Cameroon	67	Abeke	* S.c.nanus
	73	Blé E	
	74	Djaba Hosséré	
	75	Nkang	* S.c.nanus
	76	Shum Laka II c	* S.c.nanus
	77	Shum Laka II d	* S.c.nanus
C.A.R.	80	Lingbangbo	
Chad	82	Enneri Bardagué	
	84	Gabrong-d	
	85	Koyom	
Congo	86	Kawezi	
Egypt	122	Dakhleh Ouasis, Ba	
	184	Sodmein Cave - lev 5 (MP5)	
	191	Toukh	
Ethiopia	216	Moche Borago-1b	
	217	Moche Borago-1c	
	218	Moche Borago-2	
	219	Moche Borago-3	
Ghana	241	Mandara	
	242	Mankuma	
	244	Ntereso	*?
	246	Nyanga	
Kenya	258	Enkapune Ya Muto - Eb 4a	
	259	Enkapune Ya Muto - Eb 4b	
	260	Enkapune Ya Muto - Eb 5a	
	272	Gogo Falls-I	
	273	Gogo Falls-II-III	
	274	Gogo Falls-sq26-29	
	284	Kulchurdo	
	298	Prolonged Drift	
Malawi	352	Malowa-2	
Mali	368	erg Ine Sakane	
	374	Hassi el-Abiod	
	381	Kobadi	
	387	Tessalit	
	388	Tiabel Goudiodié	
	390	Windé Koroji I,III	
	391	Windé Koroji SO	
Mauritania	393	Berouâga	
	406	Khneivissa	
Mozambique	444	Manekweni-4	
Niger	465	Adrar Bous 10	
	473	Gobero-K	
	474	Gobero-T	
	477	Mentes 1	
	478	Mentes 13	
	481	Mentes 4	
	483	Mentes 6	
	484	Mentes 7	
	487	Takene Bawat 1	
	490	Takene Bawat 6	
	495	Tin Ouaffadene	
	497	TMY3	
Nigeria	505	Gajiganna B2	
Rwanda	536	Muganza	* S.c.caffer
Senegal	540	Cubalel-II	
South Africa	564	Boomplass Cave-b	
	565	Boomplass Cave-c	
	568	Buffelskloof RS-a	*?
	570	Buffelskloof RS-c	
	576	Byeneskranskop Cave 1a	
	577	Byeneskranskop Cave 1b	
	578	Byeneskranskop Cave 1c	
	579	Byeneskranskop Cave 1d	
	581	Colwinton Shelter-CW	
	586	Die Kelders	
	589	Doornfontein	
	594	Eiland Salt Works-b	
	595	Eland´s Bay Cave-Al	
	596	Eland´s Bay Cave-Rl	
	600	Eland´s Bay Cave-Wid	
	604	Equus Cave 1A	
	608	Forest Hall	*cf.
	609	Glentyre	
	614	Hapi Pan	
	616	Heuningneskrans-I/O	
	617	Heuningneskrans-R	
	618	Hill X	
	631	Klasies River Mouth-1I	
	632	Klasies River Mouth-1II	
	633	Klasies River Mouth-5II	
	638	Langsdraai	
	640	Le2	
	641	Le6	
	642	Le7a	
	643	Le7b	
	659	Marupale Hill	
	664	Mo8	
	668	Nelson Bay Cave- PW	
	669	Nelson Bay Cave- R	
	670	Nelson Bay Cave- W	
	671	Oakhurst	
	673	Ol20	
	677	Pa8.1	
	685	Ph6	
	688	Pont-Drift-LKA	
	691	Pr34	
	711	Serotwe Hill	
	713	SH16	
	716	Shankare Hill-b	
	717	SK17	
	718	Sk4	
	736	Tortoise cave-b	
	749	Witklip-3	
	750	Witklip-4	
Sudan	754	Abu Darbein	
	757	Aneibis	
	758	Awlad el Imam	

	761	Debbat Bangdit	
	762	Debbat El Eheima	
	764	El Damer	
	770	El Temeyim	
	773	Esh Shaheinab */Homoioceras	
	774	Gash Delta-AA	*?
	777	Geili	
	781	Jebel Umm Marrahi	
	782	Kabbashi	
	783	Kadero	
	786	Khabbashi Haitah	
	787	Khartoum Hopital	
	788	Khasm el Girba - Kassala	
	789	Khasm el Girba - KG14	
	791	Khasm el Girba - KG68	
	797	Mahal Teglinos K1-CG	
	798	Mahal Teglinos K1-EG	
	799	Mahal Teglinos K1-JM	
	800	Mahal Teglinos K1-LG	
	802	Mahal Teglinos K1-PG	
	805	Saggai 1	
	810	Shabona	
	815	Sheikh Mustafa	*cf.
	820	Umm Singid	
	822	Upper Atbara Valley	
Tanzania	835	Mumba	
Zaire	845	Ishango-Z-Post-Em	
	848	Kiantapo	*?
	848	Kiantapo	
	854	Matupi-I	
	855	Matupi-II	
Zambia	861	Leopard's Hill-NII	
	863	Leopard's Hill-PN	
Zimbabwe	870	Redcliff Cave	

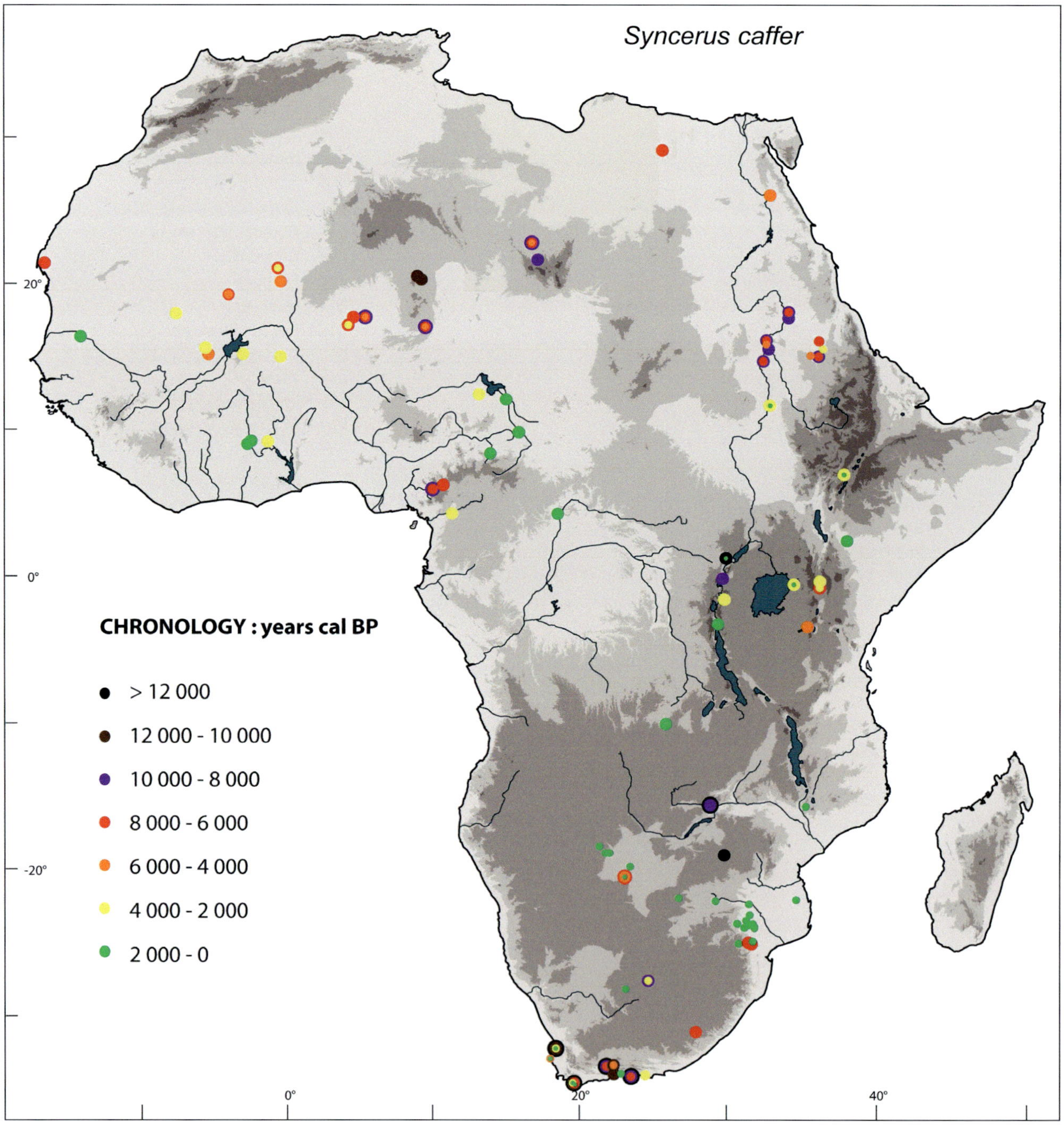

Reduncinae

Kobus sp.

Country	Site N	Site	
Ghana	244	Ntereso	
Mali	364	Dia-Shoma Horizon I	
	365	Dia-Shoma Horizon II	
	388	Tiabel Goudiodié	
Mauritania	400	Dhar Tichitt Site 38	
	406	Khneivissa	
Niger	471	Chin Tafidet	
Sudan	774	Gash Delta-AA	
	810	Shabona	
Zaire	858	Sanga	* leche/vardonii

Kobus kob

Country	Site N	Site	
Algeria	6	Columnata	*aff.
Burkina Faso	51	Cocorba-LSA	
	58	Oursi-IA	
	62	Péntènga 89/1-spits 4-10	
	63	Saouga 94/120	
Cameroon	70	Blé A	
	71	Blé B	
	72	Blé C	
	73	Blé E	
	74	Djaba Hosséré	
	75	Nkang	
Chad	85	Koyom	
Ghana	231	Daboya	
	241	Mandara	
Mali	361	Akumbu-EIA	
	369	Est-Hassi el Abiod	
	372	Gao Saney	
	374	Hassi el-Abiod	
	376	Jenné-Jeno - phase I/II	
	377	Jenné-Jeno - phase III	
	378	Jenné-Jeno - phase IV	
	381	Kobadi	
	382	Kolima Sud-I	
	383	Kolima Sud-II	
	385	Korounkorokalé-III	
	387	Tessalit	
	388	Tiabel Goudiodié	
Niger	465	Adrar Bous 10	
	467	Adrar Bous-T	*aff.
	473	Gobero-K	
Nigeria	514	Kursakata-IA	
	518	Mege I	
	519	Mege II	
	520	Mege III	
	525	Ngala I	
	526	Ngala II	
	527	Ngala IIIa	
	528	Ngala IIIb	
	529	Ngala IV	
Senegal	538	Cubalel-Ia	
	539	Cubalel-Ib	
	540	Cubalel-II	
	541	Cubalel-III	
	544	Sintiou-Bara	
	546	Siouré-Ib	
	547	Siouré-II	
	548	Siouré-IV, V	
	549	Tulel-Fobo	
	551	Waladé-W1/20-25	
Sudan	754	Abu Darbein	
	755	Abu Hugar	*cf.
	757	Aneibis	
	758	Awlad el Imam	
	761	Debbat Bangdit	
	762	Debbat El Eheima	
	764	El Damer	
	765	El Kadada	
	766	El Kenger Middle/East	
	769	El Qala´a	
	770	El Temeyim	
	771	El Zakiab	
	777	Geili	
	781	Jebel Umm Marrahi	
	782	Kabbashi	
	786	Khabbashi Haitah	
	787	Khartoum Hopital	
	805	Saggai 1	
	820	Umm Singid	
Zaire	845	Ishango-Z-Post-Em	
	854	Matupi-I	
	855	Matupi-II	

Kobus leche

Country	Site N	Site	
Botswana	28	Bosutwe-Toutswe	*cf.
	32	Divuyu	
	38	Nqoma	
	42	Toteng1-SU2	*cf.
	45	Toteng3-SU3	*cf.
	49	White Paintings RS-2b-3ab	*cf.
Sudan	810	Shabona	*cf.
Zaire	846	Kakontwe	

Kobus vardonii

Country	Site N	Site
Zaire	848	Kiantapo

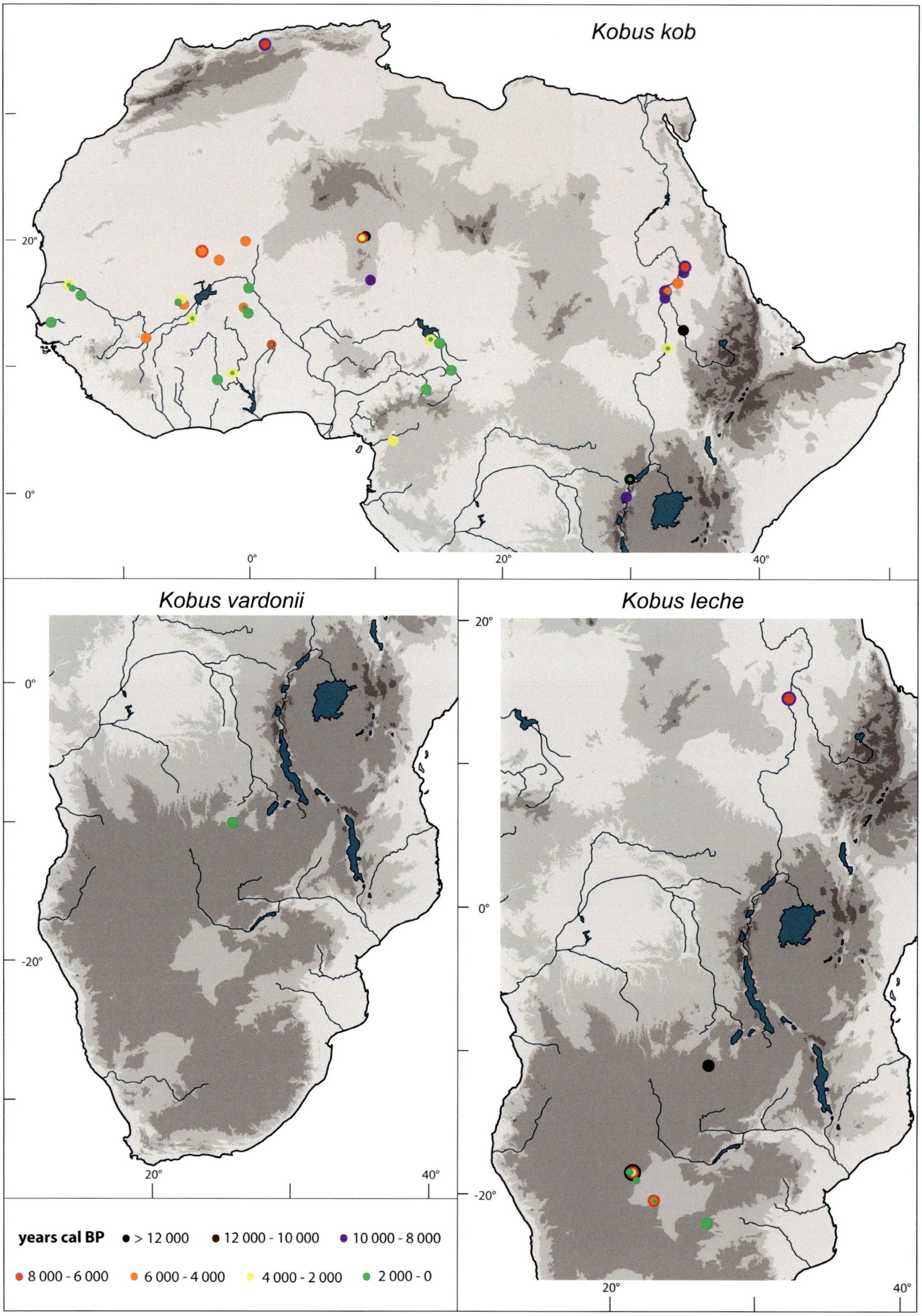
Kobus kob
20°
0°
0°
20°
40°
Kobus vardonii
0°
-20°
20°
40°
Kobus leche
20°
0°
-20°
20°
40°
years cal BP
> 12 000
12 000 - 10 000
10 000 - 8 000
8 000 - 6 000
6 000 - 4 000
4 000 - 2 000
2 000 - 0

Kobus ellipsiprymnus

Country	Site N	Site	
Botswana	27	Bosutwe-Taukome	
	28	Bosutwe-Toutswe	*cf.
	32	Divuyu	
	38	Nqoma	
Cameroon	75	Nkang	
C.A.R.	79	Mondongué	
Chad	85	Koyom	
Congo	86	Kawezi	
Ghana	227	Asantamanso-ASV	
	241	Mandara	
Kenya	273	Gogo Falls-II-III	
	292	Marula Rock Shelter	
	294	Narosura	
	297	Ngenyn	
Malawi	352	Malowa-2	
	354	Malowa-4	
Mali	371	Gao Ancien	
	382	Kolima Sud-I	
	388	Tiabel Goudiodié	
	390	Windé Koroji I,III	
	391	Windé Koroji SO	
Mozambique	442	Manekweni-2-ABC	
	444	Manekweni-4	
Niger	473	Gobero-K	
Senegal	538	Cubalel-Ia	
	540	Cubalel-II	
	541	Cubalel-III	
	546	Siouré-Ib	
	547	Siouré-II	
	549	Tulel-Fobo	
South Africa	594	Eiland Salt Works-b	
	618	Hill X	
	640	Le2	
	641	Le6	
	642	Le7a	
	643	Le7b	
	654	Mamba 1	
	659	Marupale Hill	*cf.
	673	Ol20	
	677	Pa8.1	
	714	Sh27	
	717	SK17	
	720	Sonkoanini Hill	
Sudan	787	Khartoum Hopital	
	789	Khasm el Girba - KG14	
	791	Khasm el Girba - KG68	
	805	Saggai 1	
Tanzania	835	Mumba	
Zaire	843	Ishango-AB	
	844	Ishango-C	
	845	Ishango-Z-Post-Em	

Redunca sp.

Country	Site N	Site	
Botswana	26	Bosutwe-Mapungubwe	
Kenya	258	Enkapune Ya Muto - Eb 4a	
	259	Enkapune Ya Muto - Eb 4b	
	260	Enkapune Ya Muto - Eb 5a	
	261	Enkapune Ya Muto - Eb 5b	
Lesotho	323	Tloutle-pCW	
South Africa	580	Collingham Shelter	
	621	HKLP Va	*cf.
	622	HKLP Vb	*cf.
	622	HKLP Vb	
	663	Mhlwazini Cave	
	701	Rooikrans-CW	
	718	Sk4	
Sudan	774	Gash Delta-AA	
	810	Shabona	
Tanzania	835	Mumba	

Redunca arundinum

Country	Site N	Site	
Botswana	26	Bosutwe-Mapungubwe	
	32	Divuyu	
	37	Matlapaneng	
	38	Nqoma	
	39	Qogana	
	48	White Paintings RS-2ab	*cf.
	48	White Paintings RS-1	
	49	White Paintings RS-2b-3ab	*cf.
Lesotho	321	Tloutle-CWI	
	323	Tloutle-pCW	
South Africa	564	Boomplass Cave-b	
	574	Bushman RS-LSA	
	577	Byeneskranskop Cave 1b	
	578	Byeneskranskop Cave 1c	
	579	Byeneskranskop Cave 1d	
	580	Collingham Shelter	
	589	Doornfontein	
	618	Hill X	
	622	HKLP Vb	
	642	Le7a	
	659	Marupale Hill	*cf.
	667	Ndondondwane	
	668	Nelson Bay Cave- PW	
	669	Nelson Bay Cave- R	
	670	Nelson Bay Cave- W	
	674	Ondini	
	711	Serotwe Hill	
	716	Shankare Hill-b	
	718	Sk4	
	720	Sonkoanini Hill	
	740	Twyfelpoort	
	742	Ventershoek	*cf.
	751	Wosi-early EIA	
	753	Wosi-terminal EIA	
Zaire	848	Kiantapo	
Zimbabwe	870	Redcliff Cave	

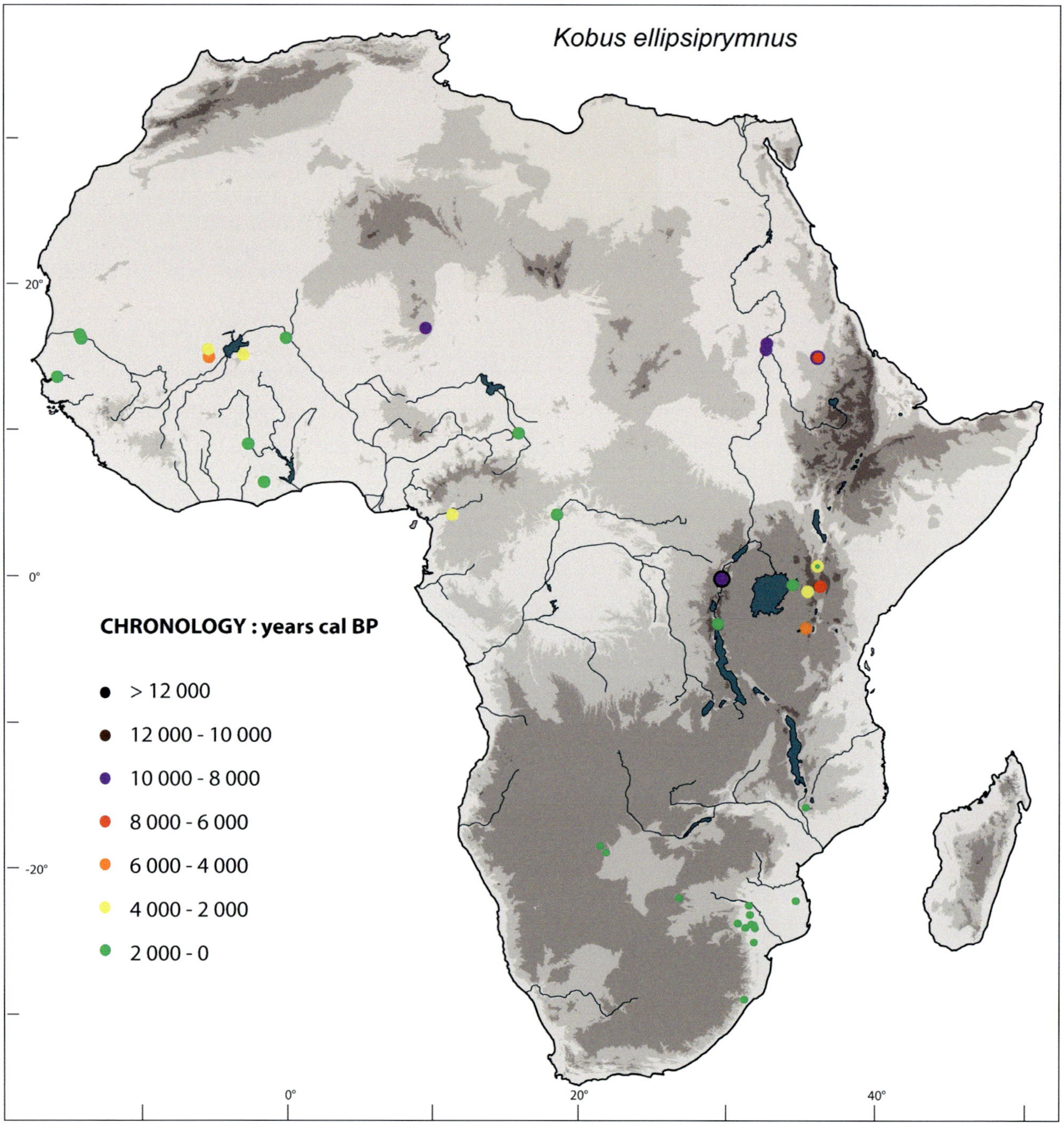

Redunca fulvorufula

Country	Site N	Site
Kenya	272	Gogo Falls-I
	275	GvJm 19-HA
	276	GvJm 19-HC
	277	GvJm 19-TLPG
	278	GvJm 22-LHC
	279	GvJm 22-LPG
	280	GvJm 46-LPG
	282	GvJm 62-LPG
Lesotho	303	Bolahla
	304	Ha Makotoko
	307	Likoaeng-III-V
	308	Likoaeng-VII-IX
	310	Likoaeng-XIII
	312	Liphofung-CW
	313	Liphofung-pCW
	314	Muela
	315	Ntloana Tsoana
	316	Sehonghong-a
	321	Tloutle-CWI
	322	Tloutle-O
	323	Tloutle-pCW
	324	Tloutle-R
South Africa	556	2627 CD7
	560	Blydefontein III-IV
	561	Blydefontein V
	562	Blydefontein VI-VII
	563	Boomplass Cave-a
	564	Boomplass Cave-b
	565	Boomplass Cave-c
	571	Buffelskloof RS-d
	574	Bushman RS-LSA

Country	Site N	Site	
	575	Bushman RS-MSA	
	578	Byeneskranskop Cave 1c	
	580	Collingham Shelter	
	603	Emambeni	
	604	Equus Cave 1A	
	616	Heuningneskrans-I/O	
	617	Heuningneskrans-R	
	619	HKLP Ia	*cf.
	620	HKLP Ib	*cf.
	621	HKLP Va	
	622	HKLP Vb	
	625	Jubilee Shelter-PW	
	626	Jubilee Shelter-W	
	631	Klasies River Mouth-1I	
	632	Klasies River Mouth-1II	
	653	Malwula	
	654	Mamba 1	
	663	Mhlwazini Cave	
	668	Nelson Bay Cave- PW	
	669	Nelson Bay Cave- R	
	670	Nelson Bay Cave- W	
	674	Ondini	
	675	Onink	
	697	Riversmead Shelter I-II	
	698	Riversmead Shelter III-IV	
	699	Riversmead Shelter VI-IX	
	701	Rooikrans-CW	
	702	Rooikrans-pCW	
	704	Rose Cottage Cave-A-Mn	
	705	Rose Cottage Cave-DB2	
	706	Rose Cottage Cave-OIC	
	707	Rose Cottage Cave-WI	
	718	Sk4	
	742	Ventershoek	
	753	Wosi-terminal EIA	
Zimbabwe	870	Redcliff Cave	

Redunca redunca

Country	Site N	Site	
Algeria	12	In Guezzam South	
	16	Meniet	
	5	Amekni	
Burkina Faso	51	Cocorba-LSA	
	53	Kissi 40	
	56	Oursi Village-LIA	
	57	Oursi Village-MIA	
	58	Oursi-IA	
Cameroon	70	Blé A	
	71	Blé B	
	72	Blé C	
	73	Blé E	
Chad	85	Koyom	
Ghana	223	Anyinam-Area A-B-C	
	226	Asantamanso-ASIV	
	227	Asantamanso-ASV	
	231	Daboya	*?
	233	Esiease - A-C-D	
	235	Esiease -recent	
Kenya	272	Gogo Falls-I	
	273	Gogo Falls-II-III	
	292	Marula Rock Shelter	
	293	Naivasha Railway Shelter	
	299	Salasun-LSA	
Libya	332	Murzuq	
Mali	361	Akumbu-EIA	
	368	erg Ine Sakane	
	369	Est-Hassi el Abiod	
	372	Gao Saney	
	374	Hassi el-Abiod	
	376	Jenné-Jeno - phase I/II	
	377	Jenné-Jeno - phase III	
	380	Karkarichinkat N-S	
	382	Kolima Sud-I	
	383	Kolima Sud-II	
	388	Tiabel Goudiodié	
	391	Windé Koroji SO	
Mauritania	393	Berouâga	
	402	DN4-BKH-S5	
Morocco	430	Kaf-taht-el-Ghar-B	
	434	Kaf-taht-el-Ghar-F	
	435	Kaf-taht-el-Ghar-G	
Niger	467	Adrar Bous-T	*cf.
	471	Chin Tafidet	
	473	Gobero-K	
	474	Gobero-T	
	488	Takene Bawat 2	*cf.
	495	Tin Ouaffadene	
Nigeria	501	Dorota	
	502	Elkido North	
	503	Gajiganna A	
	504	Gajiganna B1-lower	
	505	Gajiganna B2	
	506	Gajiganna C	
	510	Gilgila-Iic	
	511	Kariari	
	525	Ngala I	
Senegal	539	Cubalel-Ib	
	540	Cubalel-II	
	541	Cubalel-III	
	546	Siouré-Ib	
	547	Siouré-II	
	548	Siouré-IV, V	
	551	Waladé-W1/20-25	
Sudan	754	Abu Darbein	
	756	Abu Tabari 02/1	
	757	Aneibis	
	758	Awlad el Imam	
	764	El Damer	
	767	El Mahalab	
	772	Erg of Ennedi	
	781	Jebel Umm Marrahi	
	787	Khartoum Hopital	
	788	Khasm el Girba - Kassala	
	789	Khasm el Girba - KG14	
	814	Sheikh el Amin	
	815	Sheikh Mustafa	
Zaire	843	Ishango-AB	
	845	Ishango-Z-Post-Em	
	854	Matupi-I	
	855	Matupi-II	

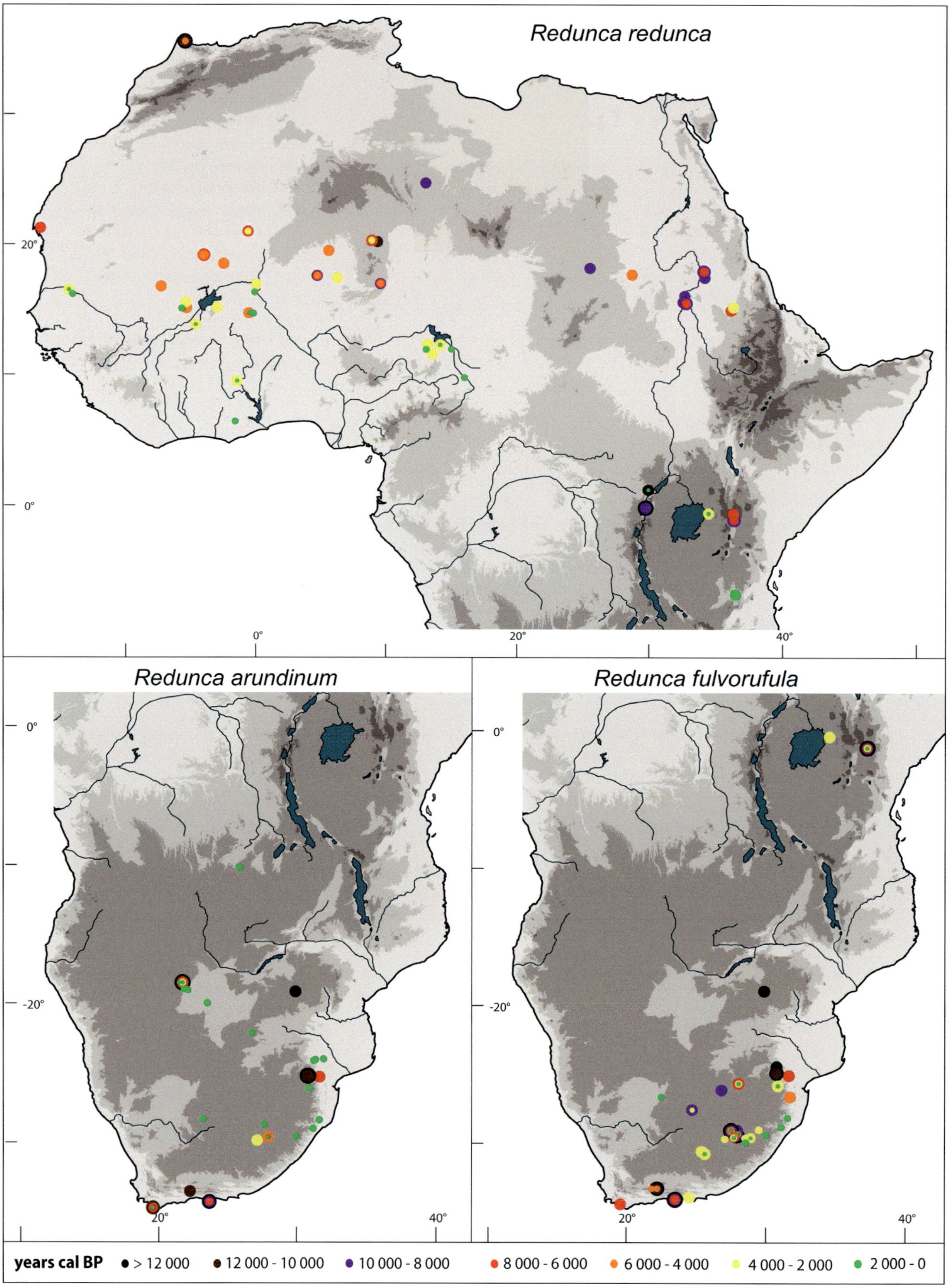
Redunca redunca
20°
0°
0°
20°
40°
Redunca arundinum
0°
-20°
20°
40°
Redunca fulvorufula
0°
-20°
20°
40°
years cal BP
> 12 000
12 000 - 10 000
10 000 - 8 000
8 000 - 6 000
6 000 - 4 000
4 000 - 2 000
2 000 - 0

Pelea capreolus

Country	Site N	Site	
Lesotho	304	Ha Makotoko	
	307	Likoaeng-III-V	
	308	Likoaeng-VII-IX	
	309	Likoaeng-XI	
	310	Likoaeng-XIII	
	312	Liphofung-CW	
	313	Liphofung-pCW	
	314	Muela	
	315	Ntloana Tsoana	
	316	Sehonghong-a	
	321	Tloutle-CWI	
	322	Tloutle-O	
South Africa	559	Blydefontein I-II	
	560	Blydefontein III-IV	
	561	Blydefontein V	
	562	Blydefontein VI-VII	
	563	Boomplass Cave-a	
	564	Boomplass Cave-b	
	565	Boomplass Cave-c	
	566	Boomplass Cave-d	
	574	Bushman RS-LSA	
	579	Byeneskranskop Cave 1d	
	580	Collingham Shelter	
	581	Colwinton Shelter-CW	
	582	Colwinton Shelter-pCW	
	586	Die Kelders	
	604	Equus Cave 1A	
	617	Heuningneskrans-R	
	620	HKLP Ib	
	622	HKLP Vb	
	626	Jubilee Shelter-W	*cf.
	631	Klasies River Mouth-1I	
	632	Klasies River Mouth-1II	
	633	Klasies River Mouth-5II	
	636	Klipspring	
	638	Langsdraai	
	639	Large Rock Shelter	
	650	Magogo-1	
	652	Magogo-3	
	661	Mgoduyanuka	
	663	Mhlwazini Cave	
	667	Ndondondwane	
	668	Nelson Bay Cave- PW	
	669	Nelson Bay Cave- R	
	670	Nelson Bay Cave- W	
	674	Ondini	
	688	Pont-Drift-LKA	*cf.
	694	Ravenscraig Shelter-pCW	
	695	Ravenscraig Shelter-R	
	698	Riversmead Shelter III-IV	
	699	Riversmead Shelter VI-IX	
	701	Rooikrans-CW	
	702	Rooikrans-pCW	
	704	Rose Cottage Cave-A-Mn	
	706	Rose Cottage Cave-OIC	
	707	Rose Cottage Cave-WI	
	719	Smitswinkelbaai Cave	
	740	Twyfelpoort	
South Africa	742	Ventershoek	

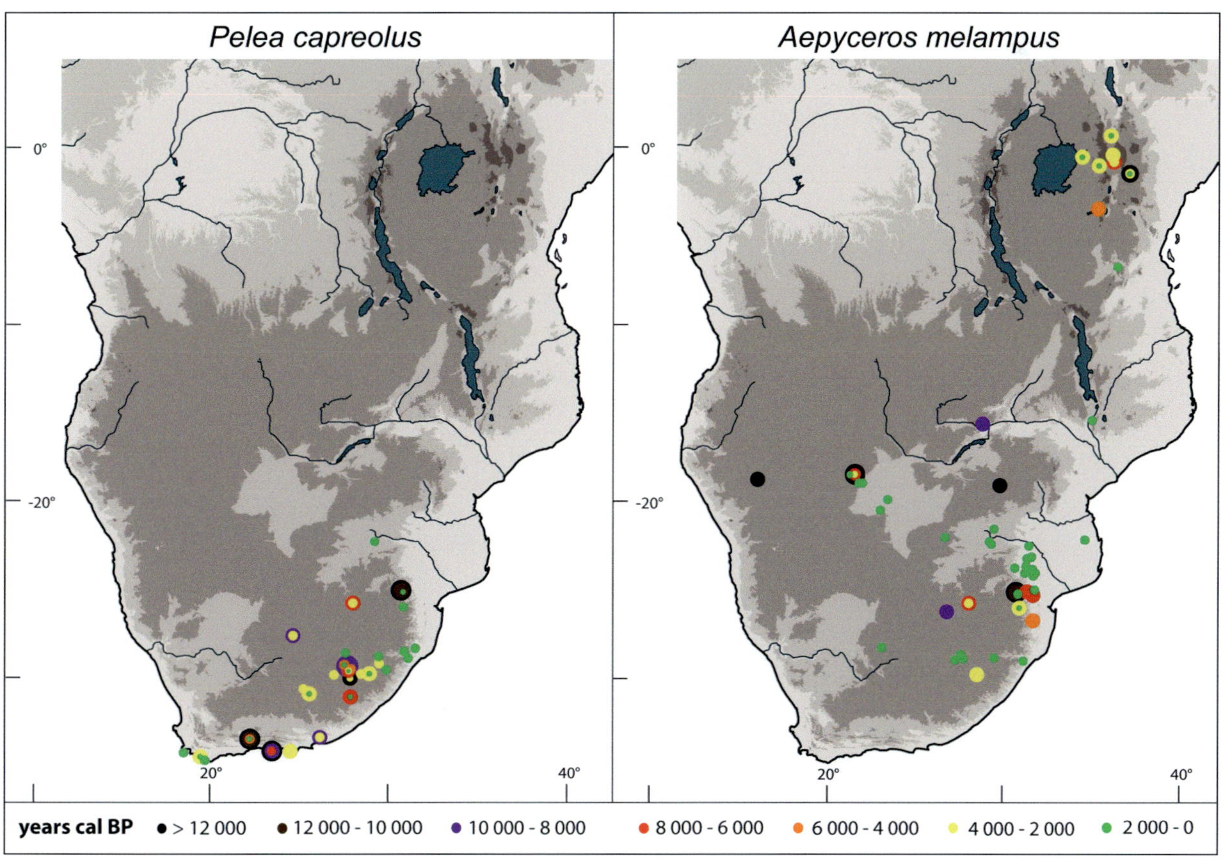

Aepycerotinae

Aepyceros melampus

Country	Site N	Site	
Botswana	26	Bosutwe-Mapungubwe	
	27	Bosutwe-Taukome	
	28	Bosutwe-Toutswe	
	28	Bosutwe-Toutswe	*cf.
	29	Bosutwe-Zimbabwe	
	32	Divuyu	
	37	Matlapaneng	
	38	Nqoma	
	39	Qogana	
	46	Toteng3-SU4	cf.
	49	White Paintings RS-2b-3ab	
Kenya	259	Enkapune Ya Muto - Eb 4b	
	260	Enkapune Ya Muto - Eb 5a	
	273	Gogo Falls-II-III	
	274	Gogo Falls-sq26-29	
	278	GvJm 22-LHC	
	280	GvJm 46-LPG	
	293	Naivasha Railway Shelter	
	296	Ngamuriak	
	297	Ngenyn	
	298	Prolonged Drift	
Lesotho	314	Muela	
Malawi	356	Matope Court-EIA	
	357	Matope Court-IA	
	358	Matope Court-LIA	
Mozambique	442	Manekweni-2-ABC	
	443	Manekweni-3	
	444	Manekweni-4	
	445	Manekweni-5	
Namibia	460	Oshigambo Peninsula	
South Africa	574	Bushman RS-LSA	
	589	Doornfontein	
	593	Eiland Salt Works-a	
	594	Eiland Salt Works-b	
	603	Emambeni	
	616	Heuningneskrans-I/O	
	617	Heuningneskrans-R	
	618	Hill X	
	619	HKLP Ia	*cf.
	620	HKLP Ib	
	620	HKLP Ib	
	622	HKLP Vb	*cf.
	622	HKLP Vb	
	623	Icon	
	626	Jubilee Shelter-W	
	638	Langsdraai	
	640	Le2	
	641	Le6	
	642	Le7a	
	643	Le7b	
	649	Ma4	
	653	Malwula	
	654	Mamba 1	
	655	Mapotini Hill	
	657	Mapungubwe-K2/TS3	
	659	Marupale Hill	
	660	Mauermanshoek	
	661	Mgoduyanuka	
	664	Mo8	
	666	Muhululu Hill	
	673	Ol20	
	677	Pa8.1	
	684	Ph1	
	685	Ph6	
	686	Ph9	
	688	Pont-Drift-LKA	
	691	Pr34	
	703	Roosfontein	
	709	Sebatini Hill	
	710	Selongwe Hill	
	711	Serotwe Hill	
	712	SH14a	
	713	SH16	
	715	Shankare Hill-a	
	716	Shankare Hill-b	
	717	SK17	
	718	Sk4	
	720	Sonkoanini Hill	
	732	Ti6	
	739	Tsh1	
	740	Twyfelpoort	
	753	Wosi-terminal EIA	
Tanzania	835	Mumba	
Zambia	861	Leopard's Hill-NII	
Zimbabwe	870	Redcliff Cave	
	872	Simamwe	

Alcelaphinae

Country	Site N	Site
Botswana	26	Bosutwe-Mapungubwe
	27	Bosutwe-Taukome
	28	Bosutwe-Toutswe
	48	White Paintings RS-1
	48	White Paintings RS-2ab
	49	White Paintings RS-2b-3ab
Burkina Faso	53	Kissi 40
	58	Oursi-IA
	61	Péntènga 89/1-spits 15-21
	63	Saouga 94/120
Chad	85	Koyom
Kenya	247	Akira-N
	248	Akira-PIA
	249	Akira-SPN
	255	Ele Bor A-C
	257	Engaruani
	272	Gogo Falls-I
	273	Gogo Falls-II-III
	273	Gogo Falls-II-III
	274	Gogo Falls-sq26-29
	277	GvJm 19-TLPG *small, extinct
	279	GvJm 22-LPG *small, extinct
	280	GvJm 46-LPG *small, extinct
	282	GvJm 62-LPG *small, extinct
	283	GvJm 62-LPNG *small, extinct
	284	Kulchurdo
	294	Narosura
	299	Salasun-LSA
	301	Salasun-PN
Lesotho	304	Ha Makotoko
	305	Lequetsoana
	312	Liphofung-CW
	315	Ntloana Tsoana
	316	Sehonghong-a
	321	Tloutle-CWI
	323	Tloutle-pCW
Mali	376	Jenné-Jeno - phase I/II
	380	Karkarichinkat N-S
	383	Kolima Sud-II
Mauritania	394	Chami-A2
Nigeria	498	Bukarkurari-I
	505	Gajiganna B2
	510	Gilgila-Iic
	512	Kelumeri
Rhodesia	534	Mabveni
Senegal	540	Cubalel-II
	546	Siouré-Ib
South Africa	559	Blydefontein I-II
	560	Blydefontein III-IV
	561	Blydefontein V
	562	Blydefontein VI-VII
	563	Boomplass Cave-a
	564	Boomplass Cave-b
	565	Boomplass Cave-c
	568	Buffelskloof RS-a
	569	Buffelskloof RS-b
	570	Buffelskloof RS-c
	576	Byeneskranskop Cave 1a
	577	Byeneskranskop Cave 1b
	578	Byeneskranskop Cave 1c
	579	Byeneskranskop Cave 1d
	580	Collingham Shelter
	581	Colwinton Shelter-CW
	604	Equus Cave 1A
	610	Green Elliott Shelter I-III
	611	Green Elliott Shelter IV-VI
	616	Heuningneskrans-I/O
	617	Heuningneskrans-R
	625	Jubilee Shelter-PW
	626	Jubilee Shelter-W
	639	Large Rock Shelter
	641	Le6
	653	Malwula
	663	Mhlwazini Cave
	676	Oudepost I
	686	Ph9
	694	Ravenscraig Shelter-pCW
	695	Ravenscraig Shelter-R
	697	Riversmead Shelter I-II
	698	Riversmead Shelter III-IV
	699	Riversmead Shelter VI-IX
	701	Rooikrans-CW
	702	Rooikrans-pCW
	704	Rose Cottage Cave-A-Mn
	705	Rose Cottage Cave-DB2
	707	Rose Cottage Cave-WI
	717	SK17
	742	Ventershoek
Sudan	774	Gash Delta-AA
	781	Jebel Umm Marrahi
	782	Kabbashi
	787	Khartoum Hopital
	788	Khasm el Girba - Kassala
	789	Khasm el Girba - KG14
	790	Khasm el Girba - KG15-16
	791	Khasm el Girba - KG68
	812	Shaqaduq-S1-B
	813	Shaqaduq-S21
Zaire	843	Ishango-AB
	844	Ishango-C
	845	Ishango-Z-Post-Em
	855	Matupi-II
Zambia	863	Leopard's Hill-PN
Zimbabwe	870	Redcliff Cave

Alcelaphus buselaphus

Country	Site N	Site	
Algeria	13	Kef Zoura D	
	15	Medjez II	
	18	Oued Telidjene A	
	2	A-88-28	
	4	Ain Mistehayia	
	5	Amekni	
	6	Columnata	
	7	Damous el Ahmar	
	9	Dra-Mta-el-Ma-el-Abiod	
Botswana	26	Bosutwe-Mapungubwe	
	27	Bosutwe-Taukome	*cf.
	28	Bosutwe-Toutswe	*cf.
	28	Bosutwe-Toutswe	*cf.
	43	Toteng1-SU3	*cf.
Cameroon	74	Djaba Hosséré	
Egypt	121	Dakhleh Oasis, SMU	
	122	Dakhleh Ouasis, Ba	
	124	Dakhleh Ouasis,OK	
	129	El Abadiya 2 North sector	
	138	Elkab-Epipaleo	
	145	Fayoum G1-H1	
	147	Fayoum G5-G4	
	148	Fayoum Site 1	
	149	Fayoum Site 2	
	150	Fayoum Site 3	
	151	Fayoum Site 4	
	152	Fayoum Site 5	
	155	Hierakonpolis HK6-Naq IIA-B	
	162	Kom Ombo	
	166	Makhadma 2	
	167	Makhadma 4	
	168	Merimde Benisâlame	
	174	Qantir	
	183	Shuwikhat 1	
	188	Tell el-Dab´a - b	
	201	Wadi Kubbaniya - E-78-3	
	202	Wadi Kubbaniya - E-78-4	
	203	Wadi Kubbaniya - E-81-1	
Ethiopia	209	Gademotta	*cf.
Ghana	241	Mandara	
Kenya	247	Akira-N	
	249	Akira-SPN	
	250	Causeway Site	
	251	Crescent Island main	
	259	Enkapune Ya Muto - Eb 4b	
	260	Enkapune Ya Muto - Eb 5a	
	272	Gogo Falls-I	
	273	Gogo Falls-II-III	
	275	GvJm 19-HA	
	276	GvJm 19-HC	
	277	GvJm 19-TLPG	
	278	GvJm 22-LHC	
	279	GvJm 22-LPG	
	280	GvJm 46-LPG	
	282	GvJm 62-LPG	
	283	GvJm 62-LPNG	
	292	Marula Rock Shelter	
	293	Naivasha Railway Shelter	
	297	Ngenyn	
	298	Prolonged Drift	
	299	Salasun-LSA	
Lesotho	314	Muela	
	316	Sehonghong-a	
	321	Tloutle-CWI	
	322	Tloutle-O	
	323	Tloutle-pCW	
Libya	325	Et Tera	
	327	Hauah Fteah-Iberom	
	328	Hauah Fteah-Lib-caps	
	329	Hauah Fteah-Neolithic	
	332	Murzuq	
	336	Uan Afuda-US1-3	
Mali	360	Akumbu-AK3	
	369	Est-Hassi el Abiod	
	374	Hassi el-Abiod	
	387	Tessalit	
	391	Windé Koroji SO	
Mauritania	394	Chami-A2	
	406	Khneivissa	
Morocco	419	Bou Saria A	
	420	Bou Saria B	
	421	Bou Saria C-D	
	422	El Harhoura 1	
	423	El Harhoura 2	
	424	El Mnasra	
	426	Ghar-Khal IIIb-V	
	427	Ghar-Khal II-IIIa	
	431	Kaf-taht-el-Ghar-C	
	432	Kaf-taht-el-Ghar-D	
	433	Kaf-taht-el-Ghar-E	
	434	Kaf-taht-el-Ghar-F	
	435	Kaf-taht-el-Ghar-G	
	437	Kehf-el-Baroud	
	438	Kehf-el-Baroud-aven	
Namibia	455	Geduld-LIA	
Niger	465	Adrar Bous 10	
	467	Adrar Bous-T	
	474	Gobero-T	
Niger	495	Tin Ouaffadene	
Senegal	539	Cubalel-Ib	
	540	Cubalel-II	
	546	Siouré-Ib	
	552	Waladé-W2	
South Africa	580	Collingham Shelter	
	586	Die Kelders	
	595	Eland´s Bay Cave-AI	
	596	Eland´s Bay Cave-RI	
	600	Eland´s Bay Cave-Wid	
	602	Eland´s Bay Open-b	
	621	HKLP Va	*cf.
	624	Jubilee Shelter-O	
	625	Jubilee Shelter-PW	
	626	Jubilee Shelter-W	
	628	Kasteelberg B	
	631	Klasies River Mouth-1I	
	632	Klasies River Mouth-1II	
	644	Leliehoek	
	660	Mauermanshoek	
	700	Rooiels	*cf.
	702	Rooikrans-pCW	
	704	Rose Cottage Cave-A-Mn	
	705	Rose Cottage Cave-DB2	
	706	Rose Cottage Cave-OIC	
	707	Rose Cottage Cave-WI	

	719	Smitswinkelbaai Cave
	730	Tandjesberg
	740	Twyfelpoort
	744	Voëlvlei
Sudan	765	El Kadada
	767	El Mahalab
	790	Khasm el Girba - KG15-16
	814	Sheikh el Amin
	815	Sheikh Mustafa
Zaire	844	Ishango-C

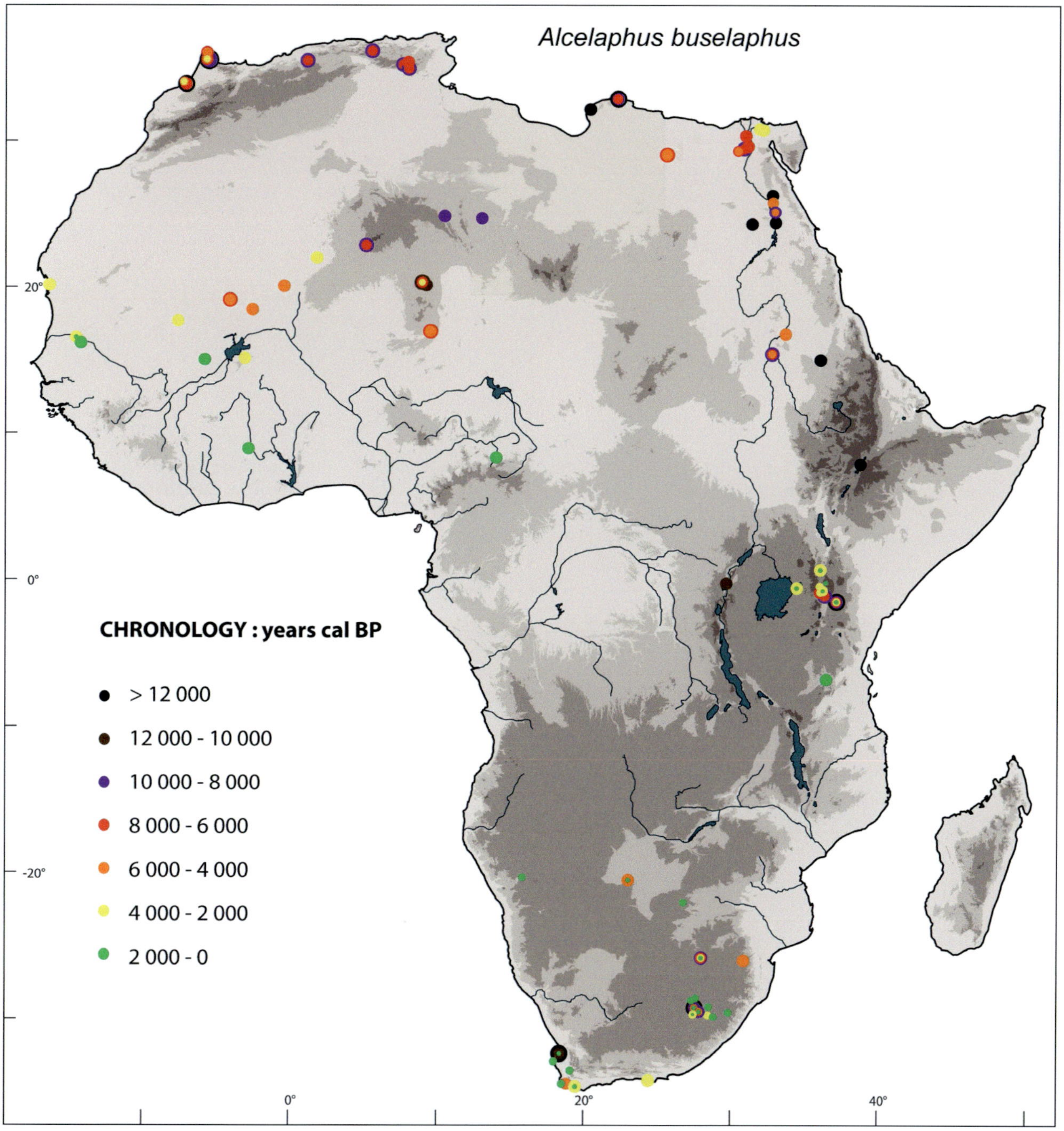

Alcelaphus caama

Country	Site N	Site	
South Africa	609	Glentyre	
	668	Nelson Bay Cave- PW	
	669	Nelson Bay Cave- R	*cf.

Alcelaphus lichtensteinii

Country	Site N	Site
Malawi	352	Malowa-2
Mozambique	442	Manekweni-2-ABC
	444	Manekweni-4
	445	Manekweni-5
Zaire	848	Kiantapo

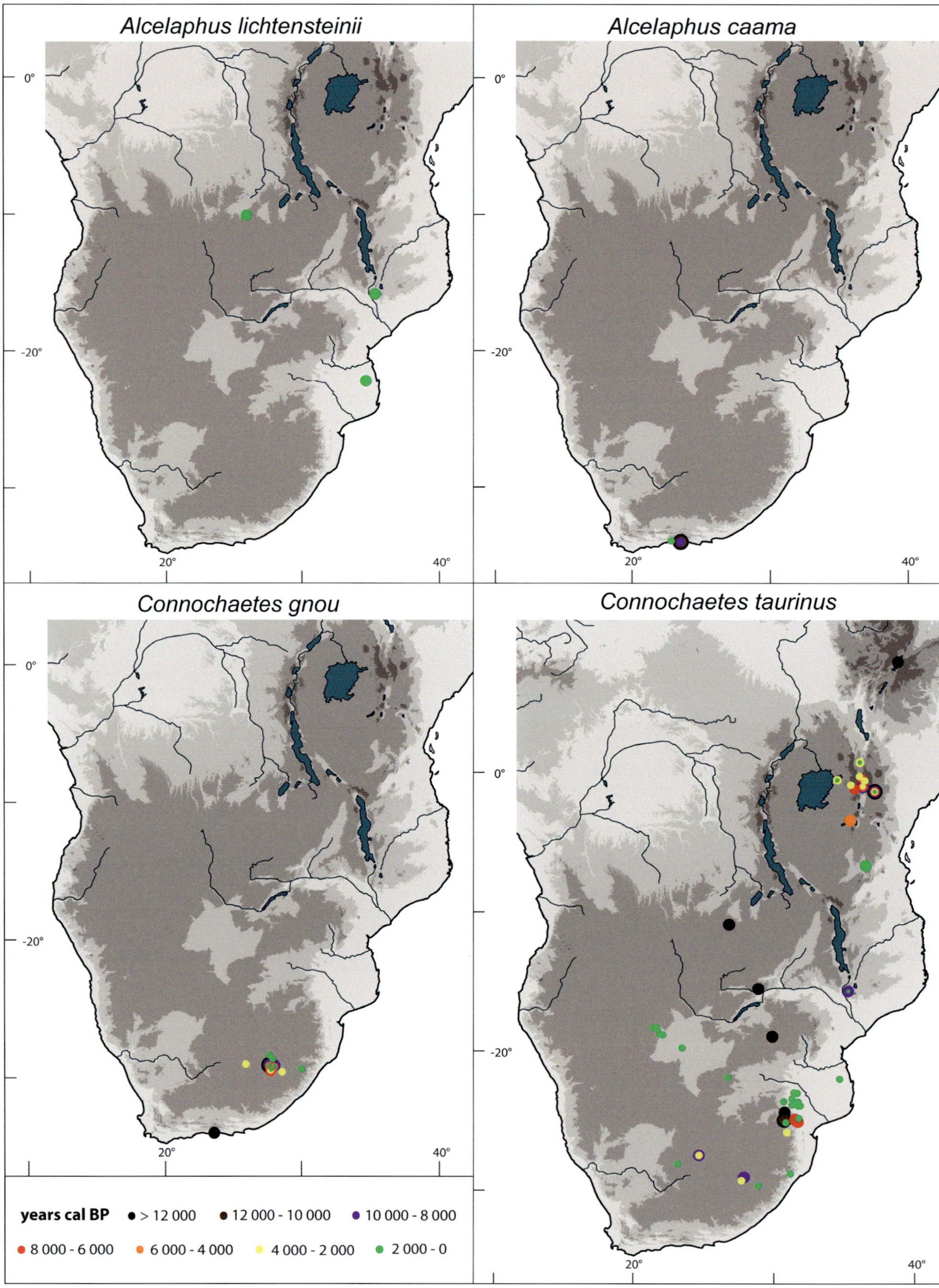

Connochaetes sp.

Country	Site N	Site	
Morocco	422	El Harhoura 1	*C. prognu
	624	Jubilee Shelter-O	
	625	Jubilee Shelter-PW	
	626	Jubilee Shelter-W	
	644	Leliehoek	

Connochaetes taurinus

Country	Site N	Site	
Botswana	26	Bosutwe-Mapungubwe	
	27	Bosutwe-Taukome	
	28	Bosutwe-Toutswe	
	28	Bosutwe-Toutswe	
	29	Bosutwe-Zimbabwe	
	32	Divuyu	
	37	Matlapaneng	
	38	Nqoma	
	39	Qogana	
	48	White Paintings RS-1	
Ethiopia	209	Gademotta	
Kenya	251	Crescent Island main	
	257	Engaruani	
	257	Engaruani	
	272	Gogo Falls-I	
	273	Gogo Falls-II-III	
	274	Gogo Falls-sq26-29	
	275	GvJm 19-HA	
	276	GvJm 19-HC	
	277	GvJm 19-TLPG	
	278	GvJm 22-LHC	
	279	GvJm 22-LPG	
	280	GvJm 46-LPG	
	282	GvJm 62-LPG	
	283	GvJm 62-LPNG	
	292	Marula Rock Shelter	
	293	Naivasha Railway Shelter	
	294	Narosura	
	297	Ngenyn	
	298	Prolonged Drift	
	299	Salasun-LSA	
	301	Salasun-PN	
Lesotho	305	Lequetsoana	
	315	Ntloana Tsoana	*cf.
	316	Sehonghong-a	
Malawi	353	Malowa-3	
	354	Malowa-4	
	355	Malowa-5	
Mozambique	442	Manekweni-2-ABC	
	444	Manekweni-4	
South Africa	574	Bushman RS-LSA	
	575	Bushman RS-MSA	
	589	Doornfontein	
	593	Eiland Salt Works-a	
	594	Eiland Salt Works-b	
	604	Equus Cave 1A	
	618	Hill X	
	619	HKLP Ia	
	636	Klipspring	
	640	Le2	
	641	Le6	
	642	Le7a	
	643	Le7b	
	648	Ma38	
	655	Mapotini Hill	
	659	Marupale Hill	
	664	Mo8	
	673	Ol20	
	685	Ph6	
	686	Ph9	
	691	Pr34	
	709	Sebatini Hill	
	712	SH14a	
	713	SH16	
	714	Sh27	
	717	SK17	
	718	Sk4	
	731	Ti3	
	732	Ti6	
	739	Tsh1	
	753	Wosi-terminal EIA	
Tanzania	835	Mumba	
Zaire	846	Kakontwe	*cf.
Zambia	860	Leopard's Hill-NI	
	863	Leopard's Hill-PN	
Zimbabwe	870	Redcliff Cave	

Connochaetes gnou

Country	Site N	Site	
Lesotho	304	Ha Makotoko	
	312	Liphofung-CW	
	313	Liphofung-pCW	
	314	Muela	
	315	Ntloana Tsoana	
	321	Tloutle-CWI	
	323	Tloutle-pCW	
South Africa	580	Collingham Shelter	
	585	Deelpan	
	669	Nelson Bay Cave- R	*cf.
	701	Rooikrans-CW	
	702	Rooikrans-pCW	
	703	Roosfontein	
	704	Rose Cottage Cave-A-Mn	
	705	Rose Cottage Cave-DB2	
	707	Rose Cottage Cave-WI	
	740	Twyfelpoort	
	745	Westbury Rock Shelter	

Damaliscus dorcas

Country	Site N	Site	
Botswana	27	Bosutwe-Taukome	*cf.
	28	Bosutwe-Toutswe	
Lesotho	312	Liphofung-CW	
	314	Muela	
	321	Tloutle-CWI	
	323	Tloutle-pCW	
	324	Tloutle-R	
South Africa	577	Byeneskranskop Cave 1b	*cf.
	578	Byeneskranskop Cave 1c	*cf.
	580	Collingham Shelter	
	581	Colwinton Shelter-CW	
	585	Deelpan	
	604	Equus Cave 1A	
	611	Green Elliott Shelter IV-VI	*?
	622	HKLP Vb	
	625	Jubilee Shelter-PW	

Country	Site N	Site	
	626	Jubilee Shelter-W	
	644	Leliehoek	
	669	Nelson Bay Cave- R	*cf.
	701	Rooikrans-CW	
	702	Rooikrans-pCW	
	704	Rose Cottage Cave-A-Mn	
	705	Rose Cottage Cave-DB2	
	707	Rose Cottage Cave-WI	
	742	Ventershoek	
Zaire	846	Kakontwe	*cf.
Zimbabwe	870	Redcliff Cave	

Damaliscus lunatus

Country	Site N	Site	
Kenya	273	Gogo Falls-II-III	
	274	Gogo Falls-sq26-29	
	270	GaJi2	
Mali	388	Tiabel Goudiodié	
Mauritania	396	Dhar Tichitt- latest phase	*cf.
	399	Dhar Tichitt- post beach	*cf.
	412	Tichitt - Arriane phase	
Namibia	460	Oshigambo Peninsula	
Senegal	540	Cubalel-II	
	541	Cubalel-III	
	543	Ogo region	*?
	548	Siouré-IV, V	
South Africa	618	Hill X	
	642	Le7a	
	685	Ph6	
	710	Selongwe Hill	
	716	Shankare Hill-b	
	717	SK17	
	718	Sk4	
	739	Tsh1	
Sudan	754	Abu Darbein	
	756	Abu Tabari 02/1	
	757	Aneibis	
	761	Debbat Bangdit	
	762	Debbat El Eheima	
	764	El Damer	
	765	El Kadada	
	771	El Zakiab	
	777	Geili	
	805	Saggai 1	
Zaire	843	Ishango-AB	
	844	Ishango-C	
	845	Ishango-Z-Post-Em	

Damaliscus pygargus

Country	Site N	Site
Lesotho	305	Lequetsoana
South Africa	584	De Hoop
	730	Tandjesberg
	740	Twyfelpoort

Megalotragus sp.

Country	Site N	Site
South Africa	669	Nelson Bay Cave- R

Megalotragus priscus

Country	Site N	Site
South Africa	705	Rose Cottage Cave-DB2

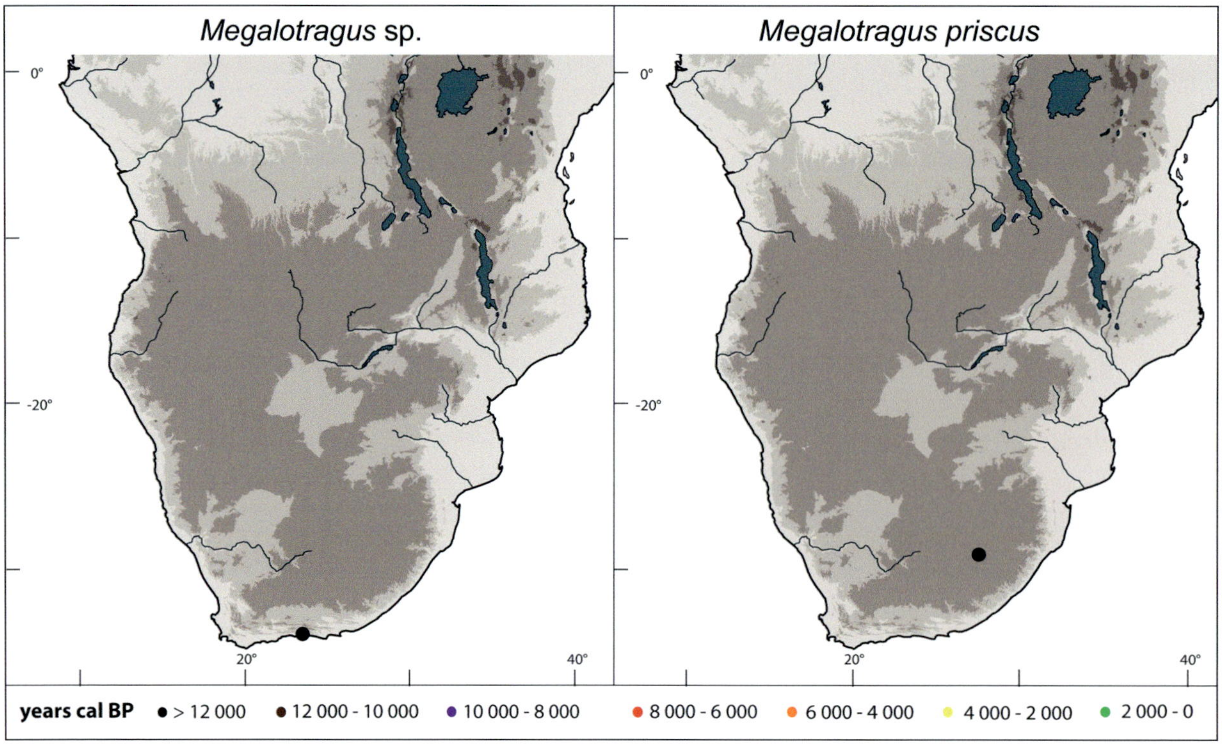

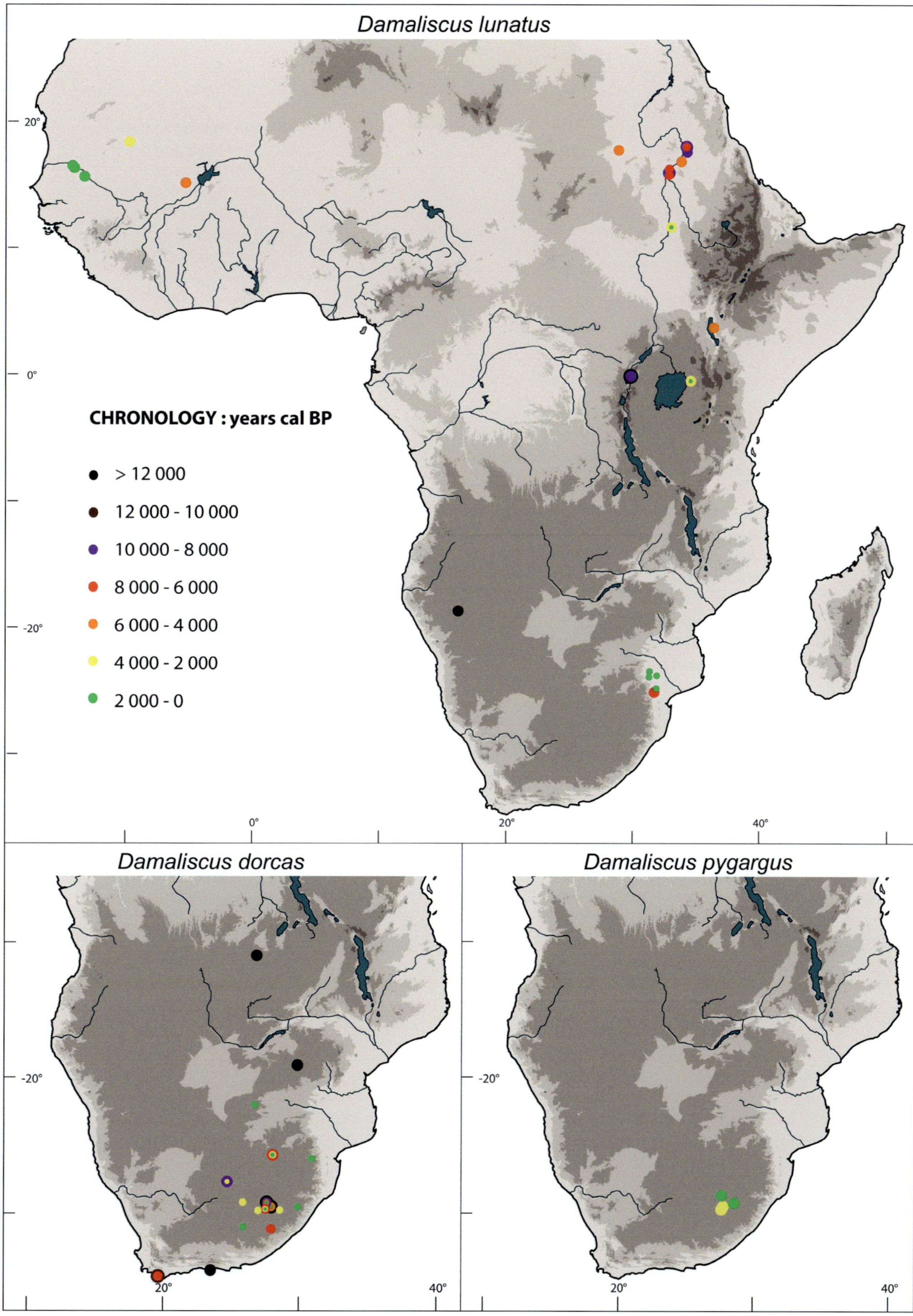
Damaliscus lunatus
20°
0°
-20°
CHRONOLOGY : years cal BP
> 12 000
12 000 - 10 000
10 000 - 8 000
8 000 - 6 000
6 000 - 4 000
4 000 - 2 000
2 000 - 0
0°
20°
40°
Damaliscus dorcas
-20°
20°
40°
Damaliscus pygargus
-20°
20°
40°

Caprinae

Ammotragus lervia

Country	Site N	Site	
Algeria	10	Grotte Capeletti	*?
	13	Kef Zoura D	
	15	Medjez II	
	20	Tamar Hat	
	21	Ti-n-Hanakaten	
	4	Ain Mistehayia	
	5	Amekni	
	6	Columnata	
	7	Damous el Ahmar	*?
	8	Djurdjura	
	9	Dra-Mta-el-Ma-el-Abiod	
Chad	83	Gabrong-b	
Egypt	100	Adaima-habitat	
	107	Berenike 95/97Late Roman	
	153	Gilf Kebir	
	162	Kom Ombo	
	193	Tree Shelter-3	
	194	Tree Shelter-4	
	195	Tree Shelter-5	
	199	Wadi el Bakht	
	99	Adaima-habitat	
Libya	325	Et Tera	
	326	Hauah Fteah- Historic	
	327	Hauah Fteah-Iberom	
	328	Hauah Fteah-Lib-caps	
	329	Hauah Fteah-Neolithic	
	333	Ti-n-Thora East	
	335	Ti-n-Thora Two Caves	
	336	Uan Afuda-US1-3	
	337	Uan Afuda-US4-7	
	338	Uan Muhuggiag (LA)	
	339	Uan Muhuggiag (LP)	
	340	Uan Muhuggiag (MP)	
	342	Uan Tabu-14-20	
	344	Uan Telocat (LP)	
Morocco	419	Bou Saria A	
	420	Bou Saria B	
	421	Bou Saria C-D	
	425	Ghar-Khal I	
	426	Ghar-Khal IIIb-V	
	427	Ghar-Khal II-IIIa	
	428	Ghar-Khal-Ib	
	430	Kaf-taht-el-Ghar-B	
	431	Kaf-taht-el-Ghar-C	
	432	Kaf-taht-el-Ghar-D	
	433	Kaf-taht-el-Ghar-E	
	434	Kaf-taht-el-Ghar-F	
	435	Kaf-taht-el-Ghar-G	
	436	Kehf el Hammar	
	437	Kehf-el-Baroud	
	438	Kehf-el-Baroud-aven	
Sudan	794	Laqiya area - 82/31	

Capra ibex

Country	Site N	Site	
Egypt	106	Berenike 95/9 Early Roman	
	137	Elephantine-Satet temple	
	164	Maadi	*C.i.nubiana
	170	Mons Claudianus	
	95	Abu Sha´ar-4C	
Sudan	785	Kerma	

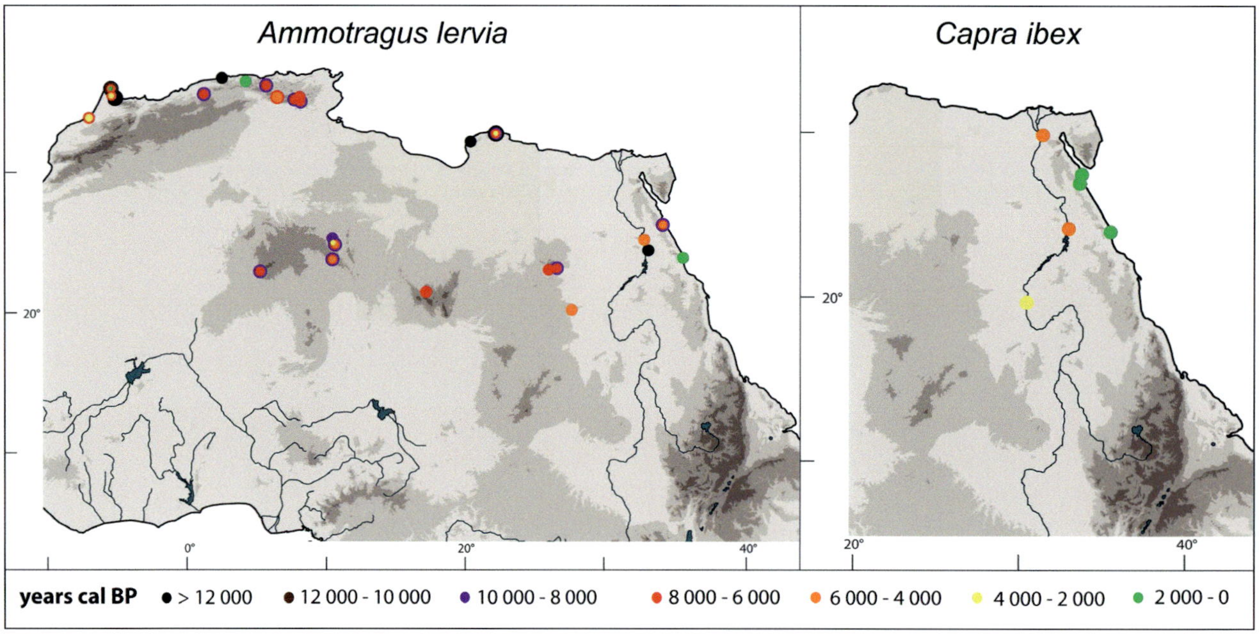

Ovis/Capra

Country	Site N	Site	
Algeria	1	A-88-15	
	10	Grotte Capeletti	
	2	A-88-28	
	7	Damous el Ahmar	
Botswana	26	Bosutwe-Mapungubwe	
	27	Bosutwe-Taukome	
	28	Bosutwe-Toutswe	
	29	Bosutwe-Zimbabwe	
	31	Commando Kop	
	32	Divuyu	
	36	Matanga	
	37	Matlapaneng	
	38	Nqoma	
	40	Taukome	
	43	Toteng1-SU3	*cf.
	44	Toteng1-SU4	
	45	Toteng3-SU3	
	46	Toteng3-SU4	*cf.
Burkina Faso	52	Kissi 22	
	53	Kissi 40	
	54	Oursi hu-beero	
	55	Oursi Village-EIA	
	56	Oursi Village-LIA	
	57	Oursi Village-MIA	
	58	Oursi-IA	
	63	Saouga 94/120	
	64	Saouga 95/7	
Cameroon	70	Blé A	
	71	Blé B	
	72	Blé C	
	73	Blé E	
	75	Nkang	
Chad	83	Gabrong-b	
	85	Koyom	
Congo	89	Tongo	
Egypt	100	Adaima-habitat	
	103	Al-Ma´abda	
	106	Berenike 95/9 Early Roman	
	107	Berenike 95/97Late Roman	
	108	Berenike 96/97 Ptolem.	
	112	Buto/1-3	
	113	Buto-6	
	114	Buto-9-12	
	121	Dakhleh Oasis, SMU	
	129	El Abadiya 2 North sector	
	130	El Abadiya 2 South sector	
	133	El Omari	
	136	Elephantine-NW Chnum T	
	139	Elkab-necropolis	
	140	El-Tell el Iswid-Early Dynastic	
	141	El-Tell el Iswid-Predynastic	
	142	El-Tell el-Iswid A	
	143	El-Tell el-Iswid B	
	146	Fayoum G3-H2	
	148	Fayoum Site 1	
	150	Fayoum Site 3	
	151	Fayoum Site 4	
	152	Fayoum Site 5	
	154	Hierakonpolis HK6-Naq IC-IIA	
	155	Hierakonpolis HK6-Naq IIA-B	
	156	Hierakonpolis HK6-Naq III	
	157	Hierakonpolis HK6-Naq I-II	
	160	Kharga dep.-graeco-roman	
	164	Maadi	
	165	Mahgar Dendera 2	
	168	Merimde Benisâlame	
	169	Minshat Abu Omar	
	172	Napta Playa 77-8	
	175	Qasr el-Sagha-1	
	176	Qasr el-Sagha-2	
	179	Quseir al-Qadim	
	182	Shenshef	
	186	Sodmein Cave y. backfill	
	200	Wadi Kalalat	
	95	Abu Sha´ar-4C	
	98	Adaima-burials	
Ethiopia	207	Axum-D	
	208	Axum-K	
	213	Lake Besala	
	214	Lalibela	
Ghana	223	Anyinam-Area A-B-C	
	224	Anyinam-Area D	
	225	Asantamanso-ASII/III	
	226	Asantamanso-ASIV	
	227	Asantamanso-ASV	
	231	Daboya	
	233	Esiease - A-C-D	
	235	Esiease -recent	
	236	Frederiksgave plantation	
	238	Kintampo RS K6-K	
	241	Mandara	
	242	Mankuma	
	244	Ntereso	
	245	Nyanawase/Akwamu Am.	
	246	Nyanga	
Kenya	251	Crescent Island main	
	260	Enkapune Ya Muto - Eb 5a	
	261	Enkapune Ya Muto - Eb 5b	
	262	Enkapune Ya Muto - Eb 5c	
	263	Enkapune Ya Muto BSS-HTL	
	272	Gogo Falls-I	
	273	Gogo Falls-II-III	
	274	Gogo Falls-sq26-29	
	275	GvJm 19-HA	
	276	GvJm 19-HC	
	278	GvJm 22-LHC	
	285	Lemek North-East	
	296	Ngamuriak	
	302	Sambo Ngige	
Lesotho	303	Bolahla	
	305	Lequetsoana	
	306	Likoaeng-I	
	307	Likoaeng-III-V	
	313	Liphofung-pCW	
	314	Muela	
	316	Sehonghong-a	
	323	Tloutle-pCW	
Libya	330	In Aghelachem-RT,SF	
	331	In Aghelachem-T1	
	339	Uan Muhuggiag (LP)	
	340	Uan Muhuggiag (MP)	
	343	Uan Telocat (EP)	
	345	Uan Telocat (MP)	
	347	Wadi Tannezzuft-96/129	
Mali	360	Akumbu-AK3	

Country	No.	Site
	361	Akumbu-EIA
	362	Akumbu-MIA
	366	Dia-Shoma Horizon III+IV
	369	Est-Hassi el Abiod
	372	Gao Saney
	373	Hamdallahi
	375	In Begaouaen
	376	Jenné-Jeno - phase I/II
	377	Jenné-Jeno - phase III
	377	Jenné-Jeno - phase III
	378	Jenné-Jeno - phase IV
	378	Jenné-Jeno - phase IV
	380	Karkarichinkat N-S
	382	Kolima Sud-I
	383	Kolima Sud-II
	384	Korounkorokalé-I
	389	Tongo Maaré Diabel
	390	Windé Koroji I,III
	392	Zampia-Tumulus 7
Mauritania	394	Chami-A2
	400	Dhar Tichitt Site 38
	401	Dhar Tichitt Site 46
	402	DN4-BKH-S5
Morocco	425	Ghar-Khal I
	426	Ghar-Khal IIIb-V
	427	Ghar-Khal II-IIIa
	437	Kehf-el-Baroud
	438	Kehf-el-Baroud-aven
Mozambique	441	Manekweni-1
	442	Manekweni-2-ABC
	443	Manekweni-3
	444	Manekweni-4
	445	Manekweni-5
Niger	471	Chin Tafidet
	478	Mentes 13
	481	Mentes 4
	483	Mentes 6
Nigeria	498	Bukarkurari-I
	499	Bukarkurari-II
	500	Daima
	501	Dorota
	502	Elkido North
	503	Gajiganna A
	504	Gajiganna B1-lower
	505	Gajiganna B2
	506	Gajiganna C
	507	Galaga
	509	Gilgila-Iib
	510	Gilgila-Iic
	511	Kariari
	512	Kelumeri
	514	Kursakata-IA
	515	Kursakata-LSA
	516	Labe Kanuri
	518	Mege I
	519	Mege II
	520	Mege III
	521	Mege IV
	522	NA 93/10
	524	NA 93/42
	525	Ngala I
	526	Ngala II
	527	Ngala IIIa
	528	Ngala IIIb
	529	Ngala IV
	530	Zilum
Rhodesia	531	Chamabvefva
	532	Harleigh Farm-1
	533	Harleigh Farm-2
	534	Mabveni
Senegal	538	Cubalel-Ia
	539	Cubalel-Ib
	540	Cubalel-II
	541	Cubalel-III
	542	Ndiamon-Badat
	544	Sintiou-Bara
	545	Siouré-Ia
	546	Siouré-Ib
	547	Siouré-II
	548	Siouré-IV, V
	549	Tulel-Fobo
	550	Walaldé-W1/1-19
	551	Walaldé-W1/20-25
	552	Walaldé-W2
Sinai	553	Abu Gedar
	554	El´Awag
South Africa	556	2627 CD7
	557	Badfontein Valley
	594	Eiland Salt Works-b
	605	Ficus A
	606	Ficus B
	607	Ficus C
	615	Happy Rest
	618	Hill X
	623	Icon
	629	Kekane
	635	Klingbeil 2530 AB 1
	636	Klipspring
	638	Langsdraai
	640	Le2
	641	Le6
	642	Le7a
	643	Le7b
	647	Lydenburg Heads site
	648	Ma38
	649	Ma4
	650	Magogo-1
	651	Magogo-2
	652	Magogo-3
	654	Mamba 1
	656	Mapungubwe-Hill
	657	Mapungubwe-K2/TS3
	658	Mapungubwe-ST
	659	Marupale Hill
	661	Mgoduyanuka
	662	Mhlopeni
	665	Msuluzi Confluence
	667	Ndondondwane
	674	Ondini
	677	Pa8.1
	685	Ph6
	688	Pont-Drift-LKA
	689	Pont-Drift-Zhizo
	692	QwaQwa Museum
	693	Randjies
	696	Rhenosterkloof
	702	Rooikrans-pCW
	704	Rose Cottage Cave-A-Mn

	708	Schroda	
	710	Selongwe Hill	
	711	Serotwe Hill	
	713	SH16	
	714	Sh27	
	715	Shankare Hill-a	
	716	Shankare Hill-b	
	717	SK17	
	720	Sonkoanini Hill	
	739	Tsh1	
	741	Umgungundlovu	
	746	Wildebeestfontein	
	751	Wosi-early EIA	
	752	Wosi-middle-late EIA	
	753	Wosi-terminal EIA	
Sudan	761	Debbat Bangdit	
	762	Debbat El Eheima	
	765	El Kadada	
	766	El Kenger Middle/East	
	768	El Nofalab	
	771	El Zakiab	
	773	Esh Shaheinab	
	775	Gash Delta-K	
	777	Geili	
	778	Gism el-Arba	
	784	Kadruka	
	793	Laqiya area - 82/29	
	797	Mahal Teglinos K1-CG	
	798	Mahal Teglinos K1-EG	
	799	Mahal Teglinos K1-JM	
	800	Mahal Teglinos K1-LG	
	801	Mahal Teglinos K1-MG	
	802	Mahal Teglinos K1-PG	
	805	Saggai 1	
	811	Shaqaduq-S1-A	
	814	Sheikh el Amin	
	817	Soba	
	819	Um Direiwa	
	824	Wadi Hariq 01/1	
	825	Wadi Hariq 01/4	
	829	Wadi Howar - 84/1	
	830	Wadi Howar - 84/12-1	
	831	Wadi Howar - 84/13	
	832	Wadi Howar - 84/19	
Zaire	853	Mashita Mbanza	
	857	Nkile	
Zambia	859	Bulila	
	865	Nanga	*cf.
	865	Nanga	
	866	Salumano	
Zanzibar	867	Kizimkazi Dimbani	
Zimbabwe	868	Mtanye	

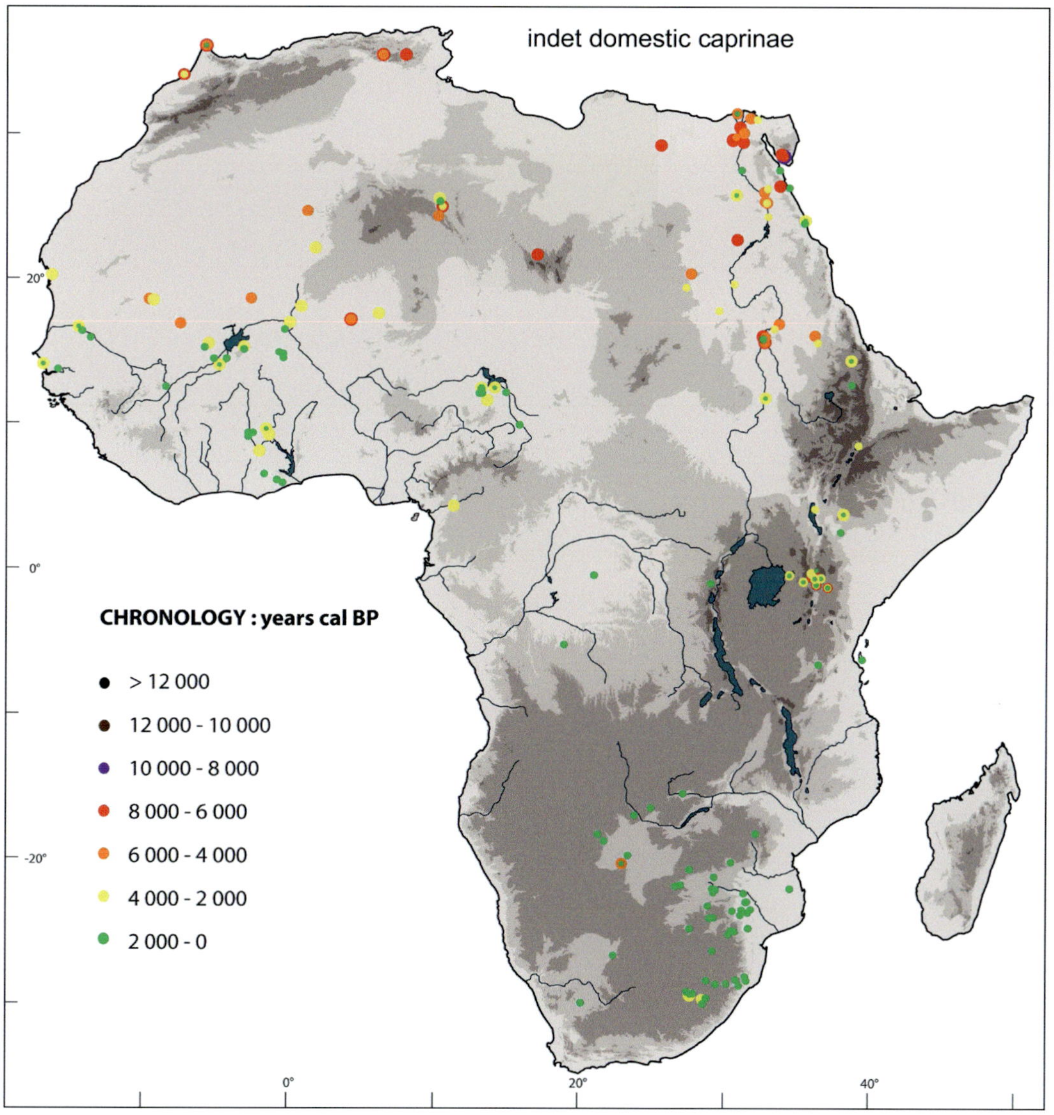

Capra sp.

Country	Site N	Site
Algeria	16	Meniet
Nigeria	498	Bukarkurari-I

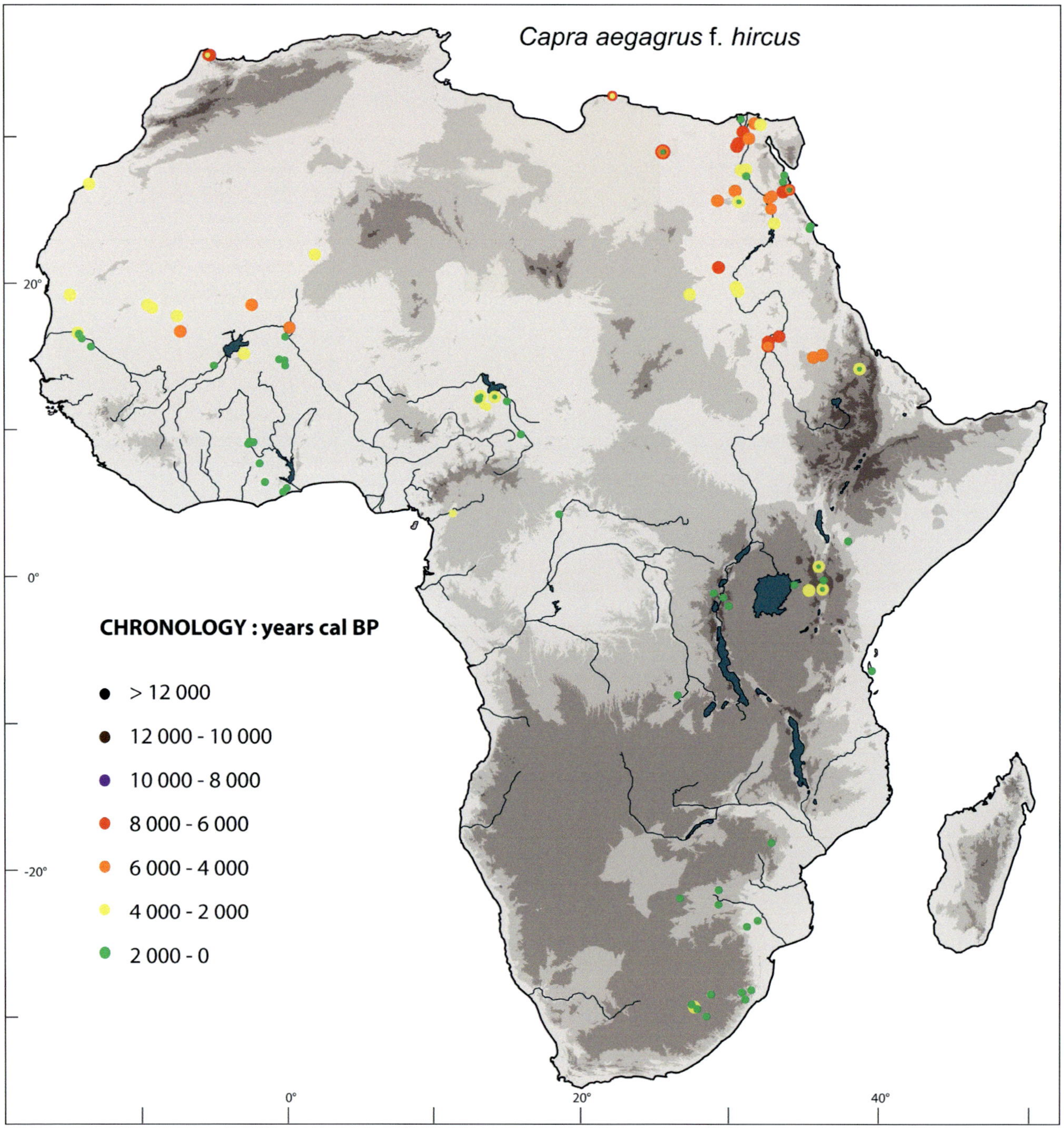

Capra aegagrus f. *hircus*

Country	Site N	Site
Algeria	2	A-88-28
Botswana	26	Bosutwe-Mapungubwe
	27	Bosutwe-Taukome
	28	Bosutwe-Toutswe
	29	Bosutwe-Zimbabwe
Burkina Faso	52	Kissi 22
	53	Kissi 40
	54	Oursi hu-beero
	55	Oursi Village-EIA
	56	Oursi Village-LIA
	57	Oursi Village-MIA
	58	Oursi-IA
	63	Saouga 94/120
	64	Saouga 95/7
Cameroon	70	Blé A
	71	Blé B

	73	Blé E	
	75	Nkang	
C.A.R.	79	Mondongué	
	80	Lingbangbo	
Chad	85	Koyom	
Congo	89	Tongo	
Egypt	102	Al-Ma´abda	
	105	Balat-Mastaba V	
	106	Berenike 95/9 Early Roman	
	107	Berenike 95/97Late Roman	
	114	Buto-9-12	
	121	Dakhleh Oasis, SMU	
	123	Dakhleh Ouasis, Roman	
	124	Dakhleh Ouasis,OK	* ?
	129	El Abadiya 2 North sector	
	130	El Abadiya 2 South sector	
	135	Elephantine-necropolis	
	136	Elephantine-NW Chnum T	
	137	Elephantine-Satet temple	
	140	El-Tell el Iswid-Early Dynastic	
	141	El-Tell el Iswid-Predynastic	
	142	El-Tell el-Iswid A	
	143	El-Tell el-Iswid B	
	146	Fayoum G3-H2	*cf.
	153	Gilf Kebir	
	154	Hierakonpolis HK6-Naq IC-IIA	
	155	Hierakonpolis HK6-Naq IIA-B	
	156	Hierakonpolis HK6-Naq III	
	160	Kharga dep.-graeco-roman	
	164	Maadi	
	168	Merimde Benisâlame	
	169	Minshat Abu Omar	
	170	Mons Claudianus	
	174	Qantir	
	175	Qasr el-Sagha-1	
	182	Shenshef	
	186	Sodmein Cave y. backfill	
	188	Tell el-Dab´a - b	
	191	Toukh	
	192	Tree Shelter-1	
	193	Tree Shelter-3	
	196	Tuna el-Gebel	
	197	Umm el-Qaab, Friedhof U	
	206	Workmen Village	
	96	Abu Sha´ar-5C	
Ethiopia	207	Axum-D	
	208	Axum-K	
Ghana	223	Anyinam-Area A-B-C	
	225	Asantamanso-ASII/III	
	227	Asantamanso-ASV	
	229	Bibease Plantation	
	230	Bono Manso	
	232	Dawu	
	233	Esiease - A-C-D	
	241	Mandara	
	242	Mankuma	
	246	Nyanga	
Kenya	248	Akira-PIA	
	273	Gogo Falls-II-III	
	284	Kulchurdo	
	294	Narosura	
	295	Ndabibi	
	297	Ngenyn	
Lesotho	303	Bolahla	
	305	Lequetsoana	
	323	Tloutle-pCW	
Libya	326	Hauah Fteah- Historic	
	328	Hauah Fteah-Lib-caps	
	329	Hauah Fteah-Neolithic	
Mali	367	Dia-Shoma Horizon V	
	369	Est-Hassi el Abiod	
	371	Gao Ancien	
	372	Gao Saney	
	379	Karkarichinkat KN05	
	390	Windé Koroji I,III	
Mauritania	402	DN4-BKH-S5	
	405	Khatt Lemaiteg	
	406	Khneivissa	
	411	Tichitt - Akjinjeir phase	
	412	Tichitt - Arriane phase	
	413	Tichitt - Chebka phase	
	414	Tichitt - Goungou phase	
	415	Tichitt - Khimiya phase	
	416	Tichitt - Naguez phase	
	417	Tichitt - Nkhal phase	
Morocco	419	Bou Saria A	
	420	Bou Saria B	
	421	Bou Saria C-D	
	429	Kaf-taht-el-Ghar-A	
	430	Kaf-taht-el-Ghar-B	
	431	Kaf-taht-el-Ghar-C	
	432	Kaf-taht-el-Ghar-D	
	433	Kaf-taht-el-Ghar-E	
Nigeria	499	Bukarkurari-II	
	501	Dorota	
	502	Elkido North	
	503	Gajiganna A	
	505	Gajiganna B2	
	506	Gajiganna C	
	507	Galaga	
	509	Gilgila-Iib	
	510	Gilgila-Iic	
	511	Kariari	
	512	Kelumeri	
	514	Kursakata-IA	
	515	Kursakata-LSA	
	518	Mege I	
	519	Mege II	
	521	Mege IV	
	524	NA 93/42	
	525	Ngala I	
	526	Ngala II	
	527	Ngala IIIa	
	528	Ngala IIIb	
	529	Ngala IV	
	530	Zilum	
Rwanda	535	Akameru-Cyinkomane	
	537	Ryamurari	
Senegal	538	Cubalel-Ia	
	539	Cubalel-Ib	
	540	Cubalel-II	
	541	Cubalel-III	
	544	Sintiou-Bara	
	545	Siouré-Ia	
	546	Siouré-Ib	
	547	Siouré-II	
	548	Siouré-IV, V	
	550	Waladé-W1/1-19	

Country	Site N	Site	
	551	Walaldé-W1/20-25	
	552	Walaldé-W2	
South Africa	615	Happy Rest	
	618	Hill X	
	623	Icon	*cf.
	623	Icon	
	650	Magogo-1	
	651	Magogo-2	
	652	Magogo-3	
	654	Mamba 1	
	659	Marupale Hill	
	667	Ndondondwane	
	674	Ondini	
	692	QwaQwa Museum	
	704	Rose Cottage Cave-A-Mn	
	709	Sebatini Hill	
	710	Selongwe Hill	
	711	Serotwe Hill	
	715	Shankare Hill-a	
	716	Shankare Hill-b	
	720	Sonkoanini Hill	
	751	Wosi-early EIA	
	752	Wosi-middle-late EIA	
	753	Wosi-terminal EIA	
Sudan	773	Esh Shaheinab	
	778	Gism el-Arba	
	783	Kadero	
	784	Kadruka	
	785	Kerma	
	808	Selima Sandsheet - 85/73-2	
	812	Shaqaduq-S1-B	
	821	Upper Atbara hinterland	
	822	Upper Atbara Valley	
	824	Wadi Hariq 01/1	
	826	Wadi Hariq 97/5	
	827	Wadi Hariq 97/7	
West. Sahara	838	Sebkha Laasailia	
Zaire	858	Sanga	
Zanzibar	867	Kizimkazi Dimbani	
Zimbabwe	868	Mtanye	
	869	Muozi	

Ovis sp.

Country	Site N	Site
Namibia	446	Aar I and II
South Africa	700	Rooiels

Ovis ammon f. *aries*

Country	Site N	Site	
Botswana	26	Bosutwe-Mapungubwe	
	27	Bosutwe-Taukome	
	28	Bosutwe-Toutswe	
	28	Bosutwe-Toutswe	*cf.
	29	Bosutwe-Zimbabwe	
	42	Toteng1-SU2	
	44	Toteng1-SU4	
	48	White Paintings RS-2ab	
Burkina Faso	52	Kissi 22	
	53	Kissi 40	
	54	Oursi hu-beero	
	56	Oursi Village-LIA	
	57	Oursi Village-MIA	
	58	Oursi-IA	
	63	Saouga 94/120	
	64	Saouga 95/7	
Cameroon	71	Blé B	
	72	Blé C	
	73	Blé E	
	75	Nkang	
Congo	89	Tongo	
Egypt	102	Al-Ma´abda	
	103	Al-Ma´abda	
	106	Berenike 95/9 Early Roman	
	107	Berenike 95/97Late Roman	
	113	Buto-6	
	114	Buto-9-12	
	127	Eastpan 95-2	
	129	El Abadiya 2 North sector	
	130	El Abadiya 2 South sector	
	135	Elephantine-necropolis	
	137	Elephantine-Satet temple	
	140	El-Tell el Iswid-Early Dynastic	
	141	El-Tell el Iswid-Predynastic	
	142	El-Tell el-Iswid A	
	143	El-Tell el-Iswid B	
	146	Fayoum G3-H2	*cf.
	147	Fayoum G5-G4	*cf.
	154	Hierakonpolis HK6-Naq IC-IIA	
	155	Hierakonpolis HK6-Naq IIA-B	
	156	Hierakonpolis HK6-Naq III	
	164	Maadi	
	168	Merimde Benisâlame	
	170	Mons Claudianus	
	175	Qasr el-Sagha-1	
	177	Qasr el-Sagha-3	
	182	Shenshef	
	188	Tell el-Dab´a - b	
	188	Tell el-Dab´a - b	
	191	Toukh * O. longipes palaeoegypticus	
	192	Tree Shelter-1	*cf.
	196	Tuna el-Gebel	
	197	Umm el-Qaab, Friedhof U	
	96	Abu Sha´ar-5C	
	99	Adaima-habitat	
Ethiopia	207	Axum-D	
	208	Axum-K	
Ghana	223	Anyinam-Area A-B-C	
	225	Asantamanso-ASII/III	
	227	Asantamanso-ASV	
	229	Bibease Plantation	
	230	Bono Manso	
	232	Dawu	
	233	Esiease - A-C-D	
	241	Mandara	

Country	No.	Site	Note
	242	Mankuma	
	246	Nyanga	
Kenya	273	Gogo Falls-II-III	
	294	Narosura	
	295	Ndabibi	
	297	Ngenyn	
Lesotho	303	Bolahla	
	307	Likoaeng-III-V	*cf., intr.
	313	Liphofung-pCW	
	314	Muela	
	316	Sehonghong-a	
Libya	326	Hauah Fteah- Historic	*?
	327	Hauah Fteah-Iberom	*?
	328	Hauah Fteah-Lib-caps	*?
	329	Hauah Fteah-Neolithic	*?
Mali	367	Dia-Shoma Horizon V	
	369	Est-Hassi el Abiod	
	372	Gao Saney	
	379	Karkarichinkat KN05	
	387	Tessalit	
Mauritania	402	DN4-BKH-S5	
	405	Khatt Lemaiteg	
	406	Khneivissa	
	407	Tegdaoust-a	
	408	Tegdaoust-b	
	409	Tegdaoust-c	
	410	Tegdaoust-d	
	413	Tichitt - Chebka phase	*?
Morocco	419	Bou Saria A	
	420	Bou Saria B	
	421	Bou Saria C-D	
	429	Kaf-taht-el-Ghar-A	
	430	Kaf-taht-el-Ghar-B	
	431	Kaf-taht-el-Ghar-C	
	432	Kaf-taht-el-Ghar-D	
	433	Kaf-taht-el-Ghar-E	
Namibia	455	Geduld-LIA	
Nigeria	498	Bukarkurari-I	
	499	Bukarkurari-II	
	502	Elkido North	
	503	Gajiganna A	
	504	Gajiganna B1-lower	
	507	Galaga	
	510	Gilgila-Iic	
	512	Kelumeri	
	514	Kursakata-IA	
	518	Mege I	
	525	Ngala I	
	526	Ngala II	
	527	Ngala IIIa	
	528	Ngala IIIb	
	529	Ngala IV	
	530	Zilum	
Rwanda	535	Akameru-Cyinkomane	
	537	Ryamurari	
Senegal	538	Cubalel-Ia	
	539	Cubalel-Ib	
	540	Cubalel-II	
	541	Cubalel-III	
	544	Sintiou-Bara	
	546	Siouré-Ib	
	547	Siouré-II	
	548	Siouré-IV, V	
	550	Waladé-W1/1-19	
	552	Waladé-W2	
South Africa	555	/Ai Tomas	
	558	Blombos Cave	
	565	Boomplass Cave-c	
	566	Boomplass Cave-d	
	579	Byeneskranskop Cave 1d	
	586	Die Kelders	
	587	Dikbosch 1	
	590	Drie Susters	
	591	Drie Susters Main	
	599	Eland´s Bay Cave-Wic	
	600	Eland´s Bay Cave-Wid	
	604	Equus Cave 1A	
	615	Happy Rest	
	622	HKLP Vb	
	623	Icon	*cf.
	627	Kasteelberg A	
	627	Kasteelberg A	
	628	Kasteelberg B	
	634	Klein Kliphuis	
	647	Lydenburg Heads site	
	650	Magogo-1	
	651	Magogo-2	
	652	Magogo-3	
	654	Mamba 1	
	661	Mgoduyanuka	
	662	Mhlopeni	
	663	Mhlwazini Cave	
	667	Ndondondwane	
	674	Ondini	
	676	Oudepost I	
	677	Pa8.1	
	702	Rooikrans-pCW	
	704	Rose Cottage Cave-A-Mn	
	717	SK17	
	719	Smitswinkelbaai Cave	
	721	Spoegrivier-a	
	722	Spoegrivier-b	
	724	Spring Cave	
	725	Steenbokfontein-1	
	726	Steenbokfontein-2	
	730	Tandjesberg	
	737	Tortoise cave-c	
	738	Tortoise cave-d	
	739	Tsh1	
	751	Wosi-early EIA	
	752	Wosi-middle-late EIA	
	753	Wosi-terminal EIA	
Sudan	773	Esh Shaheinab	
	778	Gism el-Arba	
	780	Jebel et Tomat	
	783	Kadero	
	784	Kadruka	
	785	Kerma	
	811	Shaqaduq-S1-A	
	821	Upper Atbara hinterland	
	822	Upper Atbara Valley	
	824	Wadi Hariq 01/1	
	827	Wadi Hariq 97/7	
Zaire	853	Mashita Mbanza	
Zanzibar	867	Kizimkazi Dimbani	
Zimbabwe	869	Muozi	
	869	Muozi	

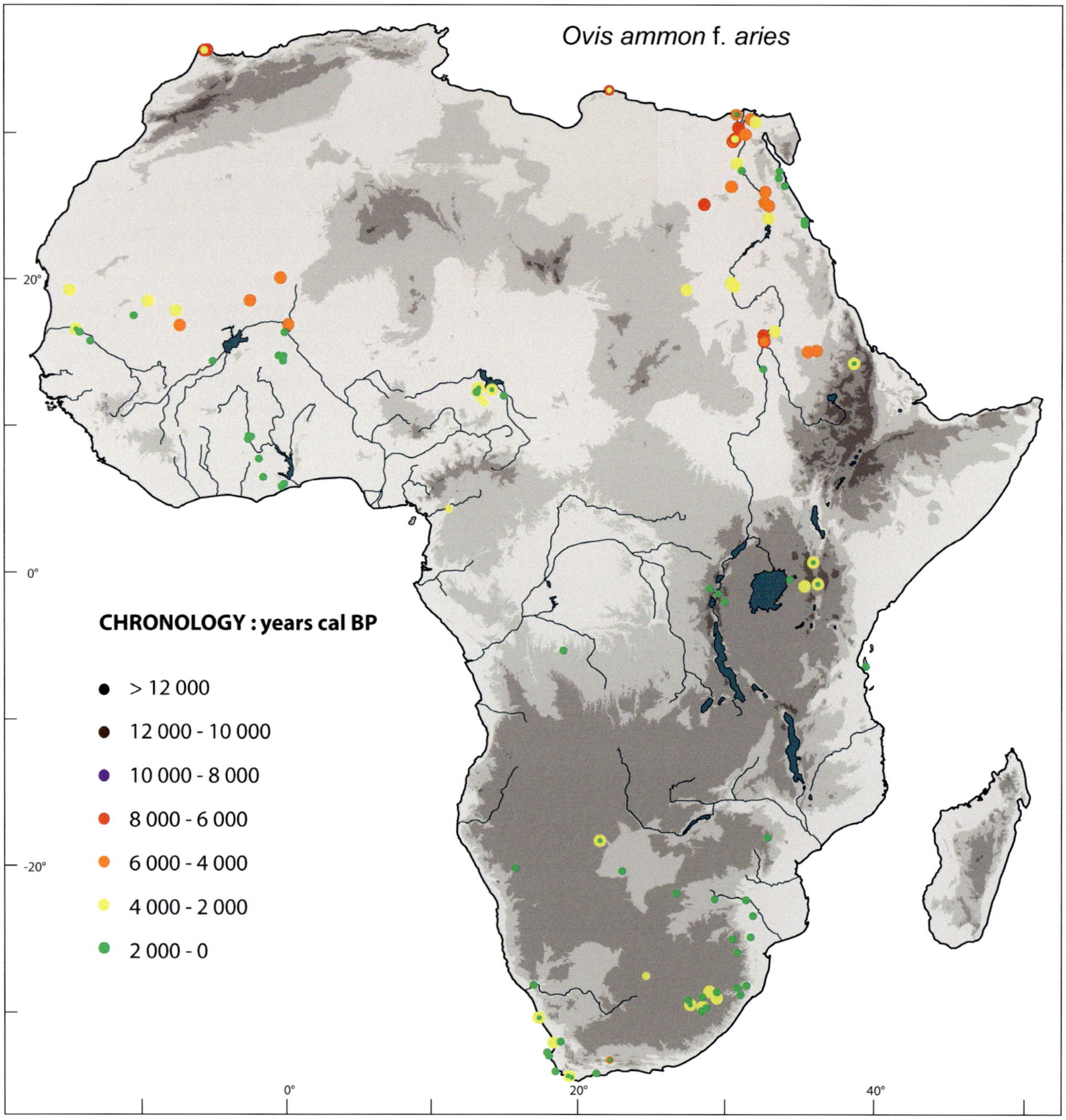
Ovis ammon f. aries
CHRONOLOGY : years cal BP
> 12 000
12 000 - 10 000
10 000 - 8 000
8 000 - 6 000
6 000 - 4 000
4 000 - 2 000
2 000 - 0
20°
0°
-20°
0°
20°
40°

Cephalophinae

Country	Site N	Site	
Botswana	32	Divuyu	
	37	Matlapaneng	
	38	Nqoma	
	39	Qogana	
Ghana	223	Anyinam-Area A-B-C	
	224	Anyinam-Area D	
	225	Asantamanso-ASII/III	
	227	Asantamanso-ASV	
	231	Daboya	
	233	Esiease - A-C-D	

Cephalophus sp.

Country	Site N	Site	
Cameroon	75	Nkang	
C.A.R.	79	Mondongué	
	80	Lingbangbo	
Congo	87	Ntadi Yom ba-level B	
	88	Ntadi Yomba-level A	
	89	Tongo	
Ghana	241	Mandara	
	242	Mankuma	
	245	Nyanawase/Akwamu Am.	
	246	Nyanga	
Zaire	854	Matupi-I	
	855	Matupi-II	*?
	855	Matupi-II	
	856	Ngovo	
	857	Nkile	
Zanzibar	867	Kizimkazi Dimbani	

Cephalophus rufilatus

Country	Site N	Site	
Ghana	223	Anyinam-Area A-B-C	
	224	Anyinam-Area D	
	225	Asantamanso-ASII/III	
	226	Asantamanso-ASIV	
	227	Asantamanso-ASV	
	233	Esiease - A-C-D	
	241	Mandara	
	242	Mankuma	
	245	Nyanawase/Akwamu Am.	
	246	Nyanga	
Mali	385	Korounkorokalé-III	
	391	Windé Koroji SO	
Mozambique	441	Manekweni-1	
	442	Manekweni-2-ABC	
	444	Manekweni-4	
	445	Manekweni-5	
Nigeria	502	Elkido North	*cf.
	504	Gajiganna B1-lower	*cf.
	509	Gilgila-lib	*cf.
Zaire	854	Matupi-I	

Cephalophus silvicultor

Country	Site N	Site	
Congo	87	Ntadi Yom ba-level B	
Gabon	221	Okala	*/T.scriptus
	235	Esiease -recent	
	244	Ntereso	
Kenya	251	Crescent Island main	
	297	Ngenyn	
Mali	380	Karkarichinkat N-S	*?
	382	Kolima Sud-I	
	383	Kolima Sud-II	
Senegal	540	Cubalel-II	
South Africa	660	Mauermanshoek	
	730	Tandjesberg	
	740	Twyfelpoort	
Ghana	225	Asantamanso-ASII/III	
	227	Asantamanso-ASV	
Zaire	842	Dimba	
	850	Malemba-Nkulu-CK	
	854	Matupi-I	
	855	Matupi-II	
	856	Ngovo	
	858	Sanga	

Cephalophus nigrifrons

Country	Site N	Site	
Ghana	237	Kintampo RS K6-IA	*?
	240	Kintampo RS K6-P	*?
Rwanda	535	Akameru-Cyinkomane	

Cephalophus natalensis

Country	Site N	Site	
Kenya	292	Marula Rock Shelter	
South Africa	580	Collingham Shelter	
	663	Mhlwazini Cave	
	718	Sk4	
	751	Wosi-early EIA	
Zaire	846	Kakontwe	*cf.

Cephalophus leucogaster

Country	Site N	Site
Zaire	854	Matupi-I

Cephalophus dorsalis

Country	Site N	Site
Ghana	223	Anyinam-Area A-B-C
	224	Anyinam-Area D
	225	Asantamanso-ASII/III
	226	Asantamanso-ASIV
	227	Asantamanso-ASV
	233	Esiease - A-C-D
	234	Esiease - H
	245	Nyanawase/Akwamu Am.

Cephalophus niger

Country	Site N	Site
Ghana	232	Dawu

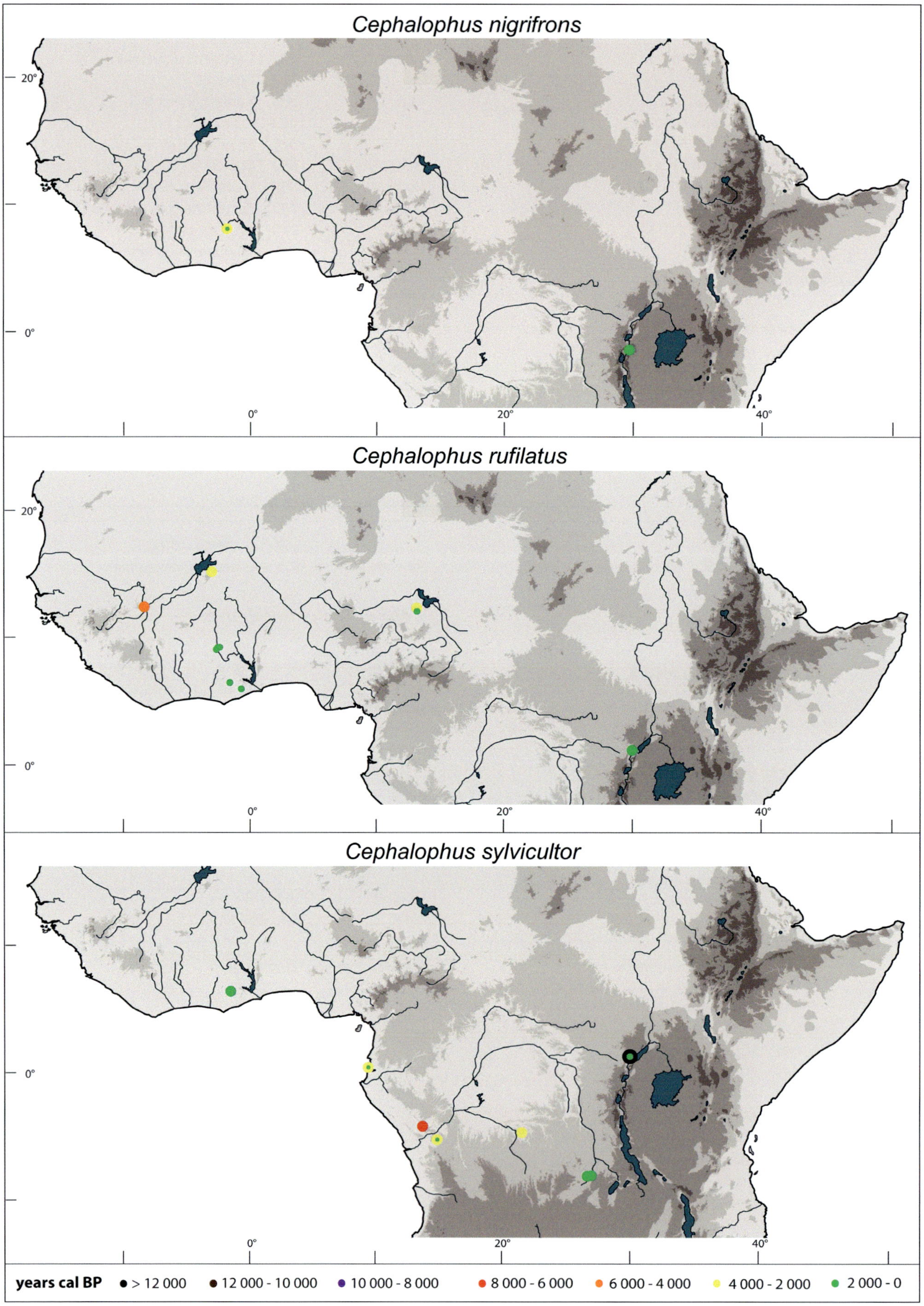
Cephalophus nigrifrons
20°
0°
0°
20°
40°
Cephalophus rufilatus
20°
0°
0°
20°
40°
Cephalophus sylvicultor
0°
0°
20°
40°
years cal BP
> 12 000
12 000 - 10 000
10 000 - 8 000
8 000 - 6 000
6 000 - 4 000
4 000 - 2 000
2 000 - 0

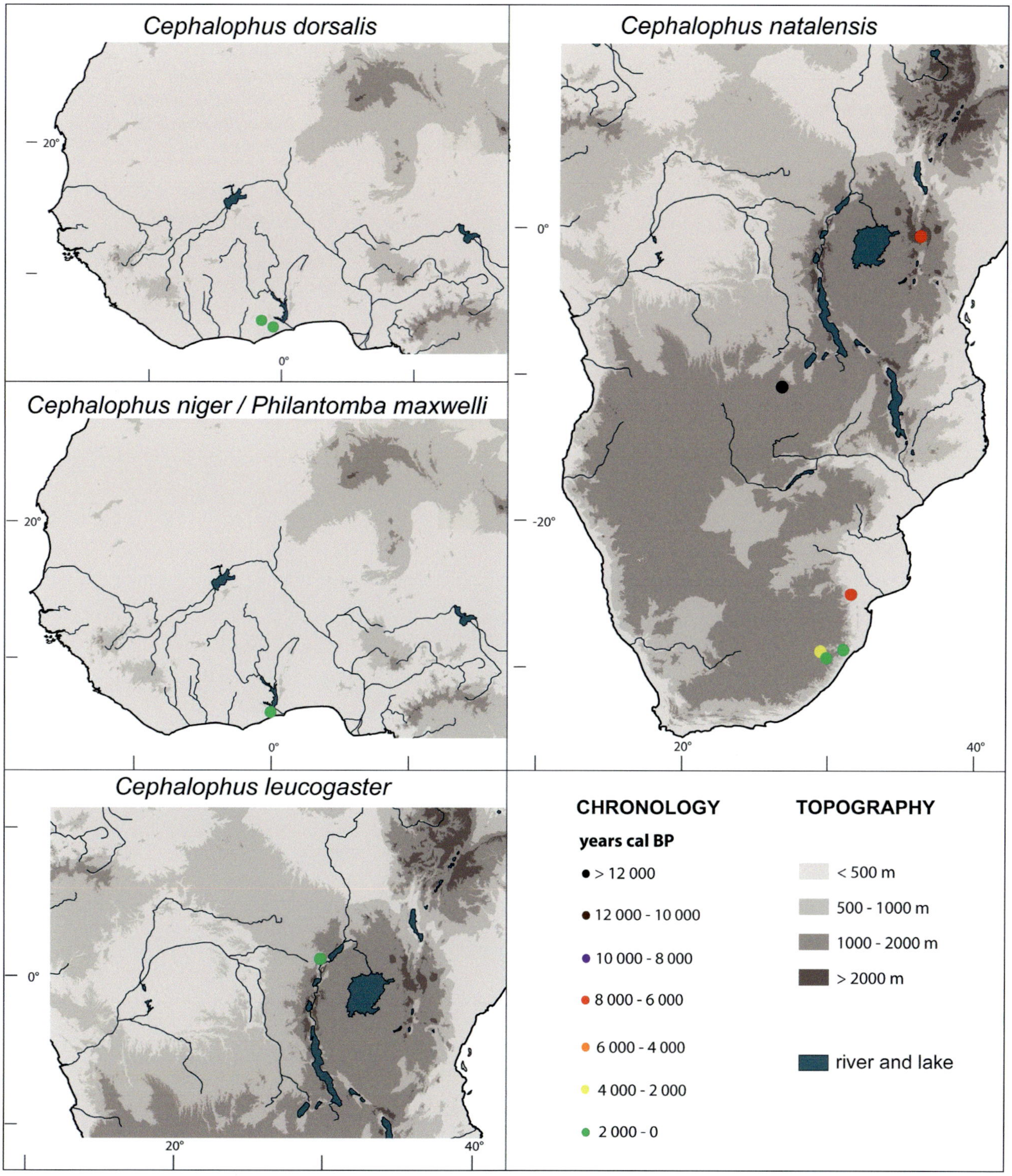

Philantomba maxwellii

Country	Site N	Site
Ghana	232	Dawu

Philantomba monticola

Country	Site N	Site
C.A.R.	80	Lingbangbo
Congo	87	Ntadi Yom ba-level B
Ghana	238	Kintampo RS K6-K
Lesotho	304	Ha Makotoko
	314	Muela
South Africa	603	Emambeni
	631	Klasies River Mouth-1I
	632	Klasies River Mouth-1II
	633	Klasies River Mouth-5II
	639	Large Rock Shelter
	653	Malwula
	663	Mhlwazini Cave
	668	Nelson Bay Cave- PW
	670	Nelson Bay Cave- W
	671	Oakhurst
	675	Onink
	751	Wosi-early EIA
	752	Wosi-middle-late EIA
	753	Wosi-terminal EIA
Zaire	848	Kiantapo
	854	Matupi-I
	855	Matupi-II
	857	Nkile

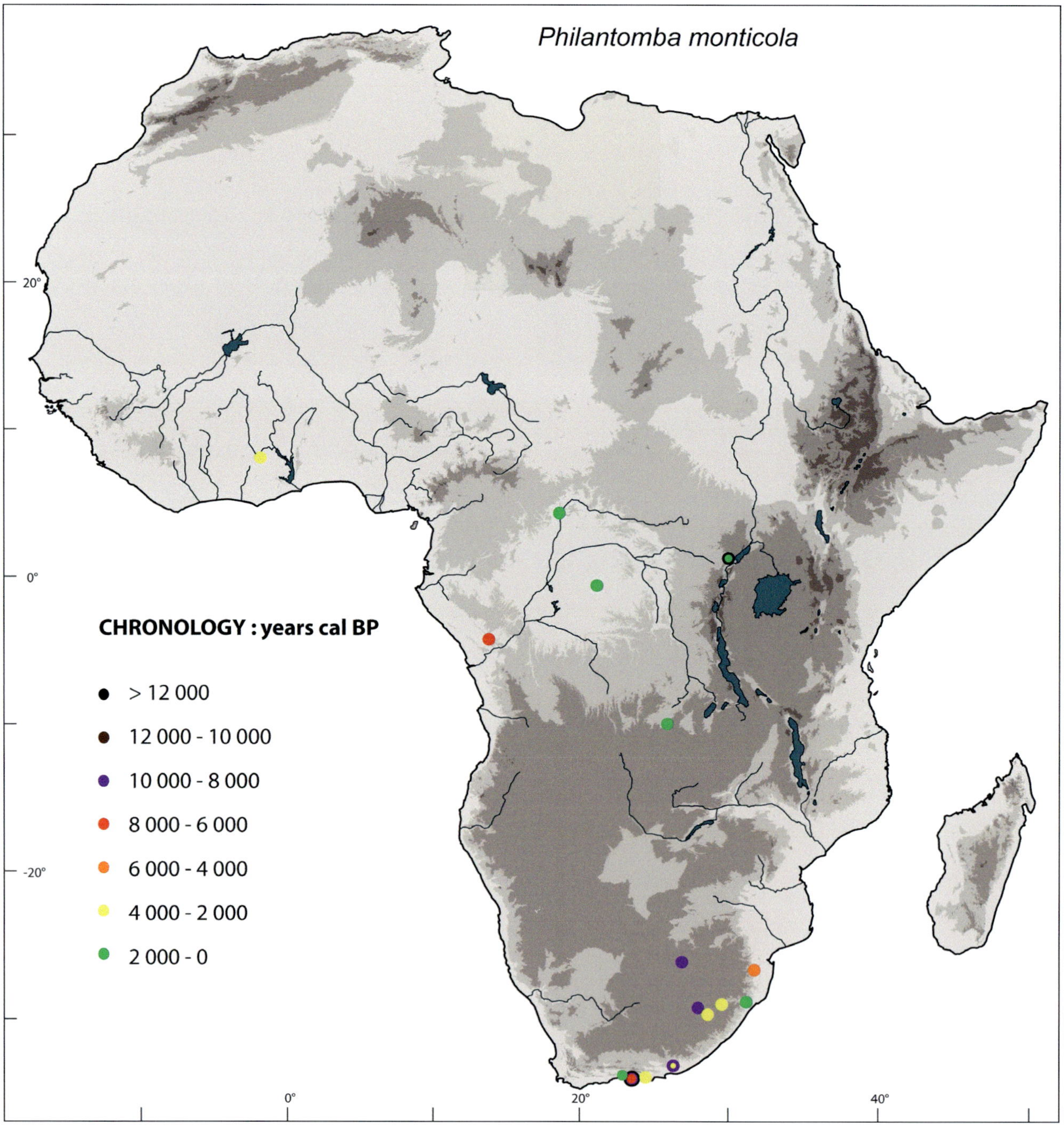

Sylvicapra grimmia

Country	Site N	Site	
Botswana	26	Bosutwe-Mapungubwe	
	27	Bosutwe-Taukome	
	28	Bosutwe-Toutswe	
	28	Bosutwe-Toutswe	*cf.
	31	Commando Kop	
	33	Drotsky's Cave-H	
	34	Drotsky's Cave-LP	
	35	Drotsky's Cave-P	
	41	Toteng1-SU1	
	41	Toteng1-SU1	*cf.
	42	Toteng1-SU2	*cf.
	44	Toteng1-SU4	
	48	White Paintings RS-1	
	48	White Paintings RS-2ab	
	49	White Paintings RS-2b-3ab	
Burkina Faso	58	Oursi-IA	
	63	Saouga 94/120	
	64	Saouga 95/7	
Chad	85	Koyom	
Ghana	223	Anyinam-Area A-B-C	
Kenya	251	Crescent Island main	
	254	Ele Bor A-B	
	255	Ele Bor A-C	
	258	Enkapune Ya Muto - Eb 4a	
	259	Enkapune Ya Muto - Eb 4b	
	260	Enkapune Ya Muto - Eb 5a	
	261	Enkapune Ya Muto - Eb 5b	
	272	Gogo Falls-I	
	273	Gogo Falls-II-III	
	273	Gogo Falls-II-III	
	275	GvJm 19-HA	
	276	GvJm 19-HC	
	277	GvJm 19-TLPG	
	279	GvJm 22-LPG	
	281	GvJm 62-H	
	282	GvJm 62-LPG	
	283	GvJm 62-LPNG	
	284	Kulchurdo	
	292	Marula Rock Shelter	
	293	Naivasha Railway Shelter	
Lesotho	303	Bolahla	
	305	Lequetsoana	
	306	Likoaeng-I	
	307	Likoaeng-III-V	
	308	Likoaeng-VII-IX	
	312	Liphofung-CW	
	314	Muela	
	316	Sehonghong-a	
	321	Tloutle-CWI	
	323	Tloutle-pCW	
Malawi	352	Malowa-2	
	356	Matope Court-EIA	
	357	Matope Court-IA	
Mali	360	Akumbu-AK3	
	361	Akumbu-EIA	
	383	Kolima Sud-II	
Mozambique	442	Manekweni-2-ABC	
	444	Manekweni-4	
	445	Manekweni-5	
Namibia	456	Geduld-LSA	
Nigeria	499	Bukarkurari-II	
	503	Gajiganna A	
	509	Gilgila-Iib	
	510	Gilgila-Iic	
	511	Kariari	
South Africa	570	Buffelskloof RS-c	
	571	Buffelskloof RS-d	
	574	Bushman RS-LSA	
	580	Collingham Shelter	
	589	Doornfontein	
	591	Drie Susters Main	
	595	Eland´s Bay Cave-AI	
	597	Eland´s Bay Cave-Wia	
	598	Eland´s Bay Cave-Wib	
	599	Eland´s Bay Cave-Wic	
	600	Eland´s Bay Cave-Wid	
	602	Eland´s Bay Open-b	
	603	Emambeni	
	604	Equus Cave 1A	
	614	Hapi Pan	
	615	Happy Rest	
	616	Heuningneskrans-I/O	
	617	Heuningneskrans-R	
	618	Hill X	
	619	HKLP Ia	*cf.
	619	HKLP Ia	
	620	HKLP Ib	
	622	HKLP Vb	
	625	Jubilee Shelter-PW	
	626	Jubilee Shelter-W	
	627	Kasteelberg A	
	628	Kasteelberg B	
	639	Large Rock Shelter	
	642	Le7a	
	644	Leliehoek	
	647	Lydenburg Heads site	*cf.
	650	Magogo-1	
	654	Mamba 1	
	657	Mapungubwe-K2/TS3	
	658	Mapungubwe-ST	
	659	Marupale Hill	
	663	Mhlwazini Cave	
	667	Ndondondwane	
	668	Nelson Bay Cave- PW	*?
	669	Nelson Bay Cave- R	
	670	Nelson Bay Cave- W	
	673	OI20	
	674	Ondini	
	676	Oudepost I	*cf.
	677	Pa8.1	
	678	Pancho's Kitchen Midden 1-2	
	679	Pancho's Kitchen Midden 3-4	
	680	Pancho's Kitchen Midden 6-7	
	682	Paternoster3-4	
	683	Paternoster5-6	
	688	Pont-Drift-LKA	
	689	Pont-Drift-Zhizo	
	692	QwaQwa Museum	
	700	Rooiels	
	701	Rooikrans-CW	
	702	Rooikrans-pCW	

Country	No.	Site
	704	Rose Cottage Cave-A-Mn
	709	Sebatini Hill
	710	Selongwe Hill
	711	Serotwe Hill
	716	Shankare Hill-b
	717	SK17
	718	Sk4
	719	Smitswinkelbaai Cave
	720	Sonkoanini Hill
	725	Steenbokfontein-1
	726	Steenbokfontein-2
	727	Steenbokfontein-3
	728	Steenbokfontein-4a
	738	Tortoise cave-d
	739	Tsh1
	741	Umgungundlovu
	749	Witklip-3
	750	Witklip-4
	751	Wosi-early EIA
	752	Wosi-middle-late EIA
	753	Wosi-terminal EIA
Sudan	773	Esh Shaheinab
	787	Khartoum Hopital
	788	Khasm el Girba - Kassala
	789	Khasm el Girba - KG14
	812	Shaqaduq-S1-B
Zaire	846	Kakontwe
	848	Kiantapo
	853	Mashita Mbanza
	858	Sanga
Zambia	860	Leopard's Hill-NI
	861	Leopard's Hill-NII
	862	Leopard's Hill-NIII
	863	Leopard's Hill-PN
Zambia	864	Namakala
Zimbabwe	870	Redcliff Cave

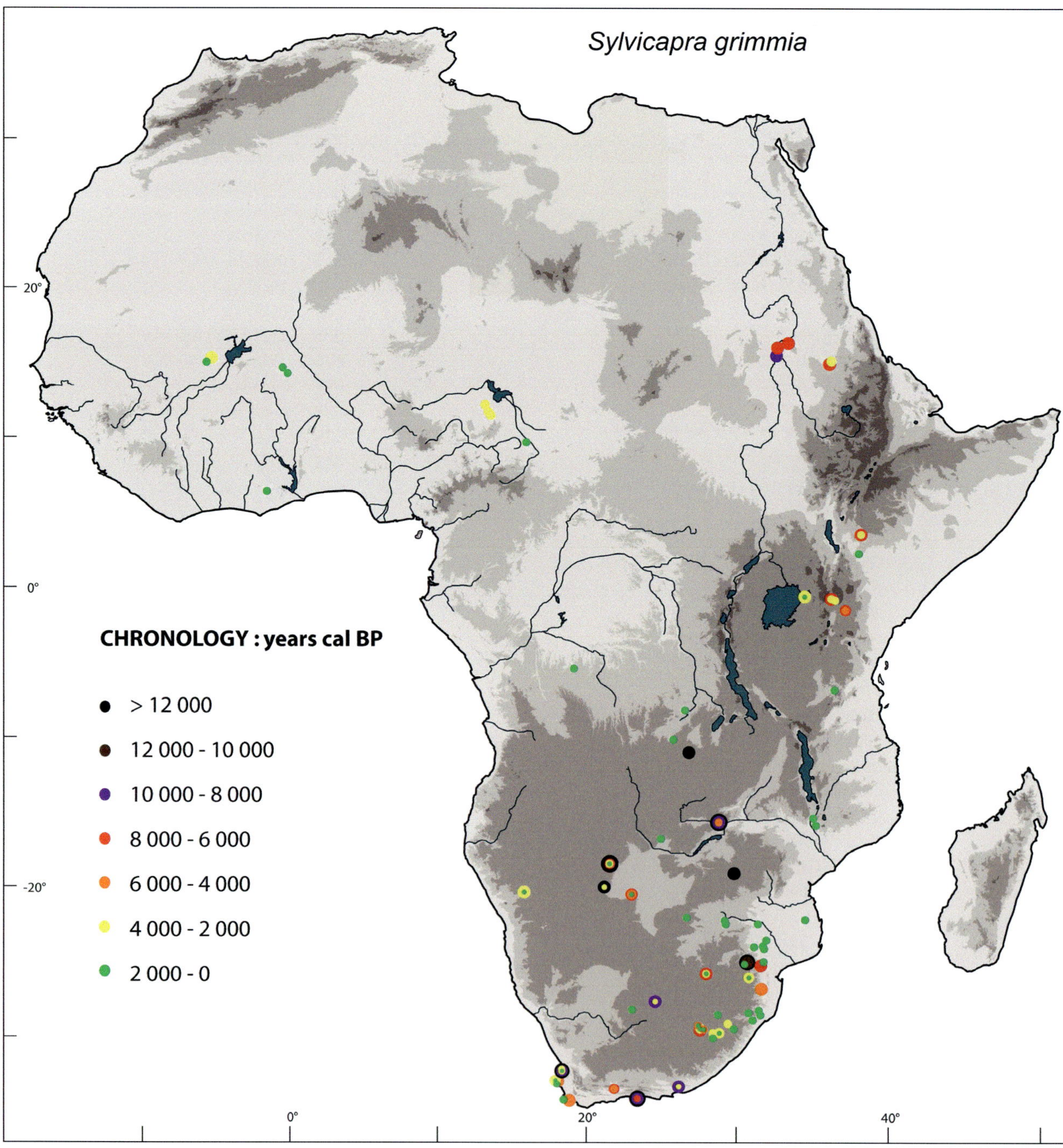

Hippotraginae

Hippotragus sp.

Country	Site N	Site	
Kenya	273	Gogo Falls-II-III	
Lesotho	304	Ha Makotoko	
	316	Sehonghong-a	
Morocco	422	El Harhoura 1	
South Africa	563	Boomplass Cave-a	
	564	Boomplass Cave-b	
	565	Boomplass Cave-c	
	566	Boomplass Cave-d	
	581	Colwinton Shelter-CW	
	604	Equus Cave 1A	
	616	Heuningneskrans-I/O	
	617	Heuningneskrans-R	
	622	HKLP Vb	
	653	Malwula	
	659	Marupale Hill	
	668	Nelson Bay Cave- PW	
	669	Nelson Bay Cave- R	
	670	Nelson Bay Cave- W	
	694	Ravenscraig Shelter-pCW	
	739	Tsh1	
Zaire	848	Kiantapo	
	855	Matupi-II	*?
	855	Matupi-II	
Zimbabwe	870	Redcliff Cave	

Hippotragus equinus

Country	Site N	Site	
Algeria	20	Tamar Hat	
Botswana	48	White Paintings RS-2ab	*cf.
	49	White Paintings RS-2b-3ab	*cf.
Burkina Faso	63	Saouga 94/120	
Cameroon	74	Djaba Hosséré	
Chad	85	Koyom	
Egypt	174	Qantir	
Ghana	230	Bono Manso	
	244	Ntereso	
Kenya	258	Enkapune Ya Muto - Eb 4a	
	259	Enkapune Ya Muto - Eb 4b	
	272	Gogo Falls-I	
	273	Gogo Falls-II-III	
	297	Ngenyn	
Lesotho	316	Sehonghong-a	
Mali	369	Est-Hassi el Abiod	
	374	Hassi el-Abiod	
	380	Karkarichinkat N-S	
	387	Tessalit	
Mauritania	400	Dhar Tichitt Site 38	
	401	Dhar Tichitt Site 46	
Niger	473	Gobero-K	
Nigeria	505	Gajiganna B2	
Senegal	540	Cubalel-II	
South Africa	574	Bushman RS-LSA	
	575	Bushman RS-MSA	
	589	Doornfontein	
	620	HKLP Ib	
	641	Le6	
	648	Ma38	
	687	Pjene Hill	
	691	Pr34	
	707	Rose Cottage Cave-WI	*cf.
	718	Sk4	
Sudan	754	Abu Darbein	
	756	Abu Tabari 02/1	
	757	Aneibis	
	764	El Damer	
	771	El Zakiab	
	773	Esh Shaheinab	
	781	Jebel Umm Marrahi	
	787	Khartoum Hopital	
	788	Khasm el Girba - Kassala	
	791	Khasm el Girba - KG68	
	804	Rahib-80/87	
	805	Saggai 1	
	812	Shaqaduq-S1-B	
	813	Shaqaduq-S21	
	822	Upper Atbara Valley	
Tanzania	835	Mumba	
Zaire	843	Ishango-AB	
	845	Ishango-Z-Post-Em	
	848	Kiantapo	

Hippotragus leucophaeus

Country	Site N	Site	
Lesotho	304	Ha Makotoko	*cf.
	315	Ntloana Tsoana	*cf.
	322	Tloutle-O	
South Africa	576	Byeneskranskop Cave 1a	
	577	Byeneskranskop Cave 1b	
	578	Byeneskranskop Cave 1c	
	579	Byeneskranskop Cave 1d	
	581	Colwinton Shelter-CW	*cf.
	595	Eland´s Bay Cave-AI	
	598	Eland´s Bay Cave-Wib	
	608	Forest Hall	*cf.
	609	Glentyre	*?
	671	Oakhurst	
	705	Rose Cottage Cave-DB2	
	706	Rose Cottage Cave-OIC	

Hippotragus niger

Country	Site N	Site	
Botswana	39	Qogana	
Malawi	356	Matope Court-EIA	*cf.
	358	Matope Court-LIA	
Mozambique	441	Manekweni-1	
	442	Manekweni-2-ABC	
	444	Manekweni-4	
	445	Manekweni-5	
South Africa	574	Bushman RS-LSA	
	589	Doornfontein	
	594	Eiland Salt Works-b	
	622	HKLP Vb	*cf.
	624	Jubilee Shelter-O	*cf.
	625	Jubilee Shelter-PW	*cf.
	626	Jubilee Shelter-W	*cf.
	642	Le7a	
	673	Ol20	
	685	Ph6	
	718	Sk4	

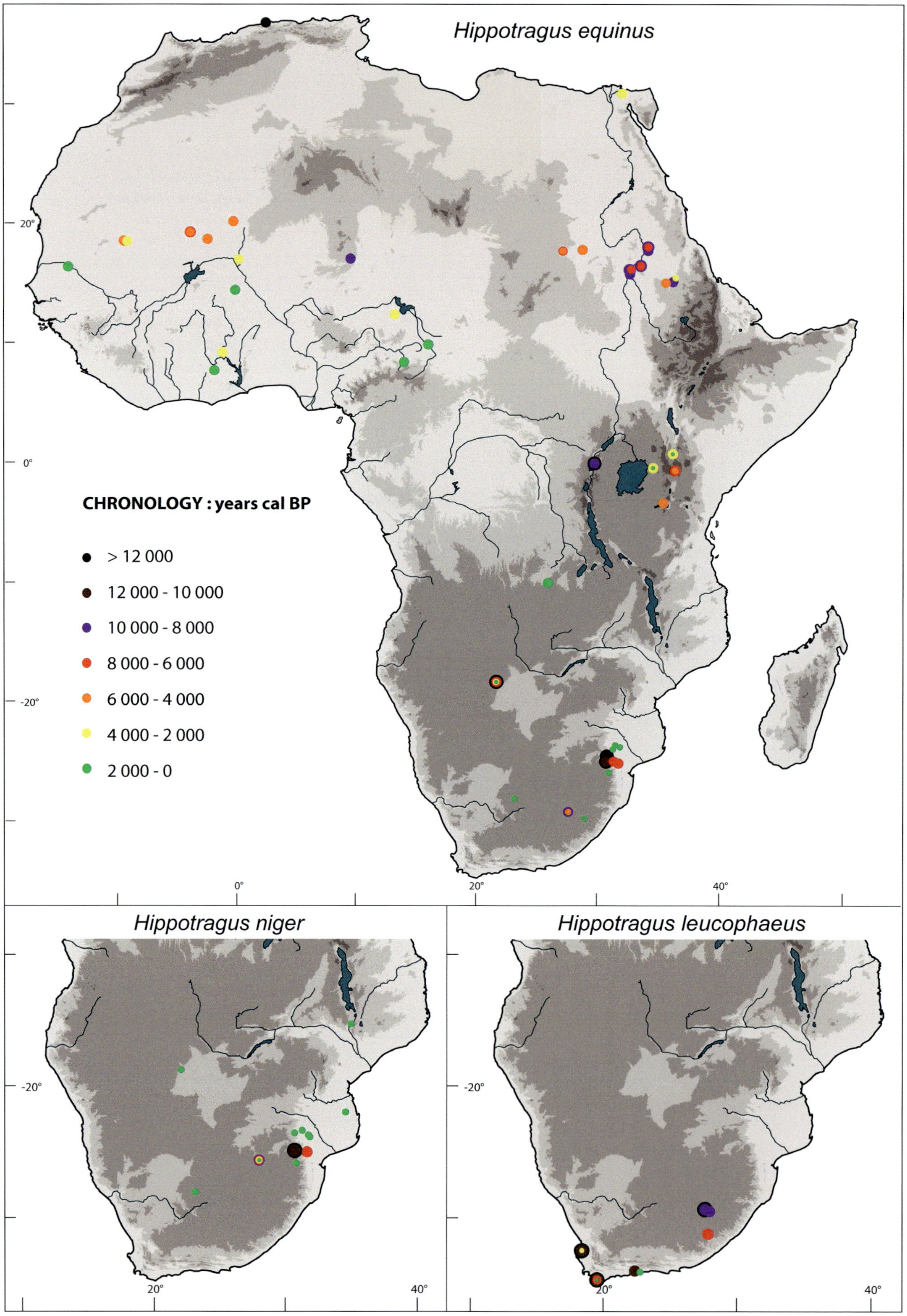
Hippotragus equinus
CHRONOLOGY : years cal BP
> 12 000
12 000 - 10 000
10 000 - 8 000
8 000 - 6 000
6 000 - 4 000
4 000 - 2 000
2 000 - 0
20°
0°
-20°
0°
20°
40°
Hippotragus niger
-20°
20°
40°
Hippotragus leucophaeus
-20°
20°
40°

Addax nasomaculatus

Country	Site N	Site	
Egypt	126	Eastpan 95-1/1	
	127	Eastpan 95-2	
	153	Gilf Kebir	
	177	Qasr el-Sagha-3	
	199	Wadi el Bakht	
Mauritania	396	Dhar Tichitt- latest phase	
	400	Dhar Tichitt Site 38	
	401	Dhar Tichitt Site 46	
	405	Khatt Lemaiteg	
	411	Tichitt - Akjinjeir phase	
Sudan	793	Laqiya area - 82/29	
	794	Laqiya area - 82/31	
	827	Wadi Hariq 97/7	

Oryx sp.

Country	Site N	Site
Algeria	22	Zmeilet el-Barka
Libya	332	Murzuq
Mauritania	394	Chami-A2
	408	Tegdaoust-b
	409	Tegdaoust-c
Morocco	422	El Harhoura 1

Oryx beisa

Country	Site N	Site	
Kenya	254	Ele Bor A-B	*cf.
	255	Ele Bor A-C	*cf.
	267	GaJi1	
	270	GaJi2	
Sudan	755	Abu Hugar	
Tanzania	835	Mumba	

Oryx dammah

Country	Site N	Site	
Algeria	6	Columnata	*aff.
	16	Meniet	*O. Leucoryx?
	21	Ti-n-Hanakaten	
Egypt	153	Gilf Kebir	
	180	Selima Sandsheet - 85/78	
	181	Selima Sandsheet - 85/79	
	205	Willmann´s camp - 81/61	
	92	Abu Ballas - 85/50-1	
	93	Abu Ballas - 85/51-3	
	94	Abu Ballas - 85/56	
Mauritania	400	Dhar Tichitt Site 38	
	401	Dhar Tichitt Site 46	
	402	DN4-BKH-S5	
Morocco	431	Kaf-taht-el-Ghar-C	
	435	Kaf-taht-el-Ghar-G	
Niger	474	Gobero-T	
	495	Tin Ouaffadene	
	497	TMY3	
Sudan	793	Laqiya area - 82/29	
	794	Laqiya area - 82/31	
	808	Selima Sandsheet - 85/73-2	
	824	Wadi Hariq 01/1	
	825	Wadi Hariq 01/4	
	826	Wadi Hariq 97/5	
	827	Wadi Hariq 97/7	

Oryx gazella

Country	Site N	Site	
Botswana	28	Bosutwe-Toutswe	
Kenya	277	GvJm 19-TLPG	
	279	GvJm 22-LPG	
	297	Ngenyn	
Namibia	446	Aar I and II	*cf.
	447	Apollo 11-a	*cf.
	449	Apollo 11-c	*cf.
	450	Brandberg-BBC	
	458	Namtib	*cf.
	461	Pokkenbank	*cf.
	463	Tiras	*cf.
South Africa	589	Doornfontein	

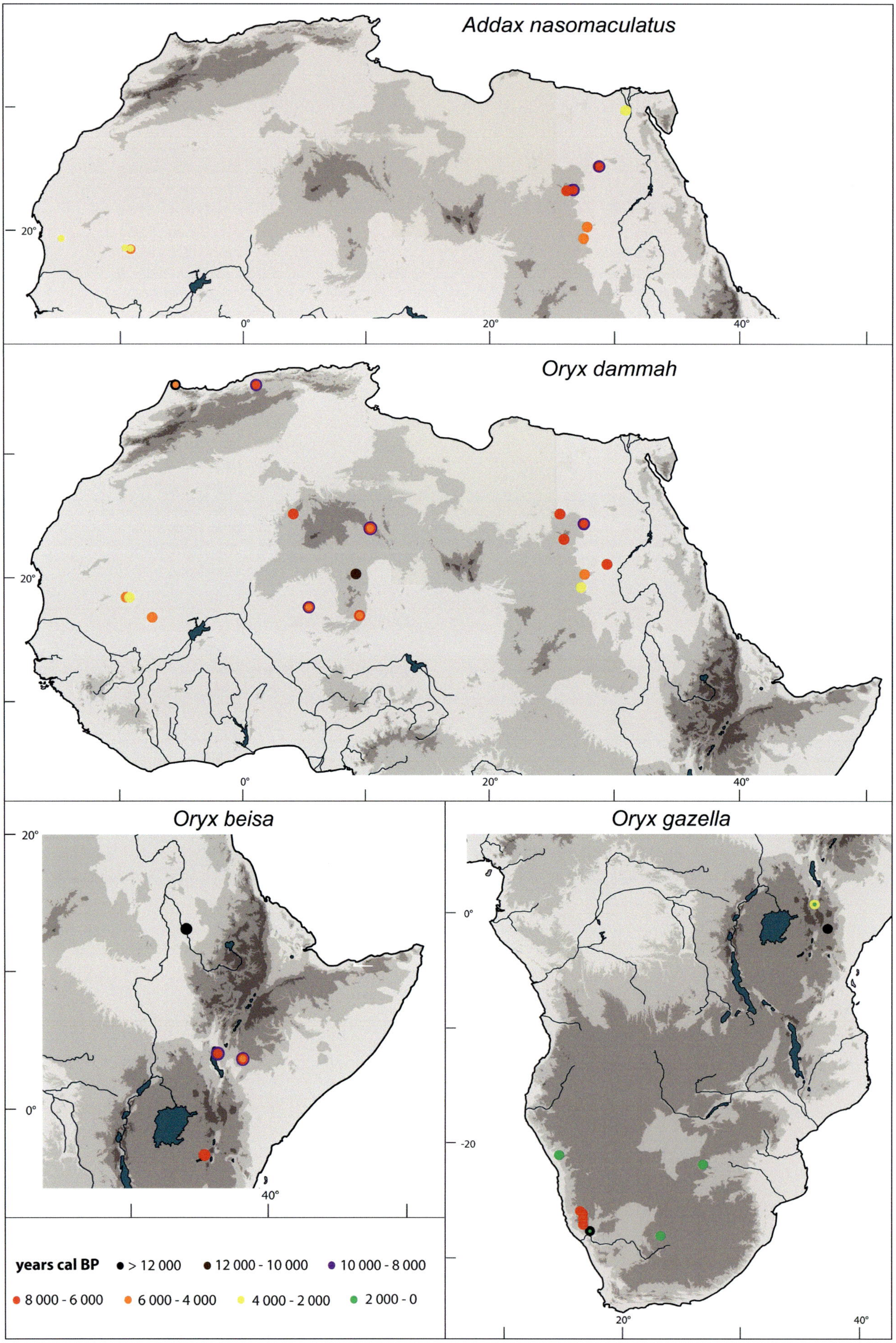
Addax nasomaculatus
20°
0°
20°
40°
Oryx dammah
20°
0°
20°
40°
Oryx beisa
20°
0°
40°
Oryx gazella
0°
-20
20°
40°
years cal BP
> 12 000
12 000 - 10 000
10 000 - 8 000
8 000 - 6 000
6 000 - 4 000
4 000 - 2 000
2 000 - 0

Cervidae

Megaceroides algericus

Country	Site N	Site
Algeria	20	Tamar Hat

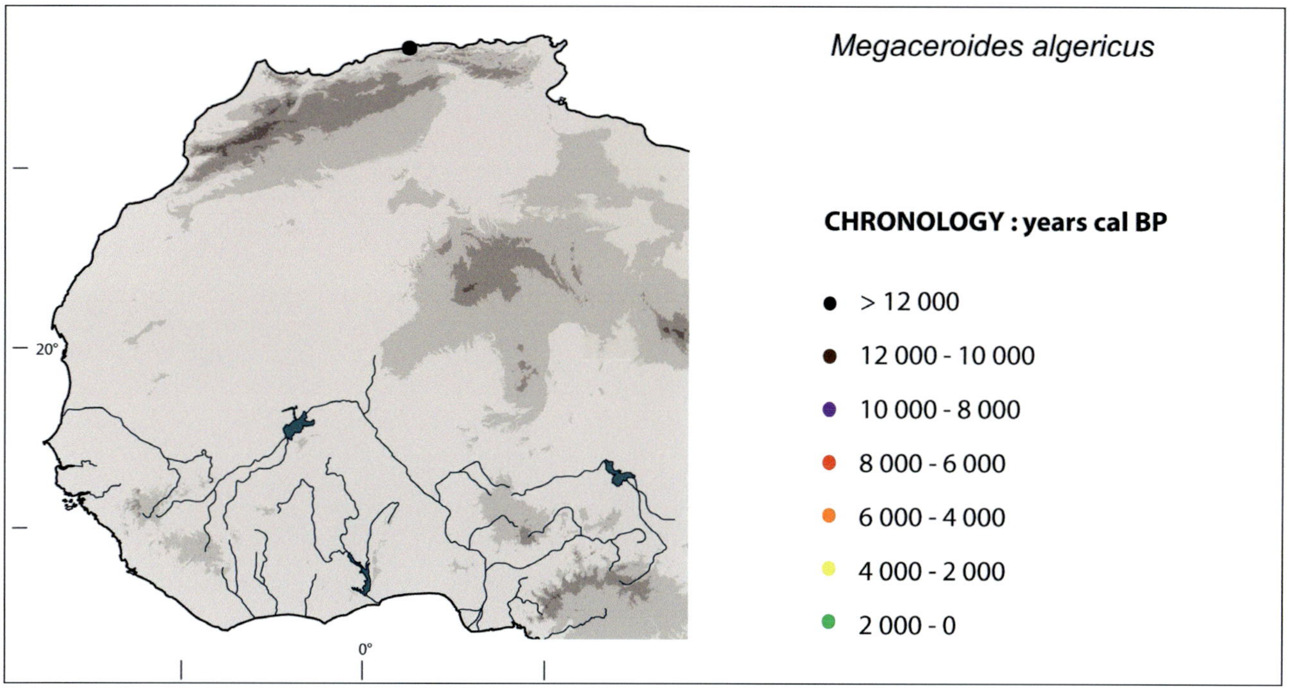

GIRAFFIDAE

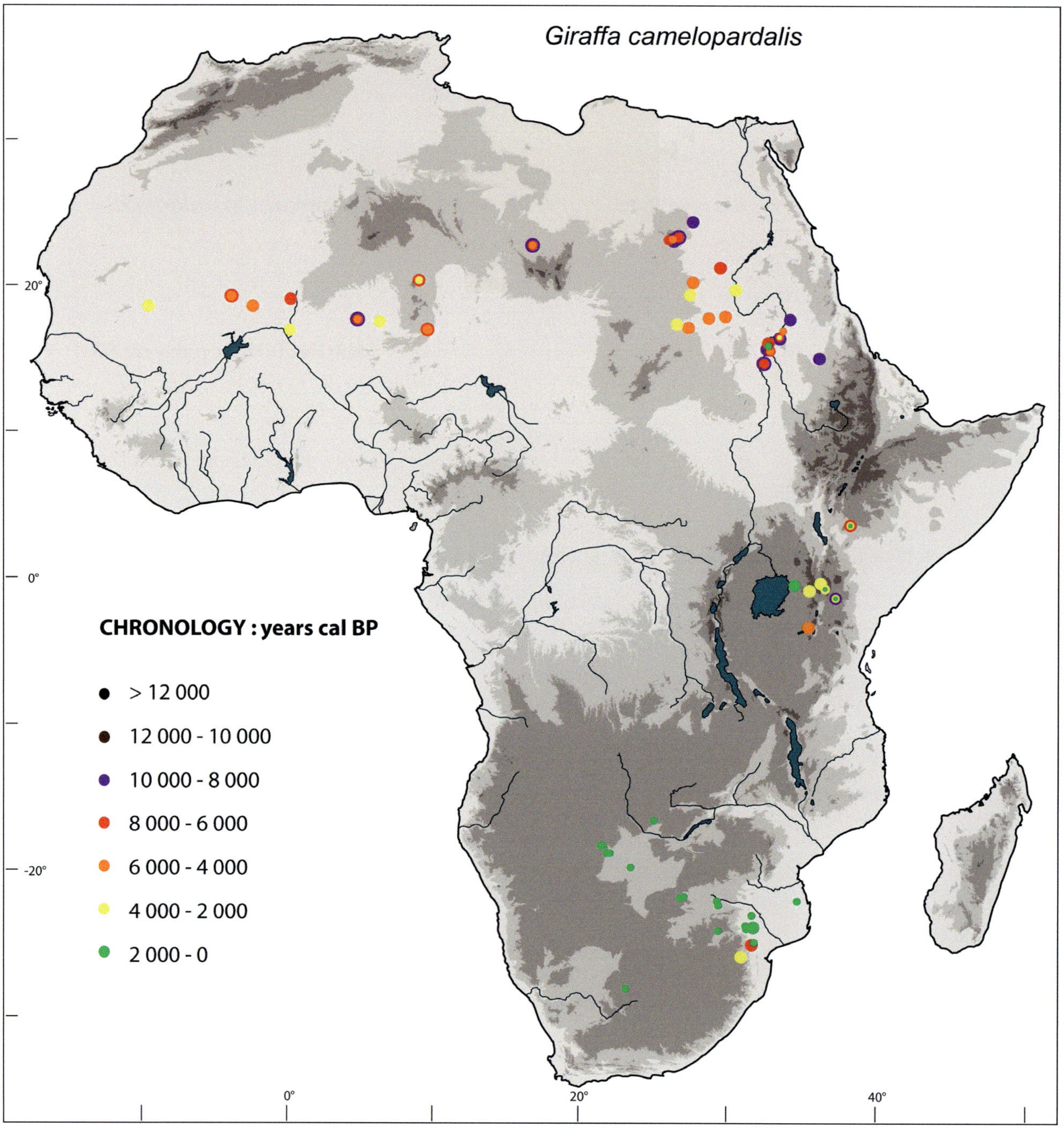

Giraffa camelopardalis

Country	Site N	Site	
Botswana	26	Bosutwe-Mapungubwe	
	28	Bosutwe-Toutswe	*cf.
	28	Bosutwe-Toutswe	
	31	Commando Kop	
	32	Divuyu	
	37	Matlapaneng	
	38	Nqoma	
	39	Qogana	
	40	Taukome	
	48	White Paintings RS-1	
Chad	82	Enneri Bardagué	
Egypt	153	Gilf Kebir	
	198	Wadi el Akhdar	
	199	Wadi el Bakht	
	91	Abu Ballas - 83/39	
Kenya	250	Causeway Site	
	251	Crescent Island main	
	253	Ele Bor A-A	
	254	Ele Bor A-B	
	255	Ele Bor A-C	
	256	Ele Bor A-D	
	273	Gogo Falls-II-III	
	275	GvJm 19-HA	
	278	GvJm 22-LHC	
	294	Narosura	
	298	Prolonged Drift	
Mali	363	Asselar	
	369	Est-Hassi el Abiod	

Country	Site N	Site	
	374	Hassi el-Abiod	
	380	Karkarichinkat N-S	
Mauritania	395	Dhar Tichitt- highest beach	
	396	Dhar Tichitt- latest phase	
	399	Dhar Tichitt- post beach	
Mozambique	441	Manekweni-1	
Niger	467	Adrar Bous-T	
	471	Chin Tafidet	
	474	Gobero-T	
	488	Takene Bawat 2	
South Africa	589	Doornfontein	
	606	Ficus B	
	607	Ficus C	
	618	Hill X	
	619	HKLP Ia	
	621	HKLP Va	
	623	Icon	*cf.
	640	Le2	
	641	Le6	
	642	Le7a	
	659	Marupale Hill	
	664	Mo8	
	673	Ol20	
	689	Pont-Drift-Zhizo	
	710	Selongwe Hill	
	713	SH16	
	714	Sh27	
	715	Shankare Hill-a	
	716	Shankare Hill-b	
	717	SK17	
	718	Sk4	
	720	Sonkoanini Hill	
	731	Ti3	
	739	Tsh1	
Sudan	756	Abu Tabari 02/1	
	763	Djarabona 84/13	
	764	El Damer	
	765	El Kadada	
	770	El Temeyim	
	773	Esh Shaheinab	
	776	Gebel Tageru	
	777	Geili	
	782	Kabbashi	
	785	Kerma	
	786	Khabbashi Haitah	
	787	Khartoum Hopital	
	791	Khasm el Girba - KG68	
	793	Laqiya area - 82/29	
	794	Laqiya area - 82/31	
	796	Lower Wadi Howar-5	
	805	Saggai 1	
	808	Selima Sandsheet - 85/73-2	
	810	Shabona	
	811	Shaqaduq-S1-A	
	812	Shaqaduq-S1-B	
	813	Shaqaduq-S21	
	814	Sheikh el Amin	
	817	Soba	
	820	Umm Singid	
	824	Wadi Hariq 01/1	
	826	Wadi Hariq 97/5	
Tanzania	835	Mumba	
Zambia	864	Namakala	

Okapia johnstoni

Country	Site N	Site
Zaire	855	Matupi-II

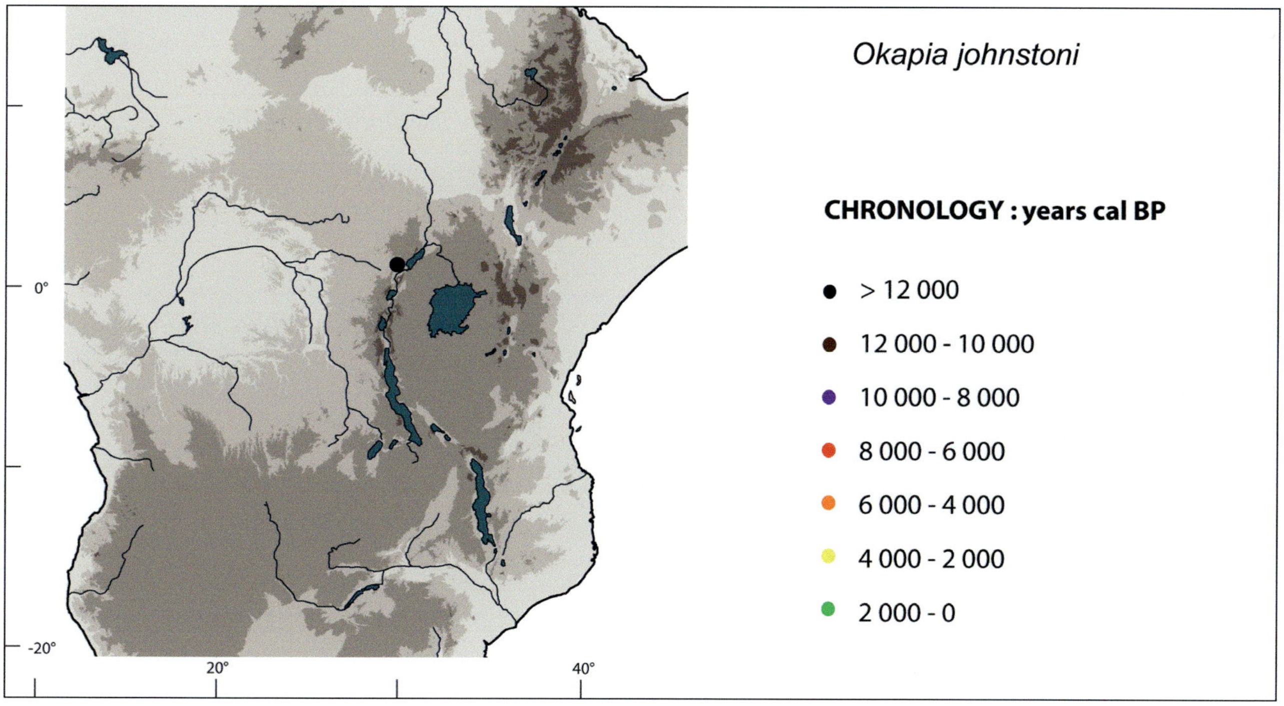

Order Cetacea

Cetacea

Country	Site N	Site
South Africa	586	Die Kelders *whale
West. Sahara	839	Sebkha Mahariat *sperm whale/seal
	841	Site H

Delphinidae

Country	Site N	Site
South Africa	586	Die Kelders
	628	Kasteelberg B
	631	Klasies River Mouth-1I
	633	Klasies River Mouth-5II
	668	Nelson Bay Cave- PW
	670	Nelson Bay Cave- W

Delphinus delphis

Country	Site N	Site
South Africa	719	Smitswinkelbaai Cave

Sousa teuszii

Country	Site N	Site
Senegal	542	Ndiamon-Badat

Tursiops sp.

Country	Site N	Site
West. Sahara	837	Sebkha Amtal

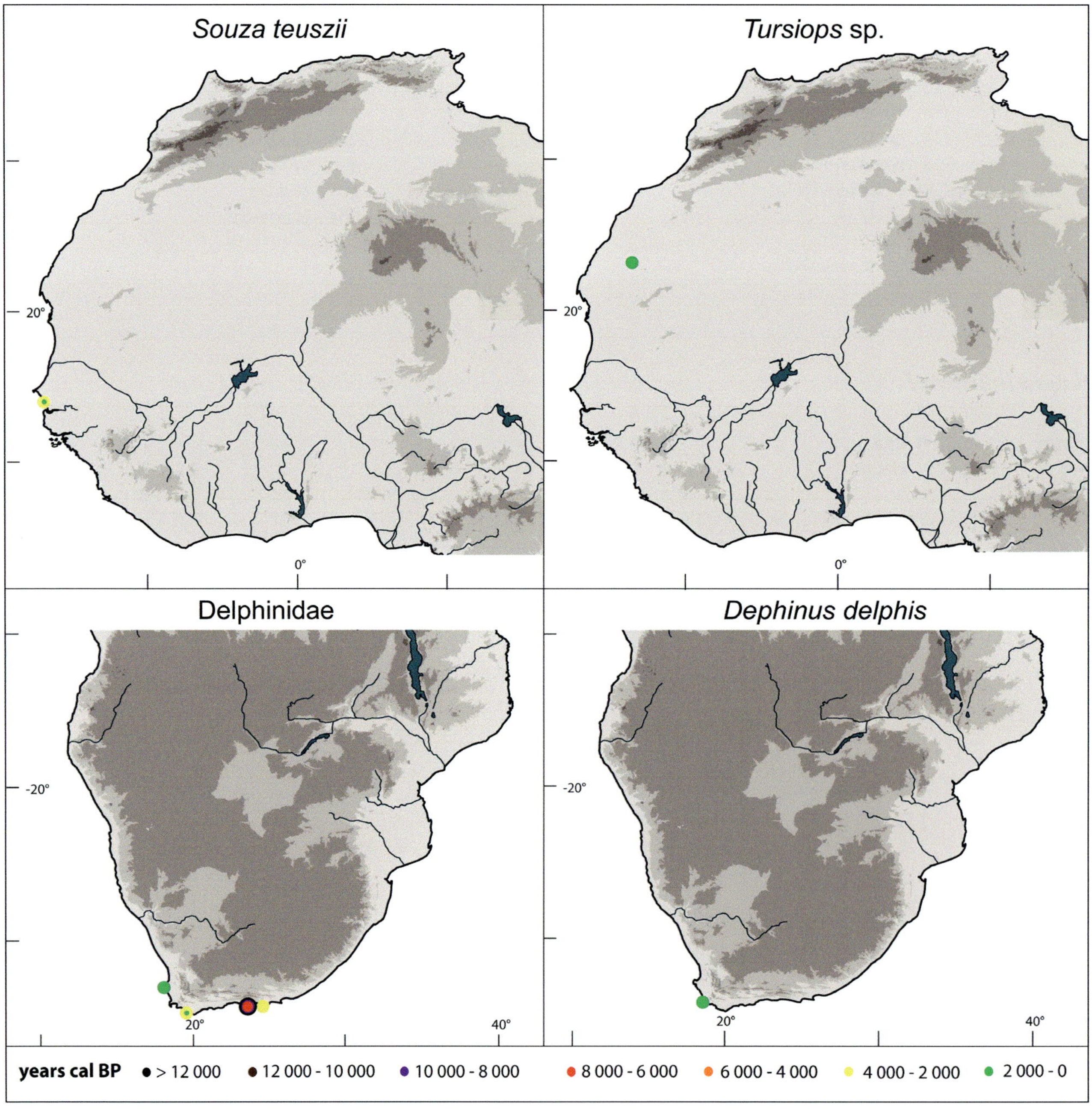

Chapter 6 References

A

1. ADAMSON, D., CLARK, J. D. & WILLIAMS, M. A. J. 1974. Barbed bone points from Central Sudan and the age of the Early Khartoum tradition. *Nature* 249:120-123.
2. ALHAIQUE, F. 2002. Archaeozoology of the funerary structures, in DI LERNIA, S. & MANZI G. (eds), *Sand, Stones and Bones. The archaeology of death in the Wadi Tannezzuft Valley (5000-2000 bp). The Archaeology of Libyan Sahara,* Arid Zone Archaeology Monographs 3, pp. 181-196.
3. ALIMEN, H., BEUCHER, F., MATEU, J. & ROUBET, C. 1980. Zmeilet el-Barka, halte saisonnière de chasseurs au VIème millénaire B.C. (Sahara algérien nord-occidental). *L'Anthropologie* 84:198-242.
4. AMBLARD-PISON, S. 2006. *Communautés villageoises néolithiques des Dhars Tichitt et Oualata (Mauritanie).* Oxford: BAR International Series 1546.
5. AOURAGHE, H. 2000. Les carnivores fossiles d'El Harhoura 1, Temara, Maroc. *L'Anthropologie* 104:147-171.
6. ARAMBOURG, C. 1948. Observations sur le Quaternaire de la région du Hoggar. *Travaux de l'Institut de Recherches Sahariennes* V:7-18.
7. AVERY, D. M. 1992. Faunal remains from Klein Kliphuis, Clanwilliam District, South Africa. *South African Archaeological Bulletin* 47:40-43.
8. AVERY, D. M. 1992. Micromammals and the environment of early pastoralists at Spoeg River, Western Cape Province, South Africa. *South African Archaeological Bulletin* 47:116-121.
9. AVERY, D. M. 1999. Holocene coastal environments in the Western Cape Province, South Africa: micromammalian evidence from Steenbokfontein, in PLUG, I. & KLEIN, R. (eds), *Archaeozoology in Africa, Archaeozoologia* X/1-2, pp. 163-180.
10. AVERY, D. M., CRUZ-URIBE, K., GOLDBERG, P., GRINE, F. E., KLEIN, R. G., LENARDI, M. J., MAREAN, C. W., RINK, W. J., SCHWARCZ, H. P., THACKERY, A. I. & WILSON, M. L. 1997. The 1992-1993 excavations at the Die Kelders Middle and Late Stone Age cave site, South Africa. *Journal of Field Archaeology* 24:263-291.

B

11. BA, M., DESCAMPS, C. & THILMANS, G. 1997. Fouille d'un tumulus à Ndiamon-Badat (Iles du Saloum, Sénégal). *Saint-Louis - Lille - Liège* 3:1-14.
12. BARKER, G. 1978. Economic models for the Manekweni, Zimbabwe, Mozambique. *Azania* 13:71-100.
13. BARTON, R. N. E., BOUZOUGGAR, A., COLLCUTT, S. N., GALE, N., HIGHAM, T. F. G., HUMPHRERY, L. T., PARFITT, S., RHODES, E., STRINGER, C. B. & MALEK, F. 2005. The Late Upper Palaeolithic occupation of the Moroccan Northwest Maghreb during the Last Glacial Maximum. *African Archaeological Review* 22:77-100.
14. BATHILY, M. S., OULD KHATTAR, M., VERNET, R., CLUZEL, C. & OTT, J. M. 1998. *Les sites néolithiques de Khat Lemaiteg (Amatlich) en Mauritanie occidentale.* Meudon: CMA, UPR 311.
15. BEAUMONT, P. & BOSHIER, A. 1974. Report on test excavations in a prehistoric pigment mine near Postmasburg, Northern Cape. *South African Archaeological Bulletin* 29:41-59.
16. BEDAUX, R., MACDONALD, K., PERSON, A., POLET, J., SANOGO, K., SCHMIDT, A. & SIDIBÉ, S. 2001. The Dia archaeological project: rescuing cultural heritage in the Inland Niger Delta (Mali). *Antiquity* 75: 837-848.
17. BOESSNECK, J. 1976. Tell el Dab'a III. Die Tierknochenfunde 1966-69. *Österreischische Akademie der Wissenschaften, Denkschriften der Gesamt-Akademie* 5.
18. BOESSNECK, J. 1988. Tierknochenfunde aus der Spätzeit (6.-4. Jh. v. Chr.) aus Minshat Abu Omar (Ägypten). *Fontes Archaeologici Posnanienses* 36:217-220.
19. BOESSNECK, J. & VON DEN DRIESCH, A. 1982. Studien an subfossilen Tierknochen aus Ägypten. *Münchner Ägyptologische Beiträge* 40.
20. BOESSNECK, J. & VON DEN DRIESCH, A. 1987. *Tuna el-Gebel I. Die Tiergalerien - Die Tierknochenfunde.* Hildesheimer Ägyptologische Beiträge 24.
21. BOESSNECK, J. & VON DEN DRIESCH, A. 1989. Vogelknochenfunde aus dem Tal der Königinnen. *Spixiana* 11:279-302.
22. BOESSNECK, J. & VON DEN DRIESCH, A. 1993. Eine Außergewöhnliche Tierknochendeponie in einem Gebäude der 25./26. Dynastie im Stadtgebiet nordwestlich des späten Chnumtemples auf Elephantine. *Mitteilungen des Deutschen Archäologischen Instituts Abteilung Kairo* 49:189-201.

23. BOESSNECK, J., VON DEN DRIESCH, A. & EISSA, A. 1992. Eine Eselbestattung der 1. Dynastie in Abusir. *Mitteilungen des Deutschen Archäologischen Instituts Abteilung Kairo* 48:1-10.
24. BOUCHUD, J. 1969. Etude des mammifères et des oiseaux d'Amekni, in CAMPS, G. (ed.) *Amekni. Néolithique ancien du Hoggar*, Mémoires du Centre de Recherches Anthropologiques Préhistoriques et Ethnographiques 10, Paris: pp. 173-177.
25. BOUCHUD, J. 1975. La faune de Medjez II, in CAMPS-FABRER, H. (ed.), *Un gisement capsien de faciès sétifien : Medjez II, el-Eulma (Algérie)*, CNRS, pp. 377-391.
26. BOUCHUD, J. 1983. Paléofaune de Tegdaoust, in BABACAR, A. D., BAH, T. H., BOUCHUD, J., COLIN, G. S., DEVISSE, J. & GHALI, N. (eds.) *Tegdaoust III - Recherches sur Aoudaghost. Campagnes 1960-65*, Enquêtes Générales Mémoires 4025, Paris, pp. 355-363.
27. BOUCHUD, J., BREBION, P. & SABAN, R. 1981. Etude des faunes holocènes provenant de la zone aride du Sahara Atlantique : les mammifères de la Mauritanie occidentale, in ROUBET, C., HUGOT, H. J. & SOUVILLE, G. (eds.), *Préhistoire africaine - Mélanges offerts au doyen L. Balout. Recherches sur les grandes civilisations 6,* Paris: A.D.P.F., pp. 237-250.
28. BOURGES, C., MACEACHERN, S. & REEVES, M. 1999. Excavations around Aissa Hardè, 1995 and 1996. *Nyame Akuma* 51:6-13.
29. BOUZOUGGAR, A., BARTON, R. N. E., BLOCKEY, S., BRONK-RAMSEY, C., COLLCUTT, S. N., GALE, R., HIGHAM, T. F. G., HUMPHRERY, L. T., PARFITT, S., TURNER, E. & WARD, S. 2008. Reevaluating the Age of the Iberomaurusian in Morocco. *African Archaeological Review* 25:3-19.
30. BREUNIG, P., NEUMANN, K. & VAN NEER, W. 1996. New research on the Holocene Settlement and Environment of the Chad Basin in Nigeria. *African Archaeological Review* 13:111-145.
31. BREWER, D. J. 1989. *Fishermens, Hunters and Herders. Zooarchaeology in the Fayum, Egypt (ca. 8200-5000 bp).* BAR International Series 478.
32. BRINK, J. & HOLT, S. 1992. A small goat, *Capra hircus*, from a Late Iron Age site in the Eastern Orange Free State. *South African Field Archaeology* 1:88-91.
33. BROOKS, A. S. & SMITH, C. 1987. Ishango revisited: new age determination and cultural interpretations. *African Archaeological Review* 5:65-78.

C

34. CAIN, C. R. 2000. *Animals at Axum: initial zooarchaeological research in the later prehistory of the northern Ethiopian Highlands*, PhD thesis, Dept. of Anthropology, Washington University.
35. CAMPS, G. 1969. *Amekni. Néolithique ancien du Hoggar.* Mémoires du Centre de Recherches Anthropologiques Préhistoriques et Ethnographiques 10. Paris.
36. CAMPS-FABRER, H. 1989. Capsien du Maghreb et Natoufien du Proche-Orient. *Travaux du LAPMO*:71-104.
37. CANEVA, I. 1983. Radiocarbon dates from Saggai 1 (Sudan). *Origini* 12:149-154.
38. CANEVA, I., GARCEA, E. A. A., GAUTIER, A. & VAN NEER, W. 1993. Pre-pastoral cultures along the Central Sudanese Nile. *Quaternaria Nova* 3:177-252.
39. CANEVA, I. & GAUTIER, A. 1994. The desert and the Nile: sixth millenium pastoral adaptions at Wadi el Kenger (Khartoum). *Archéologie du Nil Moyen* 6:65-92.
40. CARBONNEL, J. P. & BARBEY, C. 1972. Découverte de sépultures néolithiques dans le complexe dunaire du Draa Malichigdane (Mauritanie). *Notes Africaines* 136:108-111.
41. CARTER, P. L. & CLARK, J. D. 1976. Adrar Bous and African cattle. *Actes du VIIème Congrès Panafricain de Préhistoire et d'Etude du Quaternaire, Addis-Abeba, 1976*, pp. 487-493.
42. CARTER, P. L. & FLIGHT, C. 1972. A report on the fauna from the sites of Ntereso and Kintampo Rock Shelter Six in Ghana, with evidence for the practice of animal husbandry during the second millenium B.C. *Man* 7:277-282.
43. CASSOLI, F. & DURANTE, S. 1974. La fauna del Ti-n-Thora (Acacus, Libia). *Pubblicazione (seconda serie) Instituto Italiano di Paleontologica Umana*:1-3.
44. CHAÏD-SAOUDI, Y. 1987. *Les mammifères holocènes des gisements préhistoriques de Gueldaman-Akbou (Bédjaia), Columnata (Tiaret) et Ti-n-Hanakaten (Djanet) en Algérie*, PhD Thesis, Université Claude-Bernard Lyon1, Villeurbanne.
45. CHAÏD-SAOUDI, Y. 2007. Les Bovidés du site holocène de Mankhor (Sahara Central). Observations paléontologiques, archéozoologiques et taphonomiques. *ATHAR* 6:1-17.
46. CHAIX, L. 1987. Rapport préliminaire sur la faune du site de Kadruka I, Soudan Nord (Néolithique et Protohistorique), in REINOLD, J. (ed.), *Les fouilles pré- et proto-historiques de la section française de la direction des antiquités du Soudan : Les campagnes 1984-85 et 1985-86*, Archéologie du Nil Moyen 2., pp. 61-62.

47. CHAIX, L. 1990. Analyse des restes osseux, in HUYSECOM, E. (ed.), *Fanfannyégèné I, un abri sous roche à occupation néolithique du Mali - La fouille, le matériel archéologique, l'art rupestre*. Wiesbaden, F. S. Verlag, pp. 76-77.

48. CHAIX, L. 1993. The archaeozoology of Kerma (Sudan), in DAVIES, W. V. & WALKER, R. (eds.), *Biological Anthropology and the Study of ancient Egypt*. London: British Museum Press, pp. 175-185.

49. CHAIX, L. 1998. The Fauna, in Welsby, D. A. (ed.), *Soba II. Renewed excavations within the metropolis of the Kingdom of Alwa in Central Sudan*, London: Memoirs of the British Institute in Eastern Africa 15. British Museum Press, pp. 233-255.

50. CHAIX, L. 2000. An Hyksos horse from Tell Heboua (Sinaï, Egypt), in MASHKOUR, M., CHOYKE, A. M., BUITENHUIS, H. & POPLIN, F. (eds.), *Archaeozoology of the Near East*. Ivb, Groningen, pp. 177-186.

51. CHAIX, L. 2003. La faune des sites mésolithiques et néolithiques de la zone du Nil Bleu (Soudan Central). *Complutum* 14:373-396.

52. CHAIX, L. 2006. New data about rural economy in the Kerma culture: the site of Gism el-Arba (Sudan), *Archaeology of Early Northeastern Africa,* Studies in African Archaeology 9, pp. 25-37.

53. CHAIX, L. & GRATIEN, B. 2002. Un cheval du Nouvel Empire à Sai (Soudan). *Archéologie du Nil Moyen* 9:53-61.

54. CHAIX, L. & OLIVE, C. 1986. Annexe II: La faune du Mastaba V (2200 BC) à Balat (République Arabe d'Egypte), in *Balat I. Le Mastaba de Medou-Nefer*, FIFAO. vol. 31/1, pp. 201-213.

55. CHANTRET, F. & DE BAYLE DES HERMENS, R. 1968. Le gisement préhistorique de Madaouéla, République du Niger. Une nouvelle datation pour le néolithique saharien. *Bulletin de la Société préhistorique de France* 65:623-628.

56. CHAVANE, B. 1980. Extrait : *Recherches archéologiques sur la moyenne vallée du fleuve Sénégal*. PhD Thesis, Université d'Aix-en-Provence.

57. CHENAL-VÉLARDÉ, I. 1996. Etude taphonomique, observations ethnologiques et interprétations archéozoologiques : essai sur les techniques de boucherie à Hamdallahi (Mali, XIXe siècle). *Anthropozoologica* 23:85-95.

58. CHURCHER, C. S. 1993. Dogs from Ein Tirghi Cemetery, Balat, Dakhleh Oasis, Western Desert of Egypt, in CLASON, A., PAYNE, S. & UERPMANN, H.-P. (eds.), *Skeletons in her Cupboard : Festschrift fot Juliet Clutton-Brock*, Oxbow Books 34, 4, Oxford, pp. 39-60.

59. CHURCHER, C. S. 1999. Holocene faunas of the Dakhleh Oasis, in CHURCHER, C. S. & MILLS, A. J. (eds.), *Reports from the survey of the Dakleh Oasis, Western Desert of Egypt, 1977-1987*, Oxford Monograph 99, Oxford, pp. 133-170.

60. CHURCHER, C. S. 2008. Shrews from Ein el Gazzareen, Dakhleh Oasis, Western Desert, Egypt, in BADENHORST, S., MITCHELL, P. & DRIVER, J. C. (eds.), *Animals and People : Archaeozoological Papers in Honour of Ina Plug*, British Archaeological Reports 1849, Oxford, pp. 75-82.

61. CHURCHER, C. S. & SMITH, P. E. L. 1972. Kom Ombo: Preliminary report on the fauna of Late Paleolithic sites in Upper Egypt. *Science* 177:259-261.

62. CLARK, J. D. 1973. (no title). *Nyame Akuma* 3:56-64.

63. CLIST, B. 1997. Le site d'Okala, Province de l'Estuaire, Gabon et son importance pour la compréhension du passage à la sédentarisation en Afrique Centrale. *Comptes Rendus de l'Académie des Sciences* 325:151-156.

64. CLOSE, A. E. 2002. Sinai, Sahara, Sahel: the introduction of domestic caprines to Africa, in *Tides of the Desert*, Africa Praehistorica 14, Köln, pp. 459-469.

65. CLUTTON-BROCK, J. 1974. The Buhen horse. *Journal of Archaeological Science* 1:89-100.

66. COLLETT, B. P. 1982. Excavations of stone-walled ruin types in the Badfontein Valley, Eastern Transvaal, South Africa. *South African Archaeological Bulletin* 37:34-43.

67. COMMELIN, D. & PETIT-MAIRE, N. 1980. Chronologie isotopique saharienne pour les derniers 10 000 ans. Essai d'interprétation : migrations humaines et paléoclimatologie. *Bulletin du Musée d'Anthropologie Préhistorique de Monaco* 23: 37-88.

68. CORRIDI, C. 1998. Faunal remains from Holocene archaeological sites of the Tadrart Acacus and surroundings (Libyan Sahara), in CREMASCHI, M. & DI LERNIA, S. (eds.), *Wadi Teshuinat, Palaeoenvironment and Prehistory in south-western Fezzan (Libyan Sahara)*. All'Insegna del Giglio, Firenze, pp. 89-94.

69. CREMASCHI, M. 1998. Late Quaternary geological evidence for environmental changes in south-western Fezzan (Libyan Sahara), in CREMASCHI, M. & DI LERNIA, S. (eds.), *Wadi Teshuinat Palaeoenvironment and Prehistory in south-western Fezzan (Libyan Sahara)*. All'Insegna del Giglio, Firenze, pp. 13-48.

70. CRUZ-URIBE, K. 1983. The mammalian fauna from Redcliff Cave, Zimbabwe. *South African Archaeological Bulletin* 38: 7-16.

71. Cruz-Uribe, K. & Schrire, C. 1991. Analysis of faunal remains from Oudepost 1, an early outpost of the Dutch East India Company, Cape Province. *South African Archaeological Bulletin* 46: 92-106.

D

72. Daugas, J.-P. 2002. Le Néolithique du Maroc : pour un modèle d'évolution chronologique et culturelle. *Bulletin d'Archéologie Marocaine* 19:135-171.
73. Daugas, J.-P., Raynal, J.-P., El Idriss, A., Ousmoi, M., Fain, J., Miallier, D., Montret, M., Sanzelle, S., Pilleyre, T., Occhietti, S. & Rhodes, E.-J. 1998. Synthèse radiochronométrique concernant la séquence néolithique du Maroc. *Actes du colloque international C14 et Archéologie, Lyon, 1998*, pp. 349-353.
74. de Maret, P., Clist, B. & Van Neer, W. 1987. Résultats des premières fouilles dans les abris de Shum Laka et Abeke au nord-ouest du Cameroun. *L'Anthropologie* 91:559-584.
75. Deacon, H. J., Deacon, J., Brooker, M. & Wilson, M. L. 1978. The evidence for herding at Boomplaas Cave in the southern Cape, South Africa. *South African Archaeological Bulletin* 33:39-65.
76. Deacon, J. 1972. Wilton: an assessment after fifty years. *South African Archaeological Bulletin* 27:10-48.
77. Deme, A. & McIntosh, S. 2006. Excavations at Walaldé: new light on the settlement of the Middle Senegal Valley by Iron-Using people. *Journal of African Archaeology* 4:317-347.
78. Denbow, J. 1981. Broadhurst - a fourteenth century A.D. expression of the Early Iron Age in Southeastern Botswana. *South African Archaeological Bulletin* 36:66-74.
79. Denbow, J. 1990. Congo to Kalahari: data and hypothesis about the political economy of the western stream of the Early Iron Age. *African Archaeological Review* 8:139-176.
80. Di Lernia, S. & Manzi, G. 2002. *Sand, Stones and Bones. The Archaeology of Death in the Wadi Tanezzuft (5000-2000 BP)*. Arid Zone Archaeology Monographs 3.
81. Dreyer, G., von den Driesch, A., Engel, E.-M., Hartmann, R., Hartung, U., Hikade, T., Müller, V. & Peters, J. 2000. Umm el-Qaab, Nachuntersuchungen im frühzeitlichen Königsfriedhof. *Mitteilungen des Deutschen Archäologischen Instituts Abteilung Kairo* 56:43-129.
82. Durand, A., Paris, F. & Saliège, J.-F. 1999. Vallée de l'Azawagh (Sahara du Niger). I. Peuplements et environnements holocènes du bassin oriental de l'Azawagh (Niger). *Etudes nigériennes* 57.

E

83. Elamin, Y. M. & Mohammed-Ali, A. S. 2004. Umm Marrahi, An Early Holocene ceramic site, north of Khartoum (Sudan). *Sahara* 15:97-110.
84. ElMahi, A. T. 1988. *Zooarchaeology in the Middle Nile Valley*.Oxford: BAR International Series 418.
85. Esterhuysen, A. B., Behrens, J. & Harper, P. 1994. Leliehoek Shelter: a Holocene sequence from the eastern Orange Free State. *South African Archaeological Bulletin* 49:73-78.
86. Evers, T. M. 1980. Klingbell Early Iron Age sites, Lydenburg, Eastern Transvaal, South Africa. *South African Archaeological Bulletin* 35:46-57.
87. Evers, T. M., Voigt, E. A. & de Villiers, H. 1982. Excavations at the Lydenburg Heads Site, Eastern Transvall, South Africa. *South African Archaeological Bulletin* 37:16-33.

F

88. Fagan, B. M. 1960. The Glentyre Shelter and Oakhurst re-examined. *South African Archaeological Bulletin* 15:80-94.
89. Faure, M. & Guérin, C. 1990. *Hippopotamus laloumena* nov. sp., la troisième espèce d'hippopotame holocène de Madagascar. *Comptes Rendus de l'Académie des Sciences* 310:1299-1305.
90. Faye, B. 1988. Le site archéologique de Chin-Tafidet au Niger. *Bulletin mensuel de la Société linnéenne de Lyon* 57:212-213.
91. Finucane, B., Manning, K. & Touré, M. 2008. Late Stone Age subsistence in the Tilemsi Valley, Mali: Stable isotope analysis of human and animal remains from the site of Karkarichinkat Nord (KN05) and Karkarichinkat Sud (KS05). *Journal of Anthropological Archaeology* 27:82-92.
92. Frank, T., Breunig, P., Müller-HAude, P., Van Neer, W., Neumann, K., Vogelsang, R. & Wotzka, H. P. 2001. The Chaîne de Gobnangou, SE Burkina Faso: archaeological, archaeobotanical, archaeozoological and geomorphological studies. *Beiträge zur Allgemeinen und Vergleichenden Archäologie* 21:127-190.

G

93. Gaillard, C. 1934. Contribution à l'étude de la faune préhistorique de l'Egypte. *Archives du Muséum d'Histoire Naturelle de Lyon* XIV:1-143.
94. Garcea, E. A. A. 1995. New investigations in the Tadrat Acacus, Libyan Sahara. *Nyame Akuma* 44:35-37.

95. Garcea, E. A. A. 2003. Animal exploitation and pottery technology during Pastoral times: the evidence from Uan Telocat, Libyan Sahara. *Journal of African Archaeology* 1:111-126.

96. Garlake, P. S. 1973. Excavations at the Nhunguza and Ruanga ruins in Northern Mashonaland. *South African Archaeological Bulletin* 27:107-143.

97. Garth Sampson, C. 1999. Amphibians remains from a Later Stone Age Rock Shelter in the upper Karoo, South Africa, in Plug, I. & Klein, R.(eds.), *Archaeozoology in Africa*, ArchaeoZoologia X/1-2, pp. 117-154.

98. Gautier, A. 1974. Some mammalian remains from a fossil wadi near the base of the Gademotta formation, in Wendorf, F. & Schild, R. (eds.), *A Middle Stone Age sequence from the Central Rift Valley, Ethiopia*. Polska Akademia Nauk Instytut Historii Kultury Materialnej, Wroclaw, pp. 160-165.

99. Gautier, A. 1976. Animal remains from archaeological sites of Terminal Paleolithic to Old Kingdom age in the Fayum, in Wendorf, F. & Schild, R. (eds.), *Prehistory of the Nile Valley, 7*. Academic Press. pp. 369-380.

100. Gautier, A. 1978. La faune de vertébrés des sites épipaléolithiques d'Elkab, in Vermeersch, P. M. (ed.), *Elkab II. L'Elkabien, Epipaléolithique de la vallée du Nil égyptien*. Comité des fouilles belges en Afrique, Bruxelles, pp. 103-114.

101. Gautier, A. 1980. Contributions to the Archaeozoology of Egypt, in Wendorf, F. & Schild, R. (eds.), *Prehistory of the Eastern Sahara, 6*. Academic Press, New York, pp. 317-343.

102. Gautier, A. 1983. Animal life along the prehistoric Nile : The evidence from Saggai 1 and Geili (Sudan). *Origini* 12:50-115.

103. Gautier, A. 1984. Archaeozoology of the Bir Kiseiba region, eastern Sahara, in Wendorf, F., Schild, R. & Close, A. E. (eds.), *Catlle keepers of the Sahara. The Neolithic of Bir Kiseiba*, Dallas, 49-72.

104. Gautier, A. 1984. New data concerning the prehistoric fauna and domestic cattle from Ti-n-Torha (Acacus, Libya), in Coetzee, J. A. & Van Zinderen Bakker, E. M. S. (eds.), *Palaeoecology of Africa and the surrounding islands* 16. A.A. Balkema, Rotterdam, pp. 305-309.

105. Gautier, A. 1984. Quaternary mammals and archaeozoology of Egypt and the Sudan: a survey, in Kobusiewicz, L. K. M. (ed.), *Origin and early development of food-producing cultures in north-eastern Africa*. Polish Academy of Sciences, Poznan, pp. 43-56.

106. Gautier, A. 1984. The fauna of the Neolithic site of Kadero (Central Sudan). Origin and early development of food-producing cultures in north-eastern Africa. Poznan :317-319.

107. Gautier, A. 1986. La faune de l'occupation néolithique d'el Kadada (secteurs 12-22-32) au Soudan central. *Archéologie du Nil Moyen* 1:59-111.

108. Gautier, A. 1987. Prehistoric men and cattle in North Africa: a dearth of data and a surfeit of models, in Close, A. E. (ed.) *History of arid North Africa :essays in honour of Fred Wendorf*. Southern Methodist University Press, Dallas, pp. 163-187.

109. Gautier, A. 1988. Notes on the animal bone assemblage from the early Neolithic at Geili (Sudan), in Caneva, I. (ed.), *El Geili. The history of a middle Nile environment 7000 B.C.- A.D. 1500*, BAR International Series 424, pp. 57-63.

110. Gautier, A., Ballmann, P. & Van Neer, W. 1980. Molluscs fish, birds and mammals from the Late Palaeolithic sites in Wadi Kubbaniya, in Wendorf, F., Schild, R. & Close, A. E. (eds.), *Loaves and fishes : the Prehistory of Wadi Kubbaniya*. Pauls Press, New Dehli, pp. 281-294.

111. Gautier, A. & Hendrickx, S. 1999. Vultures and other animal remains from a tomb in the Elkab necropolis, Upper Egypt: an exercice in salvage archaeozoology, in *Historia Animalium ex Ossibus, Festschrift für Angela von den Driesch*, Internationale Archäologie, Studia honoraria 8, *4*, Rahden, pp. 161-179.

112. Gautier, A., Linseele, V. & Van Neer, W. 2002. The fauna of the Early Khartoum occupation on Jebel Umm Marrahi (Khartoum Province, Sudan), in *Tides of the Desert*. Africa Praehistorica 14, Köln, pp. 337-344.

113. Gautier, A. & Van Neer, W. 1982. Prehistoric fauna from Ti-n-Torha (Tadrart Acacus, Libya). *Origini* 11:87-127.

114. Gautier, A. & Van Neer, W. 1989. The animal remains from the Late Paleolithic sequence in Wadi Kubbaniya, in Wendorf, F., Schild, R. & Close, A. E. (eds.), *The Prehistory of Wadi Kubbaniya, Vol. 2, Stratigraphy, paleoeconomy and environment*. Southern Methodist University Press, Dallas, pp. 119-161.

115. Gautier, A. & Van Neer, W. 1997. Animal remains from Debbat El Eheima and Debbat Bangdit (1600-1000 B.C./400-1000 A.D., southern Sudan). *Archaeozoologia* 9:49-72.

116. Gautier, A. & Van Neer, W. 2005. The continuous exploitation of wild animal resources in the archaeozoological record of Ghana. *Journal of African Archaeology* 3:195-212.

117. Gautier, A. & Van Neer, W. 2006. Animal remains from Tahal Teglinos (Kassala, Sudan) and the arrival of cattle pastoralism in the Southern Atbai. *Journal of African Archaeology* 4:223-233.

118. Gehlen, B., Kindermann, K., Linstädter, J. & Riemer, H. 2002. The Holocene occupation of the Eastern Sahara: Regional chronologies and supra-regional developments in four areas of the absolute desert, in *Tides of the Desert*, Africa Praehistorica 14, Köln, pp. 85-116.

119. Geraads, D. 1983. Faunal remains from the Gash Delta, Sudan. *Nyame Akuma* 23:22.

120. Gifford, D. P. 1981. The faunal remains, in Philipson, D. W. (ed.), *Kulchurdo Rock Shelter and the Stone Age of Mount Marsabit*, Azania 16,167-180.

121. Gifford-Gonzales, D. P. 1984. Implication of a faunal assemblage from a Pastoral Neolithic site in Kenya : findings and a perspective on research, in Clark, J. D. & Brandt, S. (eds.), *From Hunters to Farmers: the causes and consequences of food production in Africa*. University of California Press, Berkeley, pp. 213-251.

122. Gifford-Gonzales, D. P. 1998. Early pastoralists in East Africa: ecological and social dimensions. *Journal of Anthropological Archaeology* 17:166-200.

123. Gifford-Gonzales, D.P. 2003. The fauna from Ele Bor: evidence for the persistence of foragers into the Later Holocene of arid north Kenya. *African Archaeological Review* 20:81-119.

124. Gifford-Gonzales, D. P. & Kimengich, J. 1984. Faunal evidence for early stock keeping in the central rift of Kenya: preliminary findings, in Krzyzaniak, L. & Kobusiewicz, M. (eds.), *Origin and early development of food producing cultures in northeastern Africa*. Polish Academy of Science, Poznan, pp. 457-471.

125. Gifford-Gonzales, D. & Parham, J. 2008. Fauna from Adrar Bous and surrounding areas, in Gifford-Gonzales, D. (ed.), *Adrar Bous, Archaeology of a central Saharan granitic ring complex in Niger*. Royal Museum for Central Africa, Tervuren, pp. 313-353.

126. Goudsmit, J. & Brandon-Jones, D. 1999. Mummies of Olive Baboons and Barbary Macaques in the Baboon Catacomb of the Sacred Animal Necropolis at North Saqqara. *Journal of Egyptian Archaeology* 85:45-53.

127. Goudsmit, J. & Brandon-Jones, D. 2000. Evidence from the Baboon Catacomb of the Sacred Animal Necropolis in North Saqqara for a west Mediterranean monkey trade route to Ptolemaic Alexandria. *Journal of Egyptian Archaeology* 86:111-119.

128. Green, H. S. 1975. Sudanese Radiocarbon chronology: a provisional date list. *Nyame Akuma* 6:10-24.

129. Greenwood, P. H. & Todd, E. J. 1976. Appendix J. Fish remains from Upper Palaeolithic sites near Idfu and Isna, in Wendorf, F., Schild, R. & Issawi, B. (eds.), *Prehistory of the Nile Valley*. Academic Press, New York, pp. 384-388.

130. Gronenborn, D., Van Neer, W. & Korupinski, T. 1995. Kleiner Vorbericht zur archäologischen Feldarbeit südlich des Tschad-Sees. *Berichte des Sonderforschungsbereichs* 268:27-39.

131. Guérin, C. & Faure, M. 1996. Chasse au chacal et domestication du bœuf dans le site néolithique d'Asa Koma (République de Djibouti). *Journal des Africanistes* 66:299-311.

132. Gutherz, X., Joussaume, R., Amblard, S., Guedda, M., avec la collaboration de, Bonnefille, R., Duday, H., Gouraud, R., Thiebault, S., El HAdji, T. & Van Neer, W. 1996. Le site d'Asa Koma (République de Djibouti) et les premiers producteurs dans la corne de l'Afrique. *Journal des Africanistes* 66:255-297.

H

133. Haaland, R. & Magid, A. A. 1992. Radiocarbon dates from the Mesolithic sites in the Atbara region, Sudan. *Nyame Akuma* 37:17-27.

134. Haas, H. & Haynes, C. V. 1980. Discussion of radiocarbon dates from the western desert, in Wendorf, F. & Schild, R. (eds.), *Prehistory of the Eastern Sahara*. Academic Press, New York, pp. 373-378.

135. Hamdine, W., Thévenot, M. & Michaux, J. 1998. L'histoire récente de l'ours brun au Maghreb. *Comptes Rendus de l'Académie des Sciences* 321:565-570.

136. Hartung, U., el-Gelil, M. A., von den Driesch, A., Fares, G., Hartmann, R., Hikade, T. & Ihde, C. 2003. Vorbericht über neue Untersuchungen in der prädynastischen Siedlung von Maadi. *Mitteilungen des deutschen Archäologischen Instituts Abteilung Kairo* 59:149-198.

137. Hassan, F. A. 1985. Radiocarbon chronology of Neolithic and Predynastic sites in Upper Egypt and the Delta. *African Archaeological Review* 3:95-116.

138. Hassan, F. A. 1986. Chronology of the Khartoum Mesolithic and Neolithic and related sites in the Sudan: Statistical analysis and comparison with Egypt and Predynastic sites in Upper Egypt and the Delta. *African Archaeological Review* 4:83-102.

139. HENDRICKS, S., MIDANT-REYNES, B. & VAN NEER, W. 2001. *Mahgar Dendera 2 (Haute Egypte), un site d'occupation badarien.* Egyptian Prehistory Monographs 3. Leuven University Press, Leuven.

140. HENSHILWOOD, C. 1996. A revised chronology for pastoralism in southermost Africa: new evidence for sheep at c. 2000 bp from Blombos Cave, South Africa. *Antiquity* 70:945-949.

141. HENSHILWOOD, C. 1997. Identifying the collector: evidence for human processing of the Cape Dune Mole-Rat, *Bathyergus suillus*, from Blombos Cave, Southern Cape, South Africa. *Journal of Archaeological Science* 35:659-662.

142. HIPONDOKA, M. H. T., JOUSSE, H., KEMPF, J. & BUSCHE, D. 2006. Fossil evidence for perennial lake conditions during the Holocene at Etosha Pan, Namibia. *South African Journal of Science* 102:93-95.

143. HOLL, A. 1985. Subsistence patterns of the Dhar Tichitt Neolithic, Mauritania. *African Archaeological Review* 3:151-162.

144. HOLL, A. 1986. Economie et société néolithique du Dhar Tichitt (Mauritanie), in *Recherches sur les Civilisations. Mémoire 69*, pp. 197.

145. HOLL, A. 1998. Livestock husbandry, pastoralisms, and territoriality : the West African record. *Journal of Anthropological Archaeology* 17:143-165.

146. HUFFMAN, T. N. 1975. Cattle from Mabveni. *South African Archaeological Bulletin* 30:23-24.

147. HUFFMAN, T. N. 1979. Test excavations at Chamabvefva, Southern Mashonaland. *South African Archaeological Bulletin* 34:57-70.

148. HUFFMAN, T. N. 2008. Zhizo and Leopard's Kopje: test excavations at Simamwe and Mtanye, Zimbabwe, in BADENHORST, S., MITCHELL, P. & DRIVER, J. C. (eds.), *Animals and People : Archaeozoological Papers in Honour of Ina Plug*, British Archaeological Reports 1849, Oxford, pp. 200-214.

I

149. INSOLL, T. 1994. Preliminary results of excavations at Gao, September and October 1993. *Nyame Akuma* 41:45-48.

J

150. JACKES, M. & LUBELL, D. 2008. Early and Middle Holocene environments and Capsian cultural change : evidence from the Télidjène Basin, Eastern Algeria. *African Archaeological Review*:41-55.

151. JERARDINO, A. 1998. Excavations at Pancho's kitchen middens, Western Cpae coast, South Africa: further observations into the megamidden. *South African Archaeological Bulletin* 53:16-25.

152. JERARDINO, A. & YATES, R. 1996. Preliminary results from excavations at Steenbokfontein Cave: implications for past and future research. *South African Archaeological Bulletin* 51:7-16.

153. JESSE, F., KRÖPELIN, S., LANGE, M., PÖLLATH, N. & BERKE, H. 2004. On the periphery of Kerma - The Handessi Horizon in Wadi Hariq, Northwestern Sudan. *Journal of African Archaeology* 2:123-164.

154. JOUSSE, H. 2004. Impact des variations environnementales sur la structure des communautés mammaliennes et l'anthropisation des milieux : exemple des faunes holocènes du Sahara occidental. *Documents des laboratoires de Géologie de Lyon* 160:273.

155. JOUSSE, H. & CHENAL-VÉLARDÉ, I. 2001-2002. Nouvelles données sur la faune mammalienne de Kobadi (Mali) au Néolithique : implications paléoéconomiques et paléoenvironnementales. *Préhistoire, Anthropologie méditerranéennes* 10-11:145-158.

156. JOUSSE, H., OBERMAIER, H., RAIMBAULT, M. & PETERS, J. 2008. Late Holocene economic specialisation through aquatic resource exploitation at Kobadi in the Méma, Mali. *International Journal of Osteoarchaeology*:DOI: 10.1002/oa.956.

157. JOUSSE, H., OULD MOHAMED KABER, N. & RAIMBAULT, M. 2003. Aperçu archéozoologique d'un site de chasseurs au Néolithique ancien à Berouâga (Mauritanie nord-occidentale). *Sahara* 14:81-88.

158. JUWAYEYI, Y. M. 2008. Human and animal interaction on the Shire Highlands, Malawi: The evidence from Malowa rockshelter, in BADENHORST, S., MITCHELL, P. & DRIVER, J. C. (eds.), *Animals and People : Archaeozoological Papers in Honour of Ina Plug*, British Archaeological Reports 1849, Oxford, pp. 83-93.

K

159. KAREGA-MUNENE. 2002. *Holocene foragers, fishers and herders of Western Kenya*. BAR International Series1037. Oxford.

160. KHABIR, A. R. 1985. A Note on the excavation of a Neolithic site in the Sarubab area. *Nyame Akuma* 26:40.

161. KHABIR, A. R. 1987. Mollusca remains from Sarubab site, Kharthoum Province, Sudan. *Nyame Akuma* 29:59-60.

162. KLATZOW, S. 1994. Roosfontein, a contact site in the Eastern Orange Free State. *South African Archaeological Bulletin* 49:9-15.

163. KLEIN, R. G. 1972. The Late Quaternary mammalian fauna of Nelson Bay Cave. *Quaternary Research* 2:135-142.

164. KLEIN, R. G. 1974. Environment and subsistence of prehistoric man in the Southern Cape Province. *World Archaeology* 5:249-284.

165. KLEIN, R. G. 1976. The mammalian fauna of the Klasies River Mouth sites, southern Cape Province, South Africa. *South African Archaeological Bulletin* 31:75-98.

166. KLEIN, R. G. 1977. The mammalian fauna from the Middle and Later Stone Age (Late Pleistocene) levels of Border Cave, Natal Province, South Africa. *South African Archaeological Bulletin* 32:14-27.

167. KLEIN, R. G. 1978. A preliminary report on the larger mammals from the Boomplaas Stone Age cave site, Cango Valley, Oudtshoorn District, South Africa. *South African Archaeological Bulletin* 33:66-75.

168. KLEIN, R. G. 1978. The vertebrate fauna from the Buffelskloof Rock Shelter. *South African Archaeological Bulletin* 33:35-38.

169. KLEIN, R. G. 1979. Palaeoenvironmental and cultural implications of Late Holocene archaeological faunas from the Orange Free State and North-Central Cape Province, South Africa. *South African Archaeological Bulletin* 34:34-49.

170. KLEIN, R. G. 1981. Later Stone Age subsistence at Byeneskranskop Cave, South Africa, in HARDING, R. S. O. & TELEKI, G. (eds.), *Omnivorous Primates: gathering and hunting in Human evolution*. Columbia University Press, New York, pp. 166-190.

171. KLEIN, R. G. 1984. Later Stone Age faunal samples from Heuningneskrans Shelter (Transvaal) and Leopard's Hill Cave (Zambia). *South African Archaeological Bulletin* 39:109-116.

172. KLEIN, R. G. 1986. The Prehistory of Stone Age herders in the Cape Province of South Africa. *South African Archaeology Society* 5:5-12.

173. KLEIN, R. G. & CRUZ-URIBE, K. 1987. Large mammal and tortoise bones from Eland's Bay Cave and nearby sites, Western Cape Province, South Africa, in PARKINGTON, J. & HALL, M. (eds.), *Papers in the prehistory of the Western Cape, South Africa*, BAR International Series 322, pp. 350-372.

174. KLEIN, R. G. & CRUZ-URIBE, K. 1989. Faunal evidence for prehistoric herder-forager activities at Kasteelberg, western Cape Province, South Africa. *South African Archaeological Bulletin* 44:82-97.

175. KLEIN, R. G., CRUZ-URIBE, K. & BEAUMONT, P. B. 1991. Environmental, ecological and paleoanthropological implications of the late Pleistocene mammalian fauna from Equus Cave, Northern Cape Province, South Africa. *Quaternary Research* 36:94-119.

176. KLEIN, R. G. & SCOTT, K. 1986. Re-analysis of faunal assemblages from the Haua Fteah and other Late Quaternary archaeological sites in Cyrenaican Libya. *Journal of Archaeological Science* 13:515-542.

177. KORSMAN, S. & PLUG, I. 1994. Two Later Stone Age sites on the farm Honingklip in the Eatern Transvaal. *South African Archaeological Bulletin* 49:24-32.

178. KUPER, R. 1995. Prehistoric research in the Southern Libyan Desert. A brief account and some conclusions of the B.O.S. project. *Actes de la VIIIe Conférence Internationale des Etudes Nubiennes, Lille, 11-17 Sept. 1994, 1995*, pp. 123-140.

L

179. LENOBLE, P. 1991. Chiens de paiens. Une tombe postpyramidale a doucle descenderie hors de Meroe. *Archéologie du Nil Moyen* 5:167-188.

180. LESUR, J. & LANGLOIS, O. 2002. Une communauté d'Agro-chasseurs au XIXe siècle dans la Haute Bénoué. Analyse du matériel archéozoologique du site de Djaba-Hosséré, in *Actes du XIe colloque Megatchad, Ressources vivrières et choix alimentaires dans le Bassin du lac Tchad*. IRD, Paris, pp. 113-129.

181. LESUR, J., VIGNE, J.-D. & GUTHERZ, X. 2007. Exploitation of wild mammals in South-west Ethiopia during the Holocene (4000 BC–500 AD): the finds from Moche Borago shelter (Wolayta). *Environmental Archaeology* 12:139-159.

182. LINSEELE, V. 2004. Size and change of the African Aurochs during the Pleistocene and Holocene. *Journal of African Archaeology* 2:165-185.

183. LINSEELE, V. 2005. *Domestic Livestock, subsistence strategies and environmental changes in the Sahelian West Africa during the past 4000 years: evidence from archaeofaunal remains*, PhD Thesis, Katholieke Universiteit Leuven, Leuven.

184. LOUBSER, J. H. N. 1985. Buffelshoek: an ethnoarchaeological consideration of a Late Iron Age settlement in the southern Transvaal. *South African Archaeological Bulletin* 40:81-87.

185. LUBELL, D. 2001. Late Pleistocene-Early Holocene Maghreb, in PEREGRINE, P. N. & EMBER, M. (eds.), *Encyclopedia of Prehistory*, Vol. 1, Africa. Kluwer Academic/Plenum Publishers, New York, pp. 129-149.

186. LUBELL, D. & GAUTIER, A. 1975-77. Holocene environment and Capsian subsistence in Algeria, in VAN ZINDEREN BAKKER, E. M. & COETZEE, J. A. (eds.), *Palaeoecology of Africa and the Surrounding Islands*, vol. 10-11, Rotterdam, pp. 171-178.

187. LUBELL, D., HASSAN, F. A., GAUTIER, A. & BALLAIS, J.-L. 1976. The Capsian Escargotières. An interdisciplinary study elucidates Holocene ecology and subsistence in North Africa. *Science* 191:910-920.

188. LUFF, R. M. 2001. New light on Ancient Egyptian fishing and fowling, in BUITENHUIS, H. & PRUMMEL, W. (eds.), *Animals and Man in the Past. Essays in honour of Dr. A.T. Clason*, ARC-Publicatie 41, Groningen, pp. 357-363.

M

189. MACDONALD, K. C. 1994. *Socio-economic diversity and the origins of cultural complexity along the Middle Niger (2000 BC to AD 300)*, PhD Thesis, University of Cambridge, Cambridge.

190. MACDONALD, K. C. 1995. Analysis of the mammalian, avian and reptilian remains, in MCINTOSH, S. K. (ed.), *Excavations at Jenné-Jeno, Hambarketolo, and Kaniana (Inland Delta, Mali), the 1981 season*, University of California Press Publications in Anthropology 20, Berkeley, pp. 291-318.

191. MACDONALD, K. C. 1996. The Windé Koroji complex : Evidence for the peopling of the eastern Inland Niger Delta (2100-500 BC). *Préhistoire Anthropologie Méditerranéennes* 5:147-165.

192. MACDONALD, K. C. 1997. Korounkorokalé revisited : The Pays Mande and the west African microlithic technocomplex. *African Archaeological Review* 14:161-200.

193. MACDONALD, K. C. & MACDONALD, R. H. 2000. The origins and development of domesticated animals in arid West Africa, in BLENCH, R. & MACDONALD, K. (eds.), *The origins and development of African livestock. Archaeology, genetics, linguistics and ethnography*, UCL Press, London, pp. 127-162.

194. MACDONALD, K. C. & VAN NEER, W. 1993. An Initial Report on the fauna of Akumbu (Mali), in TOGOLA, T. (ed.), *Archaeological investigations of Iron Age sites in the Méma region (Mali), Appendix 1*. Ph.D. Thesis, Rice University, pp. 215-232.

195. MACDONALD, K. C. & VAN NEER, W. 1994. Specialised fishing peoples in the Later Holocene of the Méma (Mali). *Proceedings of the 7th meeting of the ICAZ, Fish remains working group, Tervuren, 1994*, pp. 243-351.

196. MACDONALD, R. H. & MACDONALD, K. C. 1996. A Preliminary report on the faunal remains recovered from Gao Ancien and Gao Saney (1993 Season), in INSOLL, T. (ed.), *Islam, Archaeology and History, Gao Region (Mali) ca AD900-1250*, BAR International series 647, pp. 124-126.

197. MACEACHERN, S., BOURGES, C. & REEVES, M. 2001. Early horse remains from Northern Cameroon. *Antiquity* 75:62-67.

198. MAREAN, C. W. 1985. The faunal remains from Smitswinkelbaai Cave, Cape Peninsula. *South African Archaeological Bulletin* 40:100-102.

199. MAREAN, C. W. 1992. Hunter to herder : large mammal remains from the hunter-gatherer occupation at Enkapune Ya Muto rock-shelter, Central Rift, Kenya. *African Archaeological Review* 10:65-127.

200. MAREAN, C. W. 1992. Implications of the Late Quaternary mammalian fauna from Lukenya Hill (south central Kenya) for paleoenvironmental change and faunal extinctions. *Quaternary Research* 37:239-255.

201. MAREAN, C. W., MUDIDA, N. & REED, K. E. 1994. Holocene palaeoenvironmental change in the Kenyan Central Rift as indicated by micromammals from Enkapune Ya Muto Rock Shelter. *Quaternary Research* 41:376-389.

202. MARINOVA, E., LINSEELE, V. & VERMEERSCH, P. 2008. Holocene environment and subsistence patterns near the Tree Shelter, Red Sea Mountains, Egypt. *Quaternary Research* 70:392-397.

203. MARKS, A. E., PETERS, J. & VAN NEER, W. 1987. Late Pleistocene and Early Holocene Occupations in the Upper Atbara River Valley, Sudan, in CLOSE, A. E. (ed.), *Prehistory of Arid North Africa, Essays in Honor of Fred Wendorf*, S.M.U. Press, Dallas, pp. 137-161.

204. MARSHALL, F. B. 1986. *Aspect of the advent of pastoral economies in East Africa*, PhD Thesis, University of California, Berkeley.

205. MARSHALL, F. 1989. Rethinking the role of *Bos indicus* in sub-saharan Africa. *Current Anthropology* 30:235-240.

206. MARSHALL, F. 2000. The origins and spread of domestic animals in East Africa, in BLENCH, R. & MACDONALD, K. (eds.), *The origins and development of African livestock. Archaeology, genetics, linguistics and ethnography*. UCL Press, London, pp. 191-221.

207. MARSHALL, F. & STEWART, K. 1994. Hunting, fishing abd herding pastoralists of western Kenya : the fauna from Gogo Falls. *Archaeozoologia* 7:7-27.

208. MBIDA, C. M., VAN NEER, W., DOUTRELEPONT, H. & VRYDAGHS, L. 2000. Evidence for banana cultivation and animal husbandry during the first millenium BC in the forest of Southern Cameroon. *Journal of Archaeological Science* 27:151-162.

209. MEHLMAN, M. J. 1979. Mumba-Höhle revisited: the relevance of a forgotten excavation to some current issue in East African prehistory. *World Archaeology* 11:80-94.

210. MERZOUG, S. & SARI, L. 2008. Re-examination of the Zone I material from Tamar Hat (Algeria): Zooarchaeological and technofunctional analyses. *African Archaeological Review* 25:57-73.

211. MGOMEZULU, S. F. 1981. Recent archaeological research and radiocarbon dates from eastern Africa. *Journal of African History* 22:435-456.

212. MIDANT-REYNES, B., CRUBEZY, E., JANIN, T. & VAN NEER, W. 1993. Le site prédynastique d'Adaïma. Rapport préliminaire de la quatrième campagne de fouille. *Bulletin de l'Institut Français d'Archéologie Orientale* 93:349-370.

213. MITCHELL, P. J., PARKINGTON, J. & YATES, R. 1994. Recent Holocene archaeology in western and southern Lesotho. *South African Archaeological Bulletin* 49:33-52.

214. MONOD, T. 1970. A propos d'un Aulacode (*Thryonomys*) du gisement néolithique d'Amekni (Ahaggar). *Bulletin de l'IFAN* 32:531-550.

215. MUNSON, P. J. 1971. *The Tichitt Tradition. A late prehistoric occupation of the Southwestern Sahara*. Ph. D. Thesis, University of Urbana-Champaign, Illinois.

216. MUNSON, P. 1974. Late Holocene climatic chronology of the southwestern Sahara. *Paper to be presented at the 3rd Biennal Meeting of the American Quaternary Association, Madison, Wisconsin, 1974*, pp. 1-14.

N

217. NESPOULET, R., EL HAJRAOUI, M. A., AMANI, F., BEN NEER, A., DEBÉNATH, A., EL IDRISSI, A., LACOMBE, J.-P., MICHEL, P., OUJAA, A. & STOETZEL, E. 2008. Palaeolithic and Neolithic occupations in the Témara Region (Rabat, Morocco): recent data on hominin contexts and behavior. *African Archaeological Review* 25:21-39.

O

218. OPPERMAN, H. 1978. Excavations in the Buffelskloof Rock Shelter near Calitzdorp, Southern Cape. *South African Archaeological Bulletin* 33:18-38.

219. OPPERMAN, H. 1982. Some research results of excavations in the Colwinton Rock Shelter, north-eastern Cape. *South African Archaeological Bulletin* 37:51-56.

220. OUCHAOU, B. 2000. *Les faunes mammalogiques holocènes des gisements du Nord du Maroc* ? Thèse de l'Université Moulay Ismail, Meknes.

P

221. PACHUR, H. J. & HOELZMANN, P. 1991. Paleoclimatic implications of late Quaternary lacustrine sediments in Western Nubia, Sudan. *Quaternary Research* 36:257-276.

222. PACHUR, H. J. & KRÖPELIN, S. 1987. Wadi Howar. Paleoclimatic evidence from an extinct river system in the Southeastern Sahara. *Science* 237:298-300.

223. PACHUR, H. J. & PETERS, J. 2001. The position of the Murzuq Sand Sea in the palaeodrainage system of the Eastern Sahara, in BALKEMA, A.A (ed.), *Palaeoecology of Africa*, vol. 27, Rotterdam, pp. 259-290.

224. PARIS, F. 1992. Chin Tafidet, village néolithique. *Journal des Africanistes* 62:33-54.

225. PERSON, A., IBRAHIM, T., JOUSSE, H., FINCK, A., ALBARET, C., GARENNE-MAROT, L., ZEITOUN, V., SALIÈGE, J.-F. & OULD M'HEIMAM, S. 2004. Environnement et marqueurs culturels en Mauritanie sud-orientale : le site de Boû Khzâmâ (DN4), premiers résultats et approche biogéochimique, in BAZZANA, A. & BOCCOUM, A. (eds.), *Du nord au sud du Sahara : cinquante ans d'archéologie française en Afrique de l'ouest et au Magreb, bilan et perspectives*. Sepia, Paris, pp. 195-213.

226. PETERS, J. 1986. A revision of the faunal remains from two Central Sudanese sites : Khartoum Hospital and Esh Shaheinab. *Archaeozoologia* Mélanges:11-33.

227. PETERS, J. 1987. The faunal remains collected by the Bagnold-Mond expedition in the Gilf Kebir and Jebel Uweinat in 1938. *Archéologie du Nil moyen* 2:251-264.

228. PETERS, J. 1989. Faunal remains and environmental change in central and eastern Sudan from Terminal Pleistocene to Middle Holocene times. *Mededelingen van de koninklijke Academie voor Wetenschappen Letteren en Schone Kunsten van België* 4:123-147.

229. PETERS, J. 1990. Late palaeolithic ungulate fauna and landscape in the plain of Kom Ombo. *Sahara* 3:45-52.

230. PETERS, J. 1990. Late Pleistocene hunter-gatherers at Ishango (eastern Zaire) : the faunal evidence. *Revue de Paléobiologie* 9:73-112.

231. PETERS, J. 1991. The faunal remains from Shaqadud, in MARKS, A. E. & MOHAMMED-ALI, A. (eds.), *The late Prehistory of the Eastern Sahel. The Mesolithic and Neolithic of Shaqadud, Soudan.* Southern Methodist University Press, Dallas, pp. 197-235.

232. Peters, J. 1993. Animal exploitation between the fifth and the sixth Cataract ca. 8.500-7.000 B.P. : a preliminary report on the faunas from El Damer, Abu Darhein and Aneibis, in Krzyzaniak, L. & kobusiewicz, M. (eds.), *Environmental change and human culture in the Nile Basin and northern Africa until the second Millenium B.C.* Poznam Archaeological Museum, Poznan, pp. 413-419.
233. Peters, J., Pöllath, N. & von den Driesch , A. 2002. Ichthyological diversity in the Holocene Palaeodrainage Systems of Western Nubia. *Africa Praehistorica* 14:325-335.
234. Peters, J. & von den Driesch , A. 1993. Mesolithic fishing at the confluence of the Nile and the Atbara, Central Sudan, in Clason, A., Payne, S. & Uerpmann, H.-P. (eds.), *Skeletons in her Cupboard: Festschrift for Juliet Clutton-Brock*, Oxbow books, 34 Oxford, pp. 75-83.
235. Peters, J. & von den Driesch, A. 2002. Archäozoologisch-kulturhistorische Auswertung frühgeschichtlicher Krokodilknochen aus Al-Ma'abda (Mittelägypten). *Bonn. zool. Beitr.* 50:211-219.
236. Petit-Maire, N. 1979. *Le Sahara Atlantique à l'Holocène. Peuplement et écologie.* Mémoires du CRAPE 28, Alger.
237. Plug, I. 1979. Namakala and Nanga: faunal report on two early Iron Age sites, Zambia. *South African Archaeological Bulletin* 34:123-126.
238. Plug, I. 1981. Some research results on the Late Pleistocene and Early Holocene deposits of Bushman Rock Shelter, eastern Transvaal. *South African Archaeological Bulletin* 36:14-21.
239. Plug, I. 1989. Aspects of life in the Kruger National Park during the Early Iron Age. *Goodwin's Legacy Vlaeberg South African Archaeological Society, Goodwin Series* 6:62-68.
240. Plug, I. 1989. Notes on distribution and relative abundances of some animal species, and on climate in the Kruger National Park during prehistoric times. *Koedoe* 32:101-119.
241. Plug, I. 1996. Seven centuries of Iron Age traditions at Bosutswe, Botswana: a faunal perspective. *South African Journal of Science* 92:91-97.
242. Plug, I. 1997. Cattle remains in some pre- and protohistoric societies of the central cattle pattern in Southern Africa. *Anthropozoologica*:747-752.
243. Plug, I. 1997. Late Pleistocene and Holocene hunter-gatherers in the eastern highlands of South Africa and Lesotho: a faunal interpretation. *Journal of Archaeological Science* 24:715-727.
244. Plug, I. 1999. Some Early Iron Age communities of the eastern escarpment and lowveld, South Africa: a faunal perspective, in Plug, I. & Klein, R. (eds.), *Archaeozoology in Africa*, ArchaeoZoologia X/1-2, *2*, pp. 189-200.
245. Plug, I. 2000. Appendix C: The faunal remains from Westbury, Ladybrandt District, Free State, in Thorp, C. R. (ed.), *Hunter-Gatherers and Farmers. An enduring frontier in the Caledon Valley, South Africa*, BAR International Series 860, pp. 87-88.
246. Plug, I. & Brown, A. 1982. Mgoduyanuka: faunal remains. *Annals of the Natal Museum* 25:115-121.
247. Plug, I. & Dippenaar, N. J. 1979. Evidence of *Rattus rattus* (House Rat) from Pont Drift, an Iron Age Site in Northern Transvaal. *South African Journal of Science* 75:82.
248. Plug, I. & Mitchell, P. 2008. Fishing in the Lesotho Highlands: 26,000 years of fish exploitation, with special reference to Sehonghong shelter. *Journal of African Archaeology* 6:33-55.
249. Plug, I., Mitchell, P. J. & Bailey, G. N. 2003. Animal remains fronm Likoaeng, an open-air river site, and its place in the post-classic Wilton of Lesotho and the Eastern Free State, South Africa. *South African Journal of Science* 99:143-151.
250. Plug, I. & Pistorius, J. C. 1999. Animal remains from industrial Iron Age communities in Phalaborwa, South Africa. *Archaeological Review* 16:155-184.
251. Plug, I. & Roodt, F. 1990. The faunal remains from recent excavations at uMgungundlovu. *South African Archaeological Bulletin* 45:47-52.
252. Plug, I. & Skelton, P. 1991. Fish and other faunal remains from a Late Iron Age site on the Letaba River, Kruger National Park. *Koedoe* 34:1-6.
253. Plug, I., Soper, R. & Chirawu, S. 1997. Pits, tunnels and cattle in Nyanga, Zimbabwe: new light on an old problem. *South African Archaeological Bulletin* 52:89-94.
254. Plug, I. & Voigt, E. A. 1985. Archaeozoological studies of Iron Age communities in southern Africa. *Advances in World Archaeology* 4:189-238.
255. Pöllath, N. 2007. History hidden in bones: Holocene environmental change in northwestern Sudan, in *Aridity Change and Conflict in Africa*, Colloquium Africanum 2, Heinrich-Barth-Institut, Köln, pp. 91-104.

R

256. RIVALLAIN, J. & VAN NEER, W. 1983. Les fouilles de Koyom (sud du Tchad) : étude du matériel archéologique et faunique. *L'Anthropologie* 87: 221-239.
257. RIVALLAIN, J. & VAN NEER, W. 1984. Inventaire du matériel archéologique et faunique de Koyom, sud du Tchad. *L'Anthropologie* 88:441-448.
258. ROBBINS, L. H., CAMPBELL, A. C., MURPHY, M. L., BROOK, G. A., LIANG, L., SKAGGS, S. A., SRIVASTAVA, P., MABUSE, A. A. & BADENHORST, S. 2008. Recent archaeological research at Toteng, Botswana: early domesticated livestock in the Kalahari. *Journal of African Archaeology* 6:131-149.
259. ROBBINS, L. H., MURPHY, M. L., BROOK, G. A., IVESTER, A. H., CAMPBELL, A. C., KLEIN, R. G., MILO, R. G., STEWART, K. M., DOWNEY, W. S. & STEVENS, N. J. 2000. Archaeology, palaeoenvironment and chronology of the Tsodilo Hills White Paintings Rock Shelter, Northwest Kalahari Desert, Botswana. *Journal of Archaeological Science* 27:1085-1113.
260. ROBBINS, L. H., MURPHY, M. L., STEVENS, N. J., BROOK, G. A., IVESTER, A. H., HABERYAN, K. A., KLEIN, R. G., MILO, R., STEWART, K. M., MATTHIESEN, D. G. & WINKLER, A. J. 1996. Paleoenvironment and archaeology of Drotsky's Cave, Western Kalahari Desert, Botswana. *Journal of Archaeological Science* 23:7-22.
261. ROBBINS, L. H., WHITTY, A. & FAGAN, B. M. 1966. Excavations at Harleigh Farm near Rusape, Rhodesia, 1958-1962. *South African Archaeological Bulletin* 21:61-80.
262. ROBERTSHAW, P. T. 1977. Excavations at Paternoster, South Western Cape. *South African Archaeological Bulletin* 32:63-73.
263. ROBERTSHAW, P. 1984. Archaeology in the Eastern Africa; recent developments and more dates. *Journal of African History* 25:369-393.
264. ROSET, J.-P. 1987. Néolithisation, Néolithique et Post-Néolithique au Niger nord-oriental. *Bulletin de l'Association française pour l'Etude du Quaternaire* 4:203-214.
265. ROSET, J.-P., DE BROIN, F., FAURE, M., GAYET, M., GUÉRIN, C. & MOUCHET, F. 1990. La faune de Tin Ouaffadene et d'Adrar Bous 10, deux gisements archéologiques de l'Holocène ancien au Niger nord-oriental. *Géodynamique* 5:67-89.
266. ROUBET, C. 1972. Sur la définition de la chronologie du Néolithique de tradition capsienne (sensu-stricto), in VAN ZINDEREN BAKKER, E. M. & BALKEMA, A. A. (eds.), *Palaeoecology of Africa and the Surroundings Islands and Antarctica* 6, Cape Town, pp. 211-214.

S

267. SADR, K. 2008. An ageless view of first millennium AD Southern African ceramics. *Journal of African Archaeology* 6:103-129.
268. SCHWEITZER, F. R. 1974. Archaeological evidence for sheep at the Cape. *South African Archaeological Buletin.* 29:75-82.
269. SCOTT, K. & BRINK, J. S. 1992. Quaternary palaeoenvironments of pans in central South Africa: palynological and palaeontological evidence. *Suid-Afrikaanse Geograaf* 19:22-34.
270. SERENO, P., GARCEA, E. A. A., JOUSSE, H., STOJANOWSKI, C. M., SALIÈGE, J.-F., MAGA, A., IDE, O. A., KNUDSON, K., MERCURI, A. M., STAFFORD, T. W., KAYE, T. G., GIRAUDI, C., MASSAMBA N'SIALA, I., COCCA, E., MOOTS, H. M., DUTHEIL, D. & STIVERS, J. 2008. Lakeside Cemeteries in the Sahara: 5000 Years of Holocene Population and Environmental Change. *PLoS ONE* 3 (e2995).
271. SMITH, A. B. 1974. Preliminary report of excavations at Karkarichinkat Nord and Karkarichinkat Sud, Tilemsi Valley, Republic of Mali. *West African Journal of Archaeology* 4:33-55.
272. SMITH, A. B. 1975. A note on the flora and fauna from the post-paleolithic sites of Karkarichinkat Nord and Sud. *West African Journal of Archaeology* 5:201-204.
273. SMITH, A. B. 1980. The neolithic tradition in the Sahara, in WILLIAMS, M. A. J. & FAURE, H. (eds.), *The Sahara and the Nile*, A.A. Balkema, Rotterdam, pp. 451-465.
274. SMITH, A. B. 1981. An archaeological investigation of Holocene deposits at Rooeils Cave, South-western Cape. *South African Archaeological Bulletin* 36:75-83.
275. SMITH, A. B. 1998. Early domestic stock in Southern Africa: a commentary. *African Archaeological Review* 15:151-156.
276. SMITH, A. B. 2008. The Tenerian, in GIFFORD-GONZALES, D. (ed.) *Adrar Bous, Archaeology of a central Saharan granitic ring complex in Niger*. Royal Museum for Central Africa, Tervuren, pp. 201-243.
277. SMITH, A. B., SADR, K., GRIBBLE, J. & YATES, R. 1991. Excavations in the south-western Cape, South Africa, and the archaeological identity of prehistoric hunter-gatherers within the last 2000 years. *South African Archaeological Bulletin* 46:71-91.
278. SMITH, A. B., YATES, R., MILLER, D., JACOBSON, L. & EVANS, G. 1995. Excavations at Geduld and the appearance of Early domestic stock in Namibia. *South African Archaeological Bulletin* 50:3-14.

279. Stahl, A. B. 1994. Innovation, diffusion and culture contact: the Holocene archaeology of Ghana. *Journal of World Prehistory* 8:51-112.
280. Stewart, K. M. 1989. Fishing site of North and East Africa in the Late Pleistocene and Holocene: Environmental change and human adaptation. *BAR international Series* 521.

T

281. Tauveron, M., Striedter, K. H. & Ferhat, N. 2009. Neolithic domestication and pastoralism in Central Sahara: the cattle necropolis of Mankhor (Tadrart Algérienne), in Baumhauer, R. & Runge, J. (eds.), *Palaeoecology of Africa 29, Holocene Palaeoenvironemental History of the Central Sahara.* CRC Press/Balkema, Leiden, pp. 179-186.
282. Thackeray, A. I. 1979. An analysis of faunal remains from archaeological sites in southern South West Africa (Namibia). *South African Archaeological Bulletin* 34:18-33.
283. Thilmans, G. & Ravisé, E. 1980. Protohistoire du Sénégal. Recherches archéologiques. T. II. Sintiou-Bara et les sites du fleuve. *Mémoires de l'Institut fondamental d'Afrique Noire* 91:1-215.
284. Thorp, C. R. 2000. Hunter-gatherers and farmers. An enduring frontier in the Caledon Valley, South Africa. *Bar international Series* 860.
285. Togola, T. 1993. *Archaeological investigations of Iron Age sites in the Mema region, Mali (West Africa)*, PhD Thesis, Rice University, Houston.
286. Turner, G. 1986. Faunal remains from Jubilee Shelter, Transvaal. *South African Archaeological Bulletin* 41:63-68.

V

287. Van den Brink, E. C. M. 1989. A transitional Late Predynastic - Early Dynastic settlement site in the northeastern Nile Delta, Egypt. *Mitteilungen des deutschen Archäologischen Instituts Abteilung Kairo* 45:55-108.
288. Van Neer, W. 1978. Analyse de la faune trouvée dans les tombes de l'âge du Fer dans la dépression de l'Upemba, Zaïre. *Revue de Zoologie africaine* 92:703-710.
289. Van Neer, W. 1989. Contribution to the Archaeozoology of Central Africa. *Annales des Sciences zoologiques du Musée royal de l'Afrique centrale* 259:287.
290. Van Neer, W. 1990. Description of Holocene vertebrate remains from archaeological sites in the Algerian Sahara, in Fiedler, L. (ed.),*Weitere Beiträge zur Urgeschichte der Sahara.* Kleine Schriften aus dem Vorgeschichtlichen Seminar der Phillips-Universität Marburg 31, Marburg, pp. 211-215.
291. Van Neer, W. 1990. Les faunes de vertébrés quaternaires en Afrique Centrale, in Lanfranchi, R. & Schwartz, D. (eds.), *Paysages quaternaires de l'Afrique centrale atlantique*. ORSTOM, Collection Didactiques, Paris, pp. 195-220.
292. Van Neer, W. 1994. Fish remains from Late Pleistocene and Holocene archaeological sites near Khasm el Girba, Sudan. *Archaeofauna* 3:115-126.
293. Van Neer, W. 1995. Analysis of the fish remains, in McIntosh, S. K. (ed.), *Excavations at Jenné-Jeno, Hambarketolo, and Kaniana (Inland Delta, Mali), the 1981 season.*: University of California Press, Berkeley, pp. 319-347.
294. Van Neer, W. 1997. Archaeozoological data on the food provisioning of Roman settlements in the Eastern Desert of Egypt. *Archaeozoologia* 9:137-154.
295. Van Neer, W. 1997. Etude des ossements animaux de l'amas coquillier de Ndiamon-Badat (Delta du Saloum), in Thilmans, G.(ed.), *Fouille et dégradations dans les îles du Saloum,* Dakar, pp. 15-21.
296. Van Neer, W. 2000. Domestic animals from archaeological sites in Central and West-Central Africa, in Blench, R. M. & MacDonald, K. C. (eds.), *The Origins and Development of African Livestock: archaeology, genetics, linguistics and ethnography*. UCL Press, London, pp. 163-190.
297. Van Neer, W. 2000. Faunal remains from Shuwikhat 1, in Vermeersch, P. M. (ed.), *Palaeolithic living sites in Upper and Middle Egypt*. Leuven University Press, Leuven, pp. 153-154.
298. Van Neer, W. 2001. Animal remains from the medieval site of Kizimkazi Dimbani, Zanzibar, in Scarcia Amoretti, B. (ed.), *Islam in East Africa: new sources (Archives, Manuscripts and written historical sources. Oral history. Archaeology)*. Herder, Roma, pp. 385-410.
299. Van Neer, W. 2001. Restes fauniques du site Badarien de Mahgar Dendera 2 (Haute Egypte), in Hendrickx, S., Midant-Reynes, B. & Van Neer, W. (eds.), *Mahgar Dendera 2 (Haute Egypte), un site d'occupation badarien*. Leuven University Press, Leuven, pp. 91-109.

300. VAN NEER, W. 2002. Food security in western and central Africa during the late Holocene: the role of domestic stock keeping, hunting and fishing, in HASSAN, F. A. (ed.), *Droughts, food and culture, Ecological change and food security in Africa's later prehistory*, Kluwer Academy, New York, pp. 251-273.

301. VAN NEER, W. 2002. Le matériel faunique, in MIDDANT-REYNES, B. & BUCHEZ, N. (eds.), *Adaïma 1- Economie et habitat*, Institut français d'archéologie orientale 45, Le Caire, pp. 521-565.

302. VAN NEER, W. & Bocoum, H. 1991. Etude archéozoologique de Tulel-Fobo, site protohistorique (IVe-Xe siècle) de la moyenne vallée du Fleuve Sénégal (République du Sénégal). *Archaeozoologia* 4:93-114.

303. VAN NEER, W. & BREUNIG, P. 1999. Contribution to the archaeozoology of the Brandberg, Namibia. *Cimbebasia* 15:127-140.

304. VAN NEER, W. & CLIST, B. 1991. Le site de l'Age du Fer Ancien d'Oveng (Province de l'Estuaire, Gabon), analyse de sa faune et de son importance pour la problématique de l'expansion des locuteurs bantu en Afrique Centrale. *Comptes Rendus de l'Académie des Sciences* Série II, 312:105-110.

305. VAN NEER, W. & ERVYINCK, A. 1998. The faunal remains, in SIDEBOTHAM, S. E. & WENDRICH, W. Z. (eds.), *Berenike 1996- Report of the 1996 Excavations at Berenike (Egyptian Red Sea Coast) and the survey of the Eastern Desert*, CNWS Publications 3, Leiden, pp. 349-390.

306. VAN NEER, W. & ERVYNCK, A. 1999. Faunal remains from Shenshef and Kalatlat, in SIDEBOTHAM, S. E. & WENDRICH, W. Z. (eds.), *Berenike 1997- Report of the 1997 Excavations at Berenike and the survey of the Egyptian Eastern Desert, including excavations at Shenshef.* CNWS Publications 4, Leiden, pp. 431-445.

307. VAN NEER, W. & GAYET, M. 1988. Etude des poissons en provenance des sites holocènes du Bassin de Taoudenni-Araouane (Mali). *Bulletin du Muséum national d'Histoire naturelle* 10 C :343-383.

308. VAN NEER, W. & LANFRANCHI, R. 1985. Etude de la faune découverte dans l'abri Tshitolien de Ntadi Yomba (République Populaire du Congo). *L'Anthropologie* 89:351-364.

309. VAN NEER, W. & LANFRANCHI, R. 1986. Une association de faune et d'industrie du Tshitolien (Age récent de la pierre, 7000 B.P.) dans l'abri de Ntadi Yomba (Région du Niari) en R.P. du Congo. Elements nouveaux pour un essai de reconstitution du paysage à cette époque. *Comptes Rendus de l'Académie des Sciences* t. 302, série II:831-834.

310. VAN NEER, W. & LENTACKER, A. 1996. The faunal remains, in SIDEBOTHAM, S. & WENDRICH, W. (eds.), *Berenike '95. Preliminary report of the excavations at Berenike (Egyptian Red Sea Coast) and the survey of the Eastern Desert*, CNWS Publications, Leiden, pp. 337-355.

311. VAN NEER, W., LINSEELE, V. & FRIEDMAN, R. 2004. Animal burials and food offerings at the Elite Cemetery HK6 of Hierakonpolis. *Egypt at its Origins, Studies in Memory of Barbara Adams (Proceedings of the International Conference Origin of the State. Predynastic and Early Dynastic Egypt, Krakow, 28th August - 1st September 2002), Peeters Publishing, Leuven, 2004*, pp. 67-130 138.

312. VAN NEER, W., PAULISSEN, E. & VERMEERSCH, P. M. 2000. Chronology, subsistence and environment at the late palaeolithic fishing sites of Makhadma 2 and 4, in VERMEERSCH, P. M. (ed.), *Palaeolithic living sites in Upper and Middle Egypt*. Leuven University Press, Leuven, pp. 271-287.

313. VAN NEER, W. & SIDEBOTHAM, S. 2002. Animal remains from the fourth-sixth century A.D. military installations near Abu Sha'ar at the Red Sea coast, Egypt, in *Tides of the*, Africa Praehistorica14, Köln, pp. 171-195.

314. VAN NEER, W. & UERPMANN, H.-P. 1989. Palaeoecological significance of the Holocene Faunal remains, in KUPER, R. (ed.), *Forschungen zur Umweltgeschichte der Ostsahara*. Africa Praehistorica 2, Köln, pp. 307-341.

315. VAN NOTEN, F. 1977. Excavations at Matupi Cave. *Antiquity* 51:35-40.

316. VAN PEER, P., VERMEERSCH, P., MOEYERSONS, J. & VAN NEER, W. 1996. Palaeolithic sequence of Sodmein Cave, Red Sea Mountains, Egypt, in PWITI, G. & SOPER, R. (eds.), *Aspects of African Archaeology. Papers from the 10th Congress of the Panafrican Association for Prehistory and Related Studies*. University of Zimbabwe Publications, Harare, pp. 149-156.

317. VAN RIJSSEN, W. J. 1992. The Late Holocene deposits at Klein Kliphuis Shelter, Cedarberg, Western Cape Province. *South African Archaeological Bulletin* 47:34-43.

318. VAN SCHALKWYK, J. 1994. Wosi: an Early Iron Age village in the lower Thukela Basin, Natal. *Natal Museum Journal of Humanities* 6:65-117.

319. VAN WAARDEN, C. 1987. Matanga, a Late Zimbabwe cattle post. *South African Archaeological Bulletin* 42:107-124.

320. VÉLARDÉ, I. 1994. La faune de Hamdallahi (Mali, XIXème S.) et les origines du boeuf en Afrique : Etude archéozoologique et essai de synthèse critique. Diplôme d'archéologie préhistorique, Université de Genève - Faculté des Sciences - Département d'Anthropologie et d'Ecologie.

321. VERMEERSCH, P. M. 2002. The Egyptian Nile Valley during the Early Holocene, in *Tides of the Desert*, Africa Praehistorica 14, Köln,pp. 27-40.

322. VERMEERSCH, P. M., PAULISSEN, E. & VAN NEER, W. 1989. The Late Paleolithic Makhadma sites (Egypt): environment and subsistence, in KRZYZANIAK, L. & KOBUSIEWICZ, M. (eds.), *Late Prehistory of the Nile Basin and the Sahara*. Poznan, pp. 87-114.

323. VERMEERSCH, P. M., PAULISSEN, E., VAN PEER, P. & VAN NEER, W. 2000. El Abadyia, a Shuwikhatian site, in VERMEERSCH, P. M. (ed.), *Palaeolithic living sites in Upper and Middle Egypt*. Leuven University Press, Leuven, pp. 159-199.

324. VERMEERSCH, P., VAN NEER, W. & HENDRYICKX, S. 2004. El Abadiya 2, a Naqada I site near Danfiq, Upper Egypt. I. *Egypt at its Origins, Studies in Memory of Barbara Adams (Proceedings of the International Conference Origin of the State. Predynastic and Early Dynastic Egypt, Krakow, 28th August - 1st September 2002), Peeters Publishing, Leuven,* pp. 213-276.

325. VERMEERSCH, P. M., VAN PEER, P., MOYERSONS, J. & VAN NEER, W. 1994. Sodmein Cave site, Red Sea Mountains. *Sahara* 6:31-40.

326. VERMEERSCH, P. M., VAN PEER, P., MOYERSONS, J. & VAN NEER, W. 2002. The Tree Shelter, a Holocene site in the Red Sea Mountains. *Archéo-Nil* 12:123-138.

327. VERNET, R. 1998. *Inventaire des datations ^{14}C. Le Sahara et le Sahel. Paléoenvironnements et occupation humaine à la fin du Pléistocène et à l'Holocène*: Coopération française.

328. VERNET, R., HERVÉ-GRUHIER, C., SALIÈGE, J. F., LAMARCHE, B. & CHERIF, O. 2004. Le peuplement de l'île de Tidra (Banc d'Arguin, Mauritanie) à l'Holocène récent. *Actes du XIe Congrès Panafricain de Préhistoire, Bamako, 2001.*

329. VOGEL, J. C. & BEAUMONT, P. B. 1972. Revised radiocarbon chronology for the Stone Age in South Africa. *Nature* 237:50-51.

330. VOGEL, J., PLUG, I. & WEBLEY, L. 1997. New dates for the introduction of sheep into South Africa: the evidence from Spoegrivier Cave in Namaqualand. *South African Journal of Science* 93:246-248.

331. VOIGT, E. A. 1973. Faunal remains from the Iron Age site of Matope Court, Namichimba and Chikumba, southern Malawi, in ROBINSON, K. R. (ed.) *Iron Age of the Upper and Lower Shire, Malawi*. Government Press, Zomba, pp. 135-167.

332. VOIGT, E. A. 1979. The faunal remains from Icon. *South African Archaeological Society, Goodwin Series* 3:80-85.

333. VOIGT, E. A. 1980. Reconstructing Iron Age economy of the northern Transvaal : a preliminary report. *South African Archaeological Bulletin* 35:39-45.

334. VOIGT, E. A. 1982. Faunal report on the Lydenburg Heads site, 2530 AB4. Appendix to Evers, T.M. *South African Archaeological Bulletin* 135:31-32.

335. VOIGT, E. A. 1984. Faunal remains from Magogo and Mhlopeni: small stock herding in the Early Iron Age of Natal. *Annals of the Natal Museum* 26:141-163.

336. VOIGT, E. A. 1984. Iron Age herders of the northern Transvaal, South Africa, in the first millenium A.D., in CLUTTON BROCK, J. & GRIGSON, C. (eds.), *Animals and Archaeology : 3. Herders and their Flocks*, Oxford, pp. 371-393.

337. VOIGT, E. A. & PETERS, J. 1994. The faunal assemblage from Wosi in the Thukela Valley. *Natal Museum Journal of Humanities* 6:105-117.

338. VOIGT, E. A. & PETERS, J. 1994. The faunal assemblage from the Early Iron Age site of Mamba 1 in the Thukela Valley, Natal. *Natal Museum Journal of Humanities* 6:145-152.

339. VOIGT, E. A. & PLUG, I. 1984. Happy Rest: the earliest Iron Age fauna from the Soutpansberg. *South African Journal of Science* 80:221-227.

340. VOIGT, E. A. & VON DEN DRIESCH, A. 1984. Preliminary report on the faunal assemblage from Ndondondwane, Natal. *Natal. Ann. Mus.* 25:95-104.

341. VON DEN DRIESCH, A. 1983. Some archaeozoological remaks on fishes in Ancient Egypt. *Animals and Archaeology, BAR International Serie* 2.

342. VON DEN DRIESCH, A. 1986. Fische im Alten Ägypten. Eine Osteoarchäologische Untersuchung. *Documenta Naturae* 34:36.

343. VON DEN DRIESCH, A. 1986. Tierknochenfunde aus Qasr el-Sagha, Fayoum (Neolithikum und Mitteles Reich). *Mitteilungen des deutschen Archäologischen Instituts Abteilung Kairo* 42:1-8.

344. VON DEN DRIESCH, A. 1997. Tierreste aus Buto im Nildelta. *Archaeofauna* 6:23-39.

345. VON DEN DRIESCH, A. & BOESSNECK, J. 1983. A Roman cat skeleton from Quseir on the Red Sea coast. *Journal of Archaeological Science* 10:205-211.

346. VON DEN DRIESCH, A. & BOESSNECK, J. 1985. *Die Tierknochenfunde aus der neolitischen Siedlung von Merimde-Benisalâme am westlichen Delta*. Deutsches Archäologisches Institut Abteilung Kairo.

347. VON DEN DRIESCH, A. & BOESSNECK, J. 1985. Krankhaft veränderte Skelletreste von Pavianen aus altägyptischer Zeit. *Tierärztl. Prax.* 13:367-372.

348. VON DEN DRIESCH, A., KESSLER, D. & PETERS, J. 2004. Mummified Baboons and other Primates from the Saitic-Ptolemaic animal necropolis of Tuna el-Gebel, Middle Egypt. *Documenta Archaeobiologiae* 2:231-278.

349. VON DEN DRIESCH, A. & PETERS, J. 2001. Frühe Pferde- und Maultierskelette aus Auaris (Tell El-Dab'a), Östlisches Nildelta, in BIETAK, M. (ed.), *Egyptian and the Levant*, International Journal of Egyptian Archaeology and Related Disciplines 11, pp. 301-311.

W

350. WADLEY, L. 1995. Review of dated Stone Age sites recently excavated in the eastern Free State, South Africa. *South African Journal of Science* 91:574-579.

351. WATSON, D. J. 2005. Under the rocks: reconsidering the origin of the Kintampo Tradition and the development of food production in the savanna-forest/forest of West Africa. *Journal of African Archaeology* 3:3-55.

352. WATSON, E. J. & WATSON, V. 1990. Of commoners and kings: faunal remains from Ondini. *South African Archaeological Bulletin* 45:33-46.

353. WENDORF, F. & SCHILD, R. 1994. Are the early Holocene cattle in the eastern Sahara domestic or wild? *Evolutionary Anthropology* 3:118-128.

354. WENDORF, F., SCHILD, R., CLOSE, A.E., HILLMAN, G. C., GAUTIER, A., VAN NEER, W., DONAHUE, D. J., JULL, A. J. T. & LINICK, T. W. 1988. New radiocarbon dates and Late Palaeolithic diet at Wadi Kubbaniya, Egypt. *Antiquity* 62:279-283.

355. WENDT, E. W. & REED, C. A. 1966. Two prehistoric archaeological sites of Egyptian Nubia. *Postilla* 102:1-46.

356. WILSON, M. L. 1988. Forest Hall Shelter: an early excavation on the Southern Cape Coast. *South African Archaeological Bulletin* 43:53-55.

357. WILSON, D. E. AND REEDER, D. M. (1992). *Mammal species of the world. A taxonomic and geographic reference-* 2nd edition. Washington, Smithsonian Institution Press.